PARTICLES AND FIELDS

Previous Proceedings in the Series of Mexican Workshops on Particles and Fields

	Year	Held in	Publisher	ISBN
7th	1999	Mérida, Yucatán	AIP Conf. Proceedings vol. 531	1-56396-954-8
6th	1997	Morelia, Michoacán	AIP Conf. Proceedings vol. 445	1-56396-791-X
5th	1995	Puebla, Puebla	AIP Conf. Proceedings vol. 359	1-56396-548-8

Proceedings in the Series of Mexican Schools on Particles and Fields

	Year	Held in	Publisher	ISBN
IX	2000	Metepec, Puebla	AIP Conf. Proceedings vol. 562	1-56396-998-X
VIII	1998	Oaxaca, México	AIP Conf. Proceedings vol. 490	1-56396-895-9
VII	1996	Mérida, Yucatán	AIP Conf. Proceedings vol. 400	1-56396-686-7
VI	1994	Tabasco, Villahermosa	World Scientific	981-02-2121-5

Other Related Titles from AIP Conference Proceedings

619 Hadron Spectroscopy: Ninth International Conference on Hadron Spectroscopy; HADRON2001
Edited by Dmitry Amelin and Alexander M. Zaitsev, June 2002, 0-7354-0067-9

618 Heavy Flavor Physics: Ninth International Symposium on Heavy Flavor Physics
Edited by Anders Ryd and Frank C. Porter, May 2002, 0-7354-0064-4

602 QCD@Work: International Workshop on Quantum Chromodynamics: Theory and Experiment
Edited by Pietro Colangelo and Giuseppe Nardulli, December 2001, 0-7354-0046-6

601 Theoretical High Energy Physics: MRST 2001: A Tribute to Roger Migneron
Edited by V. Elias, D. G. C. McKeon, and V. A. Miransky, December 2001, 0-7354-0045-8

To learn more about these titles, or the AIP Conference Proceedings Series, please visit the webpage **http://proceedings.aip.org/proceedings**

PARTICLES AND FIELDS

Eighth Mexican Workshop

Zacatecas, México   14–20 November 2001

EDITORS
J. L. Díaz-Cruz
IF-UAP, Puebla

J. Engelfried
IF-UASLP, San Luis Potosí

M. Kirchbach
UAZ, Zacatecas

M. Mondragón
IF-UNAM, México

Melville, New York, 2002
AIP CONFERENCE PROCEEDINGS ■ VOLUME 623

Editors:

J. Lorenzo Díaz-Cruz
IF-UAP
Instituto de Fisica, BUAP
Ap. Postal J-48
Puebla, PUE 72570
MEXICO
E-mail: ldiaz@sirio.ifuap.buap.mx

Jürgen Engelfried
Instituto de Fisica
Universidad Autonoma de San Luis Potosí
Manuel Nava #6, Zona Universitaria
San Luis Potosí, S.L.P. 78240
MEXICO
E-mail: jurgen@ifisica.uaslp.mx

Mariana Kirchbach
Facultad de Fisica
Universidad Autonoma de Zacatecas
Ap. Postal C-600
Zacatecas, ZAC 98062
MEXICO
E-mail: kirchbach@chiral.reduaz.mx

Myriam Mondragón
UNAM
Instituto de Fisica
Depto. De Fisica Teorica
Ap. Postal 20-364
Mexico, D.F. 01000
MEXICO
E-mail: myriam@ft.ifisicacu.unam.mx

Authorization to photocopy items for internal or personal use, beyond the free copying permitted under the 1978 U.S. Copyright Law (see statement below), is granted by the American Institute of Physics for users registered with the Copyright Clearance Center (CCC) Transactional Reporting Service, provided that the base fee of $19.00 per copy is paid directly to CCC, 222 Rosewood Drive, Danvers, MA 01923. For those organizations that have been granted a photocopy license by CCC, a separate system of payment has been arranged. The fee code for users of the Transactional Reporting Service is: 0-7354-0072-5/02/$19.00.

© 2002 American Institute of Physics

Individual readers of this volume and nonprofit libraries, acting for them, are permitted to make fair use of the material in it, such as copying an article for use in teaching or research. Permission is granted to quote from this volume in scientific work with the customary acknowledgment of the source. To reprint a figure, table, or other excerpt requires the consent of one of the original authors and notification to AIP. Republication or systematic or multiple reproduction of any material in this volume is permitted only under license from AIP. Address inquiries to Office of Rights and Permissions, Suite 1NO1, 2 Huntington Quadrangle, Melville, N.Y. 11747-4502; phone: 516-576-2268; fax: 516-576-2450; e-mail: rights@aip.org.

L.C. Catalog Card No. 2002106592
ISBN 0-7354-0072-5
ISSN 0094-243X
Printed in the United States of America

CONTENTS

Preface ix
Organizing Committee xi
Acknowledgments xiii
Group Photo xv

COURSES

CP Violation, Rare Decays, and the CKM Matrix 3
    A. J. Buras
QCD Baryons in the $1/N_c$ Expansion 36
    E. Jenkins
Deep Inelastic Scattering at Large Energy and Momentum Transfers:
Recent Results from HERA 61
    G. Wolf

MEDAL CEREMONY

Medal Ceremony—Zacatecas, Mexico, 2001 115
    J. L. Díaz Cruz
Early Days of Particle Physics at the Institute of Physics UNAM .. 117
    A. Mondragón

INVITED TALKS

Resonant-Spin Flavour Solutions to the Solar Neutrino Problem 125
    O. G. Miranda
Scalar Mesons and Chiral Dynamics 131
    M. Napsuciale
The Structure of the Proton as seen with the H1 Detector 139
    J. G. Contreras
Electroweak Baryogenesis with Primordial Hypermagnetic Fields ... 149
    A. Ayala, J. Besprosvany, G. Pallares, and G. Piccinelli
Rare Kaon Decays 156
    P. S. Cooper
A Survey of Charm Hadroproduction Results 163
    J. S. Russ
Latest Oscillation Results from SNO 173
    R. G. Van de Water for the SNO Collaboration
Candidates for Non-Baryonic Dark Matter 182
    N. Fornengo
Results from the Mainz Neutrino Mass Experiment and Perspectives
of the Next Generation Experiment KATRIN 189
    J. Bonn

Two Phases of Neutrino Physics 195
    J. Wudka
Neutrino Oscillations in Dense Neutrino Media 202
    S. Pastor
Chiral Quark Models 209
    H. J. Weber
In-Medium Chiral Perturbation Theory 216
    U.-G. Meißner, J. A. Oller, and A. Wirzba
Chiral Symmetry and the Medium Modification of Hadrons 223
    J. Wambach
Topics on Heavy Baryon Chiral Perturbation Theory in the
Large N_c Limit 232
    R. Flores-Mendieta
Science and Technology of the TESLA Electron-Positron
Linear Collider 239
    A. Wagner
CDF at the Tevatron Collider in Run 2 245
    R. D. Erbacher
Search for the Higgs at the Tevatron in Run 2 252
    J. S. Conway
Higgs: Standard Model and Beyond 261
    C.-P. Yuan
Summary Talk 268
    G. López Castro

CONTRIBUTED TALKS

Vertex Integrals in Heavy-Particle Theories 277
    A. O. Bouzas
Stationary Effective Field Theory of Heterotic String vs.
Einstein–Maxwell Theory 281
    A. Herrera-Aguilar
Resonances in $\Lambda_c^+ \to pK^-\pi^+$ 285
    J. Medellin Z., J. Engelfried, and A. Morelos for the SELEX Collaboration
Large B-Fields and Noncommutative Solitons 289
    J. Moreno
Finite Size Effects in Colour Superconductivity 293
    P. Amore, N. R. Walet, M. C. Birse, and J. A. Mc Govern
Deconfined Matter and Λ^0 Polarization in Ultra-Relativistic
Heavy-Ion Collision 297
    A. Ayala, E. Cuautle, G. Herrera, and L. M. Montaño
Boundary and Expansion Effects on Two-Pion Correlation Function
in Relativistic Heavy-Ion Collisions 301
    A. Sánchez and A. Ayala
Ferromagnets and Antiferromagnets in the Effective
Lagrangian Perspective 305
    C. P. Hofmann

Weak Interactions Effect on the P-N Mass Splitting and the Principle of Equivalence 309
    N. Chamoun and H. Vucetich

A Nonperturbative Fermion-Boson Vertex 313
    A. Bashir and A. Raya

Higher Order Correction to the Neutrino Self-Energy 317
    S. Sahu

Ultra High Energy Neutrinos and Their Detection in the Pierre Auger Observatory 321
    A. Carreño, J. C. D'Olivo, and L. Nellen

Mixing and Instability 325
    M. Nowakowski

New Approach to the Parametrization of the Quark and Neutrino Mixing Matrices 330
    V. Gupta

Models of Flavor with Discrete Symmetries 333
    A. Aranda

On the Implications of Recent SNO Results 337
    A. A. Aguilar and J. C. D'Olivo

Systematic Study of 331 Models 341
    W. A. Ponce, Y. Giraldo, and L. A. Sánchez

Dirac Neutrino Anapole Moment 347
    L. G. Cabral-Rosetti, M. Moreno, and A. Rosado

Weinberg's Angle in Standard-Model Particles from Spin 9+1 Dimensional Space 351
    J. Besprosvany

The Condensate in QED3 355
    A. Bashir, A. Huet, and A. Raya

Pulsar Motions from VEP Neutrino Oscillations 359
    M. Barkovich, H. Casini, J. C. D'Olivo, and R. Montemayor

POSTERS

Polarization Studies of Hyperons 365
    C. J. Solano

Fundamental Measurements and Instrumentation "CKM" 369
    A. Morelos, J. Engelfried, J. Mata, I. Torres, and E. Vázquez-Jáuregui for the CKM Collaboration

Single Spin Asymmetries in p↑p→ρ+X 373
    G. Domínguez, G. Herrera, and I. León-Monzón

Numerical Analysis of the Quark Mass Matrix 379
    M. de Coss and R. Huerta

The ALICE Pixel Detector 383
    J. Mercado-Pérez

Bounds on Neutrino-Photon Interactions from Z Decays 387
    F. Larios, M. A. Pérez, and G. Tavares-Velasco

MSSM Higgs Bosons Production at e^+e^- Colliders 391
    M. A. Hernández-Ruíz, A. Gutiérrez-Rodríguez, and O. A. Sampayo
High Mountain Cosmic Ray Observatory 395
    O. Martínez, H. Salazar, L. Villaseñor, E. Ponce, E. Pérez, M. Anguiano,
    P. Bello, J. Hernández, and A. Silaev
Extensive Air Shower Detector Array at the Universidad Autonoma de Puebla 399
    J. Cotzomi, E. Moreno, S. Aguilar, B. Palma, O. Martínez, H. Salazar,
    and L. Villaseñor
Fluorescence Detector for Extensive Air Showers in the Region of $10^{17}-10^{21}$ eV 403
    M. Cuautle, E. Moreno, I. Pedraza, T. Murrieta, G. Garipov, B. Khrenov,
    H. Salazar, O. Martinez, and L. Villaseñor
Lepton Flavor Violation in the Two Higgs Doublet Model Using $g-2$ Muon Factor 407
    R. A. Diaz, R. Martinez, J.-A. Rodríguez, and E. Tuiran

List of Participants 411
Author Index 415

Preface

For the eighth time since 1985, the high-energy physics community from Mexico took the charge of organizing its workshop on particles and fields, which was held in the beautiful colonial city of Zacatecas, in the north of Mexico, November 14-20, 2001. The workshop aimed to cover, through invited lectures delivered by world renowned experts, the most recent developments in the field. The program included two courses as well, which were intended both for young researchers and graduate students. Finally, a series of short seminars and a poster session allowed the whole community to participate with their most recent research results. A special session was also dedicated to award the Division Medal to Prof. Alfonso Mondragón, from IF-UNAM. This volume is the written record of the topics and activities that were discussed at the Workshop.

The VIII Mexican Workshop was attended by more than one hundred participants, including faculty, postdocs and about 40 graduate students at different stages of their Ph.D. work. This participation would not have been possible without the generous support of various funding agencies and institutions, as well as the efforts from both the national and local organizing committees. Given the success of the meeting, which could be measured by the number of participants, the quality of the presentations, as well as the friendly atmosphere that pervaded during the event, the Mexican Division of Particles and Fields feels quite motivated to keep organizing this type of meeting for the coming years.

And just do not forget that "Penguins" will be waiting for you at the walls of "Mamá Inés" some time this century!

J. Lorenzo Díaz Cruz
President of the Division of Particles and Fields
Mexican Physical Society

Organizing Committee

Díaz Cruz, Lorenzo, IF-BUAP
Engelfried, Jürgen, IF-UASLP
Espinoza, Augusto, EF-UAZ
Gutiérrez, Alejandro, EF-UAZ
Herrera, Gerardo, CINVESTAV
Kirchbach, Mariana, EF-UAZ
Mondragón, Myriam, IF-UNAM
Nellen, Lukas, ICN-UNAM
Pérez, Miguel Angel, CINVESTAV
Rivera, Juan Manuel, EF-UAZ
Ortíz Saavedra, Juan, EF-UAZ

Acknowledgments

The VIII Mexican Workshop has been made possible by the support of the following institutions:
Centro Latinoamericano de Física (CLAF)
Centro Latinoamericano de Física (CLAF), México
Centro de Investigación y de Estudios Avanzados (CINVESTAV):
– Departamento de Física (Zacatenco)
– Departamento de Física Aplicada (Merida))
Consejo Nacional de Ciencia y Tecnología México (Conacyt)
European Laboratory for Particle Physics (CERN)
Gobierno del Estado de Zacatecas
Universidad Autónoma de San Luis Potosí:
– Instituto de Física
Universidad Autónoma de Zacatecas:
– Facultad de Física
Universidad Nacional Autónoma de México (UNAM):
– Instituto de Física
– Instituto de Ciencias Nucleares
– Coordinación de la Investigación Científica (Depto. de Intercambio Académico)
Secretaria de Relaciones Exteriores
Sociedad Mexicana de Física (SMF):
–División de Partículas y Campos
and by the generous support of our colleagues who paid some of the invited speakers from their own grants.

We thank our colleagues in the organizing committee, Agusto A. Espinoza, Alejandro Gutiérrez, Gerardo Herrera, Lukas Nellen, Miguel Angel Pérez, Juan Manuel Rivera and Juan Ortíz Saavedra for their help and support.

We are grateful to the following people in Zacatecas, who helped to bring the Workshop to a successful end: Alfredo Bañuelos, Dr. Pedro Martínez, Valerio Quintero, Josefina Rodríguez, Ricardo Rodríguez, Gabriela Cabral, José Falcón, Juan Reyes, Leonardo Quintanar, Francisco Benita, Silvia Flores, and the Chamber Choir of the City of Zacatecas and its Director Enrique Chávez. We acknowledge also the efficiency and helpfulness of our Conference Secretaries, Monika Kulova and Patricia Carranza.

Last but not least, we thank warmly all the speakers and poster presenters for their excellent talks and the careful preparation of their notes. In this proceedings we also include a lecture delivered by G. Wolf during the IX Mexican School on Particles and Fields, which for technical reasons could not be included in the proceedings of that event.

J. Lorenzo Díaz Cruz (IF-BUAP)
Jürgen Engelfried (IF-UASLP)
Mariana Kirchbach (EF-UAZ)
Myriam Mondragón (IF-UNAM)

John Conway (left) and Andrzej Buras (right)
with the Penguin in "Mama Ines".

COURSES

CP Violation, Rare Decays and the CKM Matrix

Andrzej J. Buras

Technische Universität München, Physik Department D-85748 Garching, Germany

Abstract. These lectures give a brief description of CP violation and rare decays of K and B mesons with particular emphasize put on the determination of the CKM matrix. The following topics will be discussed: i) The CKM matrix and the unitarity triangle, ii) General aspects of the theoretical framework, iii) Standard analysis of the unitarity triangle, iv) Rare decays $K^+ \to \pi^+ \nu \bar\nu$, $K_L \to \pi^0 \nu \bar\nu$ and $K_L \to \pi^0 e^+ e^-$, v) Rare B decays, vi) CP violation in B decays, vii) Models with minimal flavour violation.

INTRODUCTION

The determination of the Cabibbo-Kobayashi-Maskawa (CKM) matrix [1, 2] that parametrizes the weak charged current interactions of quarks is not surprisingly one of the central targets of particle physics. Indeed, the four parameters of this matrix govern all flavour changing transitions in the Standard Model (SM). These include in addition to tree level decays mediated by $W^\pm$-bosons, a vast number of one-loop induced flavour changing neutral current transitions involving gluons, photon, $W^\pm$, Z^0 and H^0. The latter transitions are responsible for rare decays and CP-violating decays in the SM. This important role of the CKM matrix is preserved in any extension of the SM even if more complicated extentions may contain new sources of flavour violation and CP violation.

The basic problem in the extraction of CKM parameters from hadron decays is related to strong interactions. Although due to the smallness of the effective QCD coupling at short distances, the gluonic contributions at scales $O(M_W, M_Z, m_t)$ can be calculated within the perturbative framework, the fact that hadrons are bound states of quarks and antiquarks forces us to consider QCD at long distances as well. Here we have to rely often on existing non-perturbative methods, which are not yet satisfactory at present. On the other hand, the basic experimental problem in the extraction of the CKM parameters from the relevant rare and CP-violating decays are their tiny branching ratios that are often very difficult to measure.

Fortunately there exists a handful of quantities that allow the determination of the CKM parameters essentially without any hadronic uncertainties. The prime examples are the CP asymmetry $a_{\psi K_S}$, certain strategies in B decays relevant for the angle γ in the unitarity triangle, the branching ratios for $K^+ \to \pi^+ \nu \bar\nu$ and $K_L \to \pi^0 \nu \bar\nu$, the $B^0_{d,s} - \bar B^0_{d,s}$ mixing ratio $\Delta M_d / \Delta M_s$ and suitable ratios of branching ratios for rare decays $B_{d,s} \to \mu^+ \mu^-$ and $B \to X_{d,s} \nu \bar\nu$ relevant for the determination of $|V_{td}|/|V_{ts}|$. Also the known measurement of $|V_{us}|$ could be included in this class.

The year 2001 opened a new era of theoretically clean measurements of the CKM matrix through the discovery of CP violation in B system ($a_{\psi K_S} \neq 0$) and a further evidence for the decay $K^+ \to \pi^+ \nu \bar{\nu}$. It is exciting that in the next years these two measurements will be considerably improved and $\Delta M_d/\Delta M_s$ as well as other relevant quantities measured.

These lecture notes provide a rather non-technical review of the decays that are best suited for the determination of the CKM matrix. I will not discuss the models for the CKM matrix that have been already presented here by V. Gupta and A. Mondragon. In view of considerable space limitations it is impossible to provide here the derivations and refer properly to the relevant literature. We will also omit charm decays and the ratio ε'/ε that from the present perspective are less suited for a precise determination of the CKM matrix and of the unitarity triangle. Much more detailed account of the subject with an extensive list of references can be found in our Les Houches [3], Erice [4] lectures and in the reviews [5] and [6]. On the other hand new developments until January 2002 have been taken into account, as far as the space allowed for it, and all numerical results have been updated.

CKM MATRIX AND THE UNITARITY TRIANGLE

General Remarks

The unitary CKM matrix connects the *weak eigenstates* (d', s', b') and the corresponding *mass eigenstates* d, s, b:

$$\begin{pmatrix} d' \\ s' \\ b' \end{pmatrix} = \begin{pmatrix} V_{ud} & V_{us} & V_{ub} \\ V_{cd} & V_{cs} & V_{cb} \\ V_{td} & V_{ts} & V_{tb} \end{pmatrix} \begin{pmatrix} d \\ s \\ b \end{pmatrix} \equiv \hat{V}_{CKM} \begin{pmatrix} d \\ s \\ b \end{pmatrix}. \tag{1}$$

Many parametrizations of the CKM matrix have been proposed in the literature. The classification of different parametrizations can be found in [7]. We will use two parametrizations in these lectures: the standard parametrization [8] recommended by the Particle Data Group [9] and a generalization of the Wolfenstein parametrization [10] as presented in [11].

Standard Parametrization

With $c_{ij} = \cos\theta_{ij}$ and $s_{ij} = \sin\theta_{ij}$ ($i, j = 1, 2, 3$), the standard parametrization is given by:

$$\hat{V}_{CKM} = \begin{pmatrix} c_{12}c_{13} & s_{12}c_{13} & s_{13}e^{-i\delta} \\ -s_{12}c_{23} - c_{12}s_{23}s_{13}e^{i\delta} & c_{12}c_{23} - s_{12}s_{23}s_{13}e^{i\delta} & s_{23}c_{13} \\ s_{12}s_{23} - c_{12}c_{23}s_{13}e^{i\delta} & -s_{23}c_{12} - s_{12}c_{23}s_{13}e^{i\delta} & c_{23}c_{13} \end{pmatrix}, \tag{2}$$

where δ is the phase necessary for CP violation. c_{ij} and s_{ij} can all be chosen to be positive and δ may vary in the range $0 \leq \delta \leq 2\pi$. However, the measurements of CP violation in K decays force δ to be in the range $0 < \delta < \pi$.

From phenomenological applications we know that s_{13} and s_{23} are small numbers: $O(10^{-3})$ and $O(10^{-2})$, respectively. Consequently to an excellent accuracy $c_{13} = c_{23} = 1$ and the four independent parameters are given as

$$s_{12} = |V_{us}|, \quad s_{13} = |V_{ub}|, \quad s_{23} = |V_{cb}|, \quad \delta. \tag{3}$$

The first three can be extracted from tree level decays mediated by the transitions $s \to u$, $b \to u$ and $b \to c$ respectively. The phase δ can be extracted from CP violating transitions or loop processes sensitive to $|V_{td}|$ as we will clearly see in these lectures.

For numerical evaluations the use of the standard parametrization is strongly recommended. However once the four parameters in (3) have been determined it is often useful to make a change of basic parameters in order to expose the structure of the results more transparently. This brings us to the Wolfenstein parametrization [10] and its generalization given in [11].

Wolfenstein Parameterization and its Generalization

The Wolfenstein parametrization is an approximate parametrization of the CKM matrix in which each element is expanded as a power series in the small parameter $\lambda = |V_{us}| \approx 0.22$,

$$\hat{V} = \begin{pmatrix} 1 - \frac{\lambda^2}{2} & \lambda & A\lambda^3(\rho - i\eta) \\ -\lambda & 1 - \frac{\lambda^2}{2} & A\lambda^2 \\ A\lambda^3(1 - \rho - i\eta) & -A\lambda^2 & 1 \end{pmatrix} + O(\lambda^4), \tag{4}$$

and the set (3) is replaced by

$$\lambda, \quad A, \quad \rho, \quad \eta. \tag{5}$$

Because of the smallness of λ and the fact that for each element the expansion parameter is actually λ^2, it is sufficient to keep only the first few terms in this expansion.

The Wolfenstein parametrization is certainly more transparent than the standard parametrization. However, if one requires sufficient level of accuracy, the higher order terms in λ have to be included in phenomenological applications. This can be done in many ways that correspond to different definitions of the parameters in (5) [11].

A useful definition adopted by most authors in the literature is to go back to the standard parametrization (2) and to *define* the parameters (λ, A, ρ, η) through [11, 12]

$$s_{12} = \lambda, \quad s_{23} = A\lambda^2, \quad s_{13}e^{-i\delta} = A\lambda^3(\rho - i\eta) \tag{6}$$

to *all orders* in λ. It follows that

$$\rho = \frac{s_{13}}{s_{12}s_{23}}\cos\delta, \quad \eta = \frac{s_{13}}{s_{12}s_{23}}\sin\delta. \tag{7}$$

(6) and (7) represent simply the change of variables from (3) to (5). Making this change of variables in the standard parametrization (2) we find the CKM matrix as a function

of (λ, A, ρ, η) which satisfies unitarity exactly. Expanding next each element in powers of λ we recover the matrix in (4) and in addition find explicit corrections of $O(\lambda^4)$ and higher order terms. Including $O(\lambda^4)$ and $O(\lambda^5)$ terms we find

$$V_{ud} = 1 - \frac{1}{2}\lambda^2 - \frac{1}{8}\lambda^4, \qquad V_{cs} = 1 - \frac{1}{2}\lambda^2 - \frac{1}{8}\lambda^4(1 + 4A^2) \tag{8}$$

$$V_{tb} = 1 - \frac{1}{2}A^2\lambda^4, \qquad V_{cd} = -\lambda + \frac{1}{2}A^2\lambda^5[1 - 2(\rho + i\eta)] \tag{9}$$

$$V_{us} = \lambda + O(\lambda^7), \qquad V_{ub} = A\lambda^3(\rho - i\eta), \qquad V_{cb} = A\lambda^2 + O(\lambda^8) \tag{10}$$

$$V_{ts} = -A\lambda^2 + \frac{1}{2}A\lambda^4[1 - 2(\rho + i\eta)] \qquad V_{td} = A\lambda^3(1 - \bar{\rho} - i\bar{\eta}) \tag{11}$$

where [11]

$$\bar{\rho} = \rho(1 - \frac{\lambda^2}{2}), \qquad \bar{\eta} = \eta(1 - \frac{\lambda^2}{2}). \tag{12}$$

We emphasize that by definition V_{ub} remains unchanged and the corrections to V_{us} and V_{cb} appear only at $O(\lambda^7)$ and $O(\lambda^8)$, respectively. The advantage of this generalization of the Wolfenstein parametrization over other generalizations found in the literature is the absence of relevant corrections to V_{us}, V_{cd}, V_{ub} and V_{cb} and an elegant change in V_{td} which allows a simple generalization of the so-called unitarity triangle to higher orders in λ.

Unitarity Triangle (UT)

The unitarity of the CKM-matrix implies various relations between its elements. In particular, we have

$$V_{ud}V_{ub}^* + V_{cd}V_{cb}^* + V_{td}V_{tb}^* = 0. \tag{13}$$

Phenomenologically this relation is very interesting as it involves simultaneously the elements V_{ub}, V_{cb} and V_{td} which are under extensive discussion at present.

The relation (13) can be represented as a "unitarity" triangle in the complex $(\bar{\rho}, \bar{\eta})$ plane. The invariance of (13) under any phase-transformations implies that the corresponding triangle is rotated in the $(\bar{\rho}, \bar{\eta})$ plane under such transformations. Since the angles and the sides (given by the moduli of the elements of the mixing matrix) in this triangle remain unchanged, they are phase convention independent and are physical observables. Consequently they can be measured directly in suitable experiments. One can construct additional five unitarity triangles [13] corresponding to other orthogonality relations, like the one in (13). Some of them should be useful when LHC-B and BTeV experiments will provide data. The areas (A_Δ) of all unitarity triangles are equal and related to the measure of CP violation J_{CP} [14]: $|J_{CP}| = 2 \cdot A_\Delta$.

Noting that to an excellent accuracy $V_{cd}V_{cb}^*$ is real with $|V_{cd}V_{cb}^*| = A\lambda^3 + O(\lambda^7)$ and rescaling all terms in (13) by $A\lambda^3$ we indeed find that the relation (13) can be represented as the triangle in the complex $(\bar{\rho}, \bar{\eta})$ plane as shown in fig. 1.

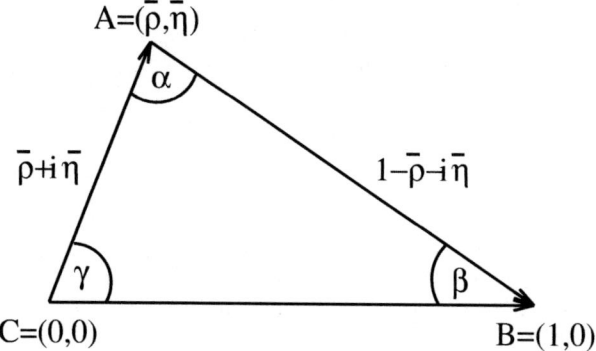

FIGURE 1. Unitarity Triangle.

Let us collect useful formulae related to this triangle:

- We can express $\sin(2\phi_i)$, $\phi_i = \alpha, \beta, \gamma$, in terms of $(\bar{\rho}, \bar{\eta})$ as follows:

$$\sin(2\alpha) = \frac{2\bar{\eta}(\bar{\eta}^2 + \bar{\rho}^2 - \bar{\rho})}{(\bar{\rho}^2 + \bar{\eta}^2)((1-\bar{\rho})^2 + \bar{\eta}^2)}, \quad (14)$$

$$\sin(2\beta) = \frac{2\bar{\eta}(1-\bar{\rho})}{(1-\bar{\rho})^2 + \bar{\eta}^2}, \quad (15)$$

$$\sin(2\gamma) = \frac{2\bar{\rho}\bar{\eta}}{\bar{\rho}^2 + \bar{\eta}^2} = \frac{2\rho\eta}{\rho^2 + \eta^2}. \quad (16)$$

- The lengths CA and BA to be denoted by R_b and R_t, respectively, are given by

$$R_b \equiv \frac{|V_{ud}V_{ub}^*|}{|V_{cd}V_{cb}^*|} = \sqrt{\bar{\rho}^2 + \bar{\eta}^2} = \left(1 - \frac{\lambda^2}{2}\right)\frac{1}{\lambda}\left|\frac{V_{ub}}{V_{cb}}\right|, \quad (17)$$

$$R_t \equiv \frac{|V_{td}V_{tb}^*|}{|V_{cd}V_{cb}^*|} = \sqrt{(1-\bar{\rho})^2 + \bar{\eta}^2} = \frac{1}{\lambda}\left|\frac{V_{td}}{V_{cb}}\right|. \quad (18)$$

- The angles β and $\gamma = \delta$ of the unitarity triangle are related directly to the complex phases of the CKM-elements V_{td} and V_{ub}, respectively, through

$$V_{td} = |V_{td}|e^{-i\beta}, \quad V_{ub} = |V_{ub}|e^{-i\gamma}. \quad (19)$$

- The angle α can be obtained through the relation

$$\alpha + \beta + \gamma = 180° \quad (20)$$

expressing the unitarity of the CKM-matrix.

The triangle depicted in fig. 1, $|V_{us}|$ and $|V_{cb}|$ give the full description of the CKM matrix. Looking at the expressions for R_b and R_t, we observe that within the SM the measurements of four CP *conserving* decays sensitive to $|V_{us}|, |V_{ub}|, |V_{cb}|$ and $|V_{td}|$ can tell us whether CP violation ($\bar{\eta} \neq 0$) is predicted in the SM. This fact is often used to determine the angles of the unitarity triangle without the study of CP-violating quantities.

Indeed, measuring the ratio $|V_{ub}/V_{cb}|$ in tree-level B decays and $|V_{td}|$ through $B_d^0 - \bar{B}_d^0$ mixing allows to determine R_b and R_t respectively. If so determined R_b and R_t satisfy

$$1 - R_b < R_t < 1 + R_b \tag{21}$$

then $\bar{\eta}$ is predicted to be non-zero on the basis of CP conserving transitions in the B-system alone without any reference to CP violation discovered in $K_L \to \pi^+\pi^-$ in 1964 [15]. Moreover one finds

$$\bar{\eta} = \pm\sqrt{R_b^2 - \bar{\rho}^2}, \qquad \bar{\rho} = \frac{1 + R_b^2 - R_t^2}{2}. \tag{22}$$

Grand Picture

The apex $(\bar{\rho}, \bar{\eta})$ of the UT can be efficiently hunted by means of rare and CP violating transitions as shown in fig. 2. Moreover the angles of this triangle can be measured in CP asymmetries in B-decays and using other strategies. This picture could describe in principle the reality in the year 2011, my retirement year, if the SM is the whole story. On the other hand in the presence of significant new physics contributions, the use of the SM expressions for rare and CP violating transitions in question, combined with future precise measurements, may result in curves which do not cross each other at a single point in the $(\bar{\rho}, \bar{\eta})$ plane. This would be truly exciting and most of us hope that this will turn out to be the case. In order to be able to draw such thin curves as in fig. 2, not only experiments but also the theory has to be under control. Let me then briefly discuss the theoretical framework for weak decays.

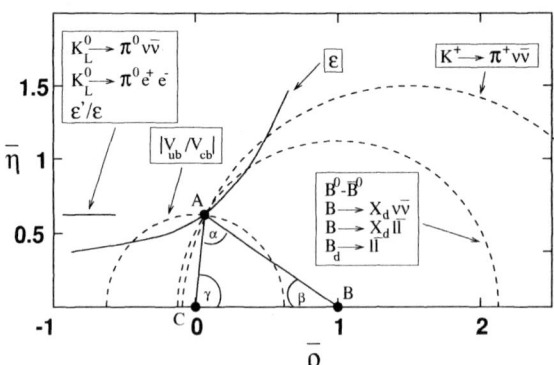

FIGURE 2. The ideal Unitarity Triangle.

THEORETICAL FRAMEWORK

The present framework for weak decays is based on the operator product expansion (OPE) that allows to separate short and long distance contributions to weak amplitudes and on the renormalization group (RG) methods that allow to sum large logarithms $\log \mu_{SD}/\mu_{LD}$ to all orders in perturbation theory. The full exposition of these methods can be found in [3, 5].

The OPE allows to write the effective weak Hamiltonian simply as follows

$$\mathcal{H}_{eff} = \frac{G_F}{\sqrt{2}} \sum_i V^i_{CKM} C_i(\mu) Q_i \, . \tag{23}$$

Here G_F is the Fermi constant and Q_i are the relevant local operators which govern the decays in question. They are built out of quark and lepton fields. The CKM factors V^i_{CKM} and the Wilson coefficients $C_i(\mu)$ describe the strength with which a given operator enters the Hamiltonian. The latter coefficients can be considered as scale dependent "couplings" related to "vertices" Q_i and as discussed below can be calculated using perturbative methods as long as μ is not too small.

An amplitude for a decay of a given meson $M = K, B, ..$ into a final state $F = \pi \nu \bar{\nu}, \pi\pi, DK$ is then simply given by

$$A(M \to F) = \langle F | \mathcal{H}_{eff} | M \rangle = \frac{G_F}{\sqrt{2}} \sum_i V^i_{CKM} C_i(\mu) \langle F | Q_i(\mu) | M \rangle, \tag{24}$$

where $\langle F | Q_i(\mu) | M \rangle$ are the matrix elements of Q_i between M and F, evaluated at the renormalization scale μ.

The essential virtue of OPE is this one. It allows to separate the problem of calculating the amplitude $A(M \to F)$ into two distinct parts: the *short distance* (perturbative) calculation of the coefficients $C_i(\mu)$ and the *long-distance* (generally non-perturbative) calculation of the matrix elements $\langle Q_i(\mu) \rangle$. The scale μ separates, roughly speaking, the physics contributions into short distance contributions contained in $C_i(\mu)$ and the long distance contributions contained in $\langle Q_i(\mu) \rangle$. Thus C_i include the top quark contributions and contributions from other heavy particles such as W-, Z-bosons and charged Higgs particles or supersymmetric particles in the supersymmetric extensions of the SM. Consequently $C_i(\mu)$ depend generally on m_t and also on the masses of new particles if extensions of the SM are considered. This dependence can be found by evaluating so-called *box* and *penguin* diagrams with full W-, Z-, top- and new particles exchanges and *properly* including short distance QCD effects. The latter govern the μ-dependence of $C_i(\mu)$.

The value of μ can be chosen arbitrarily but the final result must be μ-independent. Therefore the μ-dependence of $C_i(\mu)$ has to cancel the μ-dependence of $\langle Q_i(\mu) \rangle$. In other words it is a matter of choice what exactly belongs to $C_i(\mu)$ and what to $\langle Q_i(\mu) \rangle$. This cancellation of the μ-dependence involves generally several terms in the expansion in (24). The coefficients $C_i(\mu)$ depend also on the renormalization scheme. This scheme dependence must also be canceled by the one of $\langle Q_i(\mu) \rangle$ so that the physical amplitudes are renormalization scheme independent. Again, as in the case of the μ-dependence, the cancellation of the renormalization scheme dependence involves generally several terms in the expansion (24).

Although μ is in principle arbitrary, it is customary to choose μ to be of the order of the mass of the decaying hadron. This is $O(m_b)$ and $O(m_c)$ for B decays and D decays respectively. In the case of K decays the typical choice is $\mu = O(1-2 \text{ GeV})$ instead of $O(m_K)$, which is much too low for any perturbative calculation of the couplings C_i. Now due to the fact that $\mu \ll M_{W,Z}$, m_t, large logarithms $\ln M_W/\mu$ compensate in the evaluation of $C_i(\mu)$ the smallness of the QCD coupling constant α_s and terms $\alpha_s^n (\ln M_W/\mu)^n$, $\alpha_s^n (\ln M_W/\mu)^{n-1}$ etc. have to be resummed to all orders in α_s before a reliable result for C_i can be obtained. This can be done very efficiently by means of the renormalization group methods. The resulting *renormalization group improved* perturbative expansion for $C_i(\mu)$ in terms of the effective coupling constant $\alpha_s(\mu)$ does not involve large logarithms and is more reliable. The related technical issues are discussed in detail in [3] and [5].

Clearly, in order to calculate the amplitude $A(M \to F)$ the matrix elements $\langle Q_i(\mu) \rangle$ have to be evaluated. Since they involve long distance contributions one is forced in this case to use non-perturbative methods such as lattice calculations, the 1/N expansion (N is the number of colours), QCD sum rules, hadronic sum rules, chiral perturbation theory and so on. In the case of certain B-meson decays, the *Heavy Quark Effective Theory* (HQET) and *Heavy Quark Expansions* (HQE) also turn out to be useful tools. Needless to say, all these non-perturbative methods have some limitations. Consequently the dominant theoretical uncertainties in the decay amplitudes reside in the matrix elements $\langle Q_i(\mu) \rangle$ and non-perturbative parameters present in HQET and HQE.

The fact that in many cases the matrix elements $\langle Q_i(\mu) \rangle$ cannot be reliably calculated at present, is very unfortunate. The main goals of the experimental studies of weak decays is the determination of the CKM factors V_{CKM} and the search for the physics beyond the SM. Without a reliable estimate of $\langle Q_i(\mu) \rangle$ these goals cannot be achieved unless these matrix elements can be determined experimentally or removed from the final measurable quantities by taking suitable ratios and combinations of decay amplitudes or branching ratios. Flavour symmetries like $SU(2)_F$ and $SU(3)_F$ relating various matrix elements can be useful in this respect, provided flavour breaking effects can be reliably calculated. However, this can be achieved rarely and often one has to face directly the calculation of $\langle Q_i(\mu) \rangle$.

One of the outstanding issues in the calculation of $\langle Q_i(\mu) \rangle$ is the compatibility ("matching") of $\langle Q_i(\mu) \rangle$ with $C_i(\mu)$. $\langle Q_i(\mu) \rangle$ have to carry the correct μ and renormalization scheme dependence in order to ensure the μ and scheme independence of physical amplitudes. Most of the non-perturbative methods struggle still with this problem.

After these general remarks let me for the rest of this section discuss a master formula for weak decay amplitudes [16] that follows from OPE and RG, in particular from (24), but is more useful for phenomenological applications than the formal expressions given above. This formula while incorporating the SM contributions applies also to any extention of the SM. It reads:

$$A(\text{Decay}) = \sum_i B_i \eta_{\text{QCD}}^i V_{\text{CKM}}^i [F_{\text{SM}}^i + F_{\text{New}}^i] + \sum_k B_k^{\text{New}} [\eta_{\text{QCD}}^k]^{\text{New}} V_{\text{New}}^k [G_{\text{New}}^k]. \quad (25)$$

The non-perturbative parameters B_i represent the matrix elements of local operators present in the SM. For instance in the case of $K^0 - \bar{K}^0$ mixing, the matrix element of the operator $\bar{s}\gamma_\mu(1-\gamma_5)d \otimes \bar{s}\gamma^\mu(1-\gamma_5)d$ is represented by the parameter $\hat{B}_K$. There are other non-perturbative parameters in the SM that represent matrix elements of operators Q_i with different colour and Dirac structures. The objects η^i_{QCD} are the QCD factors resulting from RG-analysis of the corresponding operators and F^i_{SM} stand for the so-called Inami-Lim functions [17] that result from the calculations of various box and penguin diagrams. They depend on the top-quark mass. V^i_{CKM} are the CKM-factors we want to determine.

New physics can contribute to our master formula in two ways. It can modify the importance of a given operator, present already in the SM, through the new short distance functions F^i_{New} that depend on the new parameters in the extensions of the SM like the masses of charginos, squarks, charged Higgs particles and $\tan\beta = v_2/v_1$ in the MSSM. These new particles enter the new box and penguin diagrams. In more complicated extensions of the SM new operators (Dirac structures) that are either absent or very strongly suppressed in the SM, can become important. Their contributions are described by the second sum in (25) with $B^{New}_k, [\eta^k_{QCD}]^{New}, V^k_{New}, G^k_{New}$ being analogs of the corresponding objects in the first sum of the master formula. The V^k_{New} show explicitly that the second sum describes generally new sources of flavour and CP violation beyond the CKM matrix. This sum may, however, also include contributions governed by the CKM matrix that are strongly suppressed in the SM but become important in some extensions of the SM. A typical example is the enhancement of the operators with Dirac structures $(V-A) \otimes (V+A)$, $(S-P) \otimes (S \pm P)$ and $\sigma_{\mu\nu}(S-P) \otimes \sigma^{\mu\nu}(S-P)$ contributing to $K^0 - \bar{K}^0$ and $B^0_{d,s} - \bar{B}^0_{d,s}$ mixings in the MSSM with large $\tan\beta$ and in supersymmetric extensions with new flavour violation. The most recent compilation of references to existing next-to-leading (NLO) calculations of η^i_{QCD} and $[\eta^k_{QCD}]^{New}$ can be found in [4].

Clearly the new functions F^i_{New} and G^k_{New} as well as the factors V^k_{New} may depend on new CP violating phases complicating considerably phenomenological analysis. In the present lecture, that is dominantly devoted to the SM, I will only consider the simplest class of the extensions of the SM in which the second sum in (25) is absent (no new operators) and flavour changing transitions are governed by the CKM matrix. In particular there are no new complex phases beyond the CKM phase. I will call this scenario "Minimal Flavour Violation" (MFV) [18, 19] being aware of the fact that for some authors MFV means a more general framework in which also new operators can give significant contributions. In the MFV models, as defined in [18, 19], our master formula simplifies to

$$A(\text{Decay}) = \sum_i B_i \eta^i_{QCD} V^i_{CKM} [F^i_{SM} + F^i_{New}] \qquad (26)$$

with F^i_{SM} and F^i_{New} being real.

CKM FROM TREE LEVEL DECAYS

What do we know about the CKM matrix and the unitarity triangle on the basis of *tree level* decays? Here the semi-leptonic K and B decays together with chiral perturbation theory, HQET and HQE play the decisive role. The present situation can be summarized by

$$|V_{us}| = \lambda = 0.222 \pm 0.002 \qquad |V_{cb}| = 0.041 \pm 0.002, \qquad (27)$$

$$\frac{|V_{ub}|}{|V_{cb}|} = 0.085 \pm 0.018, \qquad |V_{ub}| = (3.49 \pm 0.76) \cdot 10^{-3}. \qquad (28)$$

implying

$$A = 0.83 \pm 0.04, \qquad R_b = 0.37 \pm 0.08 . \qquad (29)$$

This tells us only that the apex A of the unitarity triangle lies in the band around $(\bar{\rho}, \bar{\eta}) = (0,0)$ as described by R_b in (29). While this information appears at first sight to be rather limited, it is very important for the following reason. As $|V_{us}|$, $|V_{cb}|$, $|V_{ub}|$ and R_b are determined here from tree level decays, their values given above are to an excellent accuracy independent of any new physics contributions. That is they are universal fundamental constants valid in any extention of the SM. Therefore its precise determination is of utmost importance. Most recent discussions of the determinations of $|V_{us}|(|V_{ud}|)$ and $|V_{ub}|(|V_{cb}|)$ can be found in [20] and [21], respectively.

GOING BEYOND THE TREES

In order to answer the question where the apex A lies in the band $R_b = 0.37 \pm 0.08$, we have to look at other decays. Most promising in this respect are the so-called "loop induced" decays and transitions and CP-violating B decays which will be discussed in these lectures. They should allow us to answer already in this decade the important question, whether the Kobayashi-Maskawa picture of CP violation is correct and more generally whether the Standard Model offers a correct description of weak decays of hadrons. In the language of the unitarity triangle the question is whether the various curves in the $(\bar{\rho}, \bar{\eta})$ plane extracted from different decays and transitions using the SM formulae will cross each other at a single point as shown in fig. 2 and whether the angles (α, β, γ) in the resulting triangle will agree with those extracted from CP-asymmetries in B decays and CP-conserving B decays. The inconsistencies in the $(\bar{\rho}, \bar{\eta})$ plane will then give us some hints about the physics beyond the SM. One obvious inconsistency would be the violation of the constraint (21). Let us then summarize what we know about the unitarity triangle at present.

STANDARD ANALYSIS OF THE UNITARITY TRIANGLE

This analysis uses $|V_{us}|$, $|V_{cb}|$, $|V_{ub}/V_{cb}|$ given above and the following three constraints:

- ε_K–Hyperbola (Indirect CP Violation in $K_L \to \pi\pi$):

$$\bar{\eta}\left[(1-\bar{\rho})A^2 \eta^{tt}_{\text{QCD}} F_{tt} + P_c(\varepsilon)\right] A^2 \hat{B}_K = 0.204 , \qquad (30)$$

where $\eta^{tt}_{\text{QCD}} = 0.57 \pm 0.01$, $P_c(\varepsilon) = 0.30 \pm 0.05$ represents charm contribution and $F_{tt} = 2.38 \pm 0.11$ is the Inami-Lim (t,t) box diagram function, denoted often by $S_0(x_t)$.

- $B_d^0 - \bar{B}_d^0$-Mixing Constraint:

$$R_t = 0.85 \left[\frac{0.83}{A}\right] \sqrt{\frac{2.38}{F_{tt}}} \sqrt{\frac{\Delta M_d}{0.487/\text{ps}}} \left[\frac{230 \text{ MeV}}{\sqrt{\hat{B}_d}F_{B_d}}\right] \sqrt{\frac{0.55}{\eta_B^{\text{QCD}}}} \quad (31)$$

where $A = 0.83 \pm 0.04$, $\Delta M_d = (0.487 \pm 0.009)/\text{ps}$ and $\eta_B^{\text{QCD}} = 0.55 \pm 0.01$.

- $B_s^0 - \bar{B}_s^0$-Mixing Constraint ($\Delta M_d/\Delta M_s$):

$$R_t = 0.94 \sqrt{\frac{\Delta M_d}{0.487/\text{ps}}} \sqrt{\frac{15.0/\text{ps}}{\Delta M_s}} \left[\frac{\xi}{1.15}\right], \quad \xi = \frac{\sqrt{\hat{B}_s}F_{B_s}}{\sqrt{\hat{B}_d}F_{B_d}} \quad (32)$$

where $\Delta M_s > 15.0/\text{ps}$ from LEP experiments.

The main uncertainties in this analysis originate in the theoretical uncertainties in the non-perturbative parameters $\hat{B}_K$ and $\sqrt{\hat{B}_d}F_{B_d}$ and to a lesser extent in ξ:

$$\hat{B}_K = 0.85 \pm 0.15, \quad \sqrt{\hat{B}_d}F_{B_d} = (230 \pm 40) \text{ MeV}, \quad \xi = 1.15 \pm 0.06. \quad (33)$$

Also the uncertainties in $|V_{cb}|$ and in particular in $|V_{ub}/V_{cb}|$, are substantial. Reviews of lattice results for the parameters in question can be found in [22].

One of the important issues is the error analysis of these formulae. In the literature five different methods are used: Gaussian approach [23], Bayesian approach [24], frequentist approach [25], 95% C.L. scan method [26] and the simple (naive) scanning within one standard deviation as used by myself. Interestingly, the last method gives ranges for the output quantities that are similar to the 95% C.L. ranges obtained by the remaining methods. To this end the same input parameters have to be used and the implementation of the lower bound on ΔM_s has to be done in the same manner. I show in fig. 3 the result of my own analysis that uses naive scanning. The allowed region for $(\bar{\rho}, \bar{\eta})$ is the shaded area on the right hand side of the circle representing the lower bound for ΔM_s, that is $\Delta M_s > 15/\text{ps}$. The hyperbolas in fig. 3 give the constraint from ε and the two circles centered at $(0,0)$ the constraint from $|V_{ub}/V_{cb}|$. The circle on the right comes from $B_d^0 - \bar{B}_d^0$ mixing and excludes the region to its right. We observe that the region $\bar{\rho} < 0$ is practically excluded by the lower bound on ΔM_s. It is clear from this figure that ΔM_s is a very important ingredient in this analysis and that the measurement of ΔM_s giving also lower bound on R_t could have a large impact on the plot in fig. 3.

The ranges for various quantities found using the scanning method are compared in table 1 with the 95% C.L. ranges from Bayesian I of Ciuchini et al [24] that uses $|V_{cb}| = 0.0410 \pm 0.0016$ and $|V_{ub}/V_{cb}| = 0.086 \pm 0.009$ and Bayesian II with $|V_{cb}|$ and $|V_{ub}/V_{cb}|$ as in (27) and (28) that are used in my analysis. I thank Stocchi for providing the latter numbers. My ranges are substantially larger than Bayesian I but only slightly larger than Bayesian II. This is partly related to a different treatment of the bound on ΔM_s done by Ciuchini et al and myself. My ranges are very close to the ones obtained using the frequentist approach [25].

TABLE 1. Output of the Standard Analysis. $\lambda_t = V_{ts}^* V_{td}$.

Quantity	Scanning	Bayesian I	Bayesian II
$\bar{\rho}$	$0.07 - 0.34$	$0.14 - 0.30$	$0.13 - 0.34$
$\bar{\eta}$	$0.22 - 0.45$	$0.24 - 0.40$	$0.22 - 0.46$
$\sin(2\beta)$	$0.50 - 0.84$	$0.56 - 0.82$	$0.52 - 0.92$
$\sin(2\alpha)$	$-0.87 - 0.36$	$-0.83 - 0.04$	$-0.85 - 0.14$
γ	$37.7° - 75.7°$	$42.8° - 67.4°$	$41.8° - 67.6°$
$\mathrm{Im}\lambda_t / 10^{-4}$	$0.94 - 1.60$	$0.93 - 1.43$	$0.91 - 1.55$
$|V_{td}|/10^{-3}$	$6.7 - 9.3$	$7.0 - 8.6$	$6.8 - 8.7$

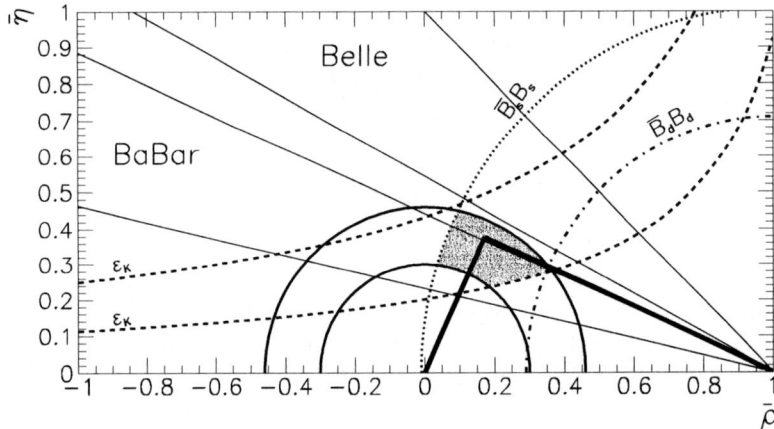

FIGURE 3. Conservative Unitarity Triangle as of January 2002.

One of the highlights of the year 2001 were the improved measurements of $\sin 2\beta$ by means of the time-dependent CP asymmetry in $B_d^0(\bar{B}_d^0) \to \psi K_S$ decays

$$a_{\psi K_S}(t) \equiv -a_{\psi K_S} \sin(\Delta M_d t) = -\sin 2\beta \sin(\Delta M_d t) \tag{34}$$

with the last relation valid in those MFV models in which as in the SM $F_{tt} > 0$ [27]. The most recent measurements of $a_{\psi K_S}$ from the BaBar and Belle Collaborations read

$$(\sin 2\beta)_{\psi K_S} = \begin{cases} 0.59 \pm 0.14 \pm 0.05 & \text{(BaBar) [28]} \\ 0.99 \pm 0.14 \pm 0.06 & \text{(Belle) [29]} \end{cases} \tag{35}$$

and establish confidently CP violation in the B system! A mile stone in the field of CP violation. Combining these results with earlier measurements by CDF ($0.79^{+0.41}_{-0.44}$) and ALEPH ($0.84^{+0.82}_{-1.04} \pm 0.16$) gives the grand average

$$a_{\psi K_S} = 0.79 \pm 0.10. \tag{36}$$

In view of the fact that the BaBar and Belle results are not fully consistent with each other, the averaging of these results and the grand average given above could be questioned. Probably a better description of the present situation is $a_{\psi K_S} = 0.80 \pm 0.20$.

In any case, these first direct measurements of the angle β are in a good agreement (see fig. 3) with the results of the standard analyses of the unitarity triangle within the SM, even if the Belle result appears a bit too high. Clearly in view of a considerable difference between BaBar and Belle results and still sizable uncertainty in the error estimates of $(\sin 2\beta)_{SM}$, there is a room for new physics contributions but the agreement of the prediction for $\sin 2\beta$ from the fits of the unitarity triangle with the measured value of $a_{\psi K_S}$ is a strong indication that the CKM matrix could turn out to be the dominant source of CP violation in flavour violating decays.

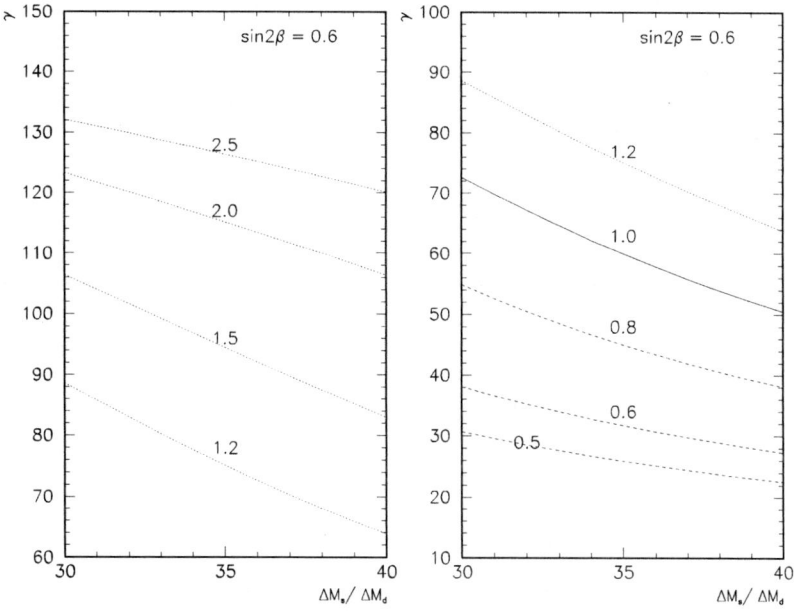

FIGURE 4. γ as a function of $\Delta M_s / \Delta M_d$ for $\sin 2\beta = 0.6$ and different R_{sd} [30].

In order to be sure whether this is indeed the case other theoretically clean quantities have to be measured. In particular the angle γ that is more sensitive to new physics contributions than β. In this context the measurement of the ratio $\Delta M_s / \Delta M_d$ will play an important role as for a fixed value of $\sin 2\beta$, the extracted value for γ is a sensitive function of $\Delta M_s / \Delta M_d$ as shown in fig. 4. The solid line, labeled by $R_{sd} = 1.0$ in the right plot, represents MFV models. The remaining lines, obtained in a general analysis in [30], represent generalized MFV models in which also significant contributions of new operators are possible. See the discussion below (25). In these models the expression for R_t in (32) receives an additional factor $\sqrt{R_{sd}}$. For $R_{sd} > 1.2$, the angle $\gamma > 90°$ is possible provided $\Delta M_s / \Delta M_d$ is not too large.

At this point I would like to stress the importance of the precise measurements of $a_{\psi K_S}$ and $\Delta M_s/\Delta M_d$ that should be available within the coming years. These two measurements taken together allow the determination of $\bar{\rho}$ and $\bar{\eta}$ through

$$\bar{\rho} \approx 1 - R_t \left[1 - \frac{a_{\psi K_S}^2}{8} \right], \qquad \bar{\eta} \approx R_t \frac{a_{\psi K_S}}{2} \left[1 + \frac{a_{\psi K_S}^2}{8} \right] \qquad (37)$$

with R_t given by (32). Exact expressions can be found in [19, 27, 30]. The only theoretical uncertainty in these formulae resides in ξ that should be known from lattice calculations within a few percent in the next years.

$K^+ \to \pi^+ \nu \bar{\nu}$, $K_L \to \pi^0 \nu \bar{\nu}$ AND $K_L \to \pi^0 e^+ e^-$

The rare decays $K^+ \to \pi^+ \nu \bar{\nu}$ and $K_L \to \pi^0 \nu \bar{\nu}$ are very promising probes of flavour physics within the SM and possible extensions, since they are governed by short distance interactions. They proceed through Z^0-penguin and box diagrams. As the required hadronic matrix elements can be extracted from the leading semileptonic decays and other long distance contributions turn out to be negligible [31], the relevant branching ratios can be computed to an exceptionally high degree of precision [32, 33]. The main theoretical uncertainty in the CP conserving decay $K^+ \to \pi^+ \nu \bar{\nu}$ originates in the value of $m_c(\mu_c)$. It has been reduced through NLO corrections down to $\pm 7\%$ [32] at the level of the branching ratio. The dominantly CP-violating decay $K_L \to \pi^0 \nu \bar{\nu}$ [34] is even cleaner as only the internal top contributions matter. The theoretical error for $Br(K_L \to \pi^0 \nu \bar{\nu})$ amounts to $\pm 2\%$ and is safely negligible.

The rare decay $K_L \to \pi^0 e^+ e^-$ is not as clean as these two decays but as discussed below could in principle also offer a useful measurement of $\text{Im}\lambda_t$.

The Decay $K^+ \to \pi^+ \nu \bar{\nu}$

The basic formula for the branching ratio is as follows

$$Br(K^+ \to \pi^+ \nu \bar{\nu}) = \kappa_+ \cdot \left[\left(\frac{\text{Im}\lambda_t}{\lambda^5} X(x_t) \right)^2 + \left(\frac{\text{Re}\lambda_c}{\lambda} P_0(X) + \frac{\text{Re}\lambda_t}{\lambda^5} X(x_t) \right)^2 \right], \qquad (38)$$

$$\kappa_+ = r_{K^+} \frac{3\alpha^2 Br(K^+ \to \pi^0 e^+ \nu)}{2\pi^2 \sin^4 \Theta_W} \lambda^8 = 4.42 \cdot 10^{-11}, \qquad (39)$$

where we have used $\lambda = 0.222$,

$$\alpha = \frac{1}{129}, \qquad \sin^2 \Theta_W = 0.23, \qquad Br(K^+ \to \pi^0 e^+ \nu) = 4.82 \cdot 10^{-2}. \qquad (40)$$

Here $\lambda_i = V_{is}^* V_{id}$ with λ_c being real to a very high accuracy. $r_{K^+} = 0.901$ summarizes isospin breaking corrections in relating $K^+ \to \pi^+ \nu \bar{\nu}$ to $K^+ \to \pi^0 e^+ \nu$ [35]. Next

$$X(x_t) = 1.51 \cdot \left[\frac{\overline{m}_t(m_t)}{166 \text{ GeV}} \right]^{1.15} \tag{41}$$

represents internal top contributions and $P_0(X) = 0.40 \pm 0.06$ results from the internal charm contributions [32].

We can next put (38) into a more transparent form [11]:

$$Br(K^+ \to \pi^+ \nu \bar{\nu}) = 4.42 \cdot 10^{-11} A^4 X^2(x_t) \frac{1}{\sigma} \left[(\sigma \bar{\eta})^2 + (\rho_0 - \bar{\rho})^2 \right], \tag{42}$$

where

$$\sigma = \left(\frac{1}{1 - \frac{\lambda^2}{2}} \right)^2, \qquad \rho_0 = 1 + \frac{P_0(X)}{A^2 X(x_t)}. \tag{43}$$

The measured value of $Br(K^+ \to \pi^+ \nu \bar{\nu})$ then determines an ellipse in the $(\bar{\rho}, \bar{\eta})$ plane centered at $(\rho_0, 0)$ and having the squared axes

$$\bar{\rho}_1^2 = r_0^2, \qquad \bar{\eta}_1^2 = \left(\frac{r_0}{\sigma} \right)^2, \qquad r_0^2 = \frac{1}{A^4 X^2(x_t)} \left[\frac{\sigma \cdot Br(K^+ \to \pi^+ \nu \bar{\nu})}{4.42 \cdot 10^{-11}} \right]. \tag{44}$$

Note that r_0 depends only on the top contribution. The departure of ρ_0 from unity measures the relative importance of the internal charm contributions. $\rho_0 \approx 1.35$.

The ellipse defined by r_0, ρ_0 and σ given above intersects with the circle (17). This allows to determine $\bar{\rho}$ and $\bar{\eta}$ with ($\bar{\eta} > 0$ assumed)

$$\bar{\rho} = \frac{1}{1 - \sigma^2} \left(\rho_0 - \sqrt{\sigma^2 \rho_0^2 + (1 - \sigma^2)(r_0^2 - \sigma^2 R_b^2)} \right), \qquad \bar{\eta} = \sqrt{R_b^2 - \bar{\rho}^2} \tag{45}$$

and consequently

$$R_t^2 = 1 + R_b^2 - 2\bar{\rho}, \qquad V_{td} = A\lambda^3 (1 - \bar{\rho} - i\bar{\eta}), \qquad |V_{td}| = A\lambda^3 R_t. \tag{46}$$

The determination of $|V_{td}|$ and of the unitarity triangle requires the knowledge of V_{cb} (or A) and of $|V_{ub}/V_{cb}|$. Both values are subject to theoretical uncertainties present in the existing analyses of tree level decays. Whereas the dependence on $|V_{ub}/V_{cb}|$ is rather weak, the very strong dependence of $Br(K^+ \to \pi^+ \nu \bar{\nu})$ on A or V_{cb} makes a precise prediction for this branching ratio difficult at present.

Scanning the input parameters of Sections 4 and 6 we find

$$Br(K^+ \to \pi^+ \nu \bar{\nu}) = (7.5 \pm 2.9) \cdot 10^{-11} \tag{47}$$

where the error comes dominantly from the uncertainties in the CKM parameters.

It is possible to derive an upper bound on $Br(K^+ \to \pi^+ \nu \bar{\nu})$ [36]:

$$Br(K^+ \to \pi^+ \nu \bar{\nu})_{\max} = \frac{\kappa_+}{\sigma} \left[P_0(X) + A^2 X(x_t) \frac{r_{sd}}{\lambda} \sqrt{\frac{\Delta M_d}{\Delta M_s}} \right]^2 \tag{48}$$

where $r_{ds} = \xi\sqrt{m_{B_s}/m_{B_d}}$ with ξ defined in (32). This equation translates a lower bound on ΔM_s into an upper bound on $Br(K^+ \to \pi^+\nu\bar{\nu})$. This bound is very clean and does not involve theoretical hadronic uncertainties except for r_{sd}. Using

$$A < 0.87, \quad P_0(X) < 0.46, \quad X(x_t) < 1.56 \tag{49}$$

we show in table 2 $Br(K^+ \to \pi^+\nu\bar{\nu})_{max}$ as a function of ξ for different values of ΔM_s. This limit could be further strengthened with improved input. However, this bound is strong enough to indicate a clear conflict with the SM if $Br(K^+ \to \pi^+\nu\bar{\nu})$ should be measured at $15 \cdot 10^{-11}$.

TABLE 2. Upper bound on $Br(K^+ \to \pi^+\nu\bar{\nu})$ in units of 10^{-11} for different values of ξ and ΔM_s.

ξ	$\Delta M_s = 15/ps$	$\Delta M_s = 18/ps$	$\Delta M_s = 21/ps$
1.25	11.9	10.4	9.3
1.20	11.2	9.8	8.8
1.15	10.5	9.3	8.3
1.10	9.9	8.7	7.9

In view of these findings the recent message from AGS E787 collaboration at Brookhaven is very exciting. In addition to the first event found in 1997 [37], this collaboration reported a second event last November [38]. The resulting branching ratio

$$Br(K^+ \to \pi^+\nu\bar{\nu}) = (15.7^{+17.5}_{-8.2}) \cdot 10^{-11} \tag{50}$$

turns out to by a factor of 2 above the SM expectation. Even if the errors are substantial and the result is compatible with the SM, the branching ratio (50) implies already a non-trivial lower bound on $|V_{td}|$ [38, 39]: $|V_{td}| > 0.0051$ (90% C.L.).

Once $Br(K^+ \to \pi^+\nu\bar{\nu}) \equiv Br(K^+)$ has been more accurately measured, $|V_{td}|$ can be extracted subject to various uncertainties:

$$\frac{\sigma(|V_{td}|)}{|V_{td}|} = \pm 0.04_{scale} \pm \frac{\sigma(|V_{cb}|)}{|V_{cb}|} \pm 0.7\frac{\sigma(\bar{m}_c)}{\bar{m}_c} \pm 0.65\frac{\sigma(Br(K^+))}{Br(K^+)} . \tag{51}$$

Taking $\sigma(|V_{cb}|) = 0.002$, $\sigma(\bar{m}_c) = 100\,\text{MeV}$ and $\sigma(Br(K^+)) = 10\%$ and adding the errors in quadrature we find that $|V_{td}|$ can be determined with an accuracy of $\pm 10\%$. This number is increased to $\pm 11\%$ once the uncertainties due to m_t, α_s and $|V_{ub}|/|V_{cb}|$ are taken into account. Clearly this determination can be improved although a determination of $|V_{td}|$ with an accuracy better than $\pm 5\%$ seems rather unrealistic in this decade.

The experimental outlook for $K^+ \to \pi^+\nu\bar{\nu}$ has been reviewed in [40]. A new experiment, AGS E949 [41] is scheduled to run in 2001-3. It is expected to reach a sensitivity of $\sim 10^{-11}$/event. In the long term, the CKM experiment at Fermilab [42] should be able to reach $\sim 10^{-12}$/event sensitivity. At this level an accurate measurement of the branching ratio should be possible.

The Decay $K_L \to \pi^0 \nu \bar\nu$

The branching ratio $Br(K_L \to \pi^0 \nu \bar\nu)$ is given simply as follows

$$Br(K_L \to \pi^0 \nu \bar\nu) = \kappa_L \cdot \left(\frac{\mathrm{Im}\lambda_t}{\lambda^5} X(x_t)\right)^2, \tag{52}$$

$$\kappa_L = \frac{r_{K_L}}{r_{K^+}} \frac{\tau(K_L)}{\tau(K^+)} \kappa_+ = 1.93 \cdot 10^{-10} \tag{53}$$

with κ_+ given in (39) and $r_{K_L} = 0.944$ summarizing isospin breaking corrections in relating $K_L \to \pi^0 \nu \bar\nu$ to $K^+ \to \pi^0 e^+ \nu$ [35].

Using the improved Wolfenstein parametrization and (41) we can rewrite (52) as

$$Br(K_L \to \pi^0 \nu \bar\nu) = 2.55 \cdot 10^{-11} \left[\frac{\bar\eta}{0.34}\right]^2 \left[\frac{\overline{m}_t(m_t)}{166\,\mathrm{GeV}}\right]^{2.3} \left[\frac{|V_{cb}|}{0.041}\right]^4. \tag{54}$$

The determination of $\bar\eta$ using $Br(K_L \to \pi^0 \nu \bar\nu)$ requires the knowledge of V_{cb} and m_t. The very strong dependence on V_{cb} makes a precise prediction for this branching ratio difficult at present.

On the other hand inverting (52) and using (41) we find

$$\mathrm{Im}\lambda_t = 1.30 \cdot 10^{-4} \left[\frac{166\,\mathrm{GeV}}{\overline{m}_t(m_t)}\right]^{1.15} \left[\frac{Br(K_L \to \pi^0 \nu \bar\nu)}{2.55 \cdot 10^{-11}}\right]^{1/2}. \tag{55}$$

without any uncertainty in $|V_{cb}|$. (55) offers the cleanest method to measure $\mathrm{Im}\lambda_t$ [43]; even better than the CP asymmetries in B decays discussed in section 9 that require the knowledge of $|V_{cb}|$ to determine $\mathrm{Im}\lambda_t$. Measuring $Br(K_L \to \pi^0 \nu \bar\nu)$ with 10% accuraccy allows to determine $\mathrm{Im}\lambda_t$ with an error of 5% [3, 43].

Scanning the input parameters of Sections 4 and 6 we find

$$Br(K_L \to \pi^0 \nu \bar\nu) = (2.6 \pm 1.2) \cdot 10^{-11} \tag{56}$$

where the error comes dominantly from the uncertainties in the CKM parameters.

The present upper bound on $Br(K_L \to \pi^0 \nu \bar\nu)$ from the KTeV experiment at Fermilab [44] reads

$$Br(K_L \to \pi^0 \nu \bar\nu) < 5.9 \cdot 10^{-7}. \tag{57}$$

This is about four orders of magnitude above the SM expectation (56). Moreover this bound is substantially weaker than the *model independent* bound [45] from isospin symmetry:

$$Br(K_L \to \pi^0 \nu \bar\nu) < 4.4 \cdot Br(K^+ \to \pi^+ \nu \bar\nu) \tag{58}$$

which through (50) gives

$$Br(K_L \to \pi^0 \nu \bar\nu) < 1.6 \cdot 10^{-9} \quad (90\%\,C.L.) \tag{59}$$

The experimental outlook for this decay has been reviewed in [40, 46]. The KEK E391a experiment [47] should reach sensitivity of $\sim 10^{-10}$/event which would give some events only in the presence of new physics contributions. Next a very interesting new experiment KOPIO at Brookhaven (BNL E926) [48] expects to reach the single event sensitivity of $2 \cdot 10^{-12}$ and to collect 50 events. Finally an experiment at the 50 GeV JHF in Japan should be able to collect 1000 events. Both KOPIO and JHF should provide very important measurements of this gold-plated decay.

Unitarity Triangle and $\sin 2\beta$ from $K \to \pi \nu \bar{\nu}$

The measurement of $Br(K^+ \to \pi^+ \nu\bar{\nu})$ and $Br(K_L \to \pi^0 \nu\bar{\nu})$ can determine the unitarity triangle completely, provided m_t and V_{cb} are known [43]. Using these two branching ratios simultaneously allows to eliminate $|V_{ub}/V_{cb}|$ from the analysis which removes a considerable uncertainty. Indeed it is evident from (38) and (52) that, given $Br(K^+ \to \pi^+ \nu\bar{\nu})$ and $Br(K_L \to \pi^0 \nu\bar{\nu})$, one can extract both $\mathrm{Im}\lambda_t$ and $\mathrm{Re}\lambda_t$. One finds [43, 5]

$$\mathrm{Im}\lambda_t = \lambda^5 \frac{\sqrt{B_2}}{X(x_t)} \qquad \mathrm{Re}\lambda_t = -\lambda^5 \frac{\frac{\mathrm{Re}\lambda_c}{\lambda} P_0(X) + \sqrt{B_1 - B_2}}{X(x_t)}, \tag{60}$$

where we have defined the "reduced" branching ratios

$$B_1 = \frac{Br(K^+ \to \pi^+ \nu\bar{\nu})}{4.42 \cdot 10^{-11}} \qquad B_2 = \frac{Br(K_L \to \pi^0 \nu\bar{\nu})}{1.93 \cdot 10^{-10}}. \tag{61}$$

Consequently with σ defined in (43) one finds to an excellent approximation

$$\bar{\rho} = 1 + \frac{P_0(X) - \sqrt{\sigma(B_1 - B_2)}}{A^2 X(x_t)}, \qquad \bar{\eta} = \frac{\sqrt{B_2}}{\sqrt{\sigma} A^2 X(x_t)} \tag{62}$$

$$\sin 2\beta = \frac{2 r_s}{1 + r_s^2}, \qquad r_s(B_1, B_2) = \sqrt{\sigma} \frac{\sqrt{\sigma(B_1 - B_2)} - P_0(X)}{\sqrt{B_2}}. \tag{63}$$

Thus, within the approximation of (62), $\sin 2\beta$ is independent of V_{cb} (or A) and m_t [43].

It should be stressed that $\sin 2\beta$ determined this way depends only on two measurable branching ratios and on the function $P_0(X)$ which is completely calculable in perturbation theory. Consequently this determination is free from any hadronic uncertainties and its accuracy can be estimated with a high degree of confidence.

An extensive numerical analysis of the formulae above has been presented in [43]. Assuming that the branching ratios are known to within $\pm 10\%$ and m_t within ± 3 GeV one finds in particular

$$\sigma(\sin 2\beta) = \pm 0.05, \quad \sigma(\mathrm{Im}\lambda_t) = \pm 5\%, \quad \sigma(\bar{\eta}) = \pm 0.03, \quad \sigma(|V_{td}|) = \pm 9\% \tag{64}$$

Of particular interest are the accurate determinations of $\sin 2\beta$ and of $\mathrm{Im}\lambda_t$. The latter quantity as seen in (55) can be obtained from $K_L \to \pi^0 \nu\bar{\nu}$ alone and does not require knowledge of V_{cb}. The importance of measuring accurately $\mathrm{Im}\lambda_t$ is evident. It plays a central role in the phenomenology of CP violation in K decays and is furthermore equivalent to the Jarlskog parameter J_{CP}, the invariant measure of CP violation in the SM, $J_{CP} = \lambda(1-\lambda^2/2)\mathrm{Im}\lambda_t$.

The accuracy to which $\sin 2\beta$ can be obtained from $K \to \pi\nu\bar{\nu}$ is, in the example discussed above, comparable to the one expected in determining $\sin 2\beta$ from $a_{\psi K_S}$ prior to LHCB and BTeV experiments. Simultaneously one finds an interesting connection between rare K decays and B physics [43]

$$\frac{2r_s(B_1,B_2)}{1+r_s^2(B_1,B_2)} = a_{\psi K_S} \qquad (65)$$

which must be satisfied in the SM and generally in MFV models. We stress that except for $P_0(X)$ all quantities in (65) can be directly measured in experiment and that this relationship is essentially independent of m_t and V_{cb}. Due to very small theoretical uncertainties in (65), this relation is particularly suited for tests of CP violation in the SM and offers a powerful tool to probe the physics beyond it. The relation (65) is one of several correlations between $K \to \pi\nu\bar{\nu}$ and observables in B physics. Another example is the upper bound in (48). A numerical analysis of such correlations can be found in [49].

$K_L \to \pi^0 e^+ e^-$

Finally we would like to discuss the rare decay $K_L \to \pi^0 e^+ e^-$. There are three contributions to this decay: CP-conserving, indirectly CP-violating and directly CP-violating. Unfortunately out of these three contributions only the directly CP-violating can be calculated reliably. This contribution is dominated by Z^0-penguin diagrams [50]. Including NLO corrections [51] and scanning the input parameters of sections 4 and 6 we find

$$Br(K_L \to \pi^0 e^+ e^-)_{\mathrm{dir}} = (4.3 \pm 2.1) \cdot 10^{-12}, \qquad (66)$$

where the errors come dominantly from the uncertainties in the CKM parameters. The calculations of indirectly CP-violating contribution are plagued by theoretical uncertainties [52, 53]. On the other hand, this contribution taken alone is given simply by

$$Br(K_L \to \pi^0 e^+ e^-)_{\mathrm{indir}} = |\varepsilon|^2 \frac{\tau(K_L)}{\tau(K_S)} Br(K_S \to \pi^0 e^+ e^-) = 3.0 \cdot 10^{-3} Br(K_S \to \pi^0 e^+ e^-), \qquad (67)$$

with $\tau(K_{L,S})$ denoting the $K_{L,S}$ life-times and $Br(K_S \to \pi^0 e^+ e^-)$ hopefully measured in the next years by KLOE at Frascati and NA48 at CERN. The two CP violating contributions will in general interfer with each other. Given the present uncertainty on $Br(K_S \to \pi^0 e^+ e^-)$ the total CP-violating contribution could be as high as few $\times 10^{-11}$ [52] but taking into account the theoretical estimates of the indirectly CP-violating contribution, one should expect it below 10^{-11} within the SM.

The upper bound on the CP-conserving contribution governed by $K_L \to \pi^0 \gamma\gamma \to \pi^0 e^+ e^-$ can be obtained with the help of chiral perturbation theory [53] and the data on $K_L \to \pi^0 \gamma\gamma$. The recent results on the latter decay [54] imply that this contribution is smaller than $2 \cdot 10^{-12}$. Consequently it is smaller than expected by some authors in the past. As this contribution does not interfere with the remaining larger CP-violating contributions, it does not present a significant problem but in order to be able to extract CKM parameters from this decay its better estimate is clearly needed.

The most recent experimental bound from KTeV [55] reads

$$Br(K_L \to \pi^0 e^+ e^-) < 5.1 \cdot 10^{-10} \ (90\% C.L.). \tag{68}$$

Considerable improvements are expected in the coming years.

We do not discuss here the decay $K_L \to \mu^+\mu^-$ as the extraction of the CKM parameters from its branching ratio is subject to very large hadronic uncertainties [56].

RARE B DECAYS

Preliminaries

The most interesting here are the decays $B \to X_{s,d}\gamma$, $B \to X_{s,d}l^+l^-$, $B \to X_{s,d}\nu\bar{\nu}$, $B_{s,d} \to l^+l^-$ and the exclusive counterparts of the first three decays with $X_{s,d}$ replaced by $K^*(\rho)$ or $K(\pi)$. There is a vast literature on these decays within the SM and its extensions. Theoretically cleanest are the decays $B \to X_{s,d}\nu\bar{\nu}$ and $B_{s,d} \to l^+l^-$ as they involve only Z^0-penguin and box diagram contributions. $B \to X_{s,d}\gamma$ and $B \to X_{s,d}l^+l^-$ involving magnetic γ-penguins and the standard photon-penguins are subject to long distance contributions which have to be taken into account. In the case of exclusive channels there are additional uncertainties in the relevant hadronic formfactors. On the experimental side the exclusive channels are easier to measure. Moreover, knowing them experimentally will help to distinguish between $B \to X_s l^+l^-$ and $B \to X_d l^+l^-$.

$B \to X_s \gamma$

A lot of efforts have been put into predicting the branching ratio for the inclusive radiative decay $B \to X_s \gamma$ including NLO-QCD corrections and higher order electroweak corrections. The relevant references can be found in [3, 4] and [57]. The final result of these efforts can be summarized by

$$Br(B \to X_s\gamma)_{\text{th}} = \begin{cases} (3.35 \pm 0.30) \cdot 10^{-4} & [58,59] \\ (3.73 \pm 0.30) \cdot 10^{-4} & [60] \end{cases} \tag{69}$$

where in [58, 59] and [60] the pole mass ratio m_c/m_b and the ratio $m_c^{\overline{MS}}/m_b^{\text{pole}}$ have been used, respectively. This result should be compared with experimental data from CLEO, ALEPH and BELLE, whose combined branching ratio reads

$$Br(B \to X_s\gamma)_{\text{exp}} = (3.22 \pm 0.40) \cdot 10^{-4} . \tag{70}$$

Clearly, the SM result agrees well with the data. In order to see whether any new physics can be seen in this decay, the theoretical and experimental errors should be reduced. This is certainly a difficult task. On the other hand already the available data put powerful constraints on the parameter space of the supersymmetric extensions of the SM [57].

It is easier to measure the exclusive branching ratios $Br(B^\pm \to K^{*\pm}\gamma)$ and $Br(B_d^0 \to K^{*0}\gamma)$. The most recent data from CLEO, BaBar and Belle are found in the ball park of $3 \cdot 10^{-5}$ and $5 \cdot 10^{-5}$ respectively. Theoretical calculations of these branching ratios are plagued by uncertainties in the relevant formfactors. The new calculations within QCD factorization approach could improve this situation [61, 62].

$B \to X_s l^+ l^-$ and $B \to K^* l^+ l^-$

These rare decays have been the subject of many theoretical studies. The NLO-QCD to corrections $B \to X_s l^+ l^-$ have been calculated long time ago in [63, 64]. As of January 2002 dominant NNLO corrections are available [65, 66]. The remaining scale dependence in the branching ratio is roughly $\pm 6.5\%$ but the total theoretical uncertainty is $\pm 13\%$ [68]. The corresponding results for $B \to K^* l^+ l^-$ have been presented in [61].

The main interest in these decays is their large sensitivity to new physics contributions [67, 68]. These contributions can be studied through the dilepton mass distribution, the leptonic forward-backward asymmetry and CP-asymmetries. Of particular interest is the leptonic forward-backward asymmetry in $B \to K^* \mu^+ \mu^-$ and $B \to K^* e^+ e^-$ which vanishes for a certain value of the dilepton mass. The position of this zero is a sensitive function of the short distance contributions with small uncertainties from the formfactors [69, 70, 71]. A nice recent review can be found in [69]. It is clear that this asymmetry will offer useful tests of the SM and of its extensions [69, 68]. Similar comments apply to CP asymmetries that are very small in the SM but can be substantial in general supersymmetric models. Within the SM the most recent analysis gives [68]

$$Br(B \to X_s \mu^+ \mu^-) = (4.15 \pm 0.70) \cdot 10^{-6}, \qquad Br(B \to K^* \mu^+ \mu^-) = (1.2 \pm 0.4) \cdot 10^{-7}, \tag{71}$$

to be compared with 90% C.L upper bounds from Belle

$$Br(B \to X_s \mu^+ \mu^-) < 19.1 \cdot 10^{-6}, \qquad Br(B \to K^* \mu^+ \mu^-) < 3.0 \cdot 10^{-6}. \tag{72}$$

Belle sees also first hints for the events of $B \to X_s \mu^+ \mu^-$ and claims the observation of $B \to K l^+ l^-$. More data on these decays are expected from Belle, BaBar and Tevatron already this year.

$B \to X_{s,d} \nu \bar{\nu}$

The decays $B \to X_{s,d} \nu \bar{\nu}$ are the theoretically cleanest decays in the field of rare B-decays. They are dominated by the same Z^0-penguin and box diagrams involving top quark exchanges which we encountered already in the case of $K^+ \to \pi^+ \nu \bar{\nu}$ and $K_L \to \pi^0 \nu \bar{\nu}$, except for the appropriate change of the external quark flavours. The charm contribution is fully negligible here and the resulting branching ratio after normalizing it to $Br(B \to X_c e\bar{\nu})$ and summing over three neutrino flavours is given as follows

$$\frac{Br(B \to X_s \nu\bar{\nu})}{Br(B \to X_c e\bar{\nu})} = \frac{3\alpha^2}{4\pi^2 \sin^4\Theta_W} \frac{|V_{ts}|^2}{|V_{cb}|^2} \frac{X^2(x_t)}{f(z)} \frac{\kappa(0)}{\kappa(z)}. \tag{73}$$

Here $f(z)$ is the phase-space factor for $B \to X_c e\bar{\nu}$ with $z = m_c^2/m_b^2$ and $\kappa(z) = 0.88$ is the corresponding QCD correction. The factor $\kappa(0) = 0.83$ represents the QCD correction to the matrix element of the $b \to s\nu\bar{\nu}$ transition due to virtual and bremsstrahlung contributions. In the case of $B \to X_d \nu\bar{\nu}$ one has to replace V_{ts} by V_{td} which results in a decrease of the branching ratio by roughly an order of magnitude.

Setting $Br(B \to X_c e\bar{\nu}) = 10.6\%$, $f(z) = 0.54$, $\kappa(z) = 0.88$ and using the values in (40) we have

$$Br(B \to X_s \nu\bar{\nu}) = 3.6 \cdot 10^{-5} \frac{|V_{ts}|^2}{|V_{cb}|^2} \left[\frac{\overline{m}_t(m_t)}{166\,\text{GeV}}\right]^{2.30}. \tag{74}$$

Taking next, $f(z) = 0.54 \pm 0.04$ and $Br(B \to X_c e\bar{\nu}) = (10.6 \pm 0.4)\%$ and scanning the input parameters we find

$$Br(B \to X_s \nu\bar{\nu}) = (3.6 \pm 0.7) \cdot 10^{-5} \tag{75}$$

to be compared with the experimental upper bound:

$$Br(B \to X_s \nu\bar{\nu}) < 6.4 \cdot 10^{-4} \quad (90\%\ \text{C.L.}) \tag{76}$$

obtained for the first time by ALEPH. This is only a factor of 20 above the SM expectation. Even if the actual measurement of this decay is very difficult, all efforts should be made to reach this goal. One should also make attempts to measure $Br(B \to X_d \nu\bar{\nu})$. Indeed

$$\frac{Br(B \to X_d \nu\bar{\nu})}{Br(B \to X_s \nu\bar{\nu})} = \frac{|V_{td}|^2}{|V_{ts}|^2} \tag{77}$$

offers the cleanest direct determination of $|V_{td}|/|V_{ts}|$ as all uncertainties related to m_t, $f(z)$ and $Br(B \to X_c e\bar{\nu})$ cancel out.

$B_{s,d} \to l^+ l^-$

The decays $B_{s,d} \to l^+ l^-$ are after $B \to X_{s,d} \nu\bar{\nu}$ the theoretically cleanest decays in the field of rare B-decays. They are dominated by the Z^0-penguin and box diagrams involving top quark exchanges which we encountered already in the case of $B \to X_{s,d} \nu\bar{\nu}$ except that due to charged leptons in the final state the charge flow in the internal lepton line present in the box diagram is reversed. This results in a different m_t dependence. The charm contributions are fully negligible here and the resulting branching ratio for $B_s \to \mu^+\mu^-$ can be written as follows [4]

$$Br(B_s \to \mu^+\mu^-) = 2.9 \cdot 10^{-9} \left[\frac{\tau(B_s)}{1.5\,\text{ps}}\right] \left[\frac{F_{B_s}}{210\,\text{MeV}}\right]^2 \left[\frac{|V_{ts}|}{0.040}\right]^2 \left[\frac{\overline{m}_t(m_t)}{166\,\text{GeV}}\right]^{3.12}. \tag{78}$$

The main uncertainty in this branching ratio results from the uncertainty in F_{B_s}. Scanning the input parameters together with $\tau(B_s) = 1.5$ ps and $F_{B_s} = (210 \pm 30)$ MeV we find

$$Br(B_s \to \mu^+\mu^-) = (3.2 \pm 1.5) \cdot 10^{-9} . \tag{79}$$

For $B_d \to \mu^+\mu^-$ a similar formula holds with obvious replacements of labels ($s \to d$). Provided the decay constants F_{B_s} and F_{B_d} will be calculated reliably by non-perturbative methods or measured in leading leptonic decays one day, the rare processes $B_s \to \mu^+\mu^-$ and $B_d \to \mu^+\mu^-$ should offer clean determinations of $|V_{ts}|$ and $|V_{td}|$. In particular the ratio

$$\frac{Br(B_d \to \mu^+\mu^-)}{Br(B_s \to \mu^+\mu^-)} = \frac{\tau(B_d)}{\tau(B_s)} \frac{m_{B_d}}{m_{B_s}} \frac{F_{B_d}^2}{F_{B_s}^2} \frac{|V_{td}|^2}{|V_{ts}|^2} \tag{80}$$

having smaller theoretical uncertainties than the separate branching ratios should offer a useful measurement of $|V_{td}|/|V_{ts}|$. Since $Br(B_d \to \mu^+\mu^-) = O(10^{-10})$ this is, however, a very difficult task.

The bounds on $B_{s,d} \to \mu\bar{\mu}$ are still many orders of magnitude away from SM expectations:

$$Br(B_s \to \mu^+\mu^-) \le 2.6 \cdot 10^{-6} \quad (95\% C.L.) \quad (CDF) \tag{81}$$

$$Br(B_d \to \mu^+\mu^-) \le 2.8 \cdot 10^{-7} \quad (90\% C.L.) \quad (Belle) \tag{82}$$

CDF should reach in Run II the sensitivity of $1 \cdot 10^{-8}$ and $4 \cdot 10^{-8}$ for $B_d \to \mu\bar{\mu}$ and $B_s \to \mu\bar{\mu}$, respectively. Thus if the SM is the whole story one will have to wait until LHC-B and BTeV to see any events. On the other hand in a Two-Higgs-Doublet-Model and in particular in the MSSM one can find substantially larger branching ratios provided $\tan\beta$ is large [72]. This means that either this decay will be measured in Run II at Fermilab or the allowed parameter space in these models will be considerably reduced.

CP VIOLATION IN B DECAYS

Preliminaries

CP violation in B decays is certainly one of the most important targets of B-factories and of dedicated B-experiments at hadron facilities. It is well known that CP-violating effects are expected to occur in a large number of channels at a level attainable experimentally in the near future. Moreover there exist channels which offer the determination of CKM phases essentially without any hadronic uncertainties.

The first results on $\sin 2\beta$ from BaBar and Belle, discussed already in Section 5, are very encouraging. These results should be improved over the coming years through the new measurements of $a_{\psi K_S}(t)$ by both collaborations and by CDF and D0 at Fermilab. An error for $\sin 2\beta$ of ± 0.08 should be achievable by the next summer. Moreover measurements of CP asymmetries in other B decays and the measurements of the angles α, β and γ by means of various strategies using two-body B decays should contribute substantially to our understanding of CP violation. In particular the KM picture of CP violation and MFV should be well tested.

Extensive discussions of CP violation in the Standard Model and its extensions can be found in the books [73, 74], in the working group reports in [75, 76] and in [6, 77].

Generally one considers three types of CP violation: a) CP violation in mixing, b) CP violation in decay and c) CP violation in the interference of mixing and decay. It turns out that the last type of CP violation is most suitable for a theoretically clean determination of the angles of the unitarity triangle. We will first recall some relevant formulae related to CP violation in the interference of mixing and decay. Subsequently we will discuss various strategies for the determination of the angles α, β and γ including the latest developments.

CP Violation in Mixing and Decay

A time dependent asymmetry $a_{CP}(t, f)$

$$a_{CP}(t,f) = \frac{\Gamma(B^0_{d,s}(t) \to f) - \Gamma(\bar{B}^0_{d,s}(t) \to f)}{\Gamma(B^0_{d,s}(t) \to f) + \Gamma(\bar{B}^0_{d,s}(t) \to f)} \tag{83}$$

with f being a CP eigenstate offers a clean measurement of the angles of the unitarity triangle provided a single mechanism dominates the decay amplitude or the contributing mechanisms have the same weak phases. Then $a_{CP}(t, f)$ is given simply by

$$a_{CP}(t,f) = \eta_f \sin(2\phi_D - 2\phi_M)\sin(\Delta M_{d,s}t) \tag{84}$$

where ϕ_D is the weak phase in the decay amplitude and ϕ_M the weak phase in the $B^0_{d,s} - \bar{B}^0_{d,s}$ mixing. $\eta_f = \pm 1$ is the CP parity of the final state. In this particular case the hadronic matrix elements drop out, the direct CP-violating contribution vanishes and the CP asymmetry is given entirely in terms of the weak phases ϕ_D and ϕ_M.

If a single tree diagram dominates, the factor $\sin(2\phi_D - 2\phi_M)$ can be calculated by using

$$\phi_D = \begin{cases} \gamma & b \to u \\ 0 & b \to c \end{cases} \qquad \phi_M = \begin{cases} -\beta & B^0_d \\ -\beta_s & B^0_s \end{cases} \tag{85}$$

where we have indicated the basic transition of the b-quark into a lighter quark. $\beta_s = O(10^{-2})$. On the other hand if the penguin diagram with internal top exchange dominates one has

$$\phi_D = \begin{cases} -\beta & b \to d \\ 0 & b \to s \end{cases} \qquad \phi_M = \begin{cases} -\beta & B^0_d \\ -\beta_s & B^0_s \end{cases}. \tag{86}$$

Let us practice these formulae. Assuming that $B_d \to \psi K_S$ and $B_d \to \pi^+\pi^-$ are dominated by tree diagrams with $b \to c$ and $b \to u$ transitions respectively we readily find

$$a_{CP}(t, \psi K_S) = -\sin(2\beta)\sin(\Delta M_d t), \tag{87}$$

$$a_{CP}(\pi^+\pi^-) = -\sin(2\alpha)\sin(\Delta M_d t). \tag{88}$$

Now in the case of $B_d \to \psi K_S$ the penguin diagrams have to a very good approximation the same phase ($\phi_D = 0$) as the tree contribution and are moreover Zweig suppressed.

Consequently (87) is very accurate. This is not the case for $B_d \to \pi^+\pi^-$ where the penguin contribution could be substantial. Heaving weak phase $\phi_D = -\beta$, that differs from the tree phase $\phi_D = \gamma$, this penguin contribution changes effectively $\sin(2\alpha)$ to $\sin(2\alpha + \theta_P)$ where θ_P is a function of β and hadronic parameters.

Classic Strategies

The Angle α

The classic determination of α by means of the time dependent CP asymmetry in the decay $B_d^0 \to \pi^+\pi^-$ as given by (88) is affected by the "QCD penguin pollution" which has to be taken care of in order to extract α. We have just mentioned this problem. The well known strategy to deal with this "penguin problem" is the isospin analysis of Gronau and London [78]. It requires however the measurement of $Br(B^0 \to \pi^0\pi^0)$ which is expected to be below 10^{-6}: a very difficult experimental task. For this reason several, rather involved, strategies have been proposed which avoid the use of $B_d \to \pi^0\pi^0$ in conjunction with $a_{CP}(\pi^+\pi^-, t)$. They are reviewed in [6, 75, 76]. It is to be seen which of these methods will eventually allow us to measure α with a respectable precision. In addition the QCD factorization approach [100] could help in calculating the penguin pollution from first principles.

At present both BaBar and Belle make efforts to measure $a_{\pi^+\pi^-}(t)$ that gives $(\sin 2\alpha)_{\text{eff}}$. The latter containing penguin contributions does not give the true angle α. Theorists are then supposed to translate $(\sin 2\alpha)_{\text{eff}}$ into the true $\sin 2\alpha$. It should be emphasized that in view of a poor knowledge of α at present (see table 1) even a rough measurement of this angle will have an important impact on the UT.

The Angle β

Here the CP-asymmetry in the decay $B_d \to \psi K_S$ remains the main tool as it allows a direct measurement of the angle β in the unitarity triangle without any theoretical uncertainties [79]. The relevant formula is given in (87). Of considerable interest [80, 81] is also the pure penguin decay $B_d \to \phi K_S$, which is expected to be sensitive to physics beyond the SM. Comparision of β extracted from $B_d \to \phi K_S$ with the one from $B_d \to \psi K_S$ should be important in this respect.

Finally we can consider the asymmetry in $B_s \to \psi\phi$, an analog of $B_d \to \psi K_s$. In the leading order of the Wolfenstein parametrization the asymmetry $a_{CP}(t, \psi\phi)$ vanishes. Including higher order terms in λ one finds

$$a_{CP}(t, \psi\phi) = 2\lambda^2 \bar{\eta} \sin(\Delta M_s t) . \qquad (89)$$

As this asymmetry is predicted to be very small, it is a useful quantity to look for the physics beyond the SM. In particular the CP violation in $B_s^0 - \bar{B}_s^0$ mixing from new sources beyond the Standard Model should be probed in this decay. Another useful channel for β is $B_d \to D^+D^-$.

The Angle γ

The two theoretically cleanest methods for the determination of γ are: i) the full time dependent analysis of $B_s \to D_s^+ K^-$ and $\bar{B}_s \to D_s^- K^+$ [82] and ii) the well known triangle construction due to Gronau and Wyler [83] which uses six decay rates $B^\pm \to D_{CP}^0 K^\pm$, $B^+ \to D^0 K^+$, $\bar{D}^0 K^+$ and $B^- \to D^0 K^-$, $\bar{D}^0 K^-$. Both methods are unaffected by penguin contributions. The first method is experimentally very challenging because of the expected large $B_s^0 - \bar{B}_s^0$ mixing. The second method is problematic because of the small branching ratios of the colour supressed channel $B^+ \to D^0 K^+$ and its charge conjugate, giving a rather squashed triangle and thereby making the extraction of γ very difficult. Variants of the latter method which could be more promising have been proposed in [84, 85]. The usefulness of $B_c \to DD_s$ for the extraction of γ was stressed in [86]. It appears that these methods will give useful results at later stages of CP-B investigations. In particular the first and the last method will be feasible only at LHC-B.

Useful strategies for γ using the U-spin symmetry have been proposed in [87, 88]. The first strategy involves the decays $B_{d,s}^0 \to \psi K_S$ and $B_{d,s}^0 \to D_{d,s}^+ D_{d,s}^-$ [87]. The second strategy involves $B_s^0 \to K^+ K^-$ and $B_d^0 \to \pi^+ \pi^-$ [88]. These strategies are unaffected by FSI and are only limited by U-spin breaking effects. They are promising for Run II at FNAL and in particular for LHC-B.

A method of determining γ, using $B^+ \to K^0 \pi^+$ and the U-spin related processes $B_d^0 \to K^+ \pi^-$ and $B_s^0 \to \pi^+ K^-$, was presented in [89]. A general discussion of U-spin symmetry in charmless B decays and more references to this topic can be found in [90].

Constraints for γ from $B \to \pi K$

The most recent developments are related to the extraction of the angle γ from the decays $B \to PP$ (P=pseudoscalar). Several of these modes have been observed by the CLEO, BaBar and Belle collaborations. In particular the results for $B \to K\pi$ branching ratios should allow us to obtain direct information on γ in the future. At present, there are only experimental results available for the combined branching ratios of these modes, i.e. averaged over decay and its charge conjugate, suffering from large uncertainties.

There has been a large activity in this field during the last four years. The main issues here are the final state interactions (FSI), SU(3) symmetry breaking effects and the importance of electroweak penguin contributions. Several interesting ideas have been put forward to extract the angle γ in spite of large hadronic uncertainties in $B \to \pi K$ decays [91, 92, 93, 94, 95, 96]. Reviews can be found in [97, 94].

Three strategies for bounding and determining γ have been proposed. The "mixed" strategy [91] uses $B_d^0 \to \pi^0 K^\pm$ and $B^\pm \to \pi^\pm K$. The "charged" strategy [96] involves $B^\pm \to \pi^0 K^\pm$, $\pi^\pm K$ and the "neutral" strategy [94] the modes $B_d^0 \to \pi^\mp K^\pm$, $\pi^0 K^0$. General parametrizations for the study of the FSI, SU(3) symmetry breaking effects and of the electroweak penguin contributions in these channels have been presented in [93, 94, 95]. Moreover, general parametrizations by means of Wick contractions [98, 99] have been proposed. They can be used for all two-body B-decays. These parametrizations should turn out to be useful when the data improve.

Parallel to these efforts an important progress has been made by Beneke, Buchalla, Neubert and Sachrajda [100] through the demonstration that in a large large class of non-leptonic two-body B-meson decays the factorization of the relevant hadronic matrix elements follows from QCD in the heavy-quark limit. The resulting QCD factorization formula incorporates elements of the naive factorization approach used in the past but allows to compute systematically non-factorizable corrections. In this approach the μ-dependence of hadronic matrix elements is under control. Moreover spectator quark effects are taken into account and final state interaction phases can be computed perturbatively. While, in my opinion, an important progress in evaluating non-leptonic amplitudes has been made in [100], the usefulness of this approach at the quantitative level has still to be demonstrated when the data improve. In particular the role of the $1/m_b$ corrections has to be considerably better understood. The techniques developed in [100] have been used for exclusive rare B decays [61, 62]. An interesting proof of factorization for $B \to D\pi$ to all orders of α_s has been presented in [101].

There is another approach, the so-called perturbative QCD approach to non-leptonic decays [102] that has been developed earlier from the QCD hard-scattering approach. The main difference between the approaches in [100] and [102] is the treatment of soft spectator contributions which are assumed to be negligible in the perturbative QCD approach. While the QCD factorization approach is more general and systematic, only time will show which of these two frameworks is more successful and whether they have to be replaced by still more powerful approaches in the future.

Finally new methods to calculate exclusive hadronic matrix elements from QCD light-cone sum rules has been developed recently in [103]. This work may shed light on the importance of $1/m_b$ and soft-gluon effects in the QCD factorization approach.

Returning to phenomenology, as demonstrated in [91, 93, 94, 95, 96], already CP-averaged $B \to \pi K$ branching ratios may imply interesting bounds on γ that may remove a large portion of the allowed range from the analysis of the unitarity triangle. In particular combining the neutral and charged strategies [94] one finds that the existing data on $B \to \pi K$ favour γ in the second quadrant, which is in conflict with the standard analysis of the unitarity triangle as we have seen in section 6. Other arguments for $\cos\gamma < 0$ using $B \to PP$, PV and VV decays were given in [104]. Also the analyses of $B \to \pi K$ in the QCD factorization approach [105] favour $\gamma > 90°$.

In view of sizable theoretical uncertainties in the analyses of $B \to \pi K$ and of large experimental errors in the corresponding branching ratios it is not yet clear whether the discrepancy in question is serious. For instance [106] sizable contributions of the so-called charming penguins to the $B \to \pi K$ amplitudes could shift γ extracted from these decays below 90° but at present these contributions cannot be calculated reliably. Similar role could be played by annihilation contributions [102] and large non-factorizable SU(3) breaking effects [94]. Also, new physics contributions in the electroweak penguin sector could shift γ to the first quadrant [94]. It should be however emphasized that the problem with the angle γ, if it persisted, would put into difficulties not only the SM but also the full class of MFV models in which the lower bound on $\Delta M_s/\Delta M_d$ implies $\gamma < 90°$. On the other hand as seen in fig. 4 for sufficiently high values of R_{sd}, the angle γ resulting from the unitarity triangle analysis in models containing new operators [30] can easily be in the second quadrant provided $\Delta M_s/\Delta M_d$ is not too large. However, this does not happen in the MSSM in the large $\tan\beta$ limit, where the presence of new operators results in $R_{sd} < 1.0$ and in γ that is generally smaller than in the SM [30].

MINIMAL FLAVOUR VIOLATION MODELS

We have defined this class of models in section 3. Here I would like just to list four interesting properties of these models that are independent of particular parameters present in these models. These are:

- There exists a universal unitarity triangle (UUT) [19] common to all these models and the SM that can be constructed by using measurable quantities that depend on the CKM parameters but are not polluted by the new parameters present in the extensions of the SM. The UUT can be constructed, for instance, by using $\sin 2\beta$ from $a_{\psi K_S}$ and the ratio $\Delta M_s / \Delta M_d$. The relevant formulae can be found in (37) and in [19, 27, 30], where also other quantities suitable for the determination of the UUT are discussed.
- There exists an absolute lower bound on $\sin 2\beta$ [107] that follows from the interplay of ΔM_d and ε_K. It depends only on $|V_{cb}|$ and $|V_{ub}/V_{cb}|$, as well as on the non-perturbative parameters $\hat{B}_K$, $F_{B_d}\sqrt{\hat{B}_d}$ and ξ entering the standard analysis of the unitarity triangle. A conservative scanning of all relevant input parameters gives [4] $(\sin 2\beta)_{min} = 0.42$. A less conservative bound of 0.52 has been found in [49]. This bound could be considerably improved when the values of $|V_{cb}|$, $|V_{ub}/V_{cb}|$, $\hat{B}_K$, $F_{B_d}\sqrt{\hat{B}_d}$, ξ and – in particular of ΔM_s – will be known better [4, 107].
- There exists an absolute upper bound on $\sin 2\beta$. It is simply given by [11]

$$(\sin 2\beta)_{max} = 2R_b^{max}\sqrt{1 - (R_b^{max})^2} \approx 0.82, \tag{90}$$

with R_b defined in (17).
- For given $a_{\psi K_S}$ and $Br(K^+ \to \pi^+ \nu \bar{\nu})$ only two values of $Br(K_L \to \pi^0 \nu \bar{\nu})$ are possible in the full class of MFV models, independently of any new parameters present in these models [27]. Consequently, measuring $Br(K_L \to \pi^0 \nu \bar{\nu})$ will either select one of these two possible values or rule out all MFV models. Taking the present experimental bound on $Br(K^+ \to \pi^+ \nu \bar{\nu})$ and (90) one finds an absolute upper bound $Br(K_L \to \pi^0 \nu \bar{\nu}) < 4.9 \cdot 10^{-10}$ (90% C.L.) [27] that is stronger than the bound in (59).

OUTLOOK

Let us then say a few words about the expectations for the coming years with regard to the CKM matrix. What follows is my personal view that may differ from the views of some of my collegues. I have divided the coming ten years in three phases:

Phase 1 (2002-2005)

In the coming three years the determination of the CKM matrix will be governed by

$$V_{us}, \quad |V_{cb}|, \quad a_{\psi K_S}, \quad \Delta M_d/\Delta M_s. \tag{91}$$

These four quantities are sufficient to determine the full CKM matrix as seen in (32) and (37). The precision of this determination will depend on the accuracy with which $a_{\psi K_S}$ and $\Delta M_d/\Delta M_s$ will be measured and the non-perturbative ratio ξ calculated by lattice and QCD sum rules methods.

An important role will also be played by

$$\varepsilon_K, \quad |V_{ub}/V_{cb}| \tag{92}$$

but in my opinion the hadronic uncertainties in these two will probably not be fully under control in spite of considerable recent progress on the lattice calculations of $\hat{B}_K$ and the determination of $|V_{ub}|$ [21]. Yet ε_K and $|V_{ub}/V_{cb}|$ combined with (91) can tell us whether the CP violation in the K-system is consistent with the one observed in the B-system.

During this phase we should also have new data on $Br(K^+ \to \pi^+ \nu \bar{\nu})$ from AGS E949. The comparison of these data with the implications of $\Delta M_d/\Delta M_s$ should be very interesting as stressed in Section 7. It would be particularly exciting if the central value did not decrease below the one in (50), that is roughly by a factor of two higher than the SM value. In any case these data should have a considerable impact on $|V_{td}|$ and the unitarity triangle.

During this phase we should also be able to get some information about α from $B \to \pi^+\pi^-$ and about γ from $B \to \pi K$ at B factories and by means of U-spin strategies in conjuction with the data from Run II at Tevatron. How useful this information will be, will depend on the relevant data and the reduction of theoretical uncertainties in the decays in question.

We will also have new data on $B_s \to \mu^+\mu^-$, $B \to X_s \nu \bar{\nu}$, $B \to X_s \gamma$, $B \to X_s \mu \bar{\mu}$ as well as on related exclusive channels. All these decays are governed by the CKM element $|V_{ts}|$ that is already well determined by the unitarity of the CKM matrix $|V_{ts}| \approx |V_{cb}|$. Consequently I do not expect that these decays will play an important role in the CKM fits. On the other hand being sensitive to new physics contributions they could give the first signals of new physics. The fact that $|V_{ts}|$ is already reasonably well known will be helpful in this context.

Phase 2 (2005-2008)

With the B-factories and Tevatron entering their mature stage and LHC-B, BTeV, Atlas and CMS beginning hopefully their operation the quantities in (91) should offer a very good determination of the CKM matrix. I expect that other decays listed in Phase 1 will become more useful in view of improved data and new theoretical ideas. The most important new developments to be expected in this phase will be clean measurements of the angle γ at LHC-B and BTeV in decays $B_s \to D_s^+ K^-$ and $\bar{B}_s \to D_s^- K^+$ and an improved measurement of $Br(K^+ \to \pi^+ \nu\bar{\nu})$ by the CKM collaboration at Fermilab. Also possible measurements of rare B-decays sensitive to both $|V_{ts}|$ and $|V_{td}|$ should play an important role. This phase should provide definite answers whether MFV is sufficient to describe the data or whether new flavour violating interactions are required.

At the end of this phase we should also witness the first events for $K_L \to \pi^0 \nu\bar{\nu}$ from KOPIO at Brookhaven and JHF in Japan.

Phase 3 (2008-2011)

Here precise measurements of $Br(K_L \to \pi^0 \nu\bar{\nu})$ from KOPIO and JHF will be among the highlights. In addition the branching ratios for most of the decays studied in phases 1 and 2 will be known with much higher precision. This will allow not only a precision test of SM but also to identify the patterns of new physics contributions that I personnally expect should show up at this level of accuracy. The combination of these studies with the results from LHC that should signal some direct signs of new physics should allow a convincing identification of this new physics.

No doubt the next ten years should be very exciting but the real progress will require extreme joined effort by theorists and experimentalists.

ACKNOWLEDGEMENTS

I would like to thank the organizers, in particular Jurgen Engelfried and Mariana Kirchbach, for inviting me to such a wonderful workshop and most enjoyable atmosphere. Special thanks go to Piotr Kielanowski for excting discussions and showing me Mexico. The work presented here has been supported in part by the German Bundesministerium für Bildung und Forschung under the contract 05HT1WOA3 and the DFG Project Bu. 706/1-1.

REFERENCES

1. N. Cabibbo, Phys. Rev. Lett. **10** (1963) 531.
2. M. Kobayashi and K. Maskawa, Prog. Theor. Phys. **49** (1973) 652.
3. A.J. Buras, hep-ph/9806471, in *Probing the Standard Model of Particle Interactions*, eds. R. Gupta, A. Morel, E. de Rafael and F. David (Elsevier Science B.V., Amsterdam, 1998), page 281.
4. A.J. Buras, hep-ph/0101336, lectures at the International Erice School, August, 2000.
5. G. Buchalla, A.J. Buras and M. Lautenbacher, Rev. Mod. Phys **68** (1996) 1125.
6. A.J. Buras and R. Fleischer, Adv. Ser. Direct. High. Energy Phys. **15** (1998) 65; hep-ph/9704376.
7. H. Fritzsch and Z.Z. Xing, Prog. Part. Nucl. Phys. **45** (2000) 1.
8. L.L. Chau and W.-Y. Keung, Phys. Rev. Lett. **53** (1984) 1802.
9. Particle Data Group, Euro. Phys. J. **C 15** (2000) 1.
10. L. Wolfenstein, Phys. Rev. Lett. **51** (1983) 1945.
11. A.J. Buras, M.E. Lautenbacher and G. Ostermaier, Phys. Rev. **D 50** (1994) 3433.
12. M. Schmidtler and K.R. Schubert, Z. Phys. **C 53** (1992) 347.
13. R. Aleksan, B. Kayser and D. London, Phys. Rev. Lett. **73** (1994) 18.
14. C. Jarlskog, Phys. Rev. Lett. **55**, (1985) 1039; Z. Phys. **C29** (1985) 491.

15. J.H. Christenson, J.W. Cronin, V.L. Fitch and R. Turlay, Phys. Rev. Lett. **13** (1964) 128.
16. A.J. Buras, hep-ph/0109197, talk given at Kaon 2001, Pisa, June, 2001.
17. T. Inami and C.S. Lim, Progr. Theor. Phys. **65** (1981) 297.
18. M. Ciuchini, G. Degrassi, P. Gambino and G.F. Giudice, Nucl. Phys. **B 534** (1998) 3.
19. A.J. Buras, P. Gambino, M. Gorbahn, S. Jäger and L. Silvestrini, Phys. Lett. **B500** (2001) 161.
20. G. Calderon and G.L. Castro, hep-ph/0111272; A. Garcia, J.L. Garcia-Luna and G.L. Castro, Phys. Lett. **B500** (2001) 66.
21. Ch.W. Bauer, Z. Ligeti and M. Luke, Phys. Lett. **B479** (2000) 395; Phys. Rev. **D64** (2001) 113004; hep-ph/0107054; Th. Becher and M. Neubert, hep-ph/0105217.
22. L. Lellouch and C.-J.D. Lin, hep-ph/0011086; J. Flynn and C.-J.D. Lin, hep-ph/0012154; C.T. Sachrajda, hep-lat/0101003.
23. A. Ali and D. London, Eur. Phys. J. **C18** (2001) 665; S. Mele, Phys. Rev. **D59** (1999) 113011; D. Atwood and A. Soni, Phys. Lett. **B508** (2001) 17.
24. M. Ciuchini et al., JHEP **0107** (2001) 013.
25. A. Höcker, H. Lacker, S. Laplace, F. Le Diberder, Eur. Phys. J. C21 (2001) 225.
26. Y. Grossman, Y. Nir, S. Plaszczynski and M.-H. Schune, Nucl. Phys. **B511** (1998) 69; S. Plaszczynski and M.-H. Schune, hep-ph/9911280.
27. A.J. Buras and R. Fleischer, Phys. Rev. **D64** (2001) 115010.
28. B. Aubert et al., BaBar Collaboration, Phys. Rev. Lett. **87**, (2001) 091801.
29. K. Abe et al., Belle Collaboration, Phys. Rev. Lett. **87**, (2001) 091802.
30. A.J. Buras, P.H. Chankowski, J. Rosiek and Ł. Sławianowska, Nucl. Phys. **B 619** (2001) 434; J. Rosiek, hep-ph/0108226.
31. D. Rein and L.M. Sehgal, Phys. Rev. **D39** (1989) 3325; J.S. Hagelin and L.S. Littenberg, Prog. Part. Nucl. Phys. **23** (1989) 1; M. Lu and M.B. Wise, Phys. Lett. **B324** (1994) 461; S. Fajfer, [hep-ph/9602322]; C.Q. Geng, I.J. Hsu and Y.C. Lin, Phys. Rev. **D54** (1996) 877; G. Buchalla and G. Isidori, Phys. Lett. **B440** (1998) 170; A.F. Falk, A. Lewandowski and A.A. Petrov, Phys. Lett. **B505** (2001) 107.
32. G. Buchalla and A.J. Buras, Nucl. Phys. **B 400** (1993) 225, Nucl. Phys. **B 412** (1994) 106, Nucl. Phys. **B 548** (1999) 309.
33. M. Misiak and J. Urban, Phys. Lett. **B541** (1999) 161.
34. L. Littenberg, Phys. Rev. **D39** (1989) 3322.
35. W. Marciano and Z. Parsa, Phys. Rev. **D53**, R1 (1996).
36. G. Buchalla and A.J. Buras, Nucl. Phys. **B 548** (1999) 309.
37. S. Adler et al., Phys. Rev. Lett. **79**, (1997) 2204; Phys. Rev. Lett. **84**, (2000) 3768.
38. S. Adler et al., hep-ex/0111091.
39. G. D'Ambrosio and G. Isidori, hep-ph/0112135.
40. L. Littenberg, hep-ex/0010048; T.K. Komatsubara, hep-ex/0112016.
41. B. Bassalleck et al., E949 Proposal, BNL 67247, TRI-PP-00-06, 1999.
42. R. Coleman et al., (CKM collaboration), Charged Kaons at the Main Injector, Fermilab-P-0905, 1998.
43. G. Buchalla and A.J. Buras, Phys. Lett. **B333** (1994) 221; Phys. Rev. **D54** (1996) 6782.
44. A. Alavi-Harati et al., Phys. Rev. **D61** (2000) 072006.
45. Y. Grossman and Y. Nir, Phys. Lett. **B398** (1997) 163.
46. A. Belyaev et al, hep-ph/0107046.
47. T. Inagaki, et al., KEK Internal 96-13, November 1996.
48. I.-H. Chiang et al., AGS Experiment Proposal 926 (1996).
49. S. Bergmann and G. Perez, Phys. Rev. **D64** (2001) 115009; JHEP **0008** (2000) 034.
50. C. Dib, I. Dunietz and F.J. Gilman, Phys. Rev. **D39** (1989) 2639; J. Flynn and L. Randall, Nucl. Phys. **B326** (1989) 31, erratum, Ibid. **B334** (1990) 580.
51. A. J. Buras, M. E. Lautenbacher, M. Misiak and M. Münz, Nucl. Phys. **B423** (1994) 349.
52. G. D'Ambrosio, G. Ecker, G. Isidori and J. Portolés, JHEP **08** (1998) 004.
53. G. Ecker, A. Pich and E. de Rafael, Nucl. Phys. **B291** (1987) 692, Nucl. Phys. **B303** (1988) 665, Phys. Lett. **B237** (1990) 481; A.G. Cohen, G. Ecker, and A. Pich, Phys. Lett. **B304** (1993) 347; L.M. Seghal, Phys. Rev. **D38** (1988) 808; P. Heiliger and L.M. Seghal, Phys. Rev. **D47** (1993) 4920; C. Bruno and J. Prades, Z. Phys. **C57** (1993) 585; J. F. Donoghue and F. Gabbiani, Phys. Rev. **D51** (1995) 2187; G. D'Ambrosio and J. Portolés, Nucl. Phys. **B492** (1997) 417.

54. A. Alavi-Harati et al. (KTeV Collaboration), Phys. Rev. Lett. **83**, (1999) 917; V.D. Kekelidze (NA48 Collaboration), talk presented at ICHEP2000, Osaka, Japan.
55. A. Alavi-Harati et al. (KTeV Collaboration), Phys. Rev. Lett. **86**, (2001) 397; Phys. Rev. Lett. **87**, (2001) 021801.
56. G. D'Ambrosio, G. Isidori and J. Portolés, Phys. Lett. **B423** (1998) 385; D. Gomez Dumm and A. Pich, Nucl. Phys. Proc. Suppl. **74** (1999) 186; G. Valencia, hep-ph/9711377; M. Knecht, S. Peris, M. Perrottet and E. de Rafael, Phys. Rev. Lett. **83**, (1999) 5230.
57. C. Greub, hep-ph/9911348; P. Gambino and U. Haisch, JHEP **0110** (2001) 020.
58. K.G. Chetyrkin, M. Misiak and M. Münz, Phys. Lett. **B400** (1997) 206; Erratum-ibid. **B425** (1998) 414.
59. A.L. Kagan and M. Neubert, Eur. Phys. J. **C7** (1999) 5.
60. P. Gambino and M. Misiak, Nucl. Phys. **B 611** (2001) 338.
61. M. Beneke and T. Feldmann, Nucl. Phys. **B592** (2001) 3. M. Beneke, T. Feldmann and D. Seidel, Nucl. Phys. **B612** (2001) 25.
62. S.W. Bosch and G. Buchalla, hep-ph/0106081.
63. M. Misiak, Nucl. Phys. **B393** (1993) 23; Erratum, Nucl. Phys. **B439** (1995) 461.
64. A.J. Buras and M. Münz, Phys. Rev. **D 52** (1995) 186.
65. Ch. Bobeth, M. Misiak and J. Urban, Nucl. Phys. **B 574** (2000) 291.
66. Ch. Greub and M. Walker, hep-ph/0110388 and references therein.
67. C. Bobeth, T. Ewerth, F. Krüger and J. Urban, Phys. Rev. **D64** (2001), 074014.
68. A. Ali, E. Lunghi, C. Greub and G. Hiller, hep-ph/0112300.
69. G. Burdman, Phys. Rev. **D57** (1998) 4254; hep-ph/0112063.
70. J. Charles et. al., Phys. Rev. **D60** (1999) 014001; Phys. Lett. **B451** (1999) 187.
71. A. Ali, P. Ball, L.T. Handoko and G. Hiller, Phys. Rev. **D61** (2000) 074024.
72. C. Hamzaoui, M. Pospelov and M. Toharia, Phys. Rev. **D59** (1999), 095005; K.S. Babu and C. Kolda, Phys. Rev. Lett. **84** (2000), 228; P.H. Chankowski and Ł. Sławianowska Phys. Rev. **D63** (2001), 054012; C. Bobeth, T. Ewerth, F. Krüger and J. Urban, Phys. Rev. **D64** (2001), 074014; C.-S. Huang, W. Liao, Q.-S. Yan and S.-H. Zhu, Phys. Rev. **D63** (2001), 114021, erratum-ibid. **D64** (2001), 059902.
73. G. Branco, L. Lavoura and J. Silva, (1999), CP Violation, Oxford Science Publications, Clarendon Press, Oxford.
74. I.I. Bigi and A.I. Sanda, (2000), CP Violation, Cambridge Monographs on Particle Physics, Nuclear Physics and Cosmology, Cambridge University Press, Cambridge.
75. The BaBar Physics Book, eds. P. Harrison and H. Quinn, (1998), SLAC report 504.
76. B Decays at the LHC, eds. P. Ball, R. Fleischer, G.F. Tartarelli, P. Vikas and G. Wilkinson, hep-ph/0003238.
77. Y. Nir, hep-ph/9911321, hep-ph/0109090.
78. M. Gronau and D. London, Phys. Rev. Lett. **65** (1990) 3381.
79. A.B. Carter and A.I. Sanda, Phys. Rev. Lett. **45** (1980) 952; Phys. Rev. **D23** (1981) 1567. I.I. Bigi and A.I. Sanda, Nucl. Phys. **B193** (1981) 85.
80. R. Fleischer, Int. J. of Mod. Phys. **A12** (1997) 2459.
81. D. London and R.D. Peccei, Phys. Lett. **B223** (1989) 257. D. London and A. Soni, Phys. Lett. **B407** (1997) 61; Y. Grossman and M.P. Worah, Phys. Lett. **B395** (1997) 241; M. Ciuchini et al., Phys. Rev. Lett. **79** (1997) 978; R. Barbieri and A. Strumia, Nucl. Phys. **B508** (1997) 3.
82. R. Aleksan, I. Dunietz and B. Kayser, Z. Phys. **C54** (1992) 653; R. Fleischer and I. Dunietz, Phys. Lett. **B387** (1996) 361.
83. M. Gronau and D. Wyler, Phys. Lett. **B265** (1991) 172.
84. M. Gronau and D. London, Phys. Lett. **B253** (1991) 483; I. Dunietz, Phys. Lett. **B270** (1991) 75.
85. D. Atwood, I. Dunietz and A. Soni, Phys. Rev. Lett. **B78** (1997) 3257.
86. R. Fleischer and D. Wyler, Phys. Rev. **D62** (2000) 057503.
87. R. Fleischer, Eur. Phys. J. **C10** (1999) 299, Phys. Rev. **D60** (1999) 073008.
88. R. Fleischer, Phys. Lett. **B459** (1999) 306.
89. M. Gronau and J.L. Rosner, Phys. Lett. **B482** (2000) 71; C.W. Chiang and L. Wolfenstein, Phys. Lett. **B493** (2000) 73.
90. M. Gronau, Phys. Lett. **B492** (2000) 297.

91. R. Fleischer, Phys. Lett. **B365** (1996) 399; R. Fleischer and T. Mannel, Phys. Rev. **D57** (1998) 2752.
92. M. Gronau and J.L. Rosner, Phys. Rev. **D57** (1998) 6843.
93. R. Fleischer, Eur. Phys. J. **C6** (1999) 451.
94. A.J. Buras and R. Fleischer, Eur. Phys. J. **C11** (1999) 93; Eur. Phys. J. **C16** (2000) 97; hep-ph/0008298.
95. M. Neubert, JHEP 9902 (1998) 014.
96. M. Neubert and J.L. Rosner, Phys. Lett. **B441** (1998) 403; Phys. Rev. Lett. **81** (1998) 5076.
97. R. Fleischer, hep-ph/9904313; M. Neubert, hep-ph/9904321.
98. M. Ciuchini, E. Franco, G. Martinelli and L. Silvestrini, Nucl. Phys. **B501** (1997) 271.
99. A.J. Buras and L. Silvestrini, Nucl. Phys. **B569** (2000) 3.
100. M. Beneke, G. Buchalla, M. Neubert and C.T. Sachrajda, Phys. Rev. Lett. **83** (1999) 1914, Nucl. Phys. **B 591** (2000) 313.
101. Ch.W. Bauer, D. Pirjol and I.W. Stewart, Phys. Rev. Lett. **87** (2001) 201806; hep-ph/0109045.
102. C.-H.V. Chang, H.-n. Li, Phys. Rev. **D55** (1997) 5577; T.-W. Yeh and H.-n. Li, Phys. Rev. **D56** (1997) 1615; H. -Y. Cheng, H.-n. Li and K.-C. Yang Phys. Rev. **D60** (1999) 094005; Y.-Y. Keum, H.-n. Li and A.I. Sanda, Phys. Lett. **B504** (2001) 6, Phys. Rev. **D63** (2001) 054008; H.-n. Li, hep-ph/0101145.
103. A. Khodjamirian, Nucl. Phys. **B 605** (2001) 558, hep-ph/0108205; R. Rückl, S. Weinzierl and O. Yakovlev, hep-ph/0105161, hep-ph/0007344.
104. D. Cronin-Hennessy et al., (CLEO), Phys. Rev. Lett. **85** (2000) 515 and 525; X.-G. He, W.-S. Hou and K-Ch. Yang, hep-ph/9902256; W.-S. Hou and K.-Ch. Yang, Phys. Rev. **D61** (2000) 073014; W.-S. Hou, J.G. Smith and F. Würthwein, hep-ex/9910014.
105. M. Beneke, G. Buchalla, M. Neubert and C.T. Sachrajda, hep-ph/0007256; Nucl. Phys. **B606** (2001) 245. D. Du, D. Yang and G. Zhu, Phys. Lett. **488** (2000) 46; T. Muta, A. Sugamoto, M.Z. Yang and Y.D. Yang, Phys. Rev. **D62** (2000) 094020;
106. M. Ciuchini, E. Franco, G. Martinelli, M. Pierini and L. Silvestrini, Phys. Lett. **515** (2001) 33.
107. A.J. Buras and R. Buras, Phys. Lett. **501** (2001) 223.
108. M. Jamin and B.O. Lange, hep-ph/0108135 and references therein.
109. Y. Nir and M.P. Worah, Phys. Lett. **423** (1998) 319. A.J. Buras, A. Romanino and L. Silvestrini, Nucl. Phys. **B520** (1998) 3. A.J. Buras and L. Silvestrini, Nucl. Phys. **B546** (1999) 299; G. Colangelo and G. Isidori, JHEP 09 (1998) 009; A.J. Buras, G. Colangelo, G. Isidori, A. Romanino and L. Silvestrini, Nucl. Phys. **B566** (2000) 3.

QCD Baryons in the $1/N_c$ Expansion

Elizabeth Jenkins

Department of Physics, 9500 Gilman Drive, University of California at San Diego, La Jolla, CA 92093-0319

Abstract. The $1/N_c$ expansion provides a theoretical method for analyzing the spin-flavor symmetry properties of baryons in QCD that is quantitative, systematic and predictive. An exact spin-flavor symmetry exists for large-N_c baryons, whereas for QCD baryons, the spin-flavor symmetry is approximate and is broken by corrections proportional to the symmetry-breaking parameter $1/N_c = 1/3$. The $1/N_c$ expansion predicts a hierarchy of spin and flavor symmetry relations for QCD baryons that is observed in nature. It provides a quantitative understanding of why some $SU(3)$ flavor symmetry relations in the baryon sector, such as the Gell-Mann–Okubo mass formula, are satisfied to a greater precision than expected from flavor symmetry-breaking suppression factors alone.

INTRODUCTION

It has been rigorously proven that spin-flavor symmetry is an approximate symmetry of baryons in QCD [1, 2]. Spin-flavor symmetry for baryons is formally an exact symmetry in the t'Hooft large-N_c limit [3]. For finite N_c, the spin-flavor symmetry of baryons is only approximate and is broken explicitly by corrections suppressed by powers of $1/N_c$. The breaking of the large-N_c baryon spin-flavor symmetry for QCD baryons is order $1/N_c = 1/3$, which is comparable to the order 30% breaking of Gell-Mann $SU(3)$ flavor symmetry. Thus, spin-flavor symmetry is as good an approximate symmetry for QCD baryons as $SU(3)$ flavor symmetry.

Spin-flavor symmetry for QCD baryons has a long history; like $SU(3)$ flavor symmetry, spin-flavor symmetry predates the formulation of QCD. Although spin-flavor symmetry was phenomenologically successful early on [4], the physical basis for spin-flavor symmetry was not understood, even after the microscopic theory of the strong interactions was known. What has been possible in recent years is to justify spin-flavor symmetry as a *bona fide* symmetry of baryons in QCD, and to classify the explicit breakings of baryon spin-flavor symmetry to all orders in the $1/N_c$ expansion in a systematic and quantitative manner. It has been shown that the quark-gluon dynamics of large-N_c QCD gives rise to a spin-flavor symmetry for baryons. For finite N_c, the symmetry is only approximate; there are subleading $1/N_c$ corrections which explicitly break the symmetry. This new insight has led to the formulation of the baryon $1/N_c$ expansion as an expansion in operators with definite transformation properties under baryon spin and flavor symmetry. Each baryon operator in the $1/N_c$ expansion occurs at a known order in $1/N_c$. Each operator is multiplied by an unknown coefficient which is a reduced matrix element that is not determined by baryon spin-flavor symmetry. Calculating these reduced baryon matrix elements is tantamount to solving QCD in the baryon sector.

All of the new results obtained for baryons in the $1/N_c$ expansion can be characterized as symmetry relations involving spin and flavor. The spin-flavor structure of the baryon $1/N_c$ expansion yields model-independent results which are valid for QCD. It has been known for some time that the spin-flavor group theory for baryons in large-N_c QCD and in the large-N_c quark and Skyrme models is the same[5]. It is now known that a stronger statement applies: the spin-flavor structure of the baryon $1/N_c$ expansion is the same in QCD as in the quark model and the Skyrme model at each order in the $1/N_c$ expansion. In other words, the spin-flavor structure of the baryon $1/N_c$ expansion is the same in QCD as in the quark model and Skyrme models. QCD and these models, however, will differ in their predictions for the reduced matrix elements of the baryon $1/N_c$ expansion. It is well known that the quark and Skyrme models are not very successful at predicting these reduced matrix elements, which leads to the conclusion that most, if not all, of the successful predictions of the non-relativistic quark and Skyrme models are actually model-independent group-theoretic predictions, and therefore should not be regarded as evidence for the validity of these models *per se*.

The spin-flavor operator analysis of the baryon $1/N_c$ expansion has resulted in significant progress in understanding of the spin-flavor structure of baryons. It explains the extraordinary accuracy of many venerable spin-flavor and flavor symmetry relations for baryons as being due to the presence of $1/N_c$ suppression factors in addition to the usual flavor symmetry breaking suppression factors. It is remarkable that, after three decades, a quantitative understanding of spin-flavor symmetry for baryons has been achieved.

The outline of these lectures is as follows. First, a brief summary is given of the $1/N_c$ expansion of large-N_c QCD. The $1/N_c$ power counting of the quark-gluon dynamics of large-N_c QCD is described for the confined quark-gluon bound states: mesons and baryons. Second, it is shown that large-N_c baryons satisfy a contracted spin-flavor symmetry in the $N_c \to \infty$ limit. Baryon states transform as irreducible representations of the spin-flavor algebra, and operators acting on a baryon spin-flavor multiplet transform as irreducible tensor operators of the algebra.

The formalism of the baryon $1/N_c$ expansion is presented in the next section. Tensor operators acting on a baryon spin-flavor multiplet have an expansion in terms of operator products of the baryon spin-flavor generators. The order in $1/N_c$ of each operator product in the $1/N_c$ expansion is known. Not all operator products of the baryon spin-flavor generators are linearly independent, so it is necessary to eliminate redundant operator products using operator identities. This operator reduction is possible in terms of the operator identities for 2-body operator products. The complete set of 2-body operator product identities for $SU(6)$ spin-flavor symmetry is given, and then used to construct operator bases for the baryon $1/N_c$ expansion.

$1/N_c$ operator-product expansions are constructed for a number of baryon tensor operators in the final section. It is necessary to incorporate $SU(3)$ flavor symmetry breaking into the baryon $1/N_c$ expansion since $SU(3)$ breaking is comparable to the expansion parameter $1/N_c$. The $1/N_c$ expansions for baryon masses, axial vector couplings and magnetic moments are presented in detail. A comparison of experimental data with the predictions of the combined $1/N_c$ and flavor-symmetry breaking expansion is given in these cases. The presence of $1/N_c$ suppression factors in the experimental data is clearly evident, and provides a quantitative understanding of the accuracy of famous symmetry relations, such as the Gell-Mann–Okubo formula, Gell-Mann's Equal Spacing Rule and

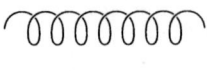

FIGURE 1. Double line notation for a gluon.

the Coleman-Glashow relation for baryon masses, as well as many others.

The focus of these lectures is on the application of the $1/N_c$ expansion to QCD baryons. The discussion is meant to be complementary to my review of large-N_c baryons [6], which is more comprehensive. The emphasis here is on baryons in QCD with $N_F = 3$ flavors of light quarks. I will try to keep the formalism introduced to a minimum throughout, although a fair amount is essential and cannot be avoided. Of necessity, a number of important topics have not been covered, even briefly. These topics include nonet symmetry of baryon amplitudes [29], exact cancellations in baryon chiral perturbation theory [44], spin-flavor symmetry of excited baryons [15], and spin-flavor symmetry of heavy quark baryons [31]. An extensive list of publications on the spin and flavor properties of baryons in the $1/N_c$ expansion is given in the references.

LARGE-N_C QCD

A recent review of large-N_c QCD can be found in Ref. [7], so the presentation here will be brief.

Large-N_c QCD is defined as the generalization of $SU(3)$ gauge theory of quarks and gluons to $SU(N_c)$ gauge theory. The naive generalization of the QCD Lagrangian is given by

$$\mathcal{L} = -\frac{1}{2}\text{Tr}\, G^{\mu\nu}G_{\mu\nu} + \sum_{f=1}^{N_F} \bar{q}_f \left(i\slashed{D} - m_f\right) q_f, \qquad (1)$$

where the gauge field strength and the covariant derivative, $D^\mu = \partial^\mu + igA^\mu$, are defined as in QCD. For $SU(N_c)$ gauge theory, the gluons appear in the adjoint representation of $SU(N_c)$ with dimension $(N_c^2 - 1)$ while the quarks appear in the fundamental representation $\mathbf{N_c}$. Thus, there are $O(N_c)$ more gluon degrees of freedom than quark degrees of freedom in large-N_c QCD.

The $1/N_c$ power counting of quark-gluon diagrams is readily obtained by introducing t'Hooft double line notation for the gluon gauge field: the adjoint index A on the gauge field $(A^\mu)^A$ is replaced by fundamental and anti-fundamental indices i and j, so that the gauge field is written as $(A^\mu)^i_j$, a substitution which is valid up to corrections which are subleading in the $1/N_c$ expansion. In double line notation, the gluon is effectively replaced by a quark line and an antiquark line, as shown in Fig. 1. When a quark-gluon Feynman diagram is rewritten in double line notation, determining the power in N_c of the diagram is equivalent to counting the number of closed quark loops with unrestricted color summations. Fig. 2 gives an explicit example of this $1/N_c$

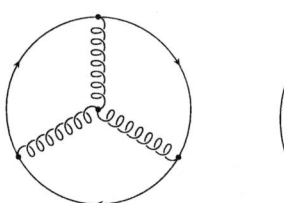 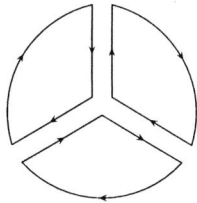

FIGURE 2.

counting. The diagram shown in Fig. 2 reduces to three quark loops, and is therefore proportional to N_c^3. In addition, the diagram is proportional to four powers of the quark-gluon coupling constant g, since the diagram contains four vertices, each of which is proportional to g. Thus, the overall diagram is proportional to $g^4 N_c^3 = (g^2 N_c)^2 N_c$. A simple analysis of other diagrams leads to the t'Hooft result that vacuum Feynman diagrams are proportional to

$$(g^2 N_c)^{\frac{1}{2}V_3 + V_4} N_c^\chi, \qquad (2)$$

where V_n is the number of n-point vertices in the diagram and χ is the Euler character of the diagram. Consequently, diagrams with arbitrary numbers of 3- and 4-point vertices grow with arbitrarily large powers of N_c unless the limit $N_c \to \infty$ is taken with $g^2 N_c$ held fixed. This limiting procedure, which is necessary to define $SU(N_c)$ gauge theory in the large-N_c limit, is known as the t'Hooft limit. The constraint $g^2 N_c$ held fixed can be implemented by rescaling the gauge coupling $g \to g/\sqrt{N_c}$ in the original Lagrangian. After this rescaling, Feynman diagrams will be proportional to N_c^χ, where the Euler character $\chi = 2 - 2H - L$ can be computed in terms of the number of handles H and quark loops L of a given diagram. The t'Hooft limit leads to the following results:

- For finite and large N_c, planar diagrams with $H = 0$ dominate the dynamics. (All planar diagrams with a given L are of the same order.)
- Diagrams with nonplanar gluon exchange ($H \neq 0$) are suppressed relative to planar diagrams by one factor of $1/N_c^2$ for each nonplanar gluon.
- Diagrams with quark loops ($L \neq 0$) are suppressed by one factor of $1/N_c$ for each quark loop.

The dynamics of large-N_c QCD is presumed to be confining. The β function of large-N_c QCD implies that the rescaled coupling gets large at some scale, let us call it Λ_{QCD}. For $E \leq O(\Lambda_{\text{QCD}})$, large-$N_c$ QCD is strongly coupled and is expected to exhibit confinement. The confined theory contains colorless bound states: mesons, baryons and glueballs. The $1/N_c$ power counting for large N_c mesons and for baryons is summarized below.

A meson in large-N_c QCD is created with unit amplitude by the operator

$$\frac{1}{\sqrt{N_c}} \sum_{i=1}^{N_c} \bar{q}_i q^i, \qquad (3)$$

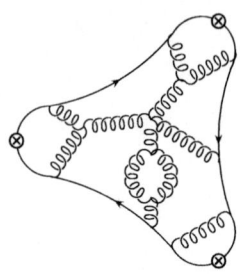

FIGURE 3. A planar diagram contributing to a meson three-point vertex at leading order.

where i is the color index of the quark. The N_c-dependence of meson amplitudes can be obtained by studying quark-gluon diagrams. The leading diagrams are planar diagrams with a single quark loop ($L = 1$) with all insertions of meson operators on the quark loop. An 3-meson diagram is shown in Fig. 3 as an example. The leading diagrams for an n-meson amplitude are $O(N_c^{1-n/2})$. For example, a meson decay constant is $O(\sqrt{N_c})$; a meson mass is $O(1)$; a 3-meson coupling is $O(1/\sqrt{N_c})$, and so forth. This power counting implies that large-N_c mesons are narrow states which are weakly coupled to one another [8, 9].

The situation for baryons is qualitatively different. A large-N_c baryon is a bound state of N_c valence quarks completely antisymmetrized in the color indices of the quarks,

$$\varepsilon_{i_1 i_2 i_3 \cdots i_{N_c}} q^{i_1} q^{i_2} q^{i_3} \cdots q^{i_{N_c}} . \tag{4}$$

The mass of the baryon is $O(N_c)$, whereas the size of the baryon is $O(1)$. The N_c-dependence of baryon-meson scattering amplitudes and couplings can be determined by studying the N_c-counting of quark-gluon diagrams. An antibaryon-baryon–n-meson vertex is $O(N_c^{1-n/2})$, as is an amplitude for the scattering baryon+meson $\to$ baryon + $(n-1)$ mesons.

Naively, this power counting is inconsistent. The amplitude for baryon + meson $\to$ baryon + meson scattering is $O(1)$, whereas an antibaryon-baryon–meson vertex is $O(\sqrt{N_c})$, and grows with N_c. A tree diagram with two different single-meson–baryon vertices produces an amplitude which is $O(N_c)$, not $O(1)$. Unless the $O(N_c)$ contributions of different tree diagrams all cancel one another exactly, the total scattering amplitude is $O(N_c)$ and will violate the $1/N_c$ power counting. Imposing the constraint that the scattering amplitude be $O(1)$ results in relations amongst single-meson–baryon-antibaryon vertices which must be satisfied for consistency of the $1/N_c$ power counting for baryon-meson scattering amplitudes. Consistency of $1/N_c$ power counting for baryon-meson scattering amplitudes and vertices results in non-trivial constraints on large-N_c baryon matrix elements at leading and subleading orders [1, 2]. Large-N_c consistency conditions also lead to the derivation of contracted spin-flavor symmetry for baryons [1, 10].

SPIN-FLAVOR SYMMETRY OF LARGE-N_C BARYONS

Large-N_c contracted spin-flavor symmetry can be derived by considering pion-baryon scattering at low energies $E \sim O(1)$. In this kinematic regime, the large-N_c baryon acts as a heavy static source for scattering the pion with no recoil. There are two tree diagrams which contribute to the scattering amplitude at $O(N_c)$, the direct and crossed diagrams. Using the N_c-independent baryon propagator of Heavy Baryon Chiral Perturbation Theory [11], it is easy to show that cancellation of the $O(N_c)$ scattering amplitude from these two diagrams is given by the large-N_c consistency condition [1]

$$N_c \left[X^{ia}, X^{jb} \right] \leq O(1), \tag{5}$$

where the baryon axial vector couplings (in the baryon rest frame) are defined by

$$A^{ia} \equiv g N_c X^{ia}. \tag{6}$$

Expanding the operator X^{ia} in a power series in $1/N_c$,

$$X^{ia} = X_0^{ia} + \frac{1}{N_c} X_1^{ia} + \frac{1}{N_c^2} X_2^{ia} + \dots, \tag{7}$$

and substituting into Eq. (5) yields the constraint

$$\left[X_0^{ia}, X_0^{jb} \right] = 0 \tag{8}$$

for the leading $O(N_c)$ matrix elements of the baryon axial vector couplings. As we will see, the matrix elements of X_0^{ia} between different baryon states are all determined relative to one another by this constraint, so the $O(N_c)$ portion of the baryon axial vector couplings A^{ia} are all related by symmetry up to an overall normalization constant given by the coupling g, which is the reduced matrix element of the axial vector couplings.

The operator X_0^{ia} is an irreducible tensor operator transforming according to the spin-1, $SU(3)$ adjoint representation of spin $\otimes$ flavor, so the commutators of X_0^{ia} with the baryon spin and flavor generators are given by

$$\left[J^i, X_0^{ja} \right] = i \varepsilon^{ijk} X_0^{ka}, \qquad \left[T^a, X_0^{ib} \right] = i f^{abc} X_0^{ic}. \tag{9}$$

The Lie algebra of the baryon spin $\otimes$ flavor generators, together with Eqs. (8) and (9), yields a contracted spin-flavor algebra [1, 10]

$$\begin{aligned}
&\left[J^i, J^j \right] = i \varepsilon^{ijk} J^k, \quad \left[T^a, T^b \right] = i f^{abc} T^c, \quad \left[J^i, I^a \right] = 0, \\
&\left[J^i, X_0^{ja} \right] = i \varepsilon^{ijk} X_0^{ka}, \quad \left[T^a, X_0^{ib} \right] = i f^{abc} X_0^{ic} \\
&\left[X_0^{ia}, X_0^{jb} \right] = 0
\end{aligned} \tag{10}$$

in the large-N_c limit.

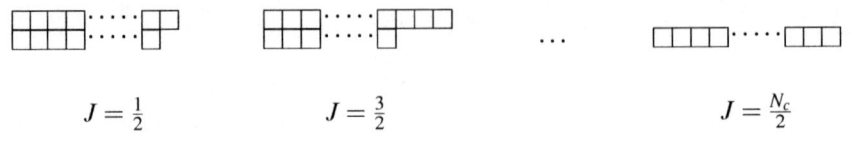

FIGURE 4. Decomposition of the $SU(6)$ baryon representation into $SU(2) \otimes SU(3)$ baryon representations. Each Young tableau has N_c boxes.

It is instructive to contrast the contracted spin-flavor algebra with the $SU(6)$ spin-flavor algebra

$$\begin{gathered}
\left[J^i, J^j\right] = i\varepsilon^{ijk}J^k, \quad \left[T^a, T^b\right] = if^{abc}T^c, \quad \left[J^i, T^a\right] = 0, \\
\left[J^i, G^{ja}\right] = i\varepsilon^{ijk}G^{ka}, \quad \left[T^a, G^{ib}\right] = if^{abc}G^{ic}, \\
\left[G^{ia}, G^{jb}\right] = \frac{i}{6}\delta^{ab}\varepsilon^{ijk}J^k + \frac{i}{4}\delta^{ij}f^{abc}T^c + \frac{i}{2}\varepsilon^{ijk}d^{abc}G^{kc}.
\end{gathered} \tag{11}$$

The contracted spin-flavor algebra can be obtained from the $SU(6)$ spin-flavor algebra with the identification

$$\lim_{N_c \to \infty} \frac{G^{ia}}{N_c} \to X_0^{ia}. \tag{12}$$

Thus, the $SU(6)$ spin-flavor algebra correctly reproduces the contracted spin-flavor algebra in the large-N_c limit. It differs from the contracted spin-flavor algebra by the inclusion of some subleading $1/N_c$ terms in the generators G^{ia}.

The contracted spin-flavor algebra in the large-N_c limit leads to baryon spin-flavor representations which are infinite-dimensional. For finite N_c, it is convenient to work with the $SU(6)$ spin-flavor algebra which leads to finite-dimensional baryon representations. Since the emphasis of these lectures is on QCD baryons with $N_c = 3$, the spin-flavor symmetry will be implemented for finite N_c. Discussion of the connection between finite-dimensional and infinite-dimensional baryon representations can be found in Ref. [6].

The lowest-lying large-N_c baryon representation of the spin-flavor algebra is given by the completely symmetric tensor product of N_c quarks in the fundamental rep of spin-flavor. Under the breakdown of spin-flavor symmetry to its spin $\otimes$ flavor subgroup, the completely symmetric spin-flavor representation decomposes into the spin and flavor representations displayed in Fig. 4. The baryon spin-flavor representation contains baryons with spins $J = \frac{1}{2}, \frac{3}{2}, \frac{5}{2}, \cdots, \frac{N_c}{2}$. The $SU(3)$ flavor representation of the baryons with a given spin J is given by the same Young tableau as its spin $SU(2)$ representation. The dimensions of the spin representations do not vary with N_c, but the dimensions of the flavor representations do. Consequently, the $SU(3)$ flavor multiplets are considerably more complicated for $N_c > 3$ than they are in QCD. The (T^3, T^8) weight diagrams for the spin-1/2 and spin-3/2 flavor multiplets for large-N_c baryons are given in Figs. 5 and 6. The numbers appearing in the weight diagrams denote the degeneracy of each weight. While there are many additional baryon states for $N_c > 3$, for $N_c = 3$ these flavor representations reduce to the usual octet and decuplet multiplets, respectively. In the

```
            1   1
          1   2   1
        1   2   2   1
      1   2   2   2   1
    1   2   2   2   2   1
  1   2   2   2   2   2   1
1   2   2   2   2   2   2   1
1   2   2   2   2   2   2   2   1
1   2   2   2   2   2   2   2   2   1
1   1   1   1   1   1   1   1   1
```

FIGURE 5. $SU(3)$ weight diagram of spin-$\frac{1}{2}$ baryons for large N_c.

```
              1   1   1   1
            1   2   2   2   1
          1   2   3   3   2   1
        1   2   3   4   3   2   1
      1   2   3   4   4   3   2   1
    1   2   3   4   4   4   3   2   1
  1   2   3   4   4   4   4   3   2   1
    1   2   3   3   3   3   3   2   1
      1   2   2   2   2   2   2   1
        1   1   1   1   1   1   1
```

FIGURE 6. $SU(3)$ weight diagram of the spin-$\frac{3}{2}$ baryons for large N_c.

end, we will apply the $1/N_c$ expansion to QCD baryons with $N_c = 3$, and there will be no unphysical baryon states in the expansion.

$1/N_C$ EXPANSION FOR BARYONS

The $1/N_c$ expansion of any baryon operator can be obtained by solving large-N_c consistency conditions. Each baryon operator has a $1/N_c$ expansion in terms of all independent operator products which can be constructed from the baryon spin-flavor generators.

The general form of the baryon $1/N_c$ expansion is given by

$$O_{\text{QCD}}^m = N_c^m \sum_{n=0}^{N_c} c_n \frac{1}{N_c^n} O^n, \qquad (13)$$

where O_{QCD}^m is an m-body quark operator in QCD which acts on baryon states. The baryon matrix elements of an m-body QCD quark operator can be at most $O(N_c^m)$, which is reflected in the factor of N_c^m in front of the $1/N_c$ operator expansion. The $1/N_c$ expansion sums over all independent operators which transform according to the same representation of spin $\otimes$ flavor symmetry as the QCD operator. The independent operators O^n which form a basis for the $1/N_c$ expansion are n^{th} degree polynomials of the baryon spin-flavor generators J^i, T^a and G^{ia}. The baryon matrix elements of these operator products can be computed in terms of the matrix elements of the baryon spin-flavor generators. The order in $1/N_c$ at which each independent operator product appears in the $1/N_c$ expansion also is known. $1/N_c$ power counting implies that an n-body operator product is multiplied by an explicit factor of $1/N_c^n$, or that each spin-flavor generator in an operator product is accompanied by a factor of $1/N_c$. Since the matrix elements of an n-body operator product are $\leq O(N_c^n)$, the matrix elements of each term in the $1/N_c$ operator expansion are manifestly $\leq O(N_c^m)$, as required. Every operator in the operator basis is accompanied by an unknown coefficient c_n, which is a reduced matrix element of the spin-flavor $1/N_c$ expansion and is not predicted by spin-flavor symmetry. The coefficients are $O(1)$ at leading order in the $1/N_c$ expansion. Finally, the summation over spin-flavor operators O^n only extends to N_c-body quark operators, since all n-body quark operators with $n > N_c$ will be redundant operators on baryons composed of N_c valence quarks.

Each operator O^n in the operator basis of the $1/N_c$ expansion can be written as an n-body quark operator using bosonic quarks which carry only spin and flavor quantum numbers. Let q^α represent an annihilation operator for a bosonic quark with spin-flavor $\alpha = 1, \cdots, 6$, and let $q_\alpha^\dagger$ represent the corresponding creation operator. In terms of these creation/annihilation operators, all n-body quark operators acting on baryon states can be catalogued. There is only a single 0-body quark operator, the baryon identity operator $\mathbb{1}$, which does not act on any of the valence quarks in the baryon. The 1-body quark operators are given by $q^\dagger q$ and the baryon spin-flavor generators

$$\begin{aligned}
J^i &= q^\dagger \left(\frac{\sigma^i}{2} \otimes \mathbb{1} \right) q, \\
T^a &= q^\dagger \left(\mathbb{1} \otimes \frac{\lambda^a}{2} \right) q, \\
G^{ia} &= q^\dagger \left(\frac{\sigma^i}{2} \otimes \frac{\lambda^a}{2} \right) q.
\end{aligned} \tag{14}$$

The notation here is compact. Each 1-body quark operator is understood to act on each quark line in the baryon. Thus,

$$G^{ia} = \sum_\ell q_\ell^\dagger \left(\frac{\sigma^i}{2} \otimes \frac{\lambda^a}{2} \right) q_\ell. \tag{15}$$

In addition, note that $q^\dagger q = N_c \mathbb{1}$ is not an independent operator and can be eliminated from the list of 1-body quark operators. The 2-body quark operators are given by products of the 1-body quark operators, the baryon spin-flavor generators. Each 2-body operator product can be written as the symmetric product (or anticommutator) of two 1-body operators since the commutator of any two 1-body operators can be replaced by

TABLE 1. $SU(6)$ Operator Identities

Identity	(J, SU(3))
$2\{J^i,J^i\} + 3\{T^a,T^a\} + 12\{G^{ia},G^{ia}\} = 5N_c(N_c+6)$	(0,0)
$d^{abc}\{G^{ia},G^{ib}\} + \frac{2}{3}\{J^i,G^{ic}\} + \frac{1}{4}d^{abc}\{T^a,T^b\} = \frac{2}{3}(N_c+3)T^c$	(0,8)
$\{T^a,G^{ia}\} = \frac{2}{3}(N_c+3)J^i$	(1,0)
$\frac{1}{3}\{J^k,T^c\} + d^{abc}\{T^a,G^{kb}\} - \varepsilon^{ijk}f^{abc}\{G^{ia},G^{jb}\} = \frac{4}{3}(N_c+3)G^{kc}$	(1,8)
$-12\{G^{ia},G^{ia}\} + 27\{T^a,T^a\} - 32\{J^i,J^i\} = 0$	(0,0)
$d^{abc}\{G^{ia},G^{ib}\} + \frac{9}{4}d^{abc}\{T^a,T^b\} - \frac{10}{3}\{J^i,G^{ic}\} = 0$	(0,8)
$4\{G^{ia},G^{ib}\} = \{T^a,T^b\}$ (27)	(0,27)
$\varepsilon^{ijk}\{J^i,G^{jc}\} = f^{abc}\{T^a,G^{kb}\}$	(1,8)
$3d^{abc}\{T^a,G^{kb}\} = \{J^k,T^c\} - \varepsilon^{ijk}f^{abc}\{G^{ia},G^{jb}\}$	(1,8)
$\varepsilon^{ijk}\{G^{ia},G^{jb}\} = f^{acg}d^{bch}\{T^g,G^{kh}\}$ $(10+\overline{10})$	$(1,10+\overline{10})$
$3\{G^{ia},G^{ja}\} = \{J^i,J^j\}$ $(J=2)$	(2,0)
$3d^{abc}\{G^{ia},G^{jb}\} = \{J^i,G^{jc}\}$ $(J=2)$	(2,8)

a linear combination of 1-body operators by the spin-flavor algebra. This observation also applies to all n-body operators with $n \geq 2$.

Not all operator products of the spin-flavor generators are linearly dependent, so it is necessary to eliminate redundant operators using operator identities. The complete set of 2-body operator identities for the completely symmetric baryon representation of $SU(6)$ are given in Table 1 along with their respective spin $\otimes$ flavor representations [25]. It is possible to eliminate all redundant n-body operators by using the 2-body operator identities, so Table 1 gives the complete set of operator identities.

The group theory behind Table 1 is interesting. Purely n-body quark operators are normal-ordered operators of the form

$$q^\dagger_{\alpha_1} \cdots q^\dagger_{\alpha_n} T^{\alpha_1 \cdots \alpha_n}_{\beta_1 \cdots \beta_n} q^{\beta_1} \cdots q^{\beta_n} \, . \tag{16}$$

For the completely symmetric $SU(6)$ representation of baryon states, the only nonvanishing normal-ordered quarks operators have tensors T which are totally symmetric in the upper and the lower spin-flavor indices. We would like to reexpress these independent normal-ordered operators as operator products of the spin-flavor generators, whose baryon matrix elements are known.

The group theory behind the operator identities is particularly elegant. The 0-body quark operator $\mathbf{1}$ is in the singlet representation of $SU(6)$, whereas the 1-body operators transform as the tensor product of a fundamental and antifundamental of $SU(6)$, which decomposes into a singlet and an adjoint of $SU(6)$. The 2-body operators are obtained from the tensor product of the symmetric 2-quark representation and its conjugate. Specifically,

$$0-\text{body}: \mathbf{1}$$
$$1-\text{body}: \left(\overline{\square} \otimes \square\right) = 1 + \text{adj} = 1 + T^\alpha_\beta \tag{17}$$

$$2 - \text{body}: \left(\square \otimes \square\right) = 1 + T^\alpha_\beta + T^{(\alpha_1\alpha_2)}_{(\beta_1\beta_2)}.$$

There are identities which relate the singlet and adjoint 2-body operators to 1-body adjoint and 0-body singlet operators, so the 2-body operators which transform as the singlet and the adjoint are not independent. The relevant identities are most easily understood keeping $SU(6)$ symmetry manifest. The generators J^i, T^a and G^{ia} form a complete set of $SU(6)$ generators Λ^A, $A = 1, \cdots, 35$. The operator identities relating the singlet 2-body operators to the 0-body operator and the adjoint 2-body operators to the 1-body operators are given by the Casimir identities

$$\Lambda^A \Lambda^A = C(R)\, \mathbb{1}$$
$$d^{ABC} \Lambda^B \Lambda^C = D(R)\, \Lambda^A, \tag{18}$$

where $C(R)$ and $D(R)$ are the quadratic and cubic Casimirs for the $SU(6)$ baryon representation R. These Casimir identities for the completely symmetric baryon spin-flavor representation produce the operator identities in the first two blocks of Table 1. The remaining 2-body operator identities arise because the completely symmetric product of two $SU(6)$ adjoints,

$$(\text{adj} \otimes \text{adj})_S = 1 + T^\alpha_\beta + T^{[\alpha_1\alpha_2]}_{[\beta_1\beta_2]} + T^{(\alpha_1\alpha_2)}_{(\beta_1\beta_2)}, \tag{19}$$

contains an additional tensor structure. The 2-body operator products corresponding to the tensor $T^{[\alpha_1\alpha_2]}_{[\beta_1\beta_2]}$ will vanish identically when acting on the completely symmetric baryon spin-flavor representation. These operator product combinations yield the vanishing operator identities given in the third part of Table 1.

The above operator identities are summarized by the following operator reduction rule: All operators in which two flavor indices are contracted using δ^{ab}, d^{abc}, or f^{abc} or two spin indices on G's are contracted using δ^{ij} or ε^{ijk} can be eliminated.

QCD BARYONS

I will now derive $1/N_c$ expansions for the masses, axial vector currents and magnetic moments of baryons in QCD. $SU(3)$ flavor breaking cannot be neglected relative to $1/N_c$, and is included in the analysis. For large-N_c baryons, the $1/N_c$ expansion extends up to N_c-body operators, so the $1/N_c$ expansion for QCD baryons goes up to third order in the generators. The $1/N_c$ expansion including flavor symmetry breaking also goes up to N_c-body operators, so perturbative $SU(3)$ breaking extends to finite order in flavor symmetry breaking. For a baryon operator with a $1/N_c$ expansion beginning with a n-body operator, the flavor symmetry breaking expansion extends to order $(N_c - n)$.

Masses

The baryon mass operator is a $J = 0$ operator. The leading operator in the $1/N_c$ expansion is the flavor singlet operator $N_c \mathbb{1}$ which gives the same $O(N_c)$ mass to all baryons in a spin-flavor representation. Since the $1/N_c$ expansion begins with a 0-body operator, the baryon mass operator can be expanded to third order in flavor symmetry breaking. Thus, the baryon mass operator decomposes into the $SU(3)$ flavor representations

$$M = M^1 + M^8 + M^{27} + M^{64}, \tag{20}$$

where the singlet, octet, **27** and **64** are zeroth, first, second and third order in $SU(3)$ flavor symmetry breaking, respectively. Each of these spin-singlet flavor representations has a $1/N_c$ operator expansion. The $1/N_c$ expansions are given by

$$M^1 = N_c \mathbb{1} + \frac{1}{N_c} J^2,$$

$$M^8 = T^8 + \frac{1}{N_c} \{J^i, G^{i8}\} + \frac{1}{N_c^2} \{J^2, T^8\},$$

$$M^{27} = \frac{1}{N_c} \{T^8, T^8\} + \frac{1}{N_c^2} \{T^8, \{J^i, G^{i8}\}\}, \tag{21}$$

$$M^{64} = \frac{1}{N_c^2} \{T^8, \{T^8, T^8\}\},$$

where it is to be understood that there is an unknown coefficient multiplying each operator in the above $1/N_c$ expansions. Note that the operators in these expansions can be derived using the operator identities in Table 1. For example, consider the octet mass expansion. $SU(3)$ flavor breaking transforms as the eighth component of an octet. There is only one 1-body operator which is $J = 0$ and the eighth component of an $SU(3)$ octet, namely T^8. From Table 1, one finds that there are three 2-body operators which transform in this manner: $d^{ab8} \{G^{ia}, G^{ib}\}$, $d^{ab8} \{T^a, T^b\}$ and $\{J^i, G^{i8}\}$. However, Table 1 shows that one linear combination of these operators is proportional to the 1-body operator T^8, and that another linear combination vanishes for the completely symmetric $SU(6)$ baryon representation. Thus, there is only one independent 2-body operator. This 2-body operator is taken to be $\{J^i, G^{i8}\}$ by the operator reduction rule. Application of the 2-body identities implies that there is a single independent 3-body operator which transforms as a spin singlet and as the eighth component of a flavor octet. Without loss of generality, this operator can be taken to be $\{J^2, T^8\}$. Similar analyses produce the other expansions in Eq. (21).

The $1/N_c$ expansion of the baryon mass operator given by Eqs. (20) and (21) contains eight independent operators, which is equal to the number of baryon masses in the **56** of $SU(6)$: the N, Λ, Σ, Ξ, Δ, Σ^*, Ξ^*, Ω. Each mass operator contributes to a unique linear combination of these eight masses. These linear combinations are given in Table 2[26]. Each mass combination occurs at specific orders in the $1/N_c$ and $SU(3)$ flavor breaking expansions; these suppression factors also appear in Table 2. The parameter $\varepsilon \sim m_s/\Lambda_{\rm QCD}$ is the suppression factor for $SU(3)$ flavor symmetry breaking. The final column gives the experimental value for the accuracy of each mass combination, which is defined by the dimensionless quantity

$$\frac{\sum B_i}{\sum |B_i|/2} \tag{22}$$

TABLE 2. Baryon Mass Hierarchy

Mass Splitting	$1/N_c$	Flavor	Expt.
$\frac{5}{8}(2N+3\Sigma+\Lambda+2\Xi) - \frac{1}{10}(4\Delta+3\Sigma^*+2\Xi^*+\Omega)$	N_c	1	*
$\frac{1}{8}(2N+3\Sigma+\Lambda+2\Xi) - \frac{1}{10}(4\Delta+3\Sigma^*+2\Xi^*+\Omega)$	$1/N_c$	1	$18.21 \pm 0.03\%$
$\frac{5}{2}(6N-3\Sigma+\Lambda-4\Xi) - (2\Delta-\Xi^*-\Omega)$	1	ε	$20.21 \pm 0.02\%$
$\frac{1}{3}(N-3\Sigma+\Lambda+\Xi)$	$1/N_c$	ε	$5.94 \pm 0.01\%$
$\frac{1}{2}(-2N-9\Sigma+3\Lambda+8\Xi) + (2\Delta-\Xi^*-\Omega)$	$1/N_c^2$	ε	$1.11 \pm 0.02\%$
$\frac{5}{4}(2N-\Sigma-3\Lambda+2\Xi) - \frac{1}{7}(4\Delta-5\Sigma^*-2\Xi^*+3\Omega)$	$1/N_c$	ε^2	$0.37 \pm 0.01\%$
$\frac{1}{2}(2N-\Sigma-3\Lambda+2\Xi) - \frac{1}{7}(4\Delta-5\Sigma^*-2\Xi^*+3\Omega)$	$1/N_c^2$	ε^2	$0.17 \pm 0.02\%$
$\frac{1}{4}(\Delta-3\Sigma^*+3\Xi^*-\Omega)$	$1/N_c^2$	ε^3	$0.09 \pm 0.03\%$

computed for each mass combination.

The $1/N_c$ and flavor symmetry breaking hierarchy predicted for the baryon masses can be tested by comparing the experimental accuracies to the $1/N_c$ and ε suppression factors. The numerical accuracies of the mass combinations are plotted in Fig. 7, except for the mass combination corresponding to the $N_c \mathbb{1}$ operator. The hierarchy predicted by the $1/N_c$ suppression factors is clearly evident. For example, there are three mass combinations that are first order in $SU(3)$ breaking, but of order $1/N_c$, $1/N_c^2$ and $1/N_c^3$ relative to the leading $O(N_c)$ singlet mass of the baryons. This pattern can be seen in Fig. 7. In addition, the two flavor **27** mass combinations which are second order in $SU(3)$ breaking are suppressed by factors of $1/N_c^2$ and $1/N_c^3$ relative to the leading $O(N_c)$ baryon mass. The Gell-Mann–Okubo flavor-**27** mass splitting of the spin-1/2 baryon octet,

$$\frac{1}{4}(2N-\Sigma-3\Lambda+2\Xi), \qquad (23)$$

and the flavor-**27** Equal Spacing Rule mass splitting of the spin-3/2 baryon decuplet,

$$\frac{1}{7}(4\Delta-5\Sigma^*-2\Xi^*+3\Omega), \qquad (24)$$

are linear combinations of the two flavor-**27** mass splittings specified by the $1/N_c$ expansion, so each is predicted to be a factor of $1/N_c^2$ more accurate than expected from flavor symmetry breaking factors alone. The most suppressed mass splitting is the flavor-**64** Equal Spacing Rule mass splitting,

$$\frac{1}{4}(\Delta-3\Sigma^*+3\Xi^*-\Omega), \qquad (25)$$

which is third order in $SU(3)$ flavor breaking and of relative order $1/N_c^3$. This mass combination is clearly suppressed by a greater factor than predicted from $SU(3)$ breaking alone. The experimental accuracy of this mass combination is consistent with the $1/N_c$ hierarchy, but a better measurement of the splitting in needed to test the $1/N_c^3$ prediction of the $1/N_c$ expansion definitively.

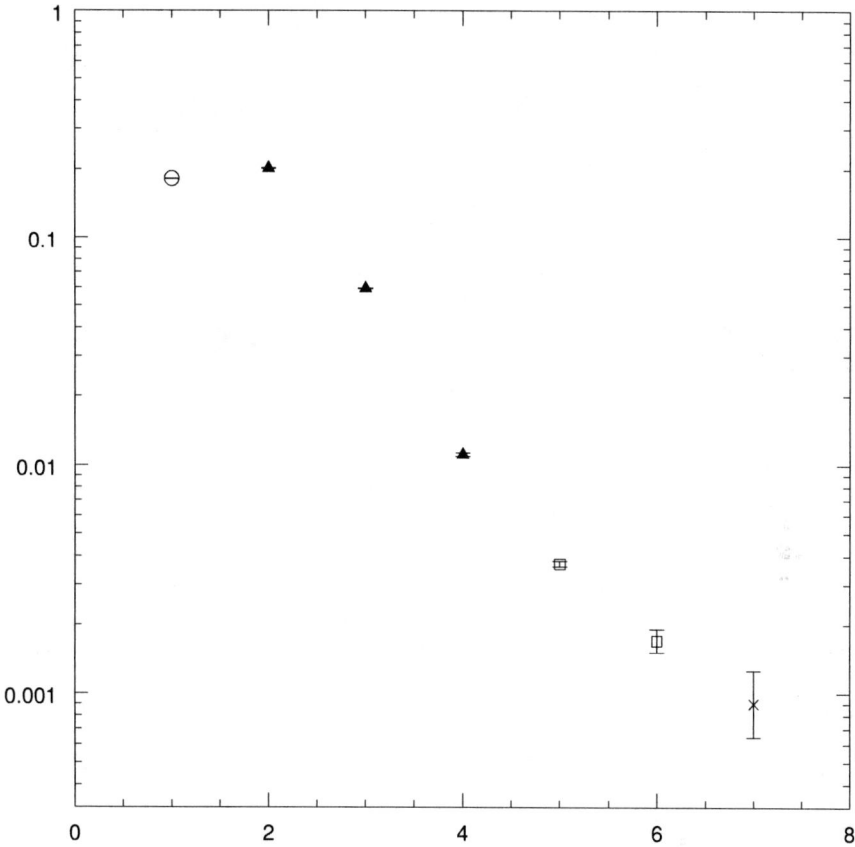

FIGURE 7. Baryon mass hierarchy. The mass combinations are of relative order $\frac{1}{N_c^2}$, $\frac{\varepsilon}{N_c}$, $\frac{\varepsilon}{N_c^2}$, $\frac{\varepsilon}{N_c^3}$, $\frac{\varepsilon^2}{N_c^2}$, $\frac{\varepsilon^2}{N_c^3}$, $\frac{\varepsilon^3}{N_c^3}$ compared to the overall $O(N_c)$ singlet mass of the baryon **56**.

In summary, the $1/N_c$ hierarchy is observed in the $I=0$ baryon mass splittings, and the presence of $1/N_c$ suppression factors explains the accuracy of baryon mass combinations quantitatively.

There also is clear evidence for the $1/N_c$ hierarchy in the $I=1$ baryon mass splittings. For example, the Coleman-Glashow mass splitting

$$[(p-n) - (\Sigma^+ - \Sigma^-) + (\Xi^0 - \Xi^-)] \tag{26}$$

has been measured to be non-zero for the first time quite recently. The measured mass splitting is more accurate than the prediction based on flavor suppression factors alone, and is consistent with an additional $1/N_c^2$ suppression predicted by the $1/N_c$ expansion[26]. It is particularly noteworthy that this prediction of the $1/N_c$ expansion was made before the Coleman-Glashow mass splitting was measured to the precision required to test the $1/N_c$ hierarchy. A more in-depth discussion of the $I = 1$ baryon mass splittings can be found in Refs. [45] and [46].

Axial Vector Couplings

The baryon axial vector current operator A^{ia} is $J = 1$ and an $SU(3)$ flavor adjoint. In the $SU(3)$ symmetry limit, the $1/N_c$ expansion of the baryon axial vector current is given by[25]

$$A^{ia} = a_1 G^{ia} + b_2 \frac{1}{N_c} J^i T^a + b_3 \frac{1}{N_c^2} \{J^i, \{J^j, G^{ja}\}\} \\ + d_3 \frac{1}{N_c^2} \left(\{J^2, G^{ia}\} - \frac{1}{2} \{J^i, \{J^j, G^{ja}\}\} \right), \qquad (27)$$

where the $1/N_c$ expansion for QCD baryons extends up to 3-body operators. The $1/N_c$ expansion involves four independent operators, so the baryon axial vector couplings are determined in terms of the four unknown coefficients a_1, b_2, b_3 and d_3. The usual $SU(3)$ flavor analysis of the axial vector couplings of the spin-1/2 baryon octet and spin-3/2 decuplet is given in terms of the four $SU(3)$ couplings D, F, C and H,

$$2D \, \mathrm{Tr} \, \bar{B} S^\mu \{\mathcal{A}_\mu, B\} + 2F \, \mathrm{Tr} \, \bar{B} S^\mu [\mathcal{A}_\mu, B] + C \left(\bar{T}^\mu \mathcal{A}_\mu B + \bar{B} \mathcal{A}_\mu T^\mu \right) + 2H \, \bar{T}^\mu S^\nu \mathcal{A}_\nu T_\mu, \quad (28)$$

where B represents the baryon octet, T^μ denotes the baryon decuplet, S^μ is a spin operator, and $\mathcal{A}^\mu$ is the axial vector current of the pion octet. The coefficients of the $1/N_c$ parametrization and the $SU(3)$ couplings are related by

$$D = \frac{1}{2}a_1 + \frac{1}{6}b_3,$$
$$F = \frac{1}{3}a_1 + \frac{1}{6}b_2 + \frac{1}{9}b_3,$$
$$C = -a_1 - \frac{1}{2}d_3, \qquad (29)$$
$$H = -\frac{3}{2}a_1 - \frac{3}{2}b_2 - \frac{5}{2}b_3.$$

The $1/N_c$ expansion for the baryon axial vector current can be truncated after the first two operators in Eq. (27) since the two 3-body operators are both suppressed relative to the 1-body operator by a factor of $1/N_c^2$. The 2-body operator $J^i T^a$ can not be neglected relative to the 1-body operator G^{ia} for all $a = 1, \cdots, 8$, so the leading order result for A^{ia} is given by

$$A^{ia} = a_1 G^{ia} + b_2 \frac{1}{N_c} J^i T^a, \qquad (30)$$

in terms of two parameters a_1 and b_2. The operator G^{ia} alone produces axial vector couplings with $SU(6)$ symmetry. If only the G^{ia} operator is retained, the four $SU(3)$ couplings satisfy

$$F/D = 2/3, \qquad C = -2D, \qquad H = -3F. \tag{31}$$

The operator $J^i T^a$ breaks the $SU(6)$ symmetry. The breaking is such that the $SU(3)$ couplings are related by

$$C = -2D, \qquad H = 3D - 9F, \tag{32}$$

which reduces to $SU(6)$ symmetry when $F/D = 2/3$.

It is worthwhile to consider the isospin decomposition of A^{ia} into the isovector, isodoublet and isosinglet axial vector currents. The $1/N_c$ expansion for the isovector axial vector current is given by

$$A^{ia} = a_1 G^{ia} + b_2 \frac{1}{N_c} J^i I^a, \qquad a = 1, 2, 3. \tag{33}$$

The 2-body operator $J^i I^a$ is suppressed relative to G^{ia} for $a = 1, 2, 3$, since the matrix elements of G^{ia} are $O(N_c)$ whereas the matrix elements of J^i and I^a are both $O(1)$. Thus, the $1/N_c$ expansion for the isovector current can be truncated to the 1-body operator G^{i3} up to a correction of order $1/N_c^2$ relative to the leading $O(N_c)$ term. A fit to the pion couplings yields $F/D = 2/3$ up to a correction of relative order $1/N_c^2$. Similar reasoning for the isodoublet axial vector current gives $F/D = 2/3$ up to a correction of relative order $1/N_c$. The 2-body operator cannot be neglected relative to the 1-body operator G^{ia} for the isosinglet axial vector current with $a = 8$.

The $1/N_c$ operator expansion for the flavor-octet baryon axial vector current can be generalized to include $SU(3)$ breaking. The $SU(3)$-symmetric expansion begins with a 1-body operator, so the baryon axial vector currents need to be expanded to second order in $SU(3)$ symmetry breaking. However, many of the baryon axial vector current observables are not measured, so it is not necessary to construct the $1/N_c$ expansion to all orders in $SU(3)$ breaking. Instead, the $1/N_c$ expansion will be considered to linear order in $SU(3)$ breaking.

The expansion at linear order in $SU(3)$ breaking involves additional spin-1 operators in different flavor representations,

$$\delta A^{ia} = A^{ia}_{\mathbf{1}} + A^{ia}_{\mathbf{8}_S} + A^{ia}_{\mathbf{8}_A} + A^{ia}_{\mathbf{27}} + A^{ia}_{\mathbf{10} + \overline{\mathbf{10}}}. \tag{34}$$

A valid truncation of the $1/N_c$ expansion to first order in $SU(3)$ breaking was constructed in Ref. [25]. The $1/N_c$ expansion is given by

$$\begin{aligned}
A^{ia} &= \left(a_1 \delta^{ab} + c_1 d^{ab8}\right) G^{ib} + \left(b_2 \delta^{ab} + c_2 d^{ab8}\right) \frac{1}{N_c} J^i T^b \\
&\quad + c_3 \frac{1}{N_c} \{G^{ia}, N_s\} + c_4 \frac{1}{N_c} \{J^i_s, T^a\} \\
&\quad + \frac{1}{3} c_5 \frac{1}{N_c} [J^2, [N_s, G^{ia}]] + \frac{1}{3} (c_1 + c_2) \delta^{a8} J^i,
\end{aligned} \tag{35}$$

TABLE 3. Axial couplings.

	Fit A
a_1	1.764 ± 0.042
b_2	-1.218 ± 0.216
d_3	0.549 ± 0.081
c_1	-0.044 ± 0.048
c_2	0.792 ± 0.228
c_3	-0.432 ± 0.036
c_4	0.096 ± 0.072
F	0.39 ± 0.02
D	0.88 ± 0.02
$3F - D$	0.27 ± 0.09

where the coefficients a_1 and b_2 are zeroth order in $SU(3)$ breaking, and the coefficients $c_1, \cdots, c_5$ are first order in the $SU(3)$ breaking. Thus, it is to be understood that the c_i are proportional to ε. Dropping the c_5 operator (since it does not contribute to any of the measured axial couplings) and adding the d_3 operator (to allow the $SU(3)$ parameters D, F and C to have arbitrary values) results in the 7-parameter formula

$$A^{ia} = a_1 G^{ia} + b_2 \frac{1}{N_c} J^i T^a + d_3 \frac{1}{N_c^2} \left(\{J^2, G^{ia}\} - \frac{1}{2} \{J^i, \{J^j, G^{ja}\}\} \right) \quad (36)$$

$$+ \Delta^a \left(c_1 G^{ia} + c_2 \frac{1}{N_c} J^i T^a \right) + c_3 \frac{1}{N_c} \{G^{ia}, N_s\} + c_4 \frac{1}{N_c} \{T^a, J_s^i\} + \frac{1}{\sqrt{3}} \delta^{a8} W^i,$$

where $\Delta_a = 1$ for $a = 4,5,6,7$ and is zero otherwise, and

$$W^i = (c_4 - 2c_1) J_s^i + \frac{1}{N_c} (c_3 - 2c_2) N_s J^i - 3 \frac{1}{N_c} (c_3 + c_4) N_s J_s^i. \quad (37)$$

A comparison of the $1/N_c$ expansion given in Eqs. (36) and (37) with the experimental data was performed in Ref. [28]. The extracted parameters from the experimental fit are tabulated in Table 3 [1]. As discussed in Ref. [28], c_1 and c_4 are anomalously small, and the character of the fit is not affected in any essential way by neglecting these parameters altogether. The coefficients c_2 and c_3 are suppressed relative to a_1 and b_2 by a factor consistent with a power of $SU(3)$ breaking ε. There is evidence for the $1/N_c$ suppression factors predicted in the $1/N_c$ expansion in the relative magnitudes of coefficients: a_1 and b_2 are comparable, as are c_2 and c_3, which is what is expected from the $1/N_c$ analysis. However, the fit of Ref. [28] is somewhat unsatisfying in that the χ^2 per d.o.f. is large, which was attributed to probable inconsistency in the experimental data.

A plot in Fig. 8 of the deviations of the baryon axial vector couplings from an $SU(3)$-symmetric fit is revealing. The $SU(3)$ breaking of the baryon octet axial vector couplings

---

[1] The parameters used in Eqs. (36) and (37) differ from those in Ref. [28] because $1/N_c$ factors have not been absorbed into the coefficients and because there is an overall factor of 2 difference in the above formula for the axial vector currents compared to the definition in Ref. [28]. The parameters given correspond to Fit A of Ref. [28].

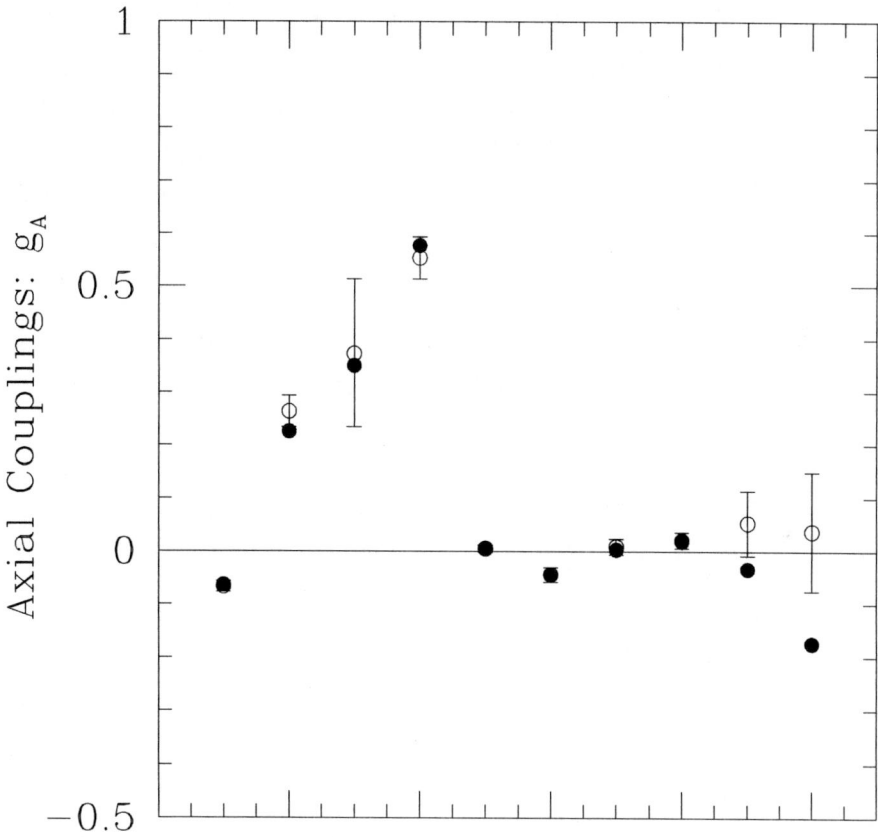

FIGURE 8. Deviation of the axial couplings from the best $SU(3)$-symmetric fit. The open circles are the experimental data, and the filled circles are the values from Fit A discussed in Ref. [28]. The points plotted are (from left to right) $\Delta \to N$, $\Sigma^* \to \Lambda$, $\Sigma^* \to \Sigma$, $\Xi^* \to \Xi$, $n \to p$, $\Sigma \to \Lambda$, $\Lambda \to p$, $\Sigma \to n$, $\Xi \to \Lambda$, and $\Xi \to \Sigma$.

obtained from hyperon β-decay measurements is very small, as is well-known. The $1/N_c$ expansion in Eq. (36) predicts the axial vector couplings of the **56** spin-flavor multiplet all together, which means that the $SU(3)$ breaking of hyperon semileptonic decay is related by spin-flavor symmetry to the $SU(3)$ breaking of nonleptonic decay decuplet → octet + pion. Thus, the pion axial vector couplings between the decuplet and octet baryons are included with the hyperon β-decay g_{AS} in the fit to $SU(3)$ breaking of the baryon axial vector couplings. Fig. 8 reveals that $SU(3)$ breaking is much larger for the decuplet-octet axial couplings than for the octet-octet couplings, and that an excellent fit is obtained with the exception of the hyperon β decays $\Xi \to \Lambda$, and $\Xi \to \Sigma$.

The experimental uncertainty of these two measurements is sizeable, and there are discrepancies between different experimental measurements for these couplings. Thus, it is likely that the experimental data for these decays is not entirely trustworthy, and may account for the large χ^2 per d.o.f. of the fit. Fig. 8 shows that the $1/N_c$ fit favors smaller $SU(3)$ breaking for these couplings.

The first six experimentally measured baryon axial vector couplings in Fig. 8 are isovector couplings, which suggests an analysis using $SU(2) \times U(1)_Y$ flavor symmetry rather than $SU(3)$ flavor symmetry. The $1/N_c$ expansion for baryon isovector axial vector couplings using $SU(2) \times U(1)_Y$ flavor symmetry is given by

$$\begin{aligned} A^{ia} &= G^{ia} + \frac{1}{N_c}\{N_s, G^{ia}\} \\ &+ \frac{1}{N_c^2}\{N_s, \{N_s, G^{ia}\}\} + \frac{1}{N_c^2}\{J^2, G^{ia}\} + \frac{1}{N_c^2}\{I^2, G^{ia}\} \\ &+ \frac{1}{N_c}J^i I^a + \frac{1}{N_c}J_s^i I^a + \frac{1}{N_c^2}\{J^i, \{G^{ka}, J_s^k\}\} \\ &+ \frac{1}{N_c^2}\{J_s^i, \{G^{ka}, J_s^k\}\} + \frac{1}{N_c^2}\{N_s, J^i I^a\} + \frac{1}{N_c^2}\{N_s, J_s^i I^a\} \end{aligned} \quad (38)$$

where $a = 1, 2, 3$ is an isovector index, and it is to be understood that each operator is multiplied by an unknown coefficient. The matrix elements of G^{ia} for baryons with strangeness of order unity are $O(N_c)$, whereas the matrix elements of J^i, I^a, and J_s^i are $O(1)$, so a valid truncation of the $1/N_c$ expansion is given by

$$A^{ia} = a_1 G^{ia} + a_2 \frac{1}{N_c}\{N_s, G^{ia}\} \quad (39)$$

up to terms which are suppressed by $1/N_c^2$ relative to the leading operator G^{ia}. Eq. (39) yields the equal spacing rule for baryon axial couplings derived in Ref. [12]. The rule implies an equal spacing of the decuplet $\to$ octet baryon non-leptonic pion couplings which is linear in strangeness,

$$\begin{aligned} g(\Sigma^* \to \Sigma\pi) - g(\Delta \to N\pi) &= g(\Xi^* \to \Xi\pi) - g(\Sigma^* \to \Sigma\pi) \\ g(\Sigma^* \to \Sigma\pi) &= g(\Sigma^* \to \Lambda\pi). \end{aligned} \quad (40)$$

This equal spacing rule is clearly evident in the experimental data, as shown in Fig. 8. Eq. (39) also implies $SU(4)$ spin-flavor symmetry for the baryon isovector axial vector couplings in each strangeness sector, so β-decay couplings $n \to p$ and $\Sigma \to \Lambda$ are related to the decuplet $\to$ octet pion couplings with strangeness zero and -1, respectively. These relations are very well-satisfied.

Magnetic Moments

The magnetic moment operator is $J = 1$ and transforms as the $Q = T^3 + T^8/\sqrt{3}$ component of an $SU(3)$ flavor **8**. The $1/N_c$ expansion of the magnetic moment operator is the same as for the axial vector couplings with

$$M^i = M^{i3} + \frac{1}{\sqrt{3}} M^{i8}. \tag{41}$$

The isovector magnetic moments are $O(N_c)$ whereas the isoscalar magnetic moments are $O(1)$ at leading order in the $1/N_c$ expansion, so it makes sense to construct $1/N_c$ expansions for the isovector and isoscalar magnetic moments separately[16]. The $1/N_c$ expansion of the isovector magnetic moments is given by

$$M^{i3} = G^{i3} + \frac{1}{N_c}\{N_s, G^{i3}\}, \tag{42}$$

up to terms which are suppressed by $1/N_c^2$ relative to leading 1-body operator G^{i3}. The $1/N_c$ expansion of the isoscalar magnetic moments is given by

$$M^{i8} = J^i + J_s^i + \frac{1}{N_c}\{N_s, J^i\} + \frac{1}{N_c}\{N_s, J_s^i\}, \tag{43}$$

up to terms of order $1/N_c^2$ compared to the two leading order 1-body operators J^i and J_s^i. It is to be understood that every operator in Eqs. (42) and (43) is multiplied by an unknown coefficient of order unity.

There are 21 independent magnetic moments of the baryon octet and decuplet, including transition magnetic moments. These 21 magnetic moments consist of 11 isovector combinations and 10 isoscalar combinations. The $1/N_c$ hierarchy of combinations of isovector and isoscalar magnetic moments is given in Table 4.

The 11 isovector magnetic moment combinations are parametrized in terms of the two operators of Eq. (42), so there are nine isovector magnetic moment relations which are satisfied to order $1/N_c$. These relations are listed as $V1-9$ in Table 4. Only one combination is measured, and the experimental accuracy $10\pm2\%$ is consistent with the $1/N_c^2$ prediction of the $1/N_c$ expansion. The $1/N_c$ expansion of the isovector magnetic moments can be truncated to the single operator G^{ia} by eliminating the subleading operator $\{N_s, G^{i3}\}$ operator. The isovector magnetic moment combination corresponding to this subleading operator is $O(1)$ in the $1/N_c$ expansion, or of relative order $1/N_c$ compared to the leading $O(N_c)$ contribution, and is listed as $V10_1$ is Table 4. The experimental accuracy of this relation is $27\pm1\%$, which is consistent with the prediction $1/N_c$ of the $1/N_c$ hierarchy. It is possible to derive a slightly different version of this mass combination by considering an $SU(3)$ analysis. In this analysis, the 2-body operator is $\{T^8, G^{i3}\}$, which is first order in $SU(3)$ breaking and order $1/N_c$ compared to the leading operator. The magnetic moment combination corresponding to this $SU(3)$ operator is listed as $V10_2$. The experimental accuracy of this relation is $13\pm2\%$, which is completely consistent with the theoretic prediction of ε/N_c of the $1/N_c$ expansion.

The 10 isoscalar magnetic moment combinations are parametrized by two 1-body operators at leading order in the $1/N_c$ expansion, so there are eight isoscalar magnetic moment relations, which appear as $S1-8$ in Table 4. The $1/N_c$ expansion of Eq. (43) contains four operators, so there are six isoscalar combinations $S1-6$ which are order $1/N_c^2$. The two subleading 2-body operators correspond to isoscalar relations $S7$ and $S8$,

TABLE 4. Baryon Magnetic Moments in the $1/N_c$ expansion. The isovector magnetic moments are $O(N_c)$ at leading order, and the isoscalar magnetic moments are $O(1)$. A $\sqrt{}$ implies that the relation is satisfied to that order in $1/N_c$ to all orders in $SU(3)$ breaking. The experimental accuracies are given for the relations whose magnetic moments have been measured.

	Isovector	$1/N_c$	Flavor	Expt.
V1	$(p-n) - 3(\Xi^0 - \Xi^-) = 2(\Sigma^+ - \Sigma^-)$	$1/N_c$	$\sqrt{}$	$10 \pm 2\%$
V2	$\Delta^{++} - \Delta^- = \frac{9}{5}(p-n)$	$1/N_c$	$\sqrt{}$	
V3	$\Lambda\Sigma^{*0} = -\sqrt{2}\Lambda\Sigma^0$	$1/N_c$	$\sqrt{}$	
V4	$\Sigma^{*+} - \Sigma^{*-} = \frac{3}{2}(\Sigma^+ - \Sigma^-)$	$1/N_c$	$\sqrt{}$	
V5	$\Xi^{*0} - \Xi^{*-} = -3(\Xi^0 - \Xi^-)$	$1/N_c$	$\sqrt{}$	
V6	$\sqrt{2}(\Sigma\Sigma^{*+} - \Sigma\Sigma^{*-}) = (\Sigma^+ - \Sigma^-)$	$1/N_c$	$\sqrt{}$	
V7	$\Xi\Xi^{*0} - \Xi\Xi^{*-} = -2\sqrt{2}(\Xi^0 - \Xi^-)$	$1/N_c$	$\sqrt{}$	
V8	$-2\Lambda\Sigma^0 = (\Sigma^+ - \Sigma^-)$	$1/N_c$	$\sqrt{}$	$11 \pm 5\%$
V9	$p\Delta^+ + n\Delta^0 = \sqrt{2}(p-n)$	$1/N_c$	$\sqrt{}$	$3 \pm 3\%$
V10$_1$	$(\Sigma^+ - \Sigma^-) = (p-n)$	1	$\sqrt{}$	$27 \pm 1\%$
V10$_2$	$(\Sigma^+ - \Sigma^-) = \left(1 - \frac{1}{N_c}\right)(p-n)$	1	ε	$13 \pm 2\%$
	Isoscalar			
S1	$(p+n) - 3(\Xi^0 + \Xi^-) = -3\Lambda + \frac{3}{2}(\Sigma^+ + \Sigma^-) - \frac{4}{3}\Omega^-$	$1/N_c^2$	$\sqrt{}$	$4 \pm 5\%$
S2	$\Delta^{++} + \Delta^- = 3(p+n)$	$1/N_c^2$	$\sqrt{}$	
S3	$\frac{2}{3}(\Xi^{*0} + \Xi^{*-}) = \Lambda + \frac{3}{2}(\Sigma^+ + \Sigma^-) - (p+n) + (\Xi^0 + \Xi^-)$	$1/N_c^2$	$\sqrt{}$	
S4	$\Sigma^{*+} + \Sigma^{*-} = \frac{3}{2}(\Sigma^+ + \Sigma^-) + 3\Lambda$	$1/N_c^2$	$\sqrt{}$	
S5	$\frac{3}{\sqrt{2}}(\Sigma\Sigma^{*+} + \Sigma\Sigma^{*-}) = 3(\Sigma^+ + \Sigma^-) - (\Sigma^{*+} + \Sigma^{*-})$	$1/N_c^2$	$\sqrt{}$	
S6	$\frac{3}{\sqrt{2}}(\Xi\Xi^{*0} + \Xi\Xi^{*-}) = -3(\Xi^0 + \Xi^-) + (\Xi^{*0} + \Xi^{*-})$	$1/N_c^2$	$\sqrt{}$	
S7	$5(p+n) - (\Xi^0 + \Xi^-) = 4(\Sigma^+ + \Sigma^-)$	$1/N_c$	$\sqrt{}$	$22 \pm 4\%$
S8	$(p+n) - 3\Lambda = \frac{1}{2}(\Sigma^+ + \Sigma^-) - (\Xi^0 + \Xi^-)$	$1/N_c$	ε	$7 \pm 1\%$
	Isoscalar/Isovector Relations			
S/V$_1$	$(\Sigma^+ + \Sigma^-) - \frac{1}{2}(\Xi^0 + \Xi^-) = \frac{1}{2}(p+n) + 3\left(\frac{1}{N_c} - \frac{2}{N_c^2}\right)(p-n)$	1	ε	$10 \pm 3\%$
	$\Delta^{++} = \frac{3}{2}(p+n) + \frac{9}{10}(p-n)$	$1/N_c^2$	$\sqrt{}$	$21 \pm 10\%$

which are order $1/N_c$. In addition, S8 is first order in $SU(3)$ breaking. The experimental accuracies of the isoscalar magnetic moment combinations are in complete accord with the $1/N_c$ hierarchy.

Finally, there are two additional relations given in Table 4. S/V_1 is a relation normalizing the isovector magnetic moments to the isoscalar magnetic moments in the $SU(3)$ flavor symmetry limit, and the last relation predicting the Δ^{++} magnetic moment is a linear combination of $V2$ and $S2$.

An alternative approach to the magnetic moments is possible using the $1/N_c$ expansion with $SU(3)$ flavor symmetry breaking. The same formula derived for the baryon axial vector currents applies since the magnetic moments also transform as $J = 1$ and as a component of an $SU(3)$ octet in the flavor symmetry limit. The analysis of flavor symmetry breaking involves the same representations analyzed for the baryon axial vec-

TABLE 5. Magnetic moments.

	Fit A
a_1	5.614 ± 0.122
b_2	0.216 ± 0.354
d_3	3.753 ± 0.639
c_1	-1.092 ± 0.230
c_2	0.612 ± 0.276
δc_2	0.066 ± 0.258
c_3	-0.522 ± 0.222
δc_3	0.024 ± 0.312
c_4	0.258 ± 0.228
δc_4	-0.288 ± 0.180

tor couplings, so Eq. (36) can be applied to the magnetic moments. A fit to the baryon magnetic moments using Eq. (36) gives the parameters listed in Table 5 taken from Ref. [28][2]. For the magnetic moments, the extracted value of b_2 is small, which implies that F/D is very close to the $SU(6)$ symmetry prediction of $2/3$. A plot of the deviations of the baryon magnetic moments from an $SU(3)$-symmetric fit, given in Fig. 9, shows that $SU(3)$ breaking is considerably larger for the magnetic moments than for the baryon axial vector couplings. Furthermore, $SU(3)$ symmetry breaking for the magnetic moments is dominated by an $O(N_c \sqrt{m_s})$ chiral loop correction, as shown in Fig 10 where the deviation of the baryon magnetic moments from an $SU(3)$-symmetric fit together with the leading chiral loop correction is plotted. Clearly, the remaining $SU(3)$ breaking in the magnetic moments is much reduced when the leading non-analytic correction is included in the fit. This result also can be seen in Table 5 in terms of the small values of the extracted parameters δc_{2-4} which measure the deviation of the extracted $SU(3)$ breaking parameters c_{2-4} from the flavor symmetry breaking structure given by the dominant chiral loop graph. There is no analogue of this chiral non-analytic correction for the baryon axial vector currents, so the $SU(3)$ breaking patterns of the baryon magnetic moments and the baryon axial vector currents are not similar.

CONCLUSIONS

The $1/N_c$ expansion for QCD baryons is both useful and predictive. In the formal large-N_c limit, there is a spin-flavor symmetry for baryons. For finite N_c, the spin and flavor structure of the baryon $1/N_c$ expansion is prescribed at each order in $1/N_c$. The $1/N_c$ expansion is given in terms of operator products of the generators of the baryon spin-flavor algebra which transform in a certain manner under spin $\otimes$ flavor symmetry. The order in $1/N_c$ of each operator structure is determined in the $1/N_c$ expansion, so the

---

[2] Again, the parameters used in Eqs. (36) and (37) differ from those in Ref. [28] because $1/N_c$ factors have not been absorbed into the coefficients and because there is an overall factor of 2 difference in the above formula for the axial vector currents compared to the definition in Ref. [28]. The parameters given correspond to Fit A of Ref. [28].

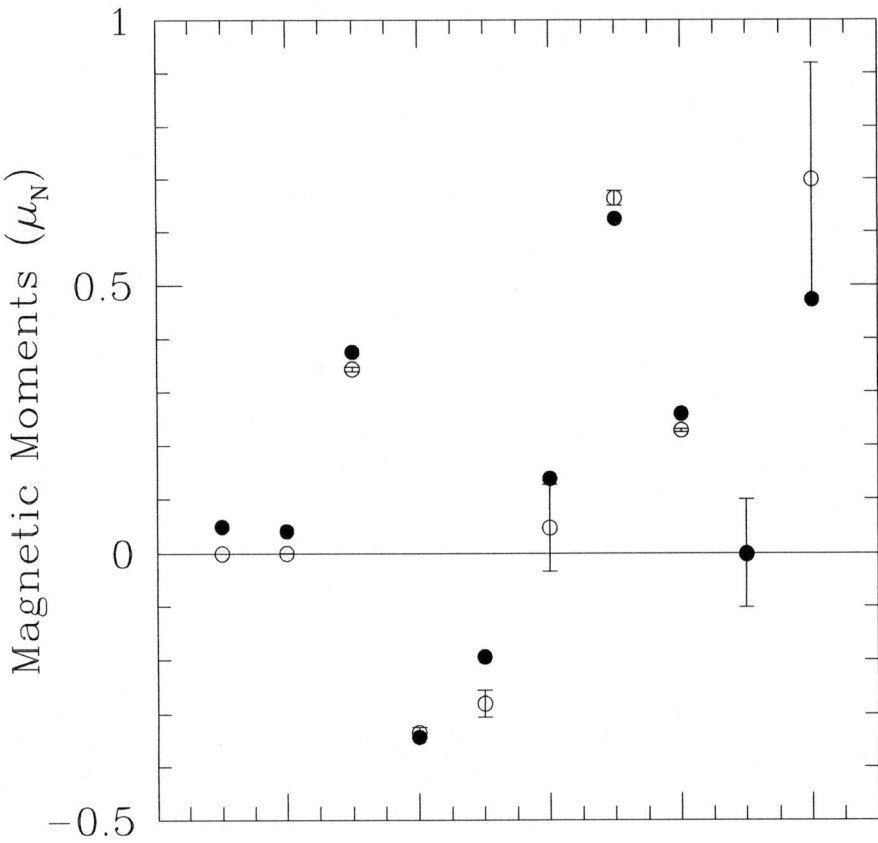

FIGURE 9. Deviation of the magnetic moments from the best $SU(3)$-symmetric fit. The open circles are the experimental data, and the filled circles are the values from Fit A discussed in the text. The order of the magnetic moments is p, n, Λ, Σ^+, Σ^-, $\Sigma^0 \Lambda$, Ξ^0, Ξ^-, $p\Delta^+$, and Ω. The Δ^{++} magnetic moment has not been plotted, since the experimental value has a very large error.

$1/N_c$ expansion predicts a hierarchy of spin and flavor relations for baryons in $1/N_c$. The predicted hierarchy of the $1/N_c$ expansion is evident in the baryon masses, axial vector currents and magnetic moments. The pattern of spin-flavor symmetry breaking is quite intricate since $1/N_c$ and $SU(3)$ flavor symmetry breaking are comparable in QCD. The presence of $1/N_c$ suppression factors explains why $SU(3)$ flavor symmetry works to a greater accuracy for baryons than predicted from an analysis of $SU(3)$ breaking alone, and gives a quantitative understanding of spin-flavor symmetry breaking for QCD baryons.

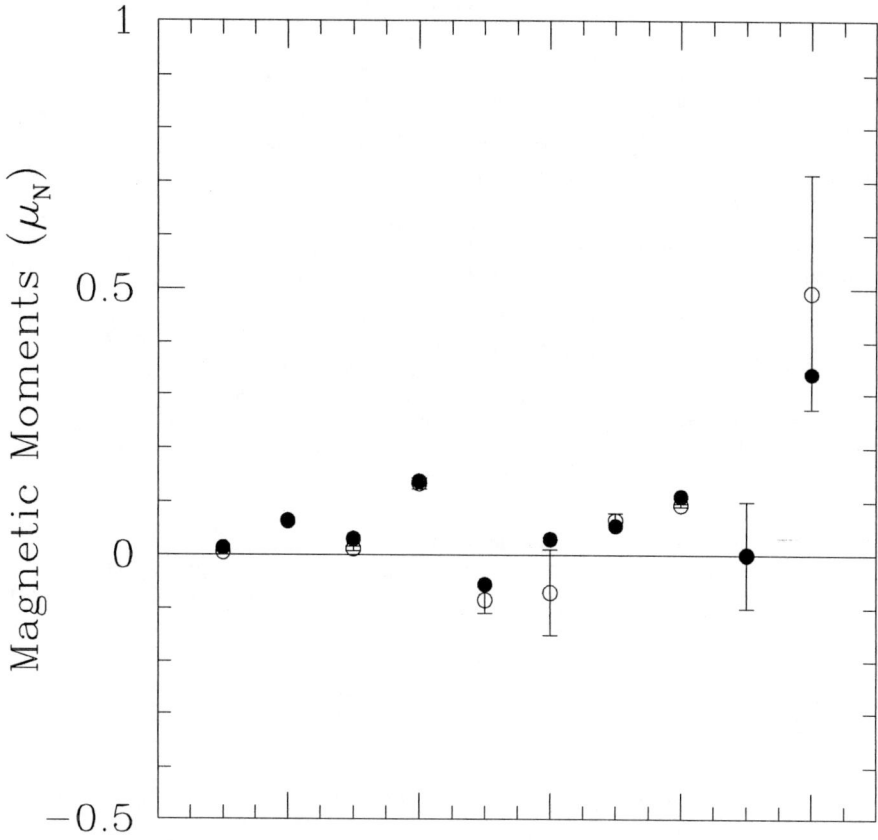

FIGURE 10. Deviation of the magnetic moments from the best $SU(3)$-symmetric fit plus the leading $O(\sqrt{m_s})$ chiral loop correction. The deviations should be compared with those in Fig. 9.

ACKNOWLEDGMENTS

I wish to thank Jürgen Engelfried and Mariana Kirchbach for organizing such a wonderful workshop. Special thanks to Ruben Flores-Mendieta for his hospitality, and to many of the participants for interesting discussions and experiences. This work was supported in part by the U.S. Department of Energy, under Grant No. DOE-FG03-97ER40546.

REFERENCES

1. Dashen, R., and Manohar, A.V., *Phys. Lett. B* **315**, 425-30 (1993); **315**, 438-40 (1993).
2. Jenkins, E., *Phys. Lett. B* **315**, 431-37 (1993); **315**, 441-46 (1993); **315**, 447-51 (1993).
3. 't Hooft, G., *Nucl. Phys. B* **72**, 461-73 (1974).
4. See Pais, A., *Rev. Mod. Phys.* **38**, 215-55 (1966), and references therein.
5. Manohar, A.V., *Nucl. Phys. B* **248**, 19-28 (1984).
6. Jenkins, E., *Annu. Rev. Nucl. Part. Sci.* **48**, 81-119 (1998).
7. Manohar, A.V., "Large N QCD," in *Les Houches Session LXVIII, Probing the Standard Model of Particle Interactions*, edited by F. David and R. Gupta, Amsterdam, Elsevier, 1998.
8. Veneziano, G., *Nucl. Phys. B* **117**, 519-45 (1976).
9. Witten, E., *Nucl. Phys. B* **160**, 57-115 (1979).
10. Gervais, J-L., and Sakita, B., *Phys. Rev. Lett.* **52**, 87-9 (1984); *Phys. Rev. D* **30**, 1795-1804 (1984).
11. Manohar, A.V., and Jenkins, E., *Phys. Lett. B* **255**, 558-62 (1991); *Phys. Lett. B* **259**, 353-8 (1991).
12. Dashen, R., Jenkins, E., and Manohar, A.V., *Phys. Rev. D* **49**, 4713-38 (1994).
13. Carone, C., Georgi, H., and Osofsky, S., *Phys. Lett. B* **322**, 227-32 (1994).
14. Luty, M., and March-Russell, J., *Nucl. Phys. B* **426**, 71-93 (1994).
15. Carone, C., Georgi, H., Kaplan, L., and Morin, D., *Phys. Rev. D* **50**, 5793-5807 (1994).
16. Jenkins, E., and Manohar, A.V., *Phys. Lett. B* **335**, 452-59 (1994).
17. Dorey, N., Hughes, J., and Mattis, M.P., *Phys. Rev. Lett.* **73**, 1211-14 (1994).
18. Manohar, A.V., *Phys. Lett. B* **336**, 502-7 (1994).
19. Broniowski, W., *Nucl. Phys. A* **580**, 429-44 (1994).
20. Wirzba, A., Kirchbach, M., and Riska, D.O., *J. Phys. G: Nucl. Part. Phys.* **20**, 1583-89 (1994).
21. Takamura, A., et al. *Prog. Theor. Phys.* **93**, 771-80 (1995); Takamura A. *Mod. Phys. Lett. A***11**, 463-70 (1996).
22. Luty, M., *Phys. Rev. D* **51**, 2322-31 (1995).
23. Luty, M., March-Russell, J., and White, M., *Phys. Rev. D* **51**, 2332-7 (1995).
24. Mattis, M.P., and Silbar, R., *Phys. Rev. D* **51**, 3267-86 (1995).
25. Dashen, R., Jenkins, E., and Manohar, A.V., *Phys. Rev. D* **51**, 3697-3727 (1995).
26. Jenkins, E., and Lebed, R.F., *Phys. Rev. D* **52**, 282-94 (1995).
27. Dorey, N., and Mattis, M.P., *Phys. Rev. D* **52**, 2891-2914 (1995).
28. Dai, J., Dashen, R., Jenkins, E., and Manohar, A.V., *Phys. Rev. D* **53**, 273-82 (1996).
29. Jenkins, E., *Phys. Rev. D* **53**, 2625-44 (1996).
30. Bedaque, P.F., and Luty, M.A., *Phys. Rev. D* **54**, 2317-27 (1996).
31. Jenkins, E., *Phys. Rev. D* **54**, 4515-31 (1996); **55**, 10-12 (1997).
32. Lam, C.S., and Liu, K.F., *Phys. Rev. Lett.* **79**, 597-600 (1997).
33. Kaplan, D.B., and Savage, M.J., *Phys. Lett. B* **365**, 244-251 (1996).
34. Kaplan, D.B., and Manohar, A.V., *Phys. Rev. C* **56**, 76-83 (1997).
35. Goity, J.L., *Phys. Lett. B* **414**, 140 (1997).
36. Pirjol, D., and Yan, T-M., *Phys. Rev. D* **57**, 1449-86 (1998).
37. Pirjol D., and Yan, T-M., *Phys. Rev. D* **57**, 5434-43 (1998).
38. Carlson, C.E., and Carone, C.D., *Phys. Rev. D* **58**, 053005 (1998).
39. Carlson, C.E., Carone, C.D., Goity, J.L., and Lebed, R.F., *Phys. Lett. B* **438**, 327-335 (1998).
40. Carlson, C.E., and Carone, C.D., *Phys. Lett. B* **441**, 363-370 (1998).
41. Carlson, C.E., Carone, C.D., Goity, J.L., and Lebed, R.F., *Phys. Rev. D* **59**, 114008 (1999).
42. Carlson, C.E., and Carone, C.D., *Phys. Lett. B* **484**, 260-266 (2000).
43. Flores-Mendieta, R., Hofmann, C.P., and Jenkins, E., *Phys. Rev. D* **61**, 116014 (2000).
44. Flores-Mendieta, R., Hofmann, C.P., Jenkins, E., and Manohar, A.V., *Phys. Rev. D* **62**, 034001 (2000).
45. Jenkins, E., and Lebed, R.F., *Phys. Rev. D* **62**, 077901 (2000).
46. Jenkins, E., *Nucl. Phys. Proc. Suppl.* **94**, 246-250 (2001).

Deep Inelastic Scattering at Large Energy and Momentum Transfers: Recent Results from HERA

Günter Wolf

Deutsches Elektronen Synchrotron DESY

Abstract. Data from H1 and ZEUS on the structure of the proton and its quark and gluon densities are discussed. A brief excursion is made into the field of inclusive diffraction by deep inelastic scattering. The comparison of e^-p and e^+p scattering at large momentum transfers demonstrates clearly the presence of weak contributions in neutral current interactions. The comparison with the corresponding charged current results shows at Q^2 values above the masses squared of the heavy vector bosons the unification of electromagnetic and weak interactions. The new data are testing the validity of the Standard Model down to spatial resolutions of the order of 10^{-16} cm.

INTRODUCTION

Ever since pointlike constituents, quarks, have been found in the nucleon [1] the question has been raised whether quarks and leptons also have substructure. Obviously, the detection of quarks and leptons as extended objects would produce a revolution in particle physics, e.g. by the possibility of constructing quarks and leptons from common subconstituents. Deep inelastic lepton nucleon scattering is particularly well suited for the study of the spatial structure of quarks and leptons. The sensitivity to substructure depends on the virtuality Q^2 of the exchanged current and hence on the energy of the scattering partners. The highest energies are provided by HERA where by now a spatial resolution of 10^{-16} cm or one thousandth of the proton radius has been achieved. The well defined initial state and the large energies and virtualities make HERA an excellent place to probe the Standard Model also in other areas such as new currents and new particles.

The first surprise encountered at HERA by H1 [2] and ZEUS [3] in deep inelastic electron proton scattering has been a rapid rise of the proton structure function F_2 as Bjorken-x tends to zero. The data are consistent with a power like rise, $F_2 \propto x^{-\lambda}$, a behaviour which for a number of reasons [4], such as overlap of partons or a Froissart like bound for the total virtual photon proton cross section or the requirement of unitarity, should not persist down to infinitely small values of x. Saturation or not is a hot research topic at HERA. The discovery of large rapidity gap events in deep inelastic scattering [5, 6], which result from diffraction, has opened another avenue on this issue. The link provided by the optical theorem between the total (F_2) and the diffractive cross section suggests that the rise should be even faster in diffraction.

This report presents new data from the two collider experiments H1 and ZEUS at

HERA with particular emphasis on the results for structure functions, diffraction and neutral and charged current processes at very large Q^2.

THE HERA COLLIDER AND THE EXPERIMENTS

The HERA collider can store electrons (positrons) of 30 GeV and protons of 920 GeV in two rings of 6.3 km circumference [7]. Table 1 lists some of the salient parameters of the machine [8]. In order to maximize the luminosity up to 210 bunches of particles can be stored for each beam. The time interval between consecutive bunches is 96 ns. The circulating e^- (e^+) beam becomes transversly polarized by the Sokolov-Ternov effect [9]. The measured specific luminosity is almost a factor of two larger compared to the design value. The maximum peak luminosity achieved so far is about 30% above the design value. The integrated luminosity per year provided by HERA for e-p collisions since 1992 is diplayed in Fig. 1. The total yearly luminosity increased by about a factor of two every year reaching 70 pb^{-1} in 2000. An upgrade program has been started which promises a factor of 3 - 5 increase in luminosity by the insertion of additional magnets close to the interaction point. Operation in the new configuration will start in the year 2001.

TABLE 1. HERA machine parameters.

parameter	electron ring	proton ring
circumference (m)	6336	
energy (GeV) design	30	820
operating energy (GeV)	27.6	920
e p c.m. energy squared (GeV^2)	10^5	
magn. bending field (T)	0.15	5.25
dipole bend. radius (m)	610	584
max. circ. curr. design (mA)	60	160
max. circ. curr. achiev.(mA)	50	110
n. bunch buckets	220	220
n. bunches (typical)	180	180
time betw. cross. (ns)	96	
max. lumi. design /achiev. $10^{31} cm^{-2} s^{-1}$	1.5/2.0	
max. yearly lumi. deliv. per expt. (pb^{-1})	70	
electron polarization	50 - 70%	

The data collected by H1 and ZEUS between 1992 and 2000 for different beam conditions correspond to integrated luminosities of about 17 pb^{-1} for e^-p and 116 pb^{-1} for e^+p collisions, per experiment.

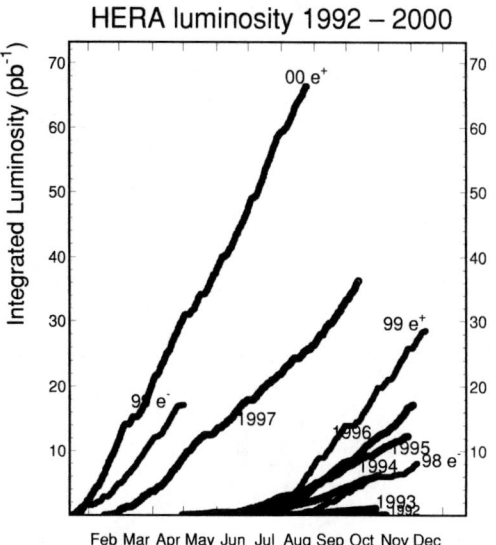

FIGURE 1. Integrated yearly luminosity delivered by HERA per experiment during 1992 - 2000.

STRUCTURE FUNCTIONS OF THE PROTON

Kinematics

Inclusive deep inelastic scattering (DIS) by neutral current (NC) exchange,

$$e(k) + p(P) \to e'(k') + anything$$

can be described as a function of Bjorken-x and Q^2 (see Fig. 2). The basic quantities, in the absence of QED radiation, are:

$$s = 4 \cdot E_e \cdot E_p \tag{1}$$

$$Q^2 = -q^2 = -(e-e')^2 \tag{2}$$

$$x = \frac{Q^2}{2P \cdot q} \tag{3}$$

$$v = (q \cdot P)/M_p \tag{4}$$

$$y = \frac{q \cdot P}{e \cdot P} \tag{5}$$

$$Q^2 = x \cdot y \cdot s \tag{6}$$

$$W^2 = \frac{Q^2(1-x)}{x} + M_p^2 \approx \frac{Q^2}{x} \text{ for } x \ll 1 \tag{7}$$

where e and e' are the four-momenta of the initial and final state electrons, P is the initial state proton four-momentum, M_p is the proton mass, s is the square of the ep c.m. energy, $-Q^2$ is the mass squared of the exchanged current, ν is the energy transfer and y the fractional energy transfer from the incident electron to the proton as measured in the proton rest frame, $y = \nu/\nu_{max}$, $\nu_{max} = s/(2M_p)$, x is the fractional momentum of the proton carried by the struck quark and W is the $\gamma^* p$ c.m. energy.

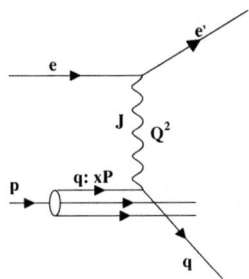

FIGURE 2. Diagram for deep inelastic ep scattering.

Definition of the structure functions

The differential cross section for deep inelastic scattering DIS can be expressed in terms of three structure functions, $\mathcal{F}_2, \mathcal{F}_L, \mathcal{F}_3$ [10]:

$$\frac{d^2\sigma}{dxdQ^2} = \frac{2\pi\alpha^2}{xQ^4}[(1+(1-y)^2)\mathcal{F}_2 - y^2\mathcal{F}_L \pm (1-(1-y)^2)x\mathcal{F}_3] \quad (8)$$

where α is the fine structure constant and the upper (lower) sign applies to e^- (e^+) p scattering. Since the contributions from $\mathcal{F}_L$ and $x\mathcal{F}_3$ are expected to be small in the measured region, the radiatively corrected NC cross section can be written as:

$$\frac{d^2\sigma}{dxdQ^2} = \frac{2\pi\alpha^2}{xQ^4}[(1+(1-y)^2)\mathcal{F}_2](1-\delta_L-\delta_3). \quad (9)$$

The structure function $\mathcal{F}_2$ receives contributions from photon and Z^o exchange and can be written as

$$\mathcal{F}_2 = F_2^{em}(1+\delta_Z). \quad (10)$$

where F_2^{em} denotes the contribution from photon exchange alone.

The corrections $\delta_{Z,L,3}$ are functions of x and Q^2 but are, to a good approximation, independent of $\mathcal{F}_2$, i.e. they are insensitive to the parton density distributions. They were calculated from theory using structure functions which gave a good representation of the data. In the measured region δ_L is small except when $y \geq 0.7$ where $\delta_L \approx 0.12$. The contributions from $\delta_{Z,3}$ are negligible for $Q^2 < 1000$ GeV^2 and small up to $Q^2 \approx 5000$ GeV^2. Whenever, in the following, Q^2 is below 1000 GeV^2 the notations $\mathscr{F}_2$, F_2^{em} and F_2 are used interchangeably.

In QCD, ignoring Z^0 exchange, $\mathscr{F}_2$ can be expressed in terms of the quark densities $q(x, Q^2)$ of the proton:

$$\mathscr{F}_2 = \sum_q e_q^2 x q(x, Q^2), \tag{11}$$

where e_q is the electric charge of quark q and the summation is performed over all quarks and antiquarks.

Structure function F_2 in the DIS regime

The proton structure function $\mathscr{F}_2$ has been measured at HERA over a wide range in x and Q^2 as shown in Fig. 3. At large x the HERA data overlap with those obtained by fixed target experiments. Until recently, most of the HERA results had come from data taken up to 1994, corresponding to about 2 - 3 pb^{-1} per experiment. H1 and ZEUS have now presented preliminary analyses of data from 1996-7 which are based on an order of magnitude higher integrated luminosity. The region covered in x, Q^2 has been enlarged substantially. At low Q^2 ZEUS and H1 have started to map out the transition region from photoproduction to deep inelastic scattering in the region of $x = 10^{-4} - 10^{-6}$. At the high Q^2 end, the structure function measurements have been extended up to $Q^2 \approx 30000$ GeV^2. Progress has also been made in the measurement of the charm contribution to F_2 which provides for a direct test of the gluon density $g(x, Q^2)$ extracted with QCD fits from the F_2 data.

The results on the x dependence of $\mathscr{F}_2$ from H1 [11] and ZEUS [12] are presented in Figs. 4, 5 for different Q^2 intervals. The error bars show the statistical and systematic uncertainties added in quadrature. For $Q^2 < 100$ GeV^2 the typical statistical errors are 2% at low Q^2 rising to 6% at $Q^2 \approx 100$ GeV^2. There is good agreement between the two HERA experiments. Also shown are the data from the fixed target experiments: BCDMS [13], E665 [14], NMC [15] and SLAC [16] which cover the region of 'large' x. In the region, where the HERA and the fixed target data overlap, good agreement is observed.

Figure 6 shows the $\mathscr{F}_2$ values as a function of Q^2 for fixed x. For $x > 0.1$ the data now span almost four decades in Q^2. Scaling violations proportional to $\ln Q^2$ are observed which decrease as x increases.

The most striking feature of the HERA data is the rapid rise of $\mathscr{F}_2$ as $x \to 0$ which is seen to persist down to Q^2 values as small as 1.5 GeV^2. The rise accelerates with increasing Q^2 as shown by Fig. 7 where ZEUS data from $Q^2 = 10, 22, 90$ and 250 GeV^2 have been overlaid. Some insight can be gained by fitting these data to the form $\mathscr{F}_2 = a + bx^{-\lambda}$. Fit results are given in Table 2. The contribution from the constant

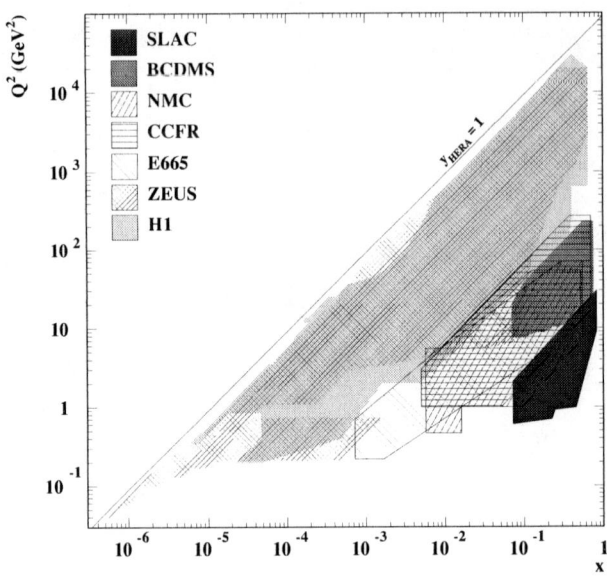

FIGURE 3. The $x - Q^2$ plane: the regions covered by H1 and ZEUS and by fixed target experiments.

term a is found to decrease rapidly with Q^2. The power λ of the x dependent term is rather constant with Q^2 while the coefficient b rises for Q^2 between 10 and 90 GeV^2 and appears to be driving the rapid rise of $\mathcal{F}_2$.

TABLE 2. Parameters from the fit of ZEUS $\mathcal{F}_2$ data to the form $\mathcal{F}_2 = a + bx^{-\lambda}$.

parameter	Q^2=10GeV^2	22GeV^2	90GeV^2	250GeV^2
a	0.31±0.03	0.14±0.05	0.01±0.11	0.06±0.11
b	0.05±0.01	0.10±0.02	0.18±0.07	0.11±0.06
λ	0.36±0.02	0.32±0.02	0.32±0.05	0.44±0.10

Fits performed without a constant term ($a \equiv 0$) result in considerably larger values of χ^2/ndf for $Q^2 < 40$ GeV^2. Thus, in this range of Q^2 the data indicate the presence of a soft term $a \neq 0$ as has been noticed before [17, 18]. If a is set to zero λ is found to rise with Q^2 as shown in Fig. 8, see e.g. [19, 20].

The curves shown in Figs. 4, 5 and 6 are the result of QCD NLO fits by H1 [11], ZEUS [12] and [21, 22, 23, 24] based on DGLAP evolution [25]. The fits show that NLO DGLAP evolution can give a consistent description of the data over the full Q^2 range.

ZEUS+H1 Preliminary 96/97

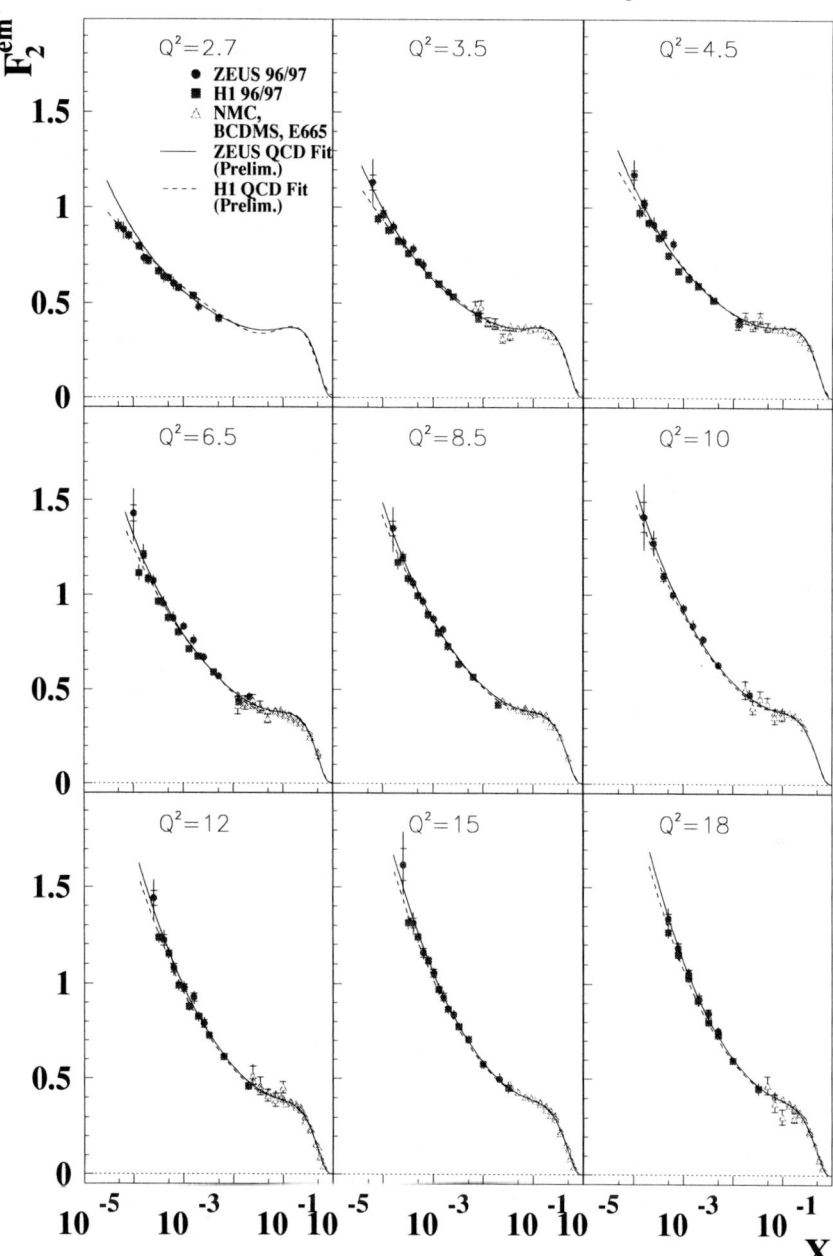

FIGURE 4. Structure function $\mathcal{F}_2$ from NC scattering as a function of x for fixed values of Q^2 between 2.7 and 18 GeV^2 as measured by H1, ZEUS. Also shown are the data from the fixed target experiments BCDMS, E665 and NMC. The lines indicate QCD NLO fits to the data by H1 and ZEUS.

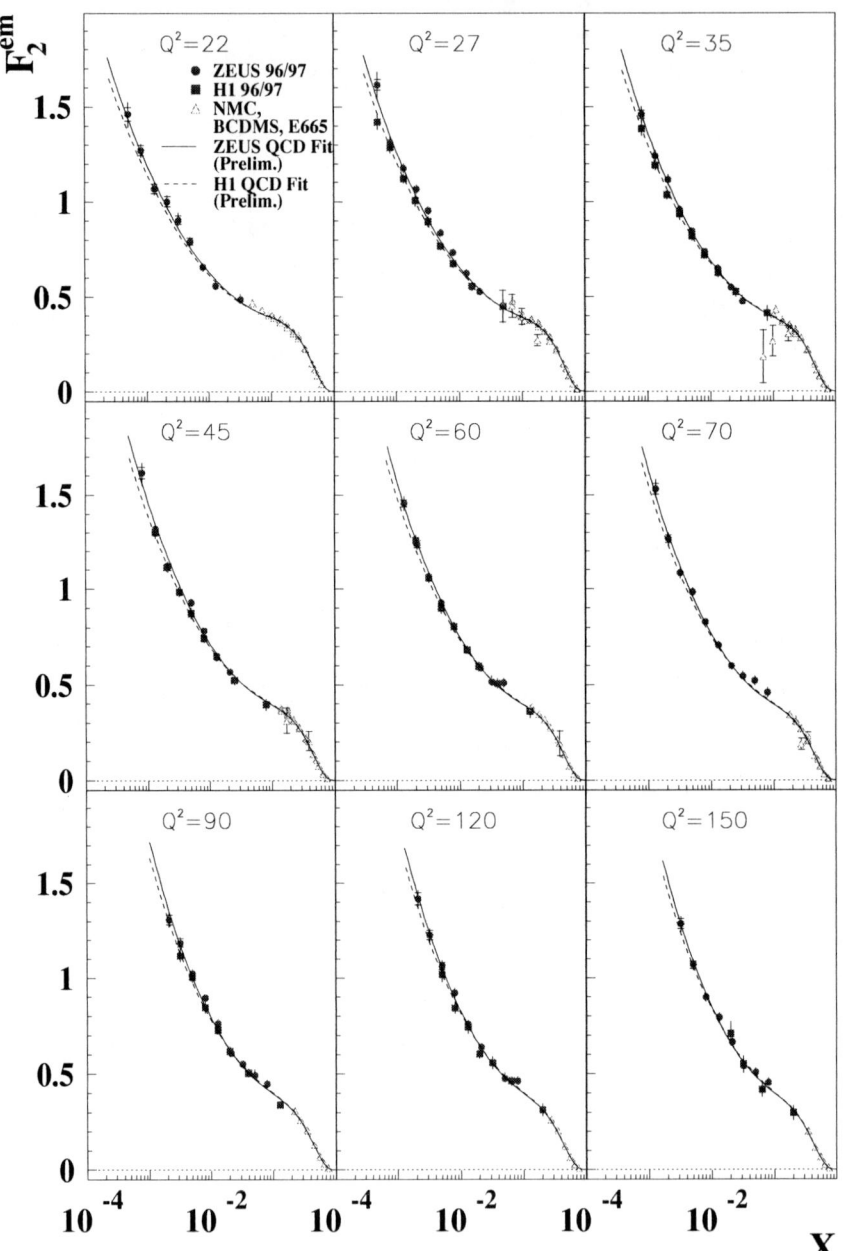

FIGURE 5. Structure function $\mathcal{F}_2$ from NC scattering as a function of x for fixed values of Q^2 between 22 and 150 GeV^2 as measured by H1, ZEUS. Also shown are the data from the fixed target experiments BCDMS, E665 and NMC. The lines indicate QCD NLO fits to the data by H1 and ZEUS.

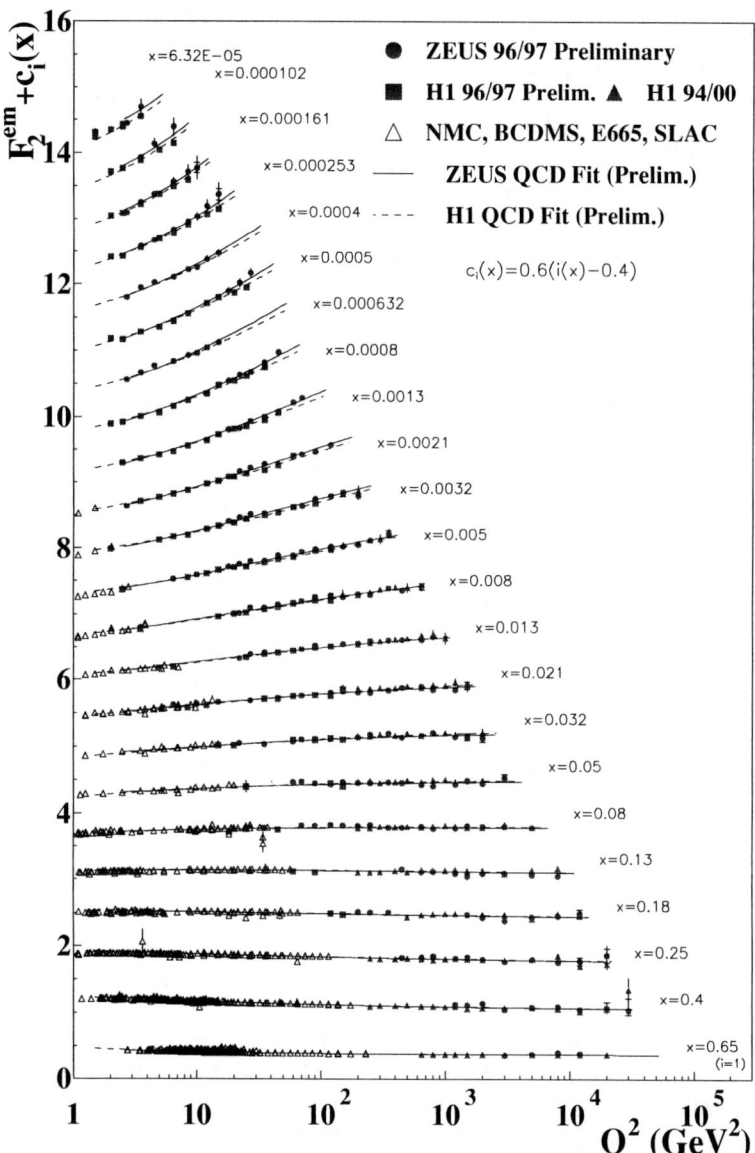

FIGURE 6. Structure function $\mathcal{F}_2$ from NC scattering as a function of Q^2 for fixed values of x as measured by H1, ZEUS. Also shown are the data from the fixed target experiments BCDMS, E665, NMC and SLAC. The dashed (solid) lines indicate QCD NLO fits to the data by H1 and ZEUS.

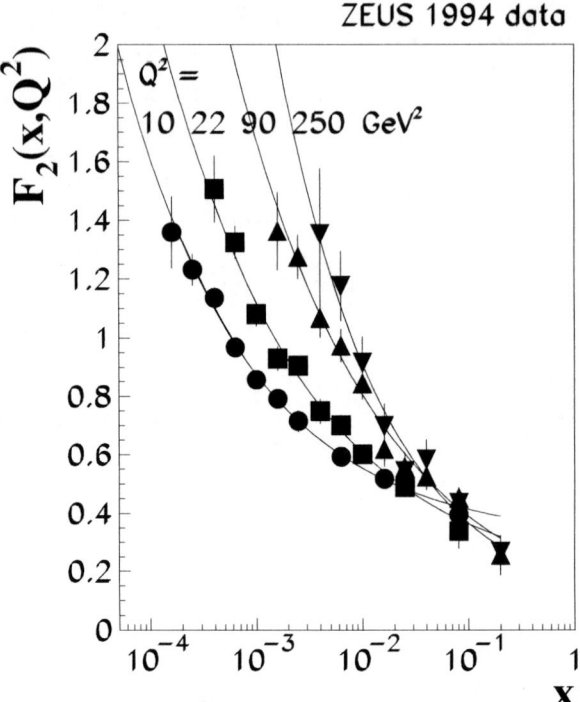

FIGURE 7. Structure function $\mathcal{F}_2$ from NC scattering as a function of x for fixed values of $Q^2 = 10$, 22, 90 and 250 GeV^2 taken from ZEUS 1994 data. The curves were obtained by fitting $\mathcal{F}_2$ to the form $\mathcal{F}_2 = a + bx^{-\lambda}$. From [18].

Structure function $\mathcal{F}_L$

In the Quark Parton Model (QPM = zero order QCD), $\mathcal{F}_L$ vanishes for spin 1/2 partons. In LO QCD $\mathcal{F}_L$ acquires a nonzero value due to the contribution from gluon radiation which is proportional to the strong coupling constant α_s. A direct determination of $\mathcal{F}_L$ requires the measurement of the DIS cross section at fixed x, Q^2 for different values of y which can be accomplished e.g. by varying the ep c.m. energy squared s.

H1 has shown that for a limited region of high y, $\mathcal{F}_L$ can be extracted from the $\mathcal{F}_2$ measurements at a single value of s if these are combined with a rather weak assumption on the validity of the DGLAP evolution [26, 11]. At high y the factors $1 + (1-y)^2$ and y^2 which multiply $\mathcal{F}_2$ and $\mathcal{F}_L$, respectively, in the expression for the DIS cross section (see Eq. 8), are of comparable magnitude. With this in mind, the following procedure was chosen. The $\mathcal{F}_2$ values measured by H1 for $y < 0.35$ and by BCDMS at larger values of x are used to extract the parton distribution functions. The DGLAP equations allow to evolve the parton distribution functions in Q^2 for fixed x and to predict $\mathcal{F}_2$ at

ZEUS 1995

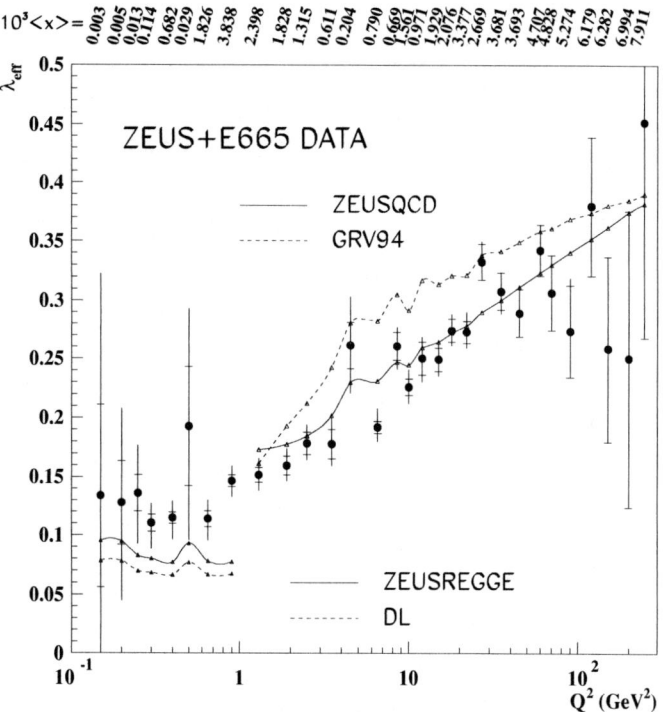

FIGURE 8. The power λ as a function of Q^2 obtained from fitting $\mathcal{F}_2$ to the form $\mathcal{F}_2 = bx^{-\lambda}$..

high y. Subtraction of the $\mathcal{F}_2$ contribution yields then $\mathcal{F}_L$. Note, as shown above, NLO DGLAP gives a good description of the $\mathcal{F}_2$ data over four orders of magnitude in x and Q^2, while for the determination of $\mathcal{F}_L$ the evolution extends the maximum Q^2 at fixed x by only a factor of two. Nevertheless, this analysis cannot strictly exclude the possibility that $\mathcal{F}_2$ behaves differently than assumed.

The longitudinal structure function $\mathcal{F}_L$ extracted by H1 is shown in Fig. 9 as a function of x. The full error bars represent the statistical and systematic uncertainties added in quadrature. The $\mathcal{F}_L$ values are significantly above zero and a factor of 2 - 3 below those of $\mathcal{F}_2$. The dashed bands, which show $\mathcal{F}_L$ as expected from the QCD NLO analysis, are consistent with the extracted $\mathcal{F}_L$ values.

This result lends also support to the procedure applied in the extraction of $\mathcal{F}_2$ from the DIS cross ection where the (small) contribution from $\mathcal{F}_L$ was taken from a QCD analysis, see above.

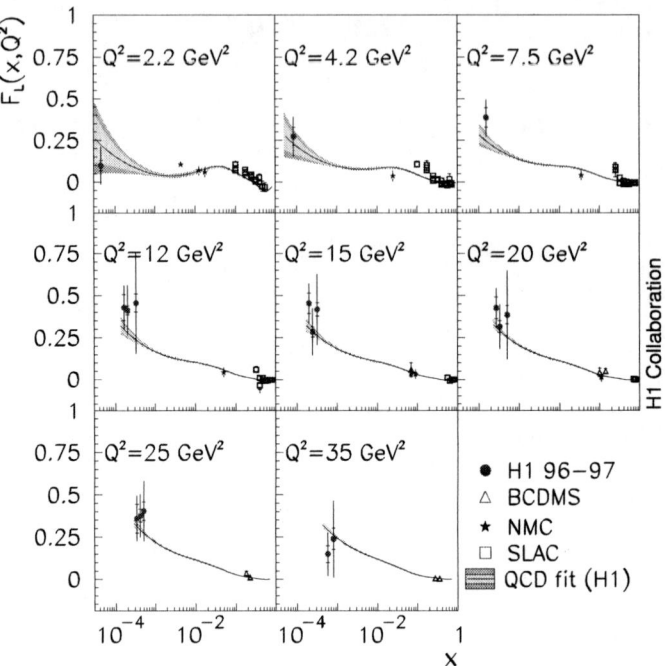

FIGURE 9. Structure function $\mathcal{F}_L$ from NC scattering as a function of x as determined by H1. The dashed band shows $\mathcal{F}_L$ as expected from the QCD NLO analysis.

Gluon density of the proton

In the QPM (diagram(a) in Fig. 10) the structure functions scale. Violations of scaling arise from QCD radiative effects which in LO (diagrams (b)-(d) in Fig. 10) have a simple mathematical expression:

$$\frac{d\mathcal{F}_2}{d\ln Q^2} = \sum_q e_q^2 \frac{\alpha_s(Q^2)}{2\pi} \int_x^1 \frac{dy}{y} [P_{qq}(\frac{x}{y})q(y,Q^2) + P_{qg}(\frac{x}{y})g(x,Q^2)]. \quad (12)$$

Here, P_{qq} and P_{qg} are the quark and gluon splitting functions and $g(x,Q^2)$ is the gluon density of the proton. At small x, $x < 10^{-2}$, the dominant contribution is expected to come from quark-pair creation by gluons (diagram (d) in Fig. 10 and second term in Eq. 12) which offers the possibility to determine the density of gluons g in the proton rather directly. It is instructive to look first at the approximate relation derived by [27] who considers only the contribution from diagram(d):

$$x \cdot g(x,Q^2) \approx \frac{27\pi}{10\alpha_s(Q^2)} \frac{dF_2(x,Q^2)}{d\ln Q^2}. \quad (13)$$

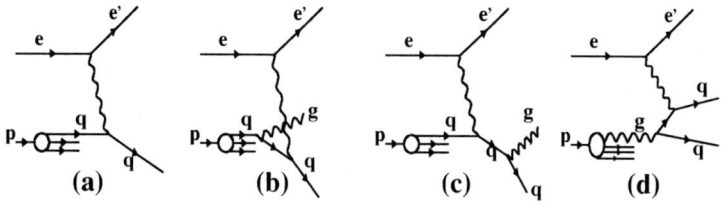

FIGURE 10. DIS diagrams: (a) no QCD radiation, (b)-(d) lowest order QCD processes.

The scaling violations of F_2, $\frac{d\mathcal{F}_2}{d\ln Q^2}$, measure directly the gluon momentum density $xg(x,Q^2)$.

H1 and ZEUS determined the gluon density by performing a global DGLAP type QCD fit to the $\mathcal{F}_2$ data such as shown by the curves in Figs. 4- 6. This method takes the contributions from the quark densities automatically into account but requires assumptions on the x dependence of the quark and gluon densities at the evolution scale Q_0^2 while for the approximate method these are not needed.

The gluon momentum density $xg(x,Q^2)$ of the proton as determined by H1 [11] at $Q^2 = 5$, 20 and 200 GeV^2 is shown in Fig. 11. Similar results have been reported by ZEUS [20, 21]. The precision now achieved for $xg(x,Q^2)$ is around 8% at $x = 5.10^{-4}$ and $Q^2 = 20$ GeV^2.

It is instructive to convert the HERA data for $\mathcal{F}_2$ and $xg(x,Q^2)$ which give the quark plus antiquark and gluon momentum densities, respectively, into numbers of partons and compare these with those obtained at large x by [24] from the parton density set MRSD0'. The result is shown in Table 3 for $Q^2 = 20$ GeV^2 which corresponds to a spatial resolution of about 5.10^{-15} cm. For high x, $x > 0.06$, one finds about 2.4 quarks which is close to the canonical number of 3 quarks; in addition there are roughly 1.8 gluons. The parton numbers increase rapidly towards small x: for $5.10^{-4} < x < 5.10^{-3}$ there are 4 times as many $q,\bar{q}$ and 15 times as many gluons as compared to the high-x regime.

Charm contribution to F_2

The structure function F_2 measures the momentum densities of quarks in the proton summed over all quark flavours. Charm quarks are expected to contribute via fusion

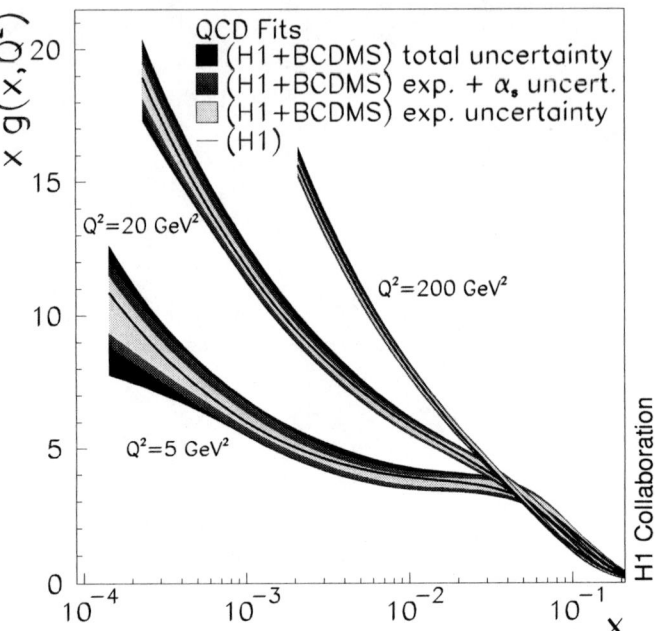

FIGURE 11. The gluon momentum density of the proton as a function of x at $Q^2 = 5, 20, 200$ GeV^2 as determined by H1.

TABLE 3. Equivalent number of partons of the proton for $Q^2 = 20$ GeV^2 determined at low x from the H1 and ZEUS data and at large x from the MRSD0' set.

	$x > 0.06$	$5 \cdot 10^{-4} < x < 5 \cdot 10^{-3}$
$N_{q,\bar{q}}$	≈ 2.4	9 ± 1
N_g	≈ 1.8	27 ± 2

of the virtual photon with a gluon from the proton (boson-gluon fusion) (diagram (d) of Fig. 10). Therefore pair production of charm quarks in DIS offers another way to measure the gluon density of the proton.

The charm contribution to DIS was determined by detecting D^* production in a limited $\eta^{D^*}, p_\perp^{D^*}$ region [28, 29, 30, 31, 32]. The measured cross section $\sigma(ep \to eD^*X)$ was extrapolated to the full range in $\eta^{D^*}, p_\perp^{D^*}$ with the help of a QCD model [33]. Using the branching ratio for $c \to D^*$ measured at LEP the cross section $\sigma(ep \to ec\bar{c}X)$ and from this the charm contribution $F_2^c(x, Q^2)$ were determined. In Figs. 12, 13 F_2^c is displayed as a function of x for fixed Q^2 values, and for fixed x as a function of Q^2. A comparison with the F_2 data shows that for $Q^2 \geq 6$ GeV^2 charm contributes about 20 - 30% of F_2. This is in broad agreement with the ratio of 4/10 expected for a democratic sea assuming massless quarks and neglecting the b quark contribution.

The curves in Fig. 12, 13 show the prediction for F_2^c using the gluon density as obtained from the QCD fit to F_2. They are in good agreement with the data providing an important test for the QCD analyses and fits performed by H1 and ZEUS on F_2.

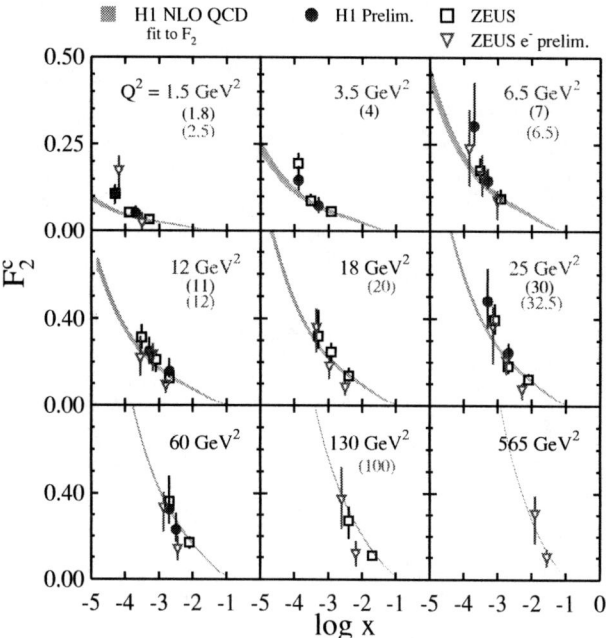

FIGURE 12. The charm contribution $F_2^c(x,Q^2)$ to the proton structure function F_2 for fixed Q^2 as a function of x as measured by H1 and ZEUS. The curves show the predictions from the H1 NLO QCD fit to the 1996-7 F_2 data.

In Fig. 14 the Q^2 dependence of F_2^c is compared for x values between 5.10^{-5} and 0.02. The bands show the predictions of GRV94HO [34] for a range of charm quark masses between 1.2 and 1.6 GeV. The agreement is impressive.

The rise of $\mathcal{F}_2$ as $x \to 0$ in the light of theory and models

The rise of $\mathcal{F}_2$ as $x \to 0$ and the W dependence of $\sigma_{\gamma^* p}^{tot}$

Neglecting contributions from Z^0 exchange, the DIS cross section can be expressed in terms of the flux of virtual photons times the total cross section for virtual - photon proton scattering, $\sigma_{\gamma^* p}^{tot}$ [35], which is written in terms of the cross sections for the scattering of transverse and longitudinal photons,

$$\sigma_{\gamma^* p}^{tot}(x,Q^2) = \sigma_T(x,Q^2) + \sigma_L(x,Q^2). \qquad (14)$$

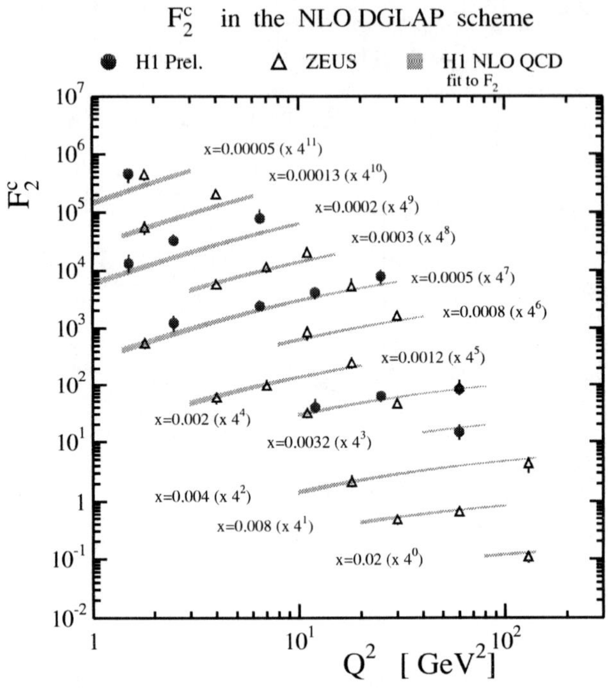

FIGURE 13. The charm contribution $F_2^c(x, Q^2)$ to the proton structure function F_2 for fixed x as a function of Q^2 as measured by H1 and ZEUS. The curves show the predictions from the H1 NLO QCD fit to the 1996-7 F_2 data.

The cross section defined in this manner can be interpreted in a way similar to the case of the interaction of real photons provided the lifetime of the virtual photon is large compared to the interaction time, which means $x \ll 1/(2M_p \cdot R_p)$, where R_p is the proton radius, $R_p = 4$ GeV^{-1} [36]. This requirement is well satisfied if $x \ll 0.1$. The expression for $\mathcal{F}_2$ in terms of σ_T and σ_L is

$$\mathcal{F}_2(x, Q^2) = \frac{Q^2(1-x)}{4\pi^2 \alpha} \sigma_{\gamma^* p}^{tot}. \tag{15}$$

At small x the expression can be rewritten in terms of the virtual-photon proton c.m. energy W, $W^2 \approx Q^2/x$ leading to

$$\sigma_{\gamma^* p}^{tot} \approx \frac{4\pi^2 \alpha}{Q^2} \cdot \mathcal{F}_2(W, Q^2) \tag{16}$$

Equation 16 was used by ZEUS [37] to determine from the 1993 $\mathcal{F}_2$ data $\sigma_{\gamma^* p}^{tot}$. Figure 15 shows for the 1994 data $Q^2 \sigma_{\gamma^* p}^{tot}$ as a function of W from 20 to 260 GeV for fixed Q^2 between 15 and 70 GeV^2 [18]. The data cluster around a narrow band rising almost

HERA PRELIMINARY 95-97

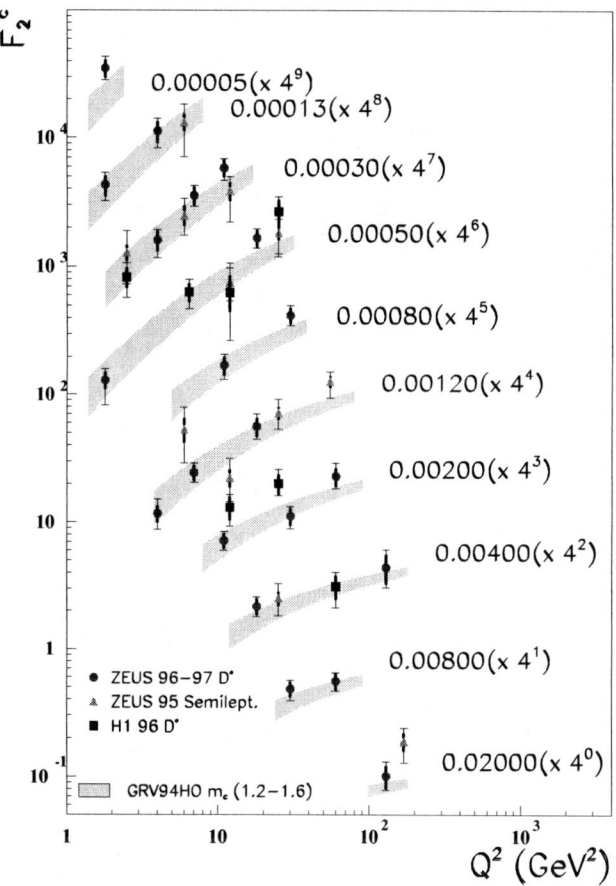

FIGURE 14. The charm contribution $F_2^c(x, Q^2)$ to the proton structure function F_2 for fixed x as a function of Q^2 as measured by H1 and ZEUS. The curves show the GRV94HO predictions for masses of the charm quark between 1.2 and 1.6 GeV.

linearly with W. A fit of the form $Q^2 \sigma_{\gamma^* p}^{tot} = a + b \cdot W^\varepsilon$ gave the value of $\varepsilon \approx 0.9$ (a fit with a=0 yielded a smaller value, $\varepsilon_{a=0} \approx 0.5$, though with considerable larger χ^2/ndf). If $\sigma_{\gamma^* p}^{tot}$ is described in terms of a pomeron trajectory, $\sigma_{\gamma^* p}^{tot} \sim (W^2)^{\alpha_P(0)-1}$, the intercept $\alpha_p(0) = 1 + \frac{\varepsilon}{2} \approx 1.45$.

The observed rise of $\sigma_{\gamma^* p}^{tot}$ is in marked contrast to the behaviour of the total cross section for antiproton-proton scattering and for *real* photon-proton scattering which are compatible with $\sigma_{\gamma^* p}^{tot} \approx W^{0.2}$, see curves in Fig. 15. Thus, the rise of $\mathscr{F}_2$ as $x \to 0$ or that of $\sigma_{\gamma^* p}^{tot}$ as $W \to \infty$ signals the presence of a new phenomenon. In QCD, the rise is

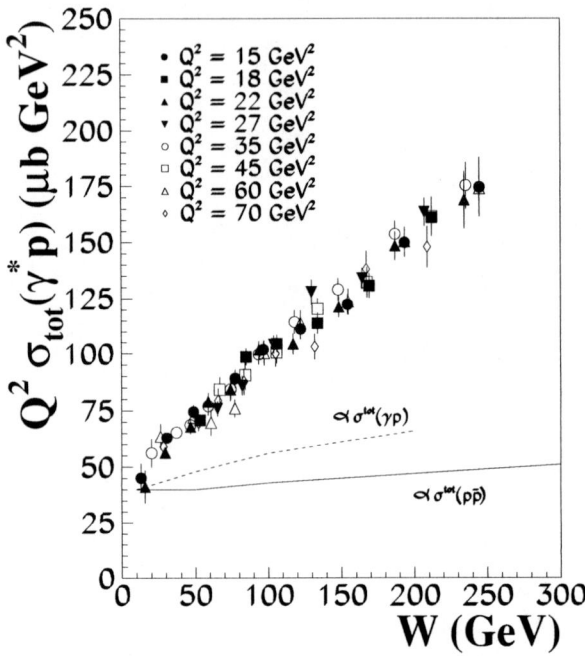

FIGURE 15. $Q^2 \cdot \sigma_{\gamma^*p}^{tot}$ as a function of W for fixed Q^2 between 15 and 70 GeV^2 as determined from the ZEUS data. The curves are proportional to the total cross sections $\sigma_{\gamma p}^{tot}$ and $\sigma_{p\bar{p}}^{tot}$.

the result of the strong increase of the number of partons at small x which blacken the proton and reduce its transparency for virtual photons of high Q^2.

DGLAP evolution

The DGLAP equations describe the evolution of the parton densities with Q^2. In order to solve these equations one must provide the parton densities as a function of x at some reference scale Q_o^2 which should be large enough for perturbative QCD to be applicable [38]. Assuming a Regge-type behavior [39, 40], the small x dependence of the valence quark (v), sea quark (s) and gluon densities is of the form:

$$xq_v(x, Q^2) \propto x^{1-\alpha_R} \tag{17}$$

$$xq_s(x, Q^2) \propto x^{1-\alpha_P} \tag{18}$$

$$xg(x, Q^2) \propto x^{1-\alpha_P} \tag{19}$$

where α_R and α_p denote the intercepts of the reggeon and pomeron trajectories. For $\alpha_R \approx 0.5$ and $\alpha_p \approx 1$ one obtains $xq_v(x,Q_o^2) \propto x^{0.5}$ and $xq_s(x,Q_o^2) \propto xg(x,Q_o^2) \propto const$.

In the leading log Q^2 approximation the $x^{0.5}$ behavior of the valence distribution remains unchanged by the Q^2 evolution while the sea quark and gluon distributions at small x become steeper. In fact, in perturbative QCD, $\mathscr{F}_2$ is expected to grow faster than any power of $\ln(1/x)$ as $x \to 0$ [41, 42]:

$$\mathscr{F}_2(x,Q^2) \approx C_o \left[\frac{33-2n_f}{576\pi^2 \ln \frac{1}{x} \ln \frac{\alpha_s(Q_o^2)}{\alpha_s(Q^2)}} \right]^{1/4} \cdot \exp\sqrt{\frac{144 \ln \frac{1}{x}}{33-2n_f} \ln \frac{\alpha_s(Q_o^2)}{\alpha_s(Q^2)}}. \quad (20)$$

where n_f is the number of quark flavors. The rise of $\mathscr{F}_2$ as $x \to 0$ can be accelerated by decreasing the reference scale Q_o^2. This can be seen by applying Eq. 20 for a specific set of parameters, e.g. $n_f = 3$ and $\alpha_s(Q^2) = \frac{4\pi}{(11-\frac{2}{3}n_f)\ln Q^2/\Lambda^2}$ with $\Lambda = 0.2$ GeV. Starting with x-independent parton distributions at Q_o^2 and parametrizing the $\mathscr{F}_2(x,Q^2)$ values obtained from Eq. 20 as $\mathscr{F}_2(x,Q^2) = b(Q^2)(1/x)^{\lambda(Q^2)}$ for $10^{-4} < x < 10^{-2}$ yields

for $Q_o^2 = 4$ GeV^2 at $Q^2 = 10$ (20, 100)GeV^2: $\lambda \approx 0.15$ (0.21, 0.29)
for $Q_o^2 = 1$ GeV^2 at $Q^2 = 10$ (20, 100)GeV^2: $\lambda \approx 0.29$ (0.32, 0.38).

Hence, theory predicted that F_2 would rise as $x \to 0$ but it could *not* predict the value of λ and therefore the speed of the rise since the starting parton distributions at Q_0^2 were unknown. We will return to this point below in a discussion of the GRV model [43, 44].

BFKL evolution

The DGLAP scheme requires angular ordering and neglects terms proportional to $\ln \frac{1}{x}$, an approximation, which may run in difficulties as $x \to 0$. The BFKL formalism [45] does not impose angular ordering and resums terms proportional to $\ln \frac{1}{x}$. Based on the BFKL formalism which performs QCD evolution for fixed Q^2 as function of x the gluon density in the proton was predicted to rise as $g(x,Q^2) \propto x^{-(1+\lambda)}$ as $x \to 0$, where $\lambda \approx \alpha_s(12/\ln 2)/\pi \approx 0.5$ for $Q^2 = 20$ GeV^2. BFKL-type calculations in NLO predict a considerably smaller value for λ, $\lambda \approx 0.15$ [46, 47, 48]. BFKL inspired fits to the data have been performed by [49].

Lopez-Yndurain model

The authors of [40] in 1980 presented a NLO QCD model which predicted the rise of $\mathscr{F}_2$ at small x observed at HERA with remarkable accuracy. According to the model, $\mathscr{F}_2$ should behave at small x as a power in x, $\mathscr{F}_2 \propto x^{-\lambda_s}$ where λ_s is independent of Q^2, except for heavy flavor thresholds. Extending the scanty data available then down to $x \approx 0.05$ led to the prediction $\lambda_s = 0.37 \pm 0.07$. Adding a constant term to $\mathscr{F}_2$ a fit of the model to the new data from H1 and ZEUS provided a good description of the measurements and yielded $\lambda_s = 0.355 \pm 0.01$ [50]. Dividing the data into different Q^2 intervals indicated a possible but small rise of λ_s with Q^2 from 0.325 ± 0.01 at $Q^2 < 10$ GeV^2 to 0.355 ± 0.01 at $Q^2 > 100$ GeV^2.

GRV model

The rapid rise of $\mathcal{F}_2$ at small x observed by the HERA experiments was anticipated in the GRV model [43] where a very small evolution scale, viz. $Q_0^2 = 0.34$ GeV^2, was chosen. The GRV predictions in NLO are compared with the experimental data on $\mathcal{F}_2$ data in Fig. 16 together with the NLO fits of the parton densities MRSA' [51] and CTEQ3 [52]. At large Q^2 both, GRV and the fits represent the data well. At $Q^2 \leq 18$ GeV^2 and $x < 10^{-3}$ GRV overshoots the data: a slight increase of Q_0^2 presumably can improve the agreement.

Haidt parametrisation

Double-logarithmic scaling of $\mathcal{F}_2$ with respect to x and Q^2 has been investigated in [53, 54]. The rise observed in the HERA data at small x is consistent with a logarithmic rise in x as well as in Q^2. An economical parametrization which decribes the $\mathcal{F}_2$ data of H1 and ZEUS for $x < 10^{-3}$, $Q^2 \geq 0.11$ GeV^2 has been obtained in [55]:

$$\mathcal{F}_2(x,Q^2) = m \log_{10}(1+\frac{Q^2}{Q_0^2}) \log_{10}(\frac{x_0}{x}) \qquad (21)$$

with $Q_0^2 = 0.55$ GeV^2, $x_0 = 0.04$ and $m = 0.45$.

Besides its simplicity, this parametrisation has the property that for fixed Q^2 the total cross section for virtual photon proton scattering, $\sigma_{tot}(\gamma^* p) \approx \frac{1}{Q^2} F_2(x,Q^2)$, does not violate the Froissart bound for hadronic total cross sections.

F_2 in the transition region between photoproduction and DIS

Virtual photon proton scattering at large Q^2 with its rapidly rising cross section as $W \to \infty$ behaves markedly different from ordinary hadron hadron or real photon proton collisions. The rise is a sign for hard scattering as described by perturbative QCD (pQCD). This leads to the question where in Q^2 does the transition from hadronic-type to hard scattering occur. When $\sqrt{Q^2}$ is below the typical hadronic transverse momenta of 200 - 500 MeV one expects confinement effects to dominate the interaction. On the other hand, the success of the GRV model may indicate that already at $Q^2 = 0.4$ GeV hard scattering is the dominant type of interaction. In order to gain further insight, a precise mapping of the proton structure functions in the region from photoproduction up to Q^2 of the order of a few GeV^2 is essential.

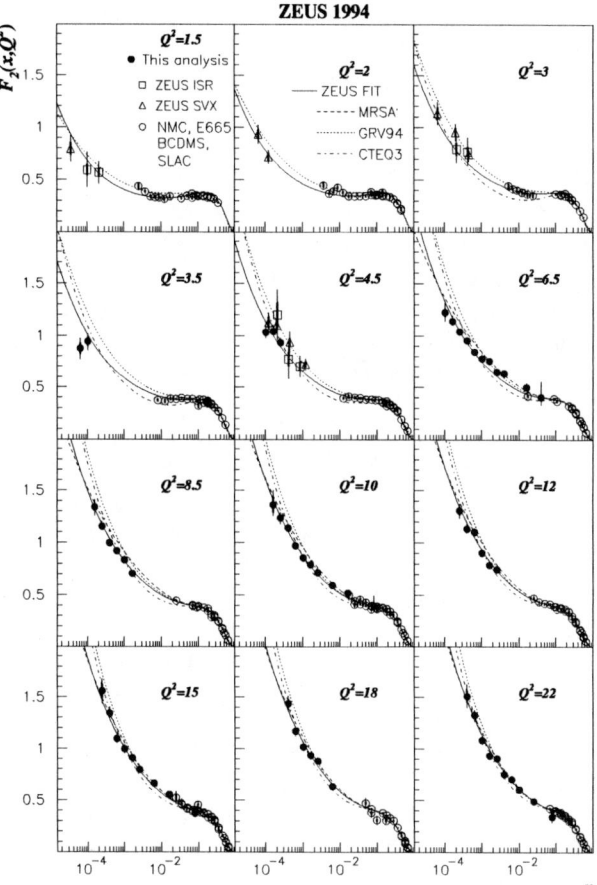

FIGURE 16. Structure function $\mathcal{F}_2$ from NC scattering as a function of x for fixed values of Q^2 as measured by ZEUS and data from the fixed target experiments BCDMS, E665, NMC and SLAC. The solid lines indicate the QCD NLO fit. Also shown are the MRSA', GRV94 and CTEQ3 parametrizations.

The structure function $F_2(x,Q^2)$ vanishes as $Q^2 \to 0$. This can be seen from its relation with the total virtual photon proton cross section, which is nonzero for $Q^2 = 0$ (Eq. 16). Measurements of F_2 at small Q^2 and small $x < 10^{-4}$ were published by H1 [56] and ZEUS [20, 57]. Figure 17 shows F_2 for fixed x as a function of Q^2. A number of phenomenological models [43, 58, 59, 60, 61, 62, 63] have been put forward to describe the behaviour at low x and low Q^2. The curves in Fig.17 show predictions of some of these models. The Vector Dominance type model DL [58] fails to reproduce the rise of $\mathcal{F}_2$ for Q^2 above ≈ 0.5 GeV^2. The BK model [59] assumes a VDM-like component which dominates the region of low Q^2 plus a hard QCD-like component for the high Q^2

regime. The predictions are somwhat above the data at low Q^2. The model CKMT [60] assumes at high Q^2 the dominance of a bare pomeron with an intercept of $\alpha_p(0) \approx 1.24$. At low Q^2 the pomeron intercept is assumed to decrease due to rescattering corrections leading to $\alpha_p(0) = 1.08$, the value obtained from hadron-hadron scattering. The CKMT predictions are found to be below the data for $Q^2 < 0.6 - 1$ GeV^2. The GRV model [43] considers only the hard scattering contribution. The GRV predictions for $\mathscr{F}_2$ are close to zero for Q^2 near the evolution scale $Q^2 = 0.34$ GeV^2. At $Q^2 = 0.44$ GeV^2 GRV accounts for about 40% of the measured $\mathscr{F}_2$ and for about 80% at $Q^2 = 0.57$ GeV^2. At $Q^2 = 0.9$ GeV^2 basically all of the DIS cross section is attributed to hard scattering. The model ABY [61], which assumes a hard plus a soft component evolved in NLO-QCD, gives a rather good descripton of the full set of data. The comparison suggests that the transition from soft to hard scattering occurs at Q^2 values somewhere between 0.5 and several GeV^2.

ZEUS studied also the scaling violations, $dF_2/d\ln Q^2$, in the transition region [64, 20, 57]. The logarithmic slope $dF_2/d\ln Q^2$ was derived from the data by fitting $F_2 = a + b\ln Q^2$ in bins of fixed x for $W^2 \simeq Q^2/x > 10$ GeV^2. Figure 18 shows $dF_2/d\ln Q^2$ as a function of x (Caldwell plot). Also shown (on top of figure) for each x bin is the weighted mean of Q^2 ($<Q^2>$) which increases as x increases due to kinematics and detector acceptance. For x values down to 3×10^{-4} the slope $dF_2/d\ln Q^2$ increases as x increases. At lower x (equivalent to lower Q^2) the slope decreases. The prediction of the GRV94 model, for which $dF_2/d\ln Q^2$ was determined in the same manner as for the data, reproduces the data for $x > 3 \times 10^{-4}$ ($Q^2 > 8$ GeV^2). For smaller x the GRV94 slope keeps on rising while in the data it decreases.

In order to gain further insight into the scaling violations at low x and Q^2 ZEUS performed QCD NLO fits using all their data with $3 \times 10^{-5} < x < 0.7$ and $Q^2 > 1$ GeV^2 together with those from NMC [65] and BCDMS [66]. A reasonable description of the data was achieved by the fits. Figure 19 shows the singlet quark momentum density ($x\Sigma \equiv x\sum_q[q(x) + \bar{q}(x)]$) and the momentum density of the gluon as a function of x for $Q^2 = 1$, 7 and 20 GeV^2. For $Q^2 \geq 7$ GeV^2 the gluon density is much larger than the singlet quark density while at $Q^2 = 1$ GeV^2 the gluon density has become equal to or lies below the singlet quark density. Also, $x\Sigma$ is seen to rise as $x \to 0$ for all three Q^2 values; xg, on the other hand, rises at $Q^2 = 7$ and 20 GeV^2 but may become constant at $Q^2 = 1$ GeV^2, or even zero. Such a behaviour was also found by [67].

One may tentatively conclude that at low Q^2, of the order of 1 GeV^2, the quark sea drives the gluon density while at higher Q^2 the gluon drives the density of the sea quarks. However, this conclusion may be premature since the fits did not allow for a contribution from soft scattering which could still be substantial at $Q^2 \geq 1$ GeV^2 (see e.g. Fig. 17).

DIFFRACTION IN DEEP INELASTIC SCATTERING

Diffraction has been studied extensively in hadron-hadron scattering at small momentum transfers [68]. An elegant parametrization of the data has been provided by the Regge formalism through the introduction of a pomeron trajectory [69]. The hypothe-

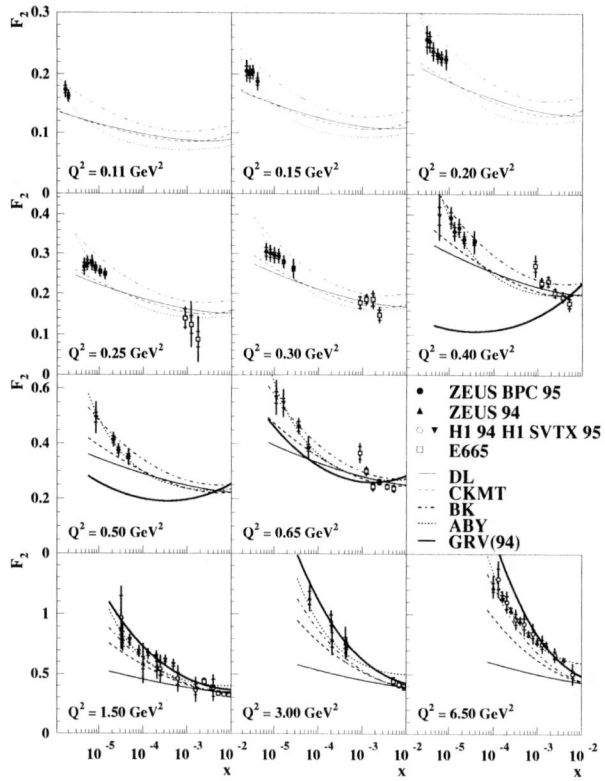

FIGURE 17. F_2 as a function of Q^2 for fixed values of x as measured by ZEUS, H1 and E665. Also shown are predictions of various models, see text.

sis that diffraction may have a partonic component [70] has been substantiated by the observation of high transverse energy jets produced in $p\bar{p}$ scattering [71]. However, in hadron-hadron scattering both collision partners are extended objects which makes extraction of the underlying partonic process(es) difficult. In DIS, on the other hand, the virtual photon has a pointlike coupling to quarks. Hence, HERA offers a unique opportunity to study the partonic structure of diffraction since it gives access to the regime of large photon virtualities and large energy transfers between the virtual photon and the target proton in its rest system, $v = Q^2/(2m_p x) = 2-20$ TeV.

Diffraction in virtual photon proton scattering has been studied at HERA in the quasielastic processes of vector meson production, $\gamma^* p \to V p$, where

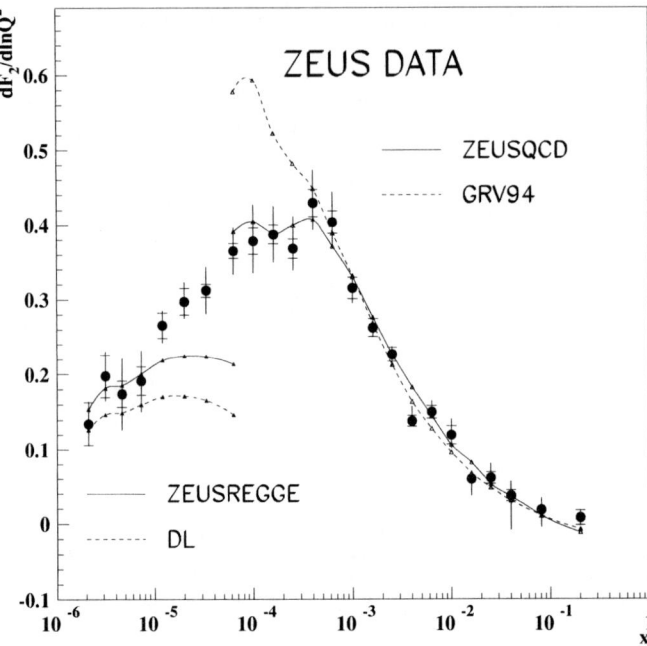

FIGURE 18. $dF_2/d\ln Q^2$ as a function of x calculated by fitting ZEUS F_2 data in bins of x to the functional form $a + b\ln Q^2$. For each x bin the average Q^2 value is indicated on top of the figure. The linked points labelled DL and GRV94 are from the Donnachie-Landshoff Regge fit and the GRV94 NLO QCD fit. In both cases, the points are obtained using the same Q^2 range as for the experimental data. From ZEUS.

$V = \rho^0, \omega, \phi, J/\Psi, \Upsilon$. While low mass V production ($V = \rho^0, \omega, \phi$) contributes more than 10% of the total cross section at $Q^2 = 0$ [72] it becomes negligible at large Q^2 [73]. However, diffractive dissociation of the virtual photon, $\gamma^* p \rightarrow XN$ (N= proton or a low mass nucleon system), into a large mass M_X, first recognized by the presence of a class of events with a large rapidity gap [5, 6] remains a substantial fraction of the total DIS cross section also at large Q^2 [74]. This has opened a window for a systematic study of diffraction in reactions initiated by a hard probe [74, 75, 76, 77, 78, 79, 80].

t-dependence of the diffractive cross section

The dependence of the diffractive cross section $d\sigma_{\gamma^* p \rightarrow Xp}/dM_X$ on the square of the four-momentum transfer t between the incoming and outgoing proton was measured

ZEUS 1995

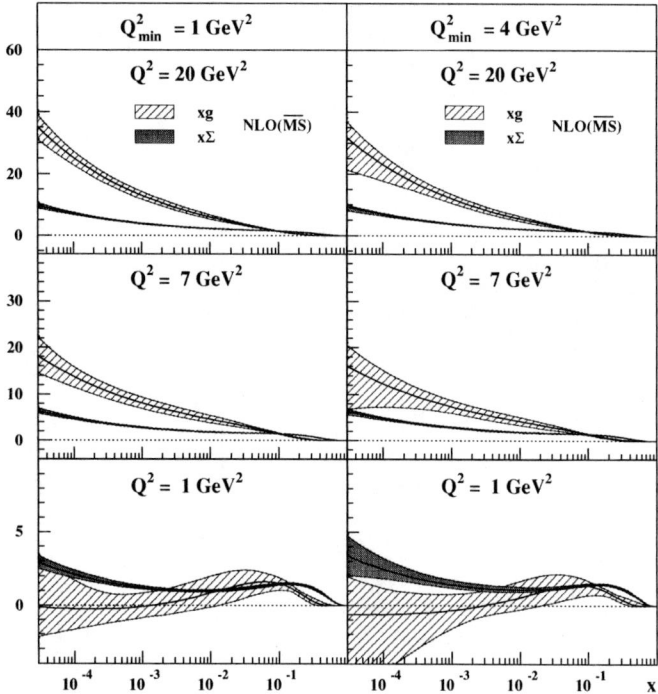

FIGURE 19. The quark singlet momentum distribution, $x\Sigma(x,Q^2)$ (shaded), and the gluon momentum density, $xg(x,Q^2)$ (hatched), as a function of x at fixed values of $Q^2 = 1, 7, 20$ GeV^2 for two different evolution scales Q^2_{min}. From ZEUS.

by ZEUS by detecting the scattered proton in the leading proton spectrometer (LPS) and the system X in the central detector [79]. The cross section is steeply falling with $-t$ as shown in Fig. 20; a fit of the form $d\sigma_{\gamma^*p \to Xp}/dM_X \propto exp(bt)$ yielded $b = 7.2 \pm 1.1(stat)^{+0.7}_{-0.9}(syst)$ GeV^{-2}. This shows that small momentum transfers between incoming and outgoing proton dominate, as expected for diffractive scattering.

Diffractive structure function and cross section

Isolation of diffractive events with the LPS is rather straightforward: detection of a proton scattered under very small angles and carrying a large fraction of the momentum of the incoming proton, $x_L = p^{LPS}/p_{p_{beam}} > 0.95$ ensures a large rapidity gap between the outgoing proton and the system X. However, the event rate is limited by the acceptance of the LPS.

In QCD, diffraction is characterized by the exchange of a colourless object, e.g. a colour singlet two-gluon system, between the incoming virtual photon and proton,

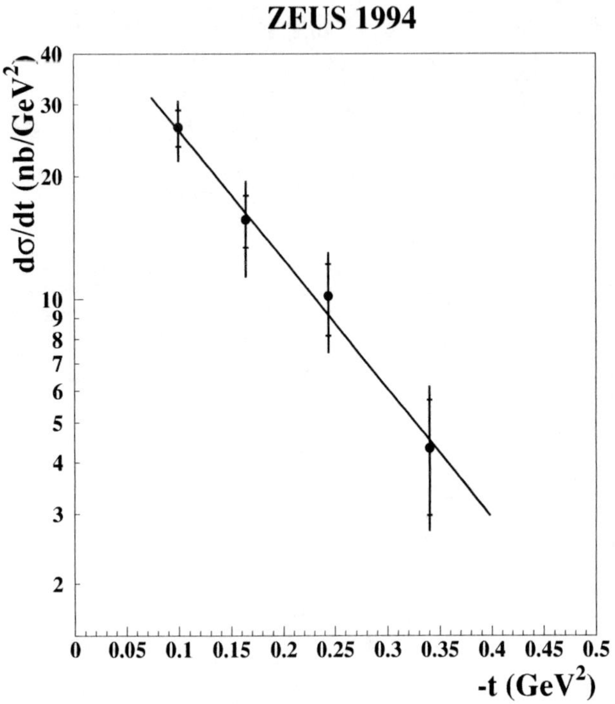

FIGURE 20. The diffractive cross section $d\sigma_{\gamma^* p \to Xp}/d|t|$ for events with a leading proton carrying more than 97% of the incoming proton momentum for $5 < Q^2 < 20$ GeV^2, $50 < W < 270$ GeV and $0.015 < \beta \approx Q^2/(M_X^2 + Q^2) < 0.5$. From ZEUS.

producing a massive photon-like state and a nucleon or a low mass excited nucleonic system, see Fig. 21. In comparison with nondiffractive scattering, the exchange of a colourless system suppresses QCD radiation and therefore the production of additional hadrons between the massive photon-like state and the nucleon or nucleonic system. Large acceptance for diffractive events has been achieved by requiring either a large rapidity gap between the nucleonic system N produced in the forward direction and the system X detected in the central detector, or by using the fact that in diffractive events most of the hadronic energy is carried away by the system N which escapes detection leaving behind, in the region of the central detector, a low mass system. Therefore, by measuring the distribution of the mass of the hadronic system observed in the central detector the diffractive contribution can be separated from the nondiffractive one (M_X method). Analyses based on the first method were performed by H1 [76, 78] and based on the M_X method by ZEUS [77, 80].

The results were presented in terms of the diffractive cross section $d\sigma_{\gamma^* p \to Xp}/dM_X$ and the diffractive structure function $F_2^{D(3)}(x_\mathbb{P}, \beta, Q^2)$ [70]. The cross section for the process $ep \to eXN$ is expressed in terms of the transverse (T) and

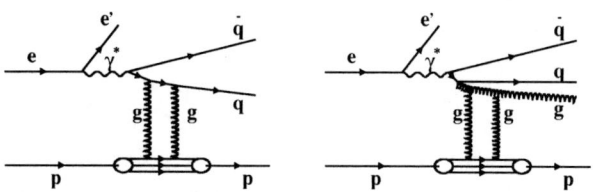

FIGURE 21. QCD diagrams for diffraction in DIS.

longitudinal (L) cross sections, σ_T^{diff} and σ_L^{diff}, for $\gamma^* p \to XN$ as:

$$\frac{d\sigma_{\gamma^* p \to XN}^{diff}(M_X,W,Q^2)}{dM_X} \equiv \frac{d(\sigma_T^{diff} + \sigma_L^{diff})}{dM_X}$$

$$\approx \frac{2\pi}{\alpha} \frac{Q^2}{(1-y)^2+1} \frac{d\sigma_{ep \to eXN}^{diff}(M_X,W,Q^2)}{dM_X d\ln W^2 dQ^2}. \quad (22)$$

The diffractive structure function of the proton can be related to the diffractive cross section in terms of the scaling variables $x_P \approx (M_X^2 + Q^2)/(W^2 + Q^2)$ and $\beta \approx Q^2/(M_X^2 + Q^2)$. In models where diffraction is described by the t-channel exchange of a system, for example the pomeron, x_P is the momentum fraction of the proton carried by this system and β is the momentum fraction of the struck quark within this system. One obtains [81]:

$$\frac{1}{2M_X} \frac{d\sigma_{\gamma^* p \to XN}^{diff}(M_X,W,Q^2)}{dM_X} = 4\pi^2 \alpha \frac{W^2}{(Q^2+W^2)^2 Q^2} F_2^{D(3)}(\beta, x_P, Q^2). \quad (23)$$

For $W^2 \gg Q^2$, Eq. 23 can be written as:

$$\frac{1}{2M_X} \frac{d\sigma_{\gamma^* p \to XN}^{diff}(M_X,W,Q^2)}{dM_X} \approx \frac{4\pi^2 \alpha}{Q^2(Q^2+M_X^2)} x_P F_2^{D(3)}(\beta, x_P, Q^2). \quad (24)$$

If $F_2^{D(3)}$ is interpreted in terms of quark densities then it specifies for a diffractive process the probability to find a quark carrying a momentum fraction $x = \beta x_P$ of the proton momentum.

Diffractive structure function measurement by H1

H1 presented their results for $\gamma^* p \to XN$ in terms of the diffractive structure function. The mass of the nucleon system N was restricted to $M_N < 1.6$ GeV. In Fig. 22 $x_P F_2^{D(3)}$ is shown as a function of x_P for fixed β values and fixed Q^2 between 4.5 and 75 GeV^2. The variation of $x_P F_2^{D(3)}$ with β and Q^2 is rather modest, indicating moderate scaling violations. In general, $x_\mathbb{P} F_2^{D(3)}$ is falling in the region $x_\mathbb{P} \leq 10^{-2}$ followed sometimes by an increase at large $x_\mathbb{P}$ values.

The $x_\mathbb{P}$ dependence of $F_2^{D(3)}$ is related to the W dependence of the diffractive cross section and, if analyzed in a Regge approach, to the Regge trajectories exchanged in the t-channel. By writing $x_\mathbb{P} F_2^{D(3)}(x_\mathbb{P}, \beta, Q^2) = (C/x_\mathbb{P}) \cdot (x_0/x_\mathbb{P})^n F_2^{D(2)}(\beta, Q^2)$ one obtains $n = 2(\overline{\alpha}_\mathbb{P} - 1)$ if only the pomeron trajectory $\alpha_\mathbb{P}$ (here averaged over t) is contributing. Because of the rise of $x_\mathbb{P} F_2^{D(3)}$ seen at large $x_\mathbb{P}$ H1 concluded that in addition to the pomeron a lower lying trajectory R is also contributing. The solid curves in Fig. 22 show the result of a two-component fit to the data. The dashed curves show the pomeron contribution alone as obtained from the fit. The fit yielded for the intercept of the pomeron trajectory $\alpha_\mathbb{P}(0) = 1.203 \pm 0.020 (stat) \pm (0.013 (syst)^{+0.030}_{-0.035}(model)$, a value which is above the results deduced from (soft) hadron-hadron scattering where $\alpha_\mathbb{P}(0) = 1.08$ [82] and $1.096^{+0.012}_{-0.009}$ [83] was found.

H1 fitted their data with a QCD motivated model, in which parton distributions are assigned to the leading and subleading exchanges. Figure 23 shows the resulting contributions to the parton densities of the pomeron as a function of the fraction z of the pomeron momentum carried by the parton. Within this model the majority of the momentum of the pomeron is found to be carried by gluons.

Diffractive cross section and structure function measured by ZEUS

ZEUS determined the diffractive cross section and structure function for $\gamma^* p \to XN$ where $M_N < 5.5$ GeV. The diffractive cross section is presented in Fig. 24 as a function of W for various M_X and Q^2 values. The diffractive cross section rises rapidly with W at all Q^2 values for M_X up to 7.5 GeV. A fit to the form

$$\frac{d\sigma^{diff}_{\gamma^* p \to XN}(M_X, W, Q^2)}{dM_X} = h \cdot W^{a^{diff}}, \tag{25}$$

where a^{diff} and the normalization constants h were treated as free parameters, yielded $a^{diff} = 0.507 \pm 0.034(stat)^{+0.155}_{-0.046}(syst)$ which corresponds to a t-averaged $\overline{\alpha}_\mathbb{P} = 1 + a^{diff}/4 = 1.127 \pm 0.009(stat)^{+0.039}_{-0.012}(syst)$. This value is consistent with the H1 result since averaging over the t-distribution gives approximately $\overline{\alpha}_\mathbb{P} = \alpha_\mathbb{P}(0) - 0.03$.

The diffractive cross section was compared with the measured total virtual-photon proton cross section. The ratio of the two cross sections,

$$r^{diff}_{tot} = \frac{\int_{M_a}^{M_b} dM_X d\sigma^{diff}_{\gamma^* p \to XN}/dM_X}{\sigma^{tot}_{\gamma^* p}}, \tag{26}$$

is displayed in Fig. 25 as a function of W for the different M_X bins and Q^2 values. The data show that, for fixed M_X, contrary to naive expectations, the diffractive cross section possesses the same W dependence as the total cross section. The rapid rise of

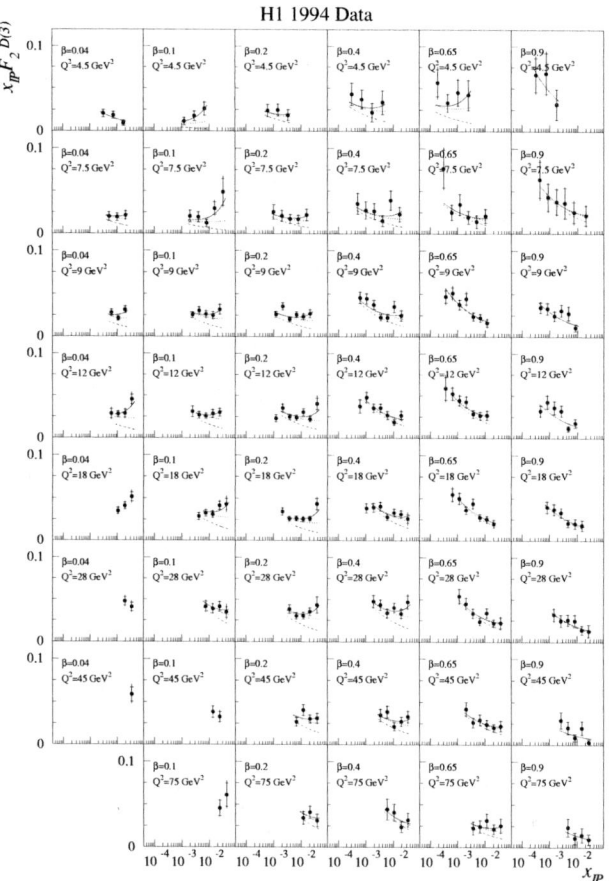

FIGURE 22. The diffractive structure function $x_\mathbb{P} F_2^{D(3)}$ as a function of $x_\mathbb{P}$ for various β and Q^2 values. The solid curves show the results from the two-component Regge fit. The dashed curves show the pomeron contribution alone. From H1.

σ_{tot} with W, which is equivalent to the rapid rise of F_2 as $x \to 0$, in QCD is attributed to the evolution of partonic processes. The observation of similar W dependences for the total and diffractive cross sections suggests, therefore, that diffraction in DIS receives sizeable contributions from hard processes. The same W dependence for the diffractive and total cross sections was predicted in [84].

The diffractive structure function, multiplied by $x_\mathbb{P}$ is shown in Fig. 26 as a function of $x_\mathbb{P}$ for different values of β and Q^2. $x_\mathbb{P} F_2^{D(3)}(x_\mathbb{P}, \beta, Q^2)$ decreases with increasing $x_\mathbb{P}$, which reflects the rapid increase of the diffractive cross section with rising W. The data are consistent with the assumption that the diffractive structure function $F_2^{D(3)}$ factorizes into a term depending only on $x_\mathbb{P}$ and a structure function $F_2^{D(2)}$ which depends

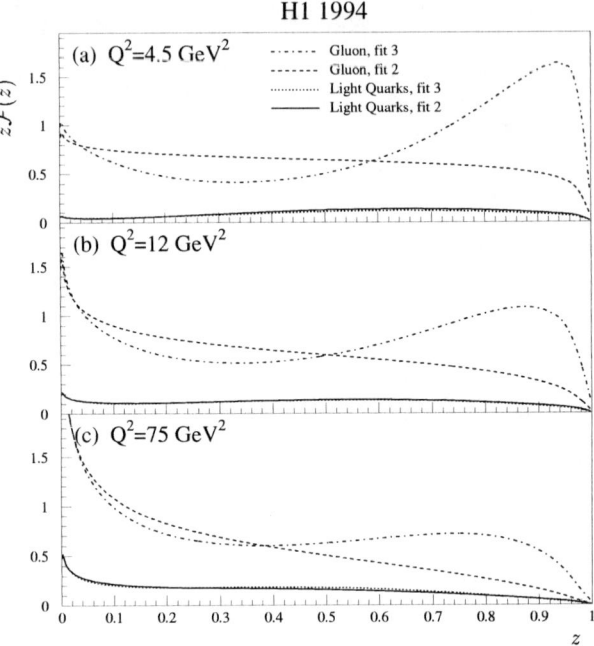

FIGURE 23. The sum of the light quark and gluon distributions as a function of the momentum fraction z of the pomeron carried by the parton at different values of Q^2. From H1.

on (β, Q^2). The rise of $x_P F_2^{D(3)}$ with x_P can be described as $x_P F_2^{D(3)} \propto (1/x_P)^n$ with $n = 0.253 \pm 0.017(stat)^{+0.077}_{-0.023}(syst)$. The data are also consistent with models which break factorization.

Figure 27 shows $F_2^{D(2)}(\beta, Q^2) = x_0 F_2^{D(3)}(x_0, \beta, Q^2)$ where $F_2^{D(3)}$ was evaluated at $x_P = x_0 = 0.0042$. The data show that $F_2^{D(2)}$ has a simple behaviour. For $\beta < 0.6$ and $Q^2 < 14$ GeV^2, $F_2^{D(2)}$ is approximately independent of β. For $\beta < 0.8$ also the data from different Q^2 values are rather similar suggesting a leading twist behaviour characterized by a slow $\ln Q^2$ type rescaling. For $\beta > 0.9$ the data show a decrease with β or Q^2. The approximate constancy of $F_2^{D(2)}$ for $\beta < 0.9$ combined with the rapid rise of $F_2^{D(3)}$ as x_P decreases can be interpreted as evidence for a substantial partonic component in DIS diffraction dissociation.

The Q^2 behaviour of $x_P F_2^{D(3)}(x_P, \beta, Q^2)$ is different from that of the proton structure function $F_2(x, Q^2)$, taken at $x = x_P$, which rises gradually with Q^2. It is in broad agreement with the conjecture [84] that

$$x_P F_2^{D(3)}(x_P, \beta, Q^2) \propto F_2(x = x_P, Q^2) / \log_{10}(Q^2/Q_0^2) \text{ where } Q_0^2 = 0.55 \text{ GeV}^2.$$

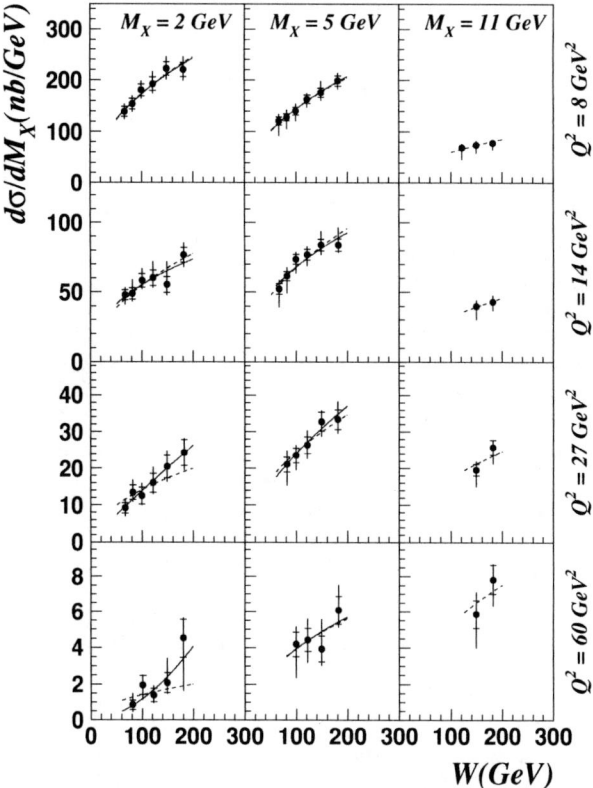

FIGURE 24. The diffractive cross section $d\sigma^{diff}_{\gamma^*p \to XN}/dM_X$, $M_N < 5.5$ GeV, as a function of W for different M_X and Q^2 values. The solid curves show the result from fitting the diffractive cross section for each (W, Q^2) bin separately using the form $d\sigma^{diff}_{\gamma^*p \to XN}/dM_X \propto W^{a^{diff}}$ where a^{diff} and the normalization constants were treated as free parameters. The dashed curves show the result from the fit where a^{diff} was assumed to be the same for all (W, Q^2) bins. From ZEUS.

Comparison with partonic models

The data were compared with several partonic models of diffraction, NZ [85], BPR [86] and BEKW [87]. Good agreement with the data can be achieved. The models provide a first glimpse of how the different components may build up the diffractive structure function.

In the BEKW model basically three components build up the diffractive structure function, $x_P F_2^{D(3)}(\beta, x_P, Q^2) = c_T \cdot F_{q\bar{q}}^T + c_L \cdot F_{q\bar{q}}^L + c_g \cdot F_{q\bar{q}g}^T$; the three terms represent the

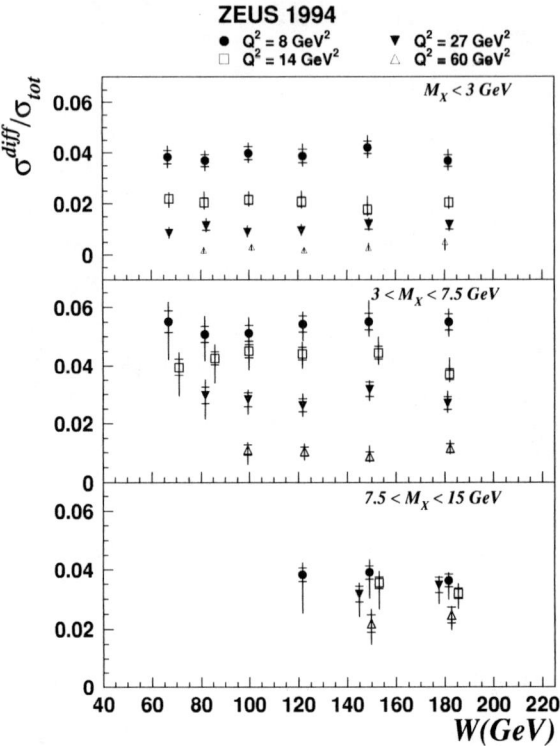

FIGURE 25. The ratio of the diffractive cross section, integrated over the M_X intervals indicated, $\sigma^{diff} = \int_{M_a}^{M_b} dM_X \sigma^{diff}_{\gamma^* p \to XN}$, for $M_N < 5.5$ GeV, to the total cross section for virtual photon proton scattering, $r^{diff}_{tot} = \sigma^{diff}/\sigma^{tot}_{\gamma^* p}$, as a function of W for the M_X intervals and Q^2 values indicated. From ZEUS.

contributions from transverse photons fluctuating into a $q\bar{q}$ or a $q\bar{q}g$ system and from longitudinal photons fluctuating into a $q\bar{q}$ system. In the model, the three terms are given in terms of x_P, β and Q^2 together with additional free parameters which had to be determined from a fit to the data.

It is instructive to compare the β and Q^2 dependences of the three components. Figure 28 shows $c_T F^T_{q\bar{q}}$ (dashed), $c_L F^L_{q\bar{q}}$ (dashed-dotted), $c_g F^T_{q\bar{q}g}$ (dotted) and their sum $x_P F_2^{D(3)}(x_P, \beta, Q^2)$ at $x_P = x_0$ (solid curves) as a function of β for $Q^2 = 8, 14, 27, 60$ GeV^2. For $\beta > 0.2$ the colourless system couples predominantly to the quarks in the virtual photon. The region $\beta \geq 0.8$ is dominated by the contributions from longitudinal photons. The contribution from coupling of the colourless system to a $q\bar{q}g$ final state becomes important for $\beta < 0.3$. The last result is in contrast to the H1 observation (see above) that the large β region is dominated by the gluon contribution. Figure 29 shows the same quantities as a function of Q^2 for $\beta = 0.1, 0.5, 0.9$. The gluon term, which

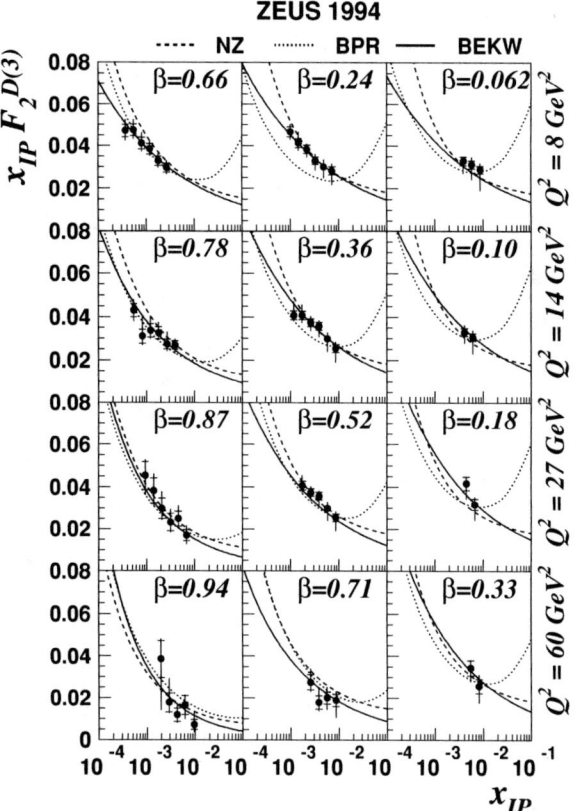

FIGURE 26. The diffractive structure function of the proton multiplied by x_P, $x_P F_2^{D(3)}$, as a function of x_P. The curves show the results from the models of Nikolaev and Zhakarov (NZ), Bialas, Peschanski and Royon (BPR) and Bartels, Ellis, Kowalski and Wüsthoff (BEKW).

dominates at $\beta = 0.1$ rises with Q^2 while the quark term, which is important at $\beta = 0.5$ shows no evolution with Q^2, i.e. is of leading twist. The contribution from longitudinal photons, which is higher twist and dominates at $\beta = 0.9$, decreases with Q^2.

In the BEKW model the x_P-dependence of the quark and gluon contributions for transverse photons is expected to be dominated by the aligned jet configuration [88] and, therefore, to be close to that given by the soft pomeron. Writing $F_{q\bar{q}}^T \propto (x_0/x_P)^{n_T}$ this implies $n_T \approx 2(\overline{\alpha}_P^{soft} - 1)$. However, perturbative admixtures in the diffractive final state are expected to have a somewhat stronger energy dependence, leading to an effective $n_T > 2(\alpha_P^{soft} - 1)$. The x_P dependence of the longitudinal contribution is driven by the square of the proton's gluon momentum density leading to $n_L > n_T$. The fit of the BEKW model to the data indicates that transverse (longitudinal) photons dominate the region $\beta < 0.8$ ($\beta > 0.8$).

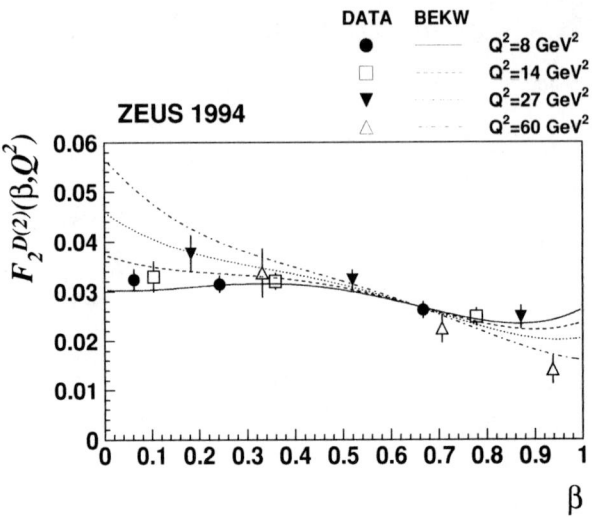

FIGURE 27. The structure function $F_2^{D(2)}(\beta,Q^2)$ for $\gamma^* p \rightarrow XN, M_N < 5.5$ GeV, for the Q^2 values indicated, as a function of β as extracted from a fit to the measured $x_p F_2^{D(3)}$ values. The curves show the fit results obtained with the BEKW model. From ZEUS.

In [89] a novel model (GBW) has been developed which links inclusive diffraction with the total cross section. In a frame where photon and proton are collinear, the total $\gamma^* p$ and the diffractive cross sections can be written as [90, 91]:

$$\sigma_{T,L}(x,Q^2) = \int d^2\mathbf{r} \int dz |\Psi_{T,L}(z,\mathbf{r})|^2 \hat{\sigma}(x,r^2) \tag{27}$$

$$\frac{d\sigma_{T,L}^{diff}(t=0)}{dt} = \frac{1}{16\pi} \int d^2\mathbf{r} \int dz |\Psi_{T,L}(z,\mathbf{r})|^2 |\hat{\sigma}(x,r^2)|^2 \tag{28}$$

where $\Psi_{T,L}(z,\mathbf{r})$ denotes the wave function for transverse (T) and longitudinal (L) photons, $\hat{\sigma}(x,r^2)$ the dipole cross section for the $q\bar{q}$ pair with the proton, z the momentum fraction of the photon carried by the quark and r the relative transverse separation between the quarks.

The wave functions are determined by the photon-$q\bar{q}$ coupling and are known in QCD [91]. The novelty of the model is an ansatz for the dipole cross section whose free parameters are determined by a comparison with the data for $F_2 \approx Q^2 \sigma_{\gamma^* p}^{tot}$. Given $\hat{\sigma}(x,r^2)$, an absolute prediction can now be made for the diffractive cross section. These predictions are found to give an almost quantitative representation of the ratio of diffractive cross section to total cross section as shown by Fig. 30. This is a further step towards a quantitative description of DIS diffraction within QCD.

An interesting discussion of DIS diffraction in terms of QCD radiation and the connection with the problem of confinement has recently been presented in [93].

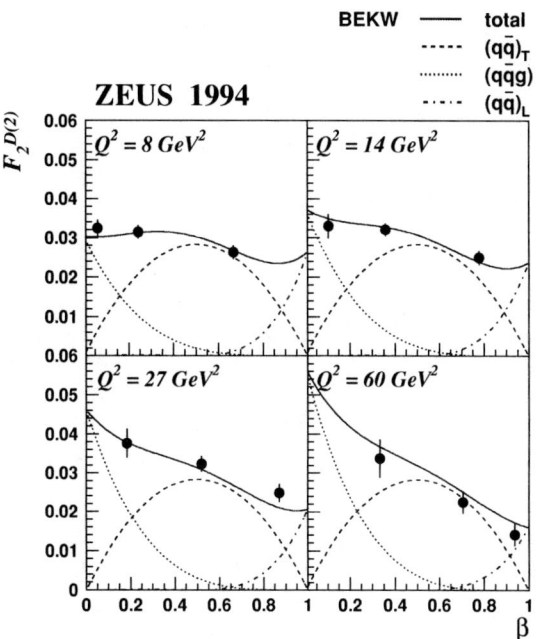

FIGURE 28. The three components $(q\bar{q})_T$, $(q\bar{q}g)$ and $(q\bar{q})_L$ of the BEKW model building up the diffractive structure function of the proton and their sum $F_2^{D(2)}(\beta, Q^2)$ as a function of β for $Q^2 = 8, 14, 27$ and 60 GeV^2, as obtained from a fit of the model to the data. From ZEUS.

WEAK INTERACTIONS BECOME STRONG: NC AND CC SCATTERING AT HIGH Q^2

In the standard model, electron proton scattering at low and medium Q^2 ($Q^2 \leq 1000$ GeV^2) proceed almost exclusively through photon exchange. At higher values of Q^2 substantial contributions are expected also from the exchange of the heavy vector bosons $W^\pm$ and Z^0. The interference between photon and Z^0 exchange contributes to e^-p and e^+p scattering with opposite sign (see xF_3 component in the expression for the cross section below) which allows for a direct detection of the weak contribution in NC scattering.

The structure functions can be expressed as sums over quark flavors of the proton's quark densities $q(x, Q^2)$ weighted according to the gauge structure of the scattering amplitudes [10], see also [94].

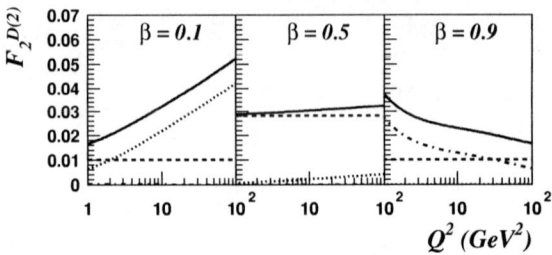

FIGURE 29. The three components $(q\bar{q})_T$, $(q\bar{q}g)$ and $(q\bar{q})_L$ of the BEKW model building up the diffractive structure function of the proton and their sum $F_2^{D(2)}(\beta, Q^2)$ as a function of Q^2 for $\beta = 0.1, 0.5$ and 0.9 as obtained from a fit of the model to the data. The notation is the same as in the previous figure. From ZEUS.

NC cross section

For the neutral current (NC) reaction, $e^{\pm}p \rightarrow e^{\pm}X$, mediated by γ and Z^o exchange, the electroweak Born–level NC DIS differential cross-section can be written as

$$\frac{d^2\sigma_{\text{Born}}^{NC}(e^{\pm}p)}{dxdQ^2} = \frac{2\pi\alpha^2}{xQ^4}\left[Y_+ F_2^{NC}(x,Q^2) \mp Y_- xF_3^{NC}(x,Q^2) - y^2 F_L^{NC}(x,Q^2)\right], \quad (29)$$

where $Y_{\pm} = 1 \pm (1-y)^2$. The structure functions F_2^{NC} and xF_3^{NC} for longitudinally unpolarized beams may be described in leading order QCD as sums over the quark flavor $f = u, ..., b$ of the product of electroweak quark couplings and quark momentum distributions in the proton

$$F_2^{NC} = \frac{1}{2}\sum_f xq_f^+ \left[(V_f^L)^2 + (V_f^R)^2 + (A_f^L)^2 + (A_f^R)^2\right],$$

$$xF_3^{NC} = \sum_f xq_f^- [V_f^L A_f^L - V_f^R A_f^R] \quad (30)$$

where $xq_f^{\pm} = xq_f(x,Q^2) \pm x\bar{q}_f(x,Q^2)$ and xq_f ($x\bar{q}_f$) are the quark (anti-quark) momentum distributions. In leading order QCD, we have $F_L^{NC} = 0$. The functions V_f and A_f can be written as

$$V_f^{L,R} = e_f - (v_e \pm a_e) v_f$$

$$A_f^{L,R} = -(v_e \pm a_e) a_f \chi_Z(Q^2), \quad (31)$$

where the weak couplings, $a_i = T_i^3$ and $v_i = T_i^3 - 2e_i\sin^2\theta_w$, are functions of the weak isospin, $T_i^3 = \frac{1}{2}$ ($-\frac{1}{2}$) for u, v (d, e), and the weak mixing angle, θ_w; e_i is the electric

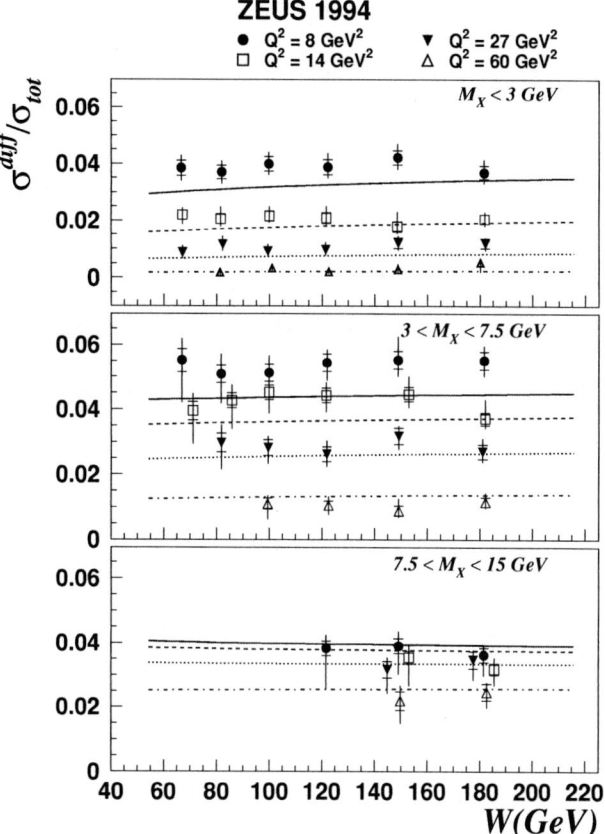

FIGURE 30. Comparison of the predictions by the GBW model with the ZEUS data for the ratio of the diffractive to total cross section as a function of W, for the M_X intervals and Q^2 values indicated. Figure taken from [92].

charge in units of the positron charge; and χ_Z is proportional to the ratio of Z^0-boson and photon propagators

$$\chi_Z = \frac{1}{4\sin^2\theta_W \cos^2\theta_W} \frac{Q^2}{Q^2 + M_Z^2}. \tag{32}$$

The contribution of F_L^{NC} to $d^2\sigma_{Born}^{NC}/dxdQ^2$ is predicted to be approximately 1.5% averaged over the kinematic range considered below. However, in the region of small x at the lower end of the Q^2 range of the data shown below the F_L^{NC} contribution to the cross-sections can be as large as 12%.

CC cross section

The electroweak Born cross section for the charged current reaction (see Fig. 31):

$$e^+ p \to \bar{\nu}_e X \tag{33}$$

can be written as

$$\frac{d^2\sigma^{CC}_{Born}(e^\pm p)}{dx dQ^2} = \frac{G_F^2}{4\pi x}\left(\frac{M_W^2}{M_W^2+Q^2}\right)^2 \left[Y_+ F_2^{CC}(x,Q^2) \mp Y_- xF_3^{CC}(x,Q^2) - y^2 F_L^{CC}(x,Q^2)\right], \tag{34}$$

where G_F is the Fermi constant and M_W is the mass of the W boson.

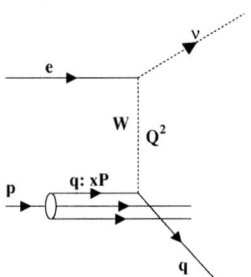

FIGURE 31. Diagram for deep inelastic ep scattering by charged current exchange.

The structure functions F_2^{CC} and xF_3^{CC}, in leading-order (LO) QCD, measure sums and differences of quark and antiquark parton momentum distributions. For longitudinally unpolarized beams, $e^- p$:

$$F_2^{CC} = x[u(x,Q^2) + c(x,Q^2) + \bar{d}(x,Q^2) + \bar{s}(x,Q^2)] \tag{35}$$
$$xF_3^{CC} = x[u(x,Q^2) + c(x,Q^2) - \bar{d}(x,Q^2) - \bar{s}(x,Q^2)] \tag{36}$$

$e^+ p$:

$$F_2^{CC} = x[d(x,Q^2) + s(x,Q^2) + \bar{u}(x,Q^2) + \bar{c}(x,Q^2)] \tag{37}$$
$$xF_3^{CC} = x[d(x,Q^2) + s(x,Q^2) - \bar{u}(x,Q^2) - \bar{c}(x,Q^2)] \tag{38}$$

where $u(x,Q^2)$ is, for example, the number density of an up quark with momentum fraction x in the proton. Since the top quark mass is large and the off-diagonal elements of the CKM matrix are small, the contribution from the third generation quarks to the structure functions may be safely ignored [95]. The chirality of the CC interaction is reflected by the factors $Y_\pm$ multiplying the structure functions. The longitudinal structure function, F_L^{CC}, is zero at leading order but is finite at next-to-leading-order (NLO) QCD. It gives a negligible contribution to the cross section except at y values close to 1, where it can be as large as 10%.

Experimental results: NC scattering

Both the H1 [96] and ZEUS [97] experiments have previously reported cross section measurements, based on data collected in 1993, which established that the Q^2 dependence of the CC cross section is consistent with the expectations from the W propagator. The data from ZEUS demonstrated also that the CC and NC cross sections are of similar magnitude for $Q^2 \geq M_W^2$.

The high precision NC data presented recently by H1 [98, 99] (e^-p data from 1998-9 with 15 pb^{-1}; e^+p data from 1994-2000 with 46 pb^{-1}) and ZEUS [100, 101] (e^-p data from 1998-9 with 16 pb^{-1}; e^+p data from 1996 with 30 pb^{-1}) allow, for the first time, to see the Z^0 contribution to the NC cross section. In Fig. 32 the e^-p and e^+p cross sections are shown in terms of $d\sigma/dQ^2$ as a function of Q^2. Both cross sections decrease by about six orders of magnitude between $Q^2 = 500$ and 40000 GeV^2, mainly governed by the photon propagator which leads to a behaviour of the form $d\sigma/dQ^2 \propto 1/Q^4$. For Q^2 values above 3000 GeV^2 there is a clear difference between the two charge states: the cross section for e^-p scattering is larger than for e^+p, which gives evidence for a weak contribution.

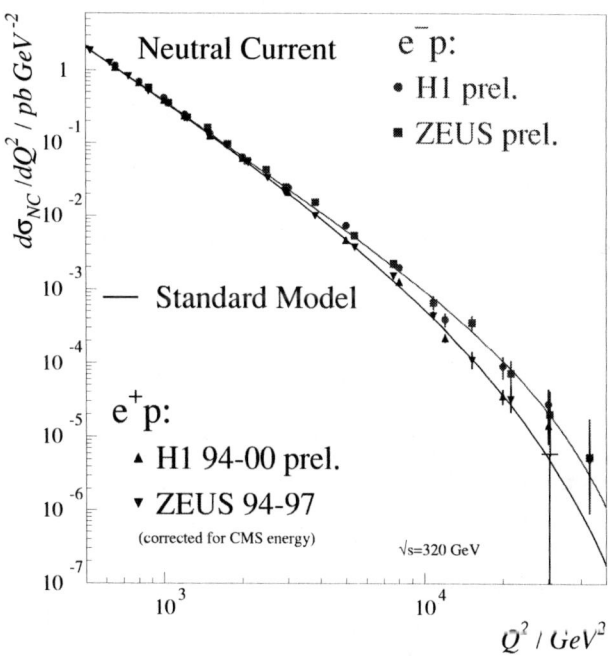

FIGURE 32. The NC cross sections for e^-p and e^+p as a function of Q^2 as measured by H1 and ZEUS. The curves show QCD-NLO predictions.

The predictions of the Standard Model (solid curves) give a good description of the data. This is also true when the model is compared with the data for different values of

x. This is shown in Figs. 33, 34 where the reduced cross sections,

$$\tilde{\sigma}_{NC}^{\mp} = \frac{xQ^4}{2\pi\alpha^2} \frac{1}{Y_+} \frac{d^2\sigma_{NC}}{dxdQ^2} = F_2^{NC} \pm \frac{Y_-}{Y_+} xF_3^{NC} \tag{39}$$

are given for fixed x as a function of Q^2.

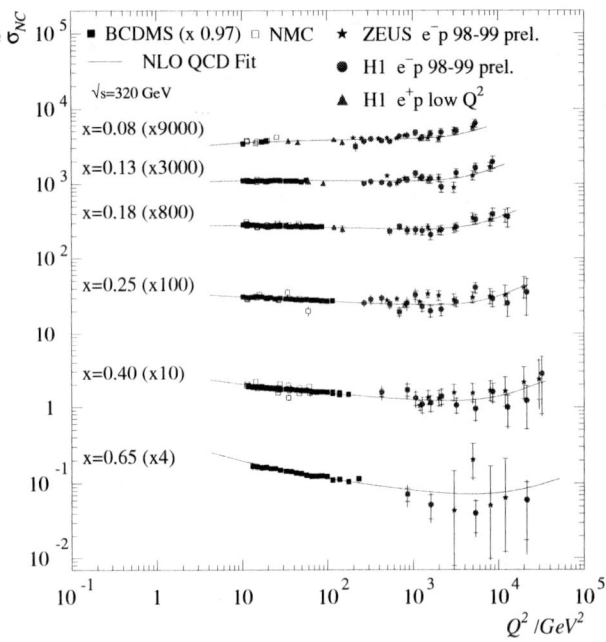

FIGURE 33. The e^-p NC reduced cross section for different values of x as a function of Q^2 as measured by H1 and ZEUS. The curves show QCD-NLO fits for $\gamma + Z^0$ exchange.

From the comparison of the e^-p and e^+p data the structure function xF_3^{NC} can be extracted. The dominant contribution comes from γZ^0 interference, which is denoted by $xF_3^{\gamma Z}$. The reduced cross sections for the two charged states and the structure function xF_3^{NC} as measured by H1 [98] are presented in Figs. 35(a,b) for different Q^2 intervals.

In Fig. 36 $xF_3^{\gamma Z}$ is shown as a function of x for Q^2 values of 1500, 5000 and 12000 GeV^2. It is remarkable that only little dependence on Q^2 is observed despite the large Q^2 range. This is in agreement with QCD where a dependence on Q^2 is expected only from scaling violations. Note also that $xF_3^{\gamma Z}$ depends on the difference between quark and antiquark densities and is therefore primarily sensitive to the valence quark contribution. The data are compared with the results of a QCD fit performed previously by H1 [99] to their NC e^+p data taken in 1994-7 combined with the data from NMC and BCDMS (called H1 97 PDF fit). The fit certainly reproduces the qualitative features of the data.

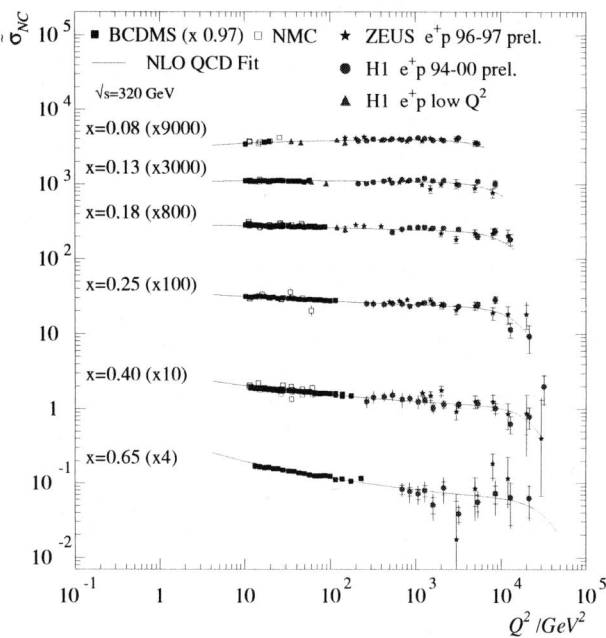

FIGURE 34. The e^+p NC reduced cross section for different values of x as a function of Q^2 as measured by H1 and ZEUS. The curves show QCD-NLO fits for $\gamma + Z^0$ exchange.

Experimental results: CC scattering

The recent evaluations of CC cross sections by H1 and ZEUS are based on an order of magnitude more data compared to the previous analyses, H1 [98] (e^-p data from 1998-9 with 16 pb^{-1}; e^+p data from 1994-7 with 36 pb^{-1}) and ZEUS [102, 103] (e^-p data from 1998-9 with 16 pb^{-1}; e^+p data from 1994-7 with 48 pb^{-1}).

The combined CC cross section data from the two experiments are shown in Fig. 37 for e^+p and e^-p scattering as a function of Q^2. For $Q^2 < M_W^2$ the cross sections show a slow decrease with Q^2 which is mainly due to the reducing phase space in x. The rapid fall at $Q^2 > M_W^2$ is mainly driven by the propagator term $\frac{1}{(M_W^2+Q^2)^2}$. At low Q^2, the cross sections for the two charge states are almost the same: the dominant contribution comes from scattering on sea quarks which is approximately flavor symmetric and contributes roughly equally in the two cases. At high Q^2 the cross section for e^-p scattering is larger than for e^+p by almost an order of magnitude. Here valence quarks dominate the cross sections and different quark flavors contribute to the two charge states. Furthermore, the xF_3 contribution is added to the e^-p cross section and subtracted for e^+p.

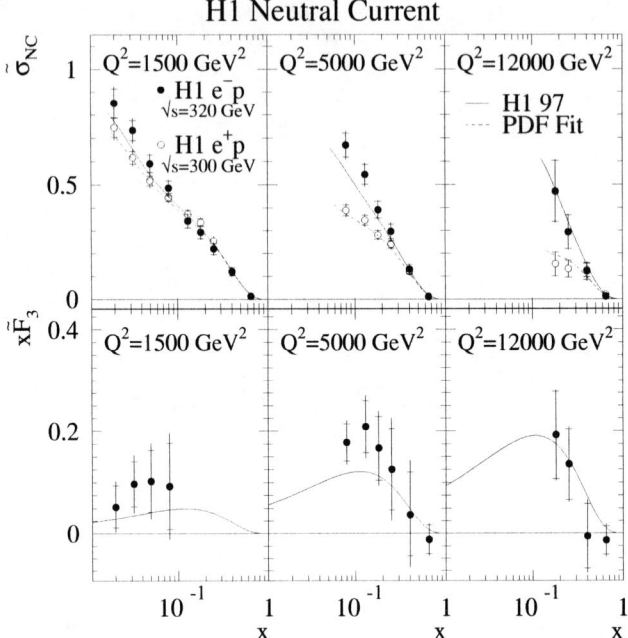

FIGURE 35. The e^+p NC reduced cross section and xF_3 for different values of Q^2 as a function of x as measured by H1. The curves show the predictions from the H1 97 PDF fit.

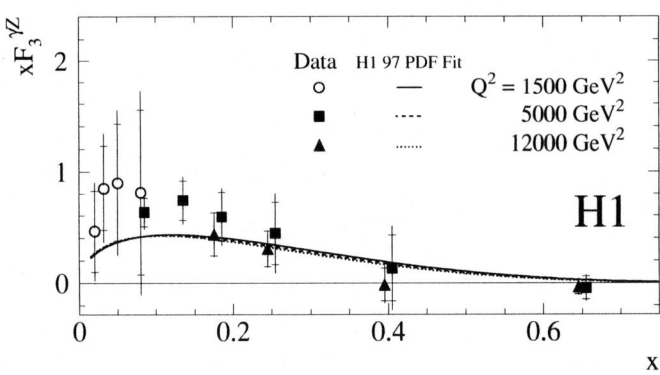

FIGURE 36. The structure function $xF_3^{\gamma Z}$ for NC scattering as a function of x for different values of Q^2 (from H1). The curves show the predictions from the H1 97 PDF fit.

In Fig. 38 the reduced cross section,

$$\tilde{\sigma}_{CC}^{\mp} = \frac{2\pi x}{G_F^2}\left[\frac{M_W^2+Q^2}{M_W^2}\right]^2 \frac{d^2\sigma_{CC}}{dxdQ^2} \qquad (40)$$

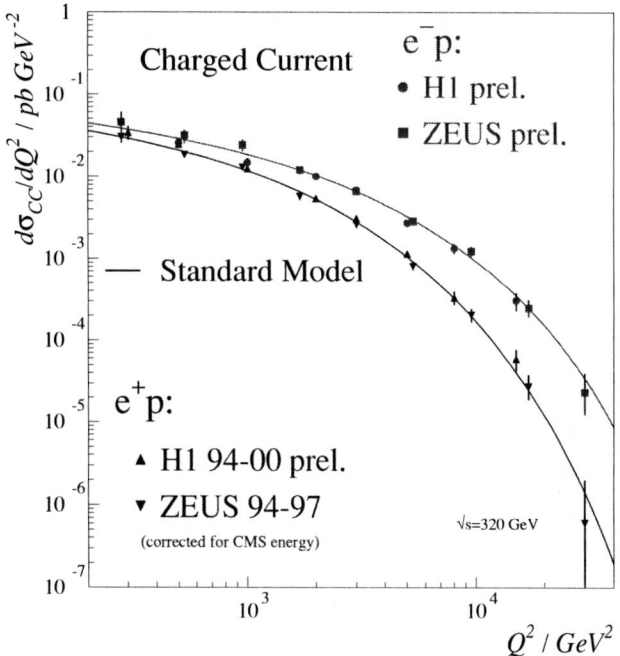

FIGURE 37. The CC cross sections for e^-p and e^+p as a function of Q^2 as measured by H1 and ZEUS. The curves show QCD NLO predictions.

as measured by ZEUS, is shown as a function of x for different Q^2 intervals. As Q^2 increases the reduced cross section for e^+p scattering becomes small with respect to the case of e^-p scattering.

While the charged vector bosons $W^\pm$ have been directly observed (CERN 1983) the effect of the W propagator on weak interaction cross sections, before HERA, had not been seen directly. As mentioned above, the Q^2 fall-off of the charged current cross sections depends primarily on the W propagator. The new data from HERA allow a significant measurement of M_W thanks to the large Q^2 range and the substantial size of the data samples. Using for G_F the measured value, the W mass was found to be:

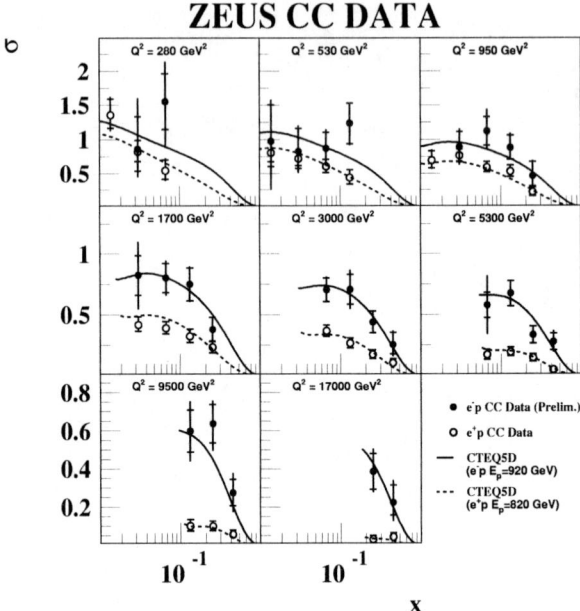

FIGURE 38. The reduced CC cross sections for $e^- p$ and $e^+ p$ scattering as a function of x for different Q^2 values (from ZEUS). The curves show the predictions using CTEQ5D.

$$M_W = 79.9 \pm 2.2(stat) \pm 0.9(syst) \pm 2.1(theor) \text{ GeV } \text{H1}$$
$$M_W = 81.4^{+2.7}_{-2.6}(stat) \pm 2.0(syst)^{+3.3}_{-3.0}(PDF) \text{ GeV } \text{ZEUS}. \quad (41)$$

The values are in good agreement with the direct measurement of $M_W = 80.419 \pm 0.056$ GeV [104].

Comparison of NC and CC cross sections

In Figs. 39, 40 the cross sections for NC and CC scattering in $e^- p$ and $e^+ p$ interactions are compared. At low Q^2, where NC scattering arises predominantly from electromagnetic interactions (= photon exchange), the NC cross section exceeds by far the CC cross section. However, at Q^2 values above 10000 GeV^2 the two processes have about the same cross sections; at these large values of Q^2, the weak force is of similar strength as the electromagnetic one: the HERA measurements show directly the unification of the electromagnetic and weak forces.

It is instructive to compare this result with typical electromagnetic and weak particle decays. There, the weak force is about ten orders of magnitude smaller than the electromagnetic one! For instance, the decay time for the electromagnetic decay $\Sigma^0 \to \Lambda\gamma$ is 7.4 ± 0.7^{-20}s while it is 4.1×10^{-10}s for the decay $\Lambda \to p\pi^-$ (note, both decays have similar c.m. momenta, viz. 0.10 GeV and 0.074 GeV, respectively, and therefore similar phase space).

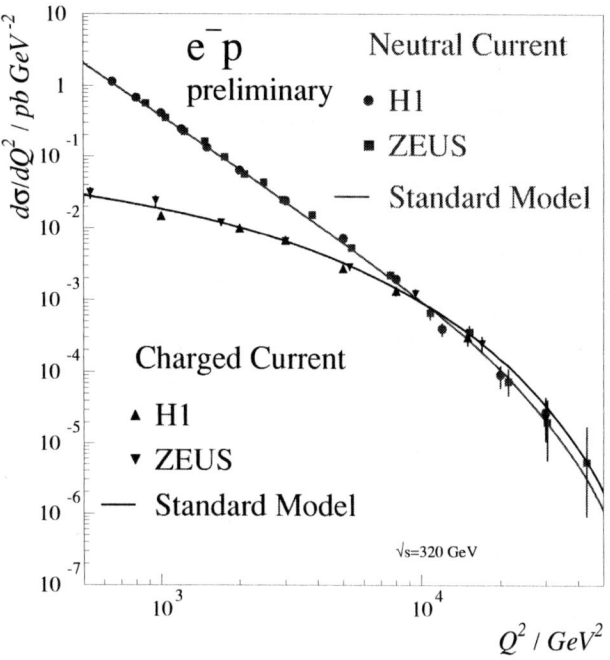

FIGURE 39. Comparison of the cross sections for e^-p scattering by NC and CC exchange as a function of Q^2, measured by H1 and ZEUS. The curves show the predictions of the Standard Model.

Testing the Standard Model

The large space covered in x and Q^2 and their high precision make the NC and CC cross section measurements of H1 and ZEUS a powerful testing ground of the Standard Model (SM). Uncertainties in the SM predictions arise from three sources: electroweak parameters, electroweak radiative corrections, and the parton momentum distributions including their higher order QCD corrections. A detailed discussion of these uncertainties can be found in [98, 99, 100, 102]

For NC scattering, the electroweak parameters have been measured to high precision by other experiments and contribute less than 0.3% uncertainty for the SM predictions. Higher order corrections for radiative corrections, vertex and propagator corrections and two-boson exchange are expected to be less than 1%. The primary source of uncertainties stem from the parton momentum distributions which have been determined from data mostly at low Q^2 and then extrapolated to higher Q^2 using DGLAP QCD evolution.

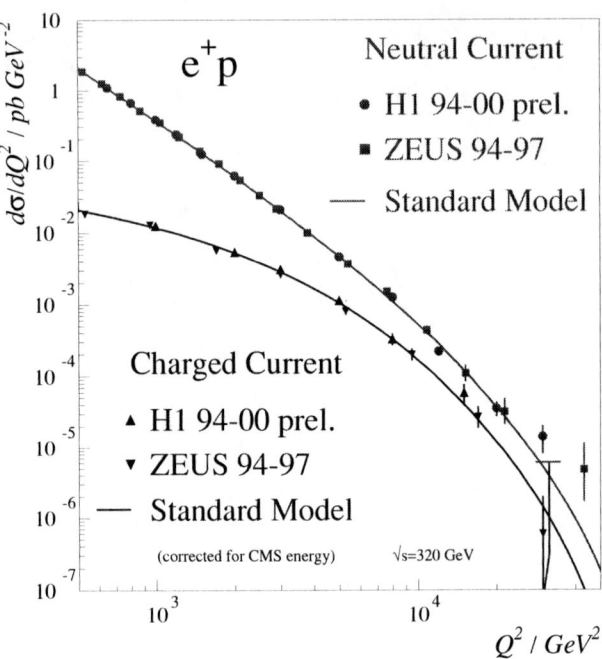

FIGURE 40. Comparison of the cross sections for e^+p scattering by NC and CC exchange as a function of Q^2, measured by H1 and ZEUS. The curves show the predictions of the Standard Model.

The HERA data included in these fits are mostly from $x < 10^{-2}$ and have little influence on the predictions for the high Q^2 regime considered here. The parton densities (PDF) give a total uncertainty on the SM predictions for $d\sigma^{NC}/dQ^2$ of 4% at $Q^2 = 400$ GeV^2 increasing to 8% at the highest Q^2 covered.

For CC scattering, the main uncertainty of the SM predictions comes also from the uncertainties of the PDF's. The resulting uncertainty in $d\sigma^{CC}/dQ^2$ ranges from 4% at $Q^2 = 400$ GeV^2 to 10% at $Q^2 = 10000$ GeV^2, and increases further at higher Q^2. The large uncertainty at high Q^2 is due to the d-quark density which is poorly constrained at high x by the experimental data.

In Fig. 41 $d\sigma^{NC}/dQ^2$ as measured for e^+p scattering is compared with the SM predictions calculated with the CTEQ4D PDF's. Taking the errors of the data and the uncertainties of the SM into account there is excellent agreement between the SM predictions and the data up to $Q^2 = 36000$ GeV^2. At higher Q^2 there are 2 events observed compared to 0.27 predicted. The corresponding data for CC scattering are presented in Fig. 42. For $Q^2 < 5000$ GeV^2 the uncertainties in the PDF's - mainly due to the uncertainty of the d-quark contribution - prevent a precise test of the SM.

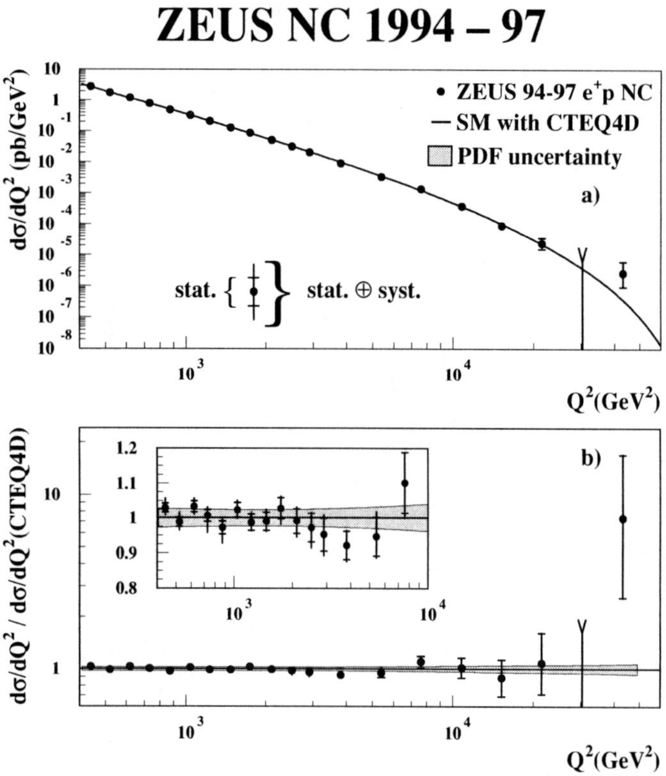

ZEUS NC 1994 – 97

FIGURE 41. The $e+p$ NC DIS cross section for data and the Standard Model (SM) predictions (a) and the ratio of data to the SM prediction (b). The shaded region gives the uncertainty in the SM prediction due to the uncertainty in the PDF's.

The Standard Model (SM) provides a constraint between G_F and M_W:

$$G_F = \frac{\pi\alpha}{\sqrt{2}} \frac{M_Z^2}{(M_Z^2 - M_W^2)M_W^2} \frac{1}{1-\Delta_r}, \tag{42}$$

where M_Z is the mass of the Z boson. The term Δ_r contains the radiative corrections to the lowest order for G_F and is a function of α and the masses of the fundamental bosons and fermions [105]. Using this constraint, assuming $M_H = 100$ GeV and treating only M_W as a free parameter ZEUS found:

$$M_W = 80.50^{+0.24}_{-0.25}(stat)^{+0.13}_{-0.16}(syst) \pm 0.31(PDF)^{+0.03}_{-0.06}(\Delta M_t, \Delta M_H, \Delta M_Z)\text{GeV} \tag{43}$$

The excellent agreement with the directly measured value of M_W indicates that the SM gives a consistent description of a variety of phenomena over a wide range of energy scales.

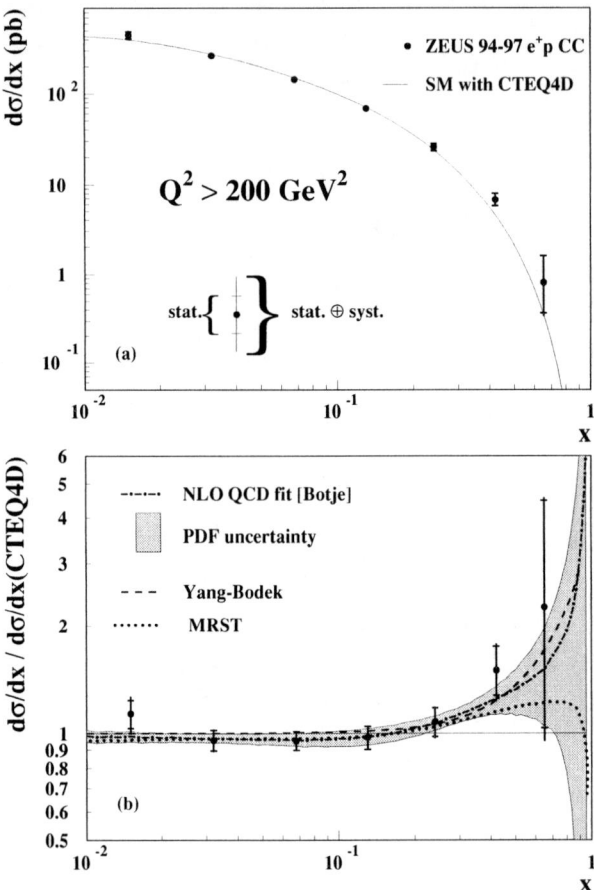

FIGURE 42. The $e + p$ CC DIS cross section for data and the Standard Model (SM) predictions (a) and the ratio of data to the SM prediction (b). The shaded region gives the uncertainty in the SM prediction due to the uncertainty in the PDF's.

OUTLOOK

The year-2000 upgrade of HERA is expected to raise the luminosity by a factor of 3 - 5 which promises for H1 and ZEUS data samples as large as 1 fb^{-1}. At the same time the interaction regions for the two collider experiments are being equiped with spin rotators in order to turn the transverse polarisation of electrons and positrons into a longitudinal polarisation. After the upgrade, the collider detectors will have improved vertexing capabilities for identifying c and b quarks with high efficiency. The combination of these improvements will increase substantially the physics capabilities [106] of the experiments.

The charm and beauty part of the proton structure function can then be measured with high precision. A special feature of c and b quarks is their heavy mass which provides a hard scale and which should suppress possible contributions from aligned jet configurations.

The structure functions $F_2, F_L, xF_3, F_2^c, F_2^b$ can be measured between $Q^2 = 10$ and 40000 GeV^2 in a single experiment. These measurements and the analyses of jet production in deep inelastic scattering are expected to determine the strong coupling with a precision of $\Delta \alpha_s = 0.0015$.

A luminosity of 1 fb^{-1} will allow to test quarks and electrons for substructure down to 4.10^{-17} cm. The measured NC and CC cross sections will be sensitive to deviations from the Standard Model predictions down to the level of 30 - 50 MeV in M_W. Additional $W's$ and $Z's$ can be detected up to 600 - 800 GeV. Perhaps most exciting will be the measurement of the charged current cross section for right (left) handed beam electrons (positrons). Any nonzero cross section contribution will be a sign for a new right handed charged current.

ACKNOWLEDGEMENTS

I want to thank Prof. G. Herrera and his staff for their kind hospitality. I am grateful to Dr. R. Yoshida for help with the data, and to Prof. E. Lohrmann for discussions and for a critical reading of the manuscript.

REFERENCES

1. SLAC - MIT Collaboration, R.R. Taylor, Proc. 4th Int. Symp. Electron Photon Interactions at High Energies, Dearsbury, England, 1969, p. 251.
2. H1 Collaboration, I. Abt et al., Nucl. Phys. B407 (1993) 515.
3. ZEUS Collaboration, M. Derrick et al., Phys. Lett. B316 (1993) 412.
4. L.V. Gribov, E.M. Levin and M.G. Ryskin, Phys. Rep. 100 (1983) 1.
5. ZEUS Collaboration, M. Derrick et al., Phys. Lett. B315 (1993) 481.
6. H1 Collaboration, T. Ahmed et al., Nucl. Phys. B429 (1994) 277.
7. *HERA, A Proposal for a Large Electron - Proton Colliding Beam Facility at DESY*, DESY HERA 81-10(1981). B.H. Wiik, *Electron - Proton Colliding Beams, The Physics Programme and the Machine*, Proc. 10th SLAC Summer Institute, ed. A. Mosher, 1982, p. 233; Proc. XXVI Int. Conf. High Energy Physics, Dallas, 1992. G.A. Voss, Proc. First Euro Acc. Conf., Rome, 1988, p. 7.
8. Private communication by F. Willeke.
9. A.A. Sokolov and M. Ternov, Sov. Phys. Doklady 8 (1964) 1203.
10. G. Ingelman and R. Rueckl, Phys. Lett. B201 (1988) 369.
11. H1 Collaboration, C. Adloff et al., DESY 00-181(2000).
12. ZEUS Collaboration, contribution to ICHEP, 2000, Osaka, Japan.
13. BCDMS Collaboration, A.C. Benvenuti et al., Phys. Lett. B223 (1989) 490.
14. E665 Collaboration, M.R. Adams et al., FERMILAB-PUB-95-396-E.
15. NMC Collaboration, M. Arneodo et al., Phys. Lett. B364 (1995) 107.
16. L.W. Whitlow et al., Phys. Lett. B282 (1992) 475.
17. G. Wolf, *HERA Physics*, DESY 94-022 (1994) and Proc. 42nd Scottish Universities Summer School in Physics, 1993.
18. G. Wolf, Proc. Int. School Subnuclear Physics, 1996, Erice, Italy, Vol 34, p. 45.

19. H1 Collaboration, C. Adloff et al., Nucl. Phys. B497 (1997) 3.
20. ZEUS Collaboration, J. Breitweg et al., Eur. Phys. J. C7 (1999) 609.
21. M. Botje, Eur. Phys. J. C14 (2000) 285.
22. A.D. Martin, R.G. Roberts, W.J. Stirling and R.S. Thorne, Eur. Phys. J. C4 (1998) 463.
23. CTEQ Collaboration, J. Huston et al., hep-ph/9801444 (1998).
24. A.D. Martin, W.J. Stirling, and R.G. Roberts, Phys. Lett. B306 (1993) 145; erratum ibid. 309 (1993) 492.
25. V.N. Gribov and L.N. Lipatov, Soc.J. Nucl. Phys. 15 (1972) 438; ibid. 675.
    L.N. Lipatov, Sov. J. Nucl. Phys. 20 (1975) 95.
    Yu.L.Dokshitzer, Sov. Phys. JETP 46 (1977) 641.
    G. Altarelli and G. Parisi, Nucl. Phys. B126 (1977) 298.
26. H1 Collaboration, C. Adloff et al., Phys. Lett. 393B (1997) 452.
27. K. Prytz, Phys. Lett. B311 (1993) 286; ibid. B332 (1994) 393.
28. H1 Collaboration, C. Adloff et al., Z. Phys. C72 (1996) 593.
29. ZEUS Collaboration, J. Breitweg et al., Phys. Lett. B407 (1997) 402.
30. ZEUS Collaboration, J. Breitweg et al., Eur. Phys. J. C12 (2000) 35.
31. ZEUS Collaboration, contribution to ICHEP, 2000, Osaka, Japan.
32. H1 Collaboration, contribution to ICHEP, 2000, Osaka, Japan.
33. B.W. Smith and J. Smith, Phys. Rev. D57 (1998) 2806.
34. M. Glück, E. Reya and A. Vogt, Z. Phys. C67 (1995) 433.
35. L.N. Hand, Phys. Rev. 129 (1963) 1834; S.D. Drell and J.D. Walecka, Ann. Phys. (N.Y.) 28 (1964) 18; F.J. Gilman, Phys. Rev. 167 (1968) 1365.
36. B.L. Ioffe, Phys. Lett. 30 (1969) 123; B.L. Ioffe, V.A. Khoze and L.N. Lipatov, "Hard Processes", North Holland (1984), p.185.
37. ZEUS Collaboration, M. Derrick et al., Z. Phys. C65 (1995) 379.
38. We follow here J. Kwiecinski, *Substructures of Matter as revealed with Electroweak Probes*, Proc. Schladming,1993, eds. L.Mathelitsch, W. Plessas, Lecture Notes in Physics 426, Springer Verlag, p.215.
39. See e.g. E. Leader and E. Predazzi, An introduction to gauge theories and the New Physics (Cambridge U.P., Cambridge 1982); R.G. Roberts, The structure of the proton (Cambridge U.P., Cambridge 1990).
40. C. Lopez and F.J. Yndurain, Nucl. Phys. B171 (1980) 231; Phys. Rev. Lett. 44 (1980) 1118.
41. For a nonasymptotically free vector theory: V.N. Gribov and L.N. Lipatov, Yad. Fiz. 15 (1972) 781; Sov. J. Nucl. Phys. 15 (1972) 438.
42. For an asymptotically free vector theory: A. De Rujula et al., Phys. Rev. D10 (1974) 1649. Detailed formulas have been given by F. Martin, Phys. Rev. D19 (1979) 1382; R.D. Ball and S. Forte, Phys. Lett. 358B (1995) 365; K. Adel, F. Barreiro and F.J. Yndurain, Nucl. Phys. B495 (1997) 221.
43. M. Glück, E. Reya and A. Vogt, Z. Phys. C67 (1995) 433; references to earlier work can also be found there.
44. M. Glück, E. Reya and A. Vogt, Eur. Phys. J. C5 (1998) 461.
45. L.N. Lipatov, Sov. J. Nucl. Phys. 23 (1976) 338; Y.Y. Balitsky and L.N. Lipatov, Sov. J. Nucl. Phys. 28 (1978) 822; E.A. Kuraev, L.N. Lipatov and V.S. Fadin, Sov. Phys. JETP 45 (1977) 199.
46. G. Bottazzi, G. Marchesini, G.P. Salam and M. Scorletti, IFUM 552-FT (1997) and hep-ph/9702418.
47. See also M. Scorletti, IFUM 574-FT (1997) and hep-ph/9707237.
48. See also S. Catani, XVIII Int. Symp. Lepton Photon Interactions, 1997, Hamburg.
49. J. Kwiecinski, A.D. Martin and A.M. Staso, hep-ph/9706455 (1997).
50. K. Adel, F. Barreiro and F.J. Yndurain, Nucl. Phys. B495 (1997) 221.
51. A.D. Martin, W.J. Stirling and R.G. Roberts, Phys. Rev. D50 (1994) 6734.
52. R. Brock et al., Rev. Mod. Phys. 67 (1995) 157.
53. R.D. Ball and R.D. Forte, Phys. Lett. B335 (1994) 77; CERN-TH/95-323(1995).
54. For a recent analysis see W. Buchmüller and D. Haidt, DESY 96-061 (1996).
55. D. Haidt, preprint, 1997.
56. H1 Collaboration, C. Adloff et al., Nucl. Phys. B497 (1997) 3.
57. ZEUS Collaboration, J. Breitweg et al., Phys. Lett. B487 (2000) 53.
58. A. Donnachie and P.V. Landshoff, Z. Phys. C61 (1994) 139.
59. B. Badelek and J. Kwiecinski, Rev. Mod. Phys. 68 (1996) 445.

60. A. Capella et al., Phys. Lett. B337 (1994) 358.
61. K. Adel, F. Barreiro and F.J. Yndurain, FTUAM 96-39 and DESY 97-088 (1997).
62. A. Donnachie and P.V. Landshoff, Phys. Lett. B296 (1992) 227.
63. H. Abramowicz and A. Levy, DESY 97-251.
64. A. Caldwell, Invited talk at the DESY Theory Workshop on *Recent Developments in QCD*, October 1997 (unpublished).
65. NMC Collaboration, M. Arneodo et al., Nucl. Phys. B483 (1997) 3.
66. BCDMS Collaboration, A.C. Benvenuti et al., Phys. Lett. B223 (1989) 485 and Phys. Lett. B237 (1990) 592.
67. A.D. Martin et al., DTP/98/10, RAL-TR-98-029 and hep-ph/9803445 (1998).
68. G. Alberi and G. Goggi, Phys. Reports 74 (1981) 1;
    K. Goulianos, Phys. Reports 101 (1983) 169; Nucl. Phys. B, Proc. Suppl. 12 (1990) 110.
    G. Giacomelli, University and INFN Bologna report DFUB 9-94 (1994).
69. N. Gribov, Sov. Phys. JETP 14 (1962) 478; ibid. 1395.
    G.F. Chew and S. Frautschi, Phys. Rev. Lett. 7 (1961) 394.
    R.D. Field and G. Fox, Nucl. Phys. B80 (1974) 367.
    A.B. Kaidalov and K.A. Ter-Martirosyan, Nucl. Phys. B75 (1974) 471.
70. G. Ingelman and P. Schlein, Phys. Lett. B152 (1985) 256.
71. UA8 Collaboration, R. Bonino et al., Phys. Lett. B211 (1988) 239; A. Brandt et al., Phys. Lett. B297 (1992) 417.
72. ZEUS Collaboration, M. Derrick et al., Z. Phys. C69 (1995) 39; Phys. Lett. B377 (1996) 259; Z. Phys. C73 (1996) 73, 253; J. Breitweg et al., Eur. Phys. J. C2 (1998) 247.
    H1 Collaboration S. Aid et al., Nucl. Phys. B463 (1996) 3.
73. ZEUS Collaboration, M. Derrick et al., Phys. Lett. B356 (1995) 601; ibid. B380 (1996) 220.
    H1 Collaboration, S. Aid et al., Nucl. Phys. B468 (1996) 3; Z. Phys. C75 (1997) 607.
74. ZEUS Collaboration, M. Derrick et al., Z. Phys. C68 (1995) 569; ibid. C70 (1996) 391.
75. ZEUS Collaboration, M. Derrick et al., Phys.lett. B332 (1994) 228; ibid. B338 (1994) 483; J. Breitweg et al., Phys. Lett. B421 (1998) 36.
76. H1 Collaboration, T. Ahmed et al., Phys. Lett. B348 (1995) 681.
77. ZEUS Collaboration, Z. Phys. C70 (1996) 391.
78. H1 Collaboration, C. Adloff et al., Z. Phys. C76 (1997) 613.
79. ZEUS Collaboration, J. Breitweg et al., Eur. Phys. J. C1 (1998) 81.
80. ZEUS Collaboration, J. Breitweg et al., Eur. Phys. J. C6 (1999) 43.
81. G. Ingelman and K. Janson-Prytz, Proc. Workshop "Physics at HERA", ed. W. Buchmüller and G. Ingelman, DESY 1992, Vol.1, 233; G. Ingelman and K. Prytz, Z. Phys. C58 (1993) 285.
82. A. Donnachie and P.V. Landshoff, Nucl. Phys. B244 (1984) 322; Phys. Lett. B296 (1992) 227.
83. J. Cudell, K. Kang and S.K. Kim, Brown-HET-1060 and hep-ph/9701312.
84. W. Buchmüller, Phys. Lett. B353 (1995) 335; W. Buchmüller and A. Hebecker, Nucl. Phys. B476 (1996) 203.
85. N.N. Nikolaev and B.G. Zakharov, Z. Phys. C53 (1992) 331; M. Genovese, N.N. Nikolaev and B.G. Zakharov, JETP81 (1995) 625.
86. A. Bialas and R. Peschanski, Phys. Lett. B387 (1996) 405; A. Bialas, R. Peschanski and Ch. Royon, Phys. Rev. D57 (1998) 6899.
87. J. Bartels, J. Ellis, H. Kowalski and M. Wüsthoff, CERN-TH/98-67, DESY 98-034, DTP 98-02 and hep-ph/9803497.
88. J.D. Bjorken, Proc. Int. Symp. Electron and Photon Interactions at High Energies, Cornell, 1971, p. 282.
    J.D. Bjorken, J. Kogut and D. Soper, Phys. Rev. D3 (1971) 1382.
    J.D. Bjorken and J. Kogut, Phys. Rev. D8 (1973) 1341.
89. K. Golec-Biernat and M. Wüsthoff, Phys. Rev. D59 (1999) 014017; Phys. Rev. D60 (1999) 114023.
90. N.N. Nikolaev and B.G. Zakharov, Z. Phys. C49 (1990) 607.
91. J.R. Forshaw and D.A. Ross, *QCD and the Pomeron*, Cambridge, University Press, 1997.
92. H. Kowalski, DESY 99-141.
93. J. Bartels and H. Kowalski, DESY 00-154 and hep-ph/0010345.
94. A.M. Cooper-Sarkar, R. Devenish and A. de Roeck, *Int'l J. Mod. Phys.* A13 (1998) 3385.

95. U. Katz, *Deep-Inelastic Positron-Proton Scattering in the High Momentum-Transfer Regime of HERA*, Springer Tracts in Modern Physics (1999) (to be published).
96. H1 Collaboration, T. Ahmed et al., Phys. Lett. B324 (1994) 241; S. Aid et al., Z. Phys. C67 (1995) 565.
97. ZEUS Collaboration, M. Derrick et al., Phys. Rev. Lett. 75 (1995) 1006.
98. H1 Collaboration, C. Adloff et al., DESY 00-187.
99. H1 Collaboration, C. Adloff et al., Eur. Phys. J. C13 (2000) 609.
100. ZEUS Collaboration, J. Breitweg et al., Eur. Phys. J. C11 (1999) 427.
101. ZEUS Collaboration, contribution to ICHEP, 2000, Osaka, Japan.
102. ZEUS Collaboration, J. Breitweg et al., Eur. Phys. J. C12 (2000) 411.
103. ZEUS Collaboration, contribution to ICHEP, 2000, Osaka, Japan.
104. Particle Data Group, D.E. Groom et al., Eur. Phys. J. C15 (2000)1.
105. D. Bardin et al, *Z. Phys.* **C44** (1989) 149;
    M. Böhm and H. Spiesberger, *Nucl. Phys.* **B304** (1988) 749;
    W.F.L. Hollik, *Fortsch. Phys.* **38** (1990) 165;
    H. Spiesberger, *Nucl. Phys.* **B349** (1991) 109;
    H. Spiesberger, in *Proc. of the Workshop "Future Physics at HERA"*, vol. 1, Eds. G. Ingelman, A. De Roeck and R. Klanner, DESY (1996) 227.
106. Future Physics at HERA, Proc. Workshop 1995/6, Eds. G. Ingelman, A. De Roeck and R. Klanner, DESY 1996, Vol.1 - 3.

MEDAL CEREMONY

Medal Ceremony – Zacatecas, Mexico, 2001

J. Lorenzo Díaz Cruz

Instituto de Física, Benemérita Universidad Autónoma de Puebla

Abstract. Laudatio in honor of Dr. Alfonso Mondragón, recipient of the 2001 Medal of the Division of Particles and Fields, of the Mexican Physical Society.

Greetings to all attendants of the VIII Mexican Workshop on Particles and Fields !
Buenos Dias Señoras y Señores, distinguidos participantes y acompañantes.

Each Year, since 1999 the council of our Division of Particles and Fields (DPyC) of the Mexican Physical Society, asks the national institutions and research groups to propose candidates for the medal, which has the purpose of recognising the work of those who have made an important impact on our community. Namely, the DPYC Medal is aimed to acknowledge:

- The generous effort of foreign scientists, who have helped the development of High-Energy Physics in Mexico, either through their support to research groups or collaborations, or through their dedication to advice Mexican Graduate students.
- The work of physicists established in Mexico, foreign or national, whose research and influence has helped the progress of our community.
- The creativity of Mexican scientists working in the field of high-energy physics anywhere, whose contributions to the field have achieved international recognition.

On this occasion, it is a pleasure for me to announce the decision of the DPyC, to award its medal to a distinguished Mexican scientist, Dr. Alfonso Mondragón from Instituto de Física at UNAM, for his pioneering work in High-Energy Physics in Mexico.

A few weeks ago, when I was asked to prepare this speech, I asked myself what could be the best words to describe the work of Prof. Mondragón. In the first place I thought I should mention its many qualities as a scientist, and its warm and kind personality. I also had in mind his long career, devoted so enthusiastically to research and teaching activities, which has benefited so many students that have taken his classes at UNAM or who have worked with him for their bachelor, master or doctoral thesis. Equally important, and impressive, is his prolific career in various aspects of theoretical physics, which started in 1960. In our particular field, during the period from 1965 to 1970, he wrote a series of articles with Germinal Cocho, on spinorial representations of the Galilean Group, crossed channel in the non-relativistic limit, Regge poles and trajectories and S-matrix theory in dual models. Both with Cocho and Jorge Flores, he wrote also a paper on the calculation of form factors with relativistic harmonic oscillators.

When one realizes that this work was done during the turbulent sixties, one may wonder about the so many changes that Prof. Mondragón witnessed, as a professor at UNAM, as "universitario", and as a Mexican scientist working in Mexico. I just can not imagine how difficult it should had been to do research before CONACYT, when in fact there were only two research groups in the whole country. Things have certainly changed for good, mainly thanks to the efforts of pioneers like Prof. Mondragón.

But the light came to my mind, after Prof. Mondragón came to Puebla, to deliver a seminar at my place, IFUAP, and after sharing lunch-time both with him and his wife, Myriam, we spend quite a few hours talking about physics, about the old and new problems in flavour physics, really enjoying our time. Therefore, I concluded that it was preferable to forget about a story of sacrifice and heroism, because I was missing the most important aspect about Prof. Mondragón. Namely, that he is a person that truly loves his work, someone who has really enjoyed "the privilege of being a physicist", "el privilegio de ser un físico".

And with this words, I ask all of you to give a warm applause to Dr. Alfonso Mondragón, the recipient of the 2001 DPyC Medal !

Early Days of Particle Physics at the Institute of Physics UNAM.

A. Mondragón

Instituto de Física, UNAM, Apdo. Postal 20-364, 01000 México, D.F., México

Abstract. Some reminiscences of the early days of particle physics at the Institute of Physics UNAM are made.

In accepting the 2001 Medal of the Division of Particles and Fields of the Sociedad Mexicana de Física I am highly honoured. I receive this award in my name and in the name of the other scientists who participated in the initial efforts to start and develop a research programme in Physics of Particles and Fields in a systematic and professional way in Mexico. The many mexican scientists now active in this field and the high quality of their published works, of which those presented at this Workshop are a small but representative sample, bear witness to the advancement and progress made in this undertaking during the last four decades. We all have participated in the efforts leading to the formation of a Mexican tradition in the Physics of Particles and Fields whose importance, we hope, will be gradually recognized in wider and wider circles.

I am also highly honoured by the invitation of the Organizing Committee to speak on the early efforts to initiate and carry out research on Particle Physics at the University of Mexico (Universidad Nacional Autónoma de México or UNAM). This I will do with pleasure, but from the start I must disclaim any attempt to settle questions of personal merit or priority and I must add that my account of past events will be heavily biased in the sense that I shall only refer to those of which I was a personal witness or had a first hand account from others who were personal witnesses.

To begin with, let me recall some necessary background. At the end of the Mexican Revolution, around 1918-20, the University of Mexico had lost a large number of its best scholars. Many university professors had been thrown into an inner exile or had just been exiled. Among them were the few scientists who had made the initial efforts to do research in physics and mathematics in Mexico during the last third of the XIXth century and the first decade of the XXth century. Urged by more pressing demands, it was not until the 1930's that the university authorities made a serious effort to upgrade higher education and organize scientific research in physics and mathematics. The subject and orientation of those initial research programmes were determined in great measure by the international prestige and high standing of Manuel Sandoval Vallarta, a mexican physicist who was then a young Professor of Physics at the Massachussetts Institute of Technology.

In scientific circles, M.S. Vallarta was well known for his many contributions to the development of Einstein's theory of general relativity, atomic physics and the foundations of quantum mechanics, but specially, since 1932, he was known for the theory of the geomagnetic effects in cosmic radiation that he formulated and developed in collaboration with Georges Lemaître. In 1933, Vallarta [1] proposed to A. H. Compton the crucial experiment on the East-West asymmetry of cosmic radiation that would allow the determination of the nature of the primary cosmic rays. The experiment was made in Mexico City by Louis W. Alvarez [2] and almost at the same time but independently by Thomas H. Johnson [3], who, in this way determined for the first time that the main component of the primary cosmic radiation is protons. The wide publicity given to this important discovery by the mexican press had a considerable impact on the mexican governement and reinforced the determination of the university authorities who set in motion a number of actions that culminated in 1935 with the foundation of a School of Advanced Studies which later (1939) became the School of Sciences and an Institute of Physics and Mathematics which in 1938 was split into an Institute of Physics and another of Mathematics. The first three directors of the Institute of Physics were Alfredo Baños (1938-1943), M.S. Vallarta (1944) and Carlos Graef-Fernández (1945-1957). Both Baños and Graef-Fernández had been students of M.S. Vallarta at MIT and had written Ph.D. dissertations on different problems related to the geomagnetic effects in cosmic rays.

In the decade from 1938 to 1947 research in physics in Mexico progressed a good deal. It was mostly done in two institutions: The University of Mexico and the Polytechnic Institute.

At the University of Mexico, basic research in theoretical physics was mostly done in two fields, the theory of gravitation and cosmic rays, while experimental work was carried out in the latter field and some work on X-rays analysis of the structure of matter had been started. The research team on cosmic rays was led by M.S. Vallarta and included A. Baños, M. Perusquía, J. Lifshitz and several others. There was also a small number of students working as assistants, among them were Juan de Oyarzabal, Marcos Moshinsky and Gustavo del Castillo.

On the theoretical side, they were interested mainly in the theory of the geomagnetic effects in cosmic radiation, that is, in the paths cosmic rays take in the magnetic field of the Earth, the stability of periodic orbits, the theory of ring currents circulating around the Earth, the consequences of a permanent magnetic field of the Sun and so forth. A good synthesis of these works and references to the original papers may be found in a long review paper on the "Theory of the Geomagnetic Effects of Cosmic Radiation" which appeared as Chapter XLVI/I in Handbuch der Physik [4], see also the Collected Scientific Works of Manuel Sandoval Vallarta [5].

On the experimental side, research was carried out on the determination of the sign and the energy spectrum of primary cosmic radiation. For this purpose, in 1941-42, a counter telescope was designed, built and calibrated at the Institute of Physics UNAM by A. Baños and M. Perusquía. It was completely automatic and registered photographycally the number of events every 32 minutes. With this instrument, lectures were taken in continuous runs lasting al least 100 days from July 1943 to May 1946. The statistical analysis of the data was made by Juan de Oyarzabal and the theoretical interpretation was made by M.S. Vallarta [6]. It was probably in connection with these measurements and their analysis that a professional interest in doing research in particle physics arose for the first time in the group working at the Institute of Physics. Although, the experiment had been wholly succesful and their results confirmed and extended the results obtained by other groups in other laboratories, they had found that, on the high energy side of the cosmic ray spectrum, their results were not conclusive, since they depended on meson production in the high altitude layers of the atmosphere. Motivated by his interest in this experiment, in 1949, G. del Castillo went to Purdue University where he got a Ph.D. degree in physics with a thesis work on a "Study of penetrating showers produced in Lead, Carbon and Beryllium" [7, 8, 9]. He came back to Mexico in 1953 to work at the National Institute for Scientific Research (Instituto Nacional de la Investigación Científica or INIC) in Mexico City on the construction of radiation detection instruments. In 1955 he left INIC for the University of San Luis Potosí, where he founded a Physics Department where he built and operated succesfully a Wilson cloud chamber which allowed him to make some experiments on high energy nuclear interactions of cosmic ray particles with carbon and lead [10].

At the Polytechnic Institute, research on methods of accelerating charged particles had been going on at the Research Laboratory. Manuel Cerrillo and a number of workers were concerned with the development in Mexico of well tried methods to obtain high energy particles, such as the van de Graaf electrostatic generator and the linear accelerator. M. Mireles-Malpica also tried some new methods such as the use of a variety of the Tesla coil and a new device, which for lack of a better name, was called the electrostatic transformer of which a small model was built and some tests were carried out. These works were interrupted in 1942, when M. Cerrillo left for MIT where he stayed until his retirement 35 years later as full Professor of Electronics.

In spite of the fact that the School of Sciences and the Institute of Physics had been working continuously since 1938, when I first came to the University as a college student, in 1950, the material resources for teaching and research in physics as a basic science, that is, not in association with the applied disciplines of Civil, Chemical and Electrical Engineering, were so few and scarce that it is fair to say that they were almost non-existent. The very few students of physics gathered with their teachers in a large classroom in the ground floor of the beautiful Palace of Mining which housed the School of Engineering. We could also use a very small room in the roof of the same building which had previously been used to keep the buckets and brooms of the cleansing staff of the School of Engineering. The offices and library of the Institute of Physics were in one large room in the first floor of the Palace of Mining. It had six desks, one for the director, who was Graef-Fernández, one for his secretary and one for each of the four senior people in the research staff who had a desk. But, the results of more than a decade of formative efforts were by now apparent. There was a small group of bright, young physicist who had recently graduated at the School of Sciences and were now very active teaching and starting a number of new research projects. In 1950, Marcos Moshinsky came back from Princeton and started teaching graduate courses, he also started a very active research programme on Theorctical Nuclear Physics. It was the living example of Marcos Moshinsky what inspired and encouraged us to work hard. So, even if the material resources were scarce, the intelectual activity was intense and, besides, we had some interesting visitors from abroad. I still remember with pleasure when I first met Richard Feynmann in 1951. He was in his way to or back from Brazil, and had come to the Institute looking for Carlos Graef. They had been schoolmates at MIT and Feynmann had written his first research paper with M.S. Vallarta on the effect of the intergalactic magnetic field in cosmic radiation. When Feynmann came, Graef was not at the Institute, so the secretary asked me to take care of the visitor who, she said, was an american scientist who wanted a cup of strong coffee. I took him to a Café around the corner where we had two cups of strong expresso coffee and I had the rare privilege of having the Feynmann graphs explained to me by Feynmann himself.

In 1953, the Institute of Physics, the Institute of Mathematics and the School of Sciences were the first university dependencies to move to the beautiful new university buildings in Ciudad Universitaria. In the same year, President Adolfo Ruiz Cortines inaugurated the new laboratories of nuclear physics, atomic and molecular physics, X-rays and cristallography, solid state, gravitation and cosmic rays. The Nuclear Physics laboratory had a 2 MeV van de Graaf accelerator with which a small group led by M. Mazari started an active program of research in low energy nuclear collisions and reactions. This initial effort in experimental nuclear physics was done in collaboration with the MIT group headed by William Buechner.

In those years, research in theoretical physics was mostly done at the Institute of Physics UNAM, but some of its younger members who were active in research received a small stipend from the National Institute for Scientific Research (INIC). This economic help was conditioned only to the publication of research papers in refereed international journals and the participation in the weekly Physics Colloquium organized and presided by M.S. Vallarta, where almost all research papers in physics were read and discussed prior to publication. It was at INIC that Alejandro Medina, who was not a member of the Institute of Physics, gave courses on Classical and Quantum Electrodynamics. He also organized a research seminar on Particle Physics and Field Theory. The participants were some of the younger physicists of the Institute of Physics and some of the more advanced students of physics, among them were Juan de Oyarzabal, Fernando Prieto, Francisco Medina and Juan Manuel Lozano. Very soon, this small group produced some papers "On the field of the Bhabha equation" by F.E. Prieto [11, 12], "Pion-pair production" by J. de Oyarzabal [13] and "Divergencies in Quantum Field Theory" by A. Medina [14] which were presented at the Joint Meeting of the Mexican and American Physical Societies in Mexico City in August 1955. At that meeting sixteen papers by mexican authors were read, among them was my first research paper on "Scattering length and effective range for velocity dependent tensor forces in nucleon-nucleon collisions" [15] which contained the results of my B.Sc. thesis which I had written under the advice of Marcos Moshinsky.

It was under the advice of A. Medina that F.E. Prieto wrote the dissertation on "Differential and total cross sections for nucleon-antinucleon scattering" that he presented as a thesis at the School of Sciences to obtain the degree of Ph.D. in 1957. The main results of this work were published in Physical Reviews [16] in 1957. I believe this was the first Ph.D. thesis written in Mexico on theoretical aspects of Particle Physics. Later, Prieto published also a paper on "Parity non-conservation in hyperon decays" [17].

In this same year, 1956, Moshinsky had organized the first Mexican Summer School in Physics, which later became the Latin American School of Physics, where I met Prof. R.E. Peierls. Other lecturers were Joseph Levinger, Berthold Stech, Robert G. Thomas and Marcos Moshinsky. It was attended by many mexican students and some foreign students. Among the students from abroad were Charles Summerfield and Sheldon Glashow, now a Nobel laureate for his work in particle physics, Martin Blume, who later became a solid state expert at Brookhaven and André Martin, a cheerful young Frenchman who became well known for his many contributions to field theory. In collaboration with these last two, I wrote and edited the notes of Peierls' course on Nuclear Forces.

In 1957, I went to the University of Birmingham, in England, as a graduate student. I took courses on Quantum Field Theory with Sam F. Edwards, later Sir Sam Edwards, on Nuclear Physics with Prof. R. E. Peierls, later Sir Rudolf Peierls, and many more. After two years of hard work I wrote a Ph.D. dissertation "On the Pauli principle in the optical model of nuclear reactions" under Peierls' advice.

I came back to Mexico at the end of 1960 and started teaching in 1961. At first I could not find easely my way back into research. In 1963, Germinal Cocho came back to Mexico from Princeton University where he had written a thesis on a "Resonance Model in nucleon-antinucleon annihilation" [18] under the advice of Prof. R. Blanckenbecler. Cocho was well versed in the S-matrix theory of particle interactions proposed by W. Heisenberg and developed by G.F. Chew and many others. It seemed natural that we tried to combine what I had learned from Peierls about the S-matrix theory of nuclear reactions with the S-matrix theory of elementary particles he had learned from Blanckenbecler in Princeton. Soon we realized that the optical model of the scattering of a nucleon by a nucleus and the Regge-pole model of hadron-hadron interactions at high energy and low values of the momentum transfer had in common the following feature: the total scattering amplitude is naturally decomposed into the sum of a term which is a smoothly varying function of the energy and describes the prompt response of the system plus a term with a rich structure in energy which is due to the contribution of many sharp resonances and describes the time delayed response of the system. This analogy lead us, first, to the definition of a non-relativistic analogue of the crossed-channel Regge-Joos representation used in relativistic particle physics [19, 20]. It was defined as a decomposition of the scattering amplitude into partial amplitudes, each one corresponding to the exchange of a set of pseudostates transforming irreducibly under the Galilei group. We gave explicit formulae for the elastic scattering of particles with arbitrary spin for the non-forward and forward scattering, where the little groups are $E(3)$ and $E(2)$. We applied this formalism to the scattering of nucleons by alpha particles and, parametrizing the scattering amplitude with only two effective multipoles, we obtained a very good fit for the differential cross section and polarization for values of the energy in the range from 66 to 320 MeV [20, 21] . Then, we turned our attention to the application of concepts and ideas from low energy nuclear physics to the interpretation of the Regge-Joos decomposition of high energy and low momentum transfer hadron-hadron collisions. In this mathematical model the amplitude is decomposed as a sum of terms, namely, the contribution of the Regge poles and the remaining term or subtracted amplitude which is given by the Regge background integral. In the early 1960s, there was a major attempt to identify poles with resonances: Pomeron, $\rho, \omega, \ldots$ and $\pi, N, \ldots$ Searches were made for "Regge recurrences", resonances with spins two higher than the original resonances, i.e., a spin 3 ρ, etc. A few meson recurrences were found and candidates for recurrences of the proton and Δ(1 2 3 2) poles were identified. The trajectories tended to lie on straight lines. These trajectories were then used in elastic and exchange scattering to try to understand the Q^2 dependence of the scattering. Thus, in principle, Regge trajectories could be used for understanding both scattering and resonances. While most people were looking at the Regge poles, we occupied ourselves with the background Regge integral. We were able to show that the integration parameter in that integral may be understood as a time delay variable. From this result, we could also show that models with linearly rising trajectories have a simple interpretation in terms of multiscattering processes in the time delay representation [22]. Furthemore, we also found that in pseudopotential theory, linear rising trajectories correspond to potentials linear in r, i.e. forces independent of r [22]. Encouraged by the success of the analogy found, at the end of 1968 we proposed a completely relativistic formulation of nuclear reactions in which the dynamics of the interaction process was parametrized in terms of the exchange of representations of some non-invariance group [23]. Two conditions had to be met: First, relativistic invariance of the formalism required that the non-invariance group had a subgroup isomorphic to the Lorentz group. Second, since the harmonic oscillator symmetry group $U(3)$ had proved so succesful in low energy nuclear physics, the non invariance group had to contain $U(3)$ as a compact subgroup. The simplest group that satisfies these conditions is the non-compact group $U(3,1)$ which has the additional advantage of being simple to deal with mathematically. Although many properties could, in principle, be computed, the physical picture was somewhat abstract and the actual applications somewhat limited. When the quark theory became succesful and Quantum Chromodynamics became established as the right field theory of strong interactions, much of the attention shifted to that theory. By then, a small group of students had become interested in particle physics, among them were Augusto García and Clicerio Avilés, and we could say that a systematic research programme in Particle Physics had been started succesfully at the Institute of Physics UNAM.

REFERENCES

1. Vallarta, M.S., *Phys. Rev.*, **44**, 1 (1934).
2. Alvarez, L.W. and Compton, A.H., *Phys. Rev.*, **43**, 835 (1933).
3. Johnson, T.H., *Phys. Rev.*, **43**, 834 (1933).
4. Vallarta, M.S., "Theory of the Geomagnetic Effects of Cosmic Radiation", in *Cosmic Rays I*, Edited by, S. Flügge, Handbuch der Physik XLVI/I, Springer. Berlin, 1961, pp 88.
5. "Manuel Sandoval Vallarta: Obra Científica", edited by D. Barnes and A. Mondragón. Universidad Nacional Autónoma de México, México City 1978.
6. Vallarta, M.S., Perusquía, M.L. and De Oyarzábal, J., *Phys. Rev.*, **71**, 393 (1947).
7. Chang, W.Y. and Del Castillo, G., *Phys. Rev.*, **81**, 584 (1951).
8. Chang, W.Y., Del Castillo, G. and Grodzins, L., *Phys. Rev.*, **84**, 582 (1951).
9. Chang, W., Del Castillo, G. and Grodzins, L., *Phys. Rev.*, **89**, 408 (1953).
10. Del Castillo, G., *Phys. Rev.*, **98**, 1163 (1955).
11. Prieto, F.E., *Rev. Mex. Fis.*, **3**, 24 (1954).
12. Prieto, F.E., *Phys. Rev.*, **100**, 974 (1955).
13. De Oyarzabal, J., *Phys. Rev.*, **100**, 974 (1955).
14. Medina A., *Phys. Rev.*, **100**, 978 (1955).
15. Mondragón A., *Phys. Rev.*, **100**, 956 (1955).
16. Prieto, F.E., *Phys. Rev.*, **107**, 1439 (1957).
17. Prieto, F.E. *Nucl. Phys.*, **13**, 456 (1959).
18. Cocho G., *Rev. Mex. Fis.*, **11**, 215 (1962).
19. Cocho, G., Colón-Vela , M. and Mondragón, A., *Rev. Mex. Fis.*, **17**, 59 (1968).
20. Cocho, G. and Mondragón, A., *Nucl. Phys.*, **A125**, 417 (1969).
21. Cocho, G., Colón-Vela, M. and Mondragón, A., *Nucl. Phys.*, **A125**, 425 (1969).
22. Cocho, G. and Mondragón, A., *Phys. Rev.*, **D1**, 3484 (1970).
23. Cocho, G., Flores J. and Mondragón, A., *Nucl. Phys.*, **A128**, 110 (1969).

INVITED TALKS

Resonant-Spin Flavour solutions to the Solar neutrino Problem

O. G. Miranda

Depto. de Fisica, CINVESTAV-IPN, A P 14-740, Mexico D F, 07000 Mexico

Abstract. Resonant spin-flavour solutions to the solar neutrino problem are introduced, in the framework of analytic solutions to the solar magneto-hydrodynamics equations. We study these solutions in a scheme with 3 effective parameters: the neutrino magnetic moment, the neutrino mass difference and mixing. We perform a fit of the solar neutrino data, including the recent SNO CC. We show how a rates-only analysis slightly flavours spin-flavour precession solutions over oscillations. In addition to the resonant solution, there is a new non-resonant solution in the dark-side. Both solutions lead to flat recoil energy spectra in excellent agreement with the latest SuperKamiokande data.

INTRODUCTION

We consider the Solar neutrino problem in the case of spin-flavour precession solutions (RSFP), based on non-zero transition magnetic moments of neutrinos [1].We include in our analysis the recent charged current measurement at the Sudbury Neutrino Observatory (SNO) [2]. Such a RSFP solution is especially attractive for several reasons: (i) on general theoretical grounds [3] neutrinos are expected to be Majorana particles; (ii) such conversions induced by transition magnetic moments may be resonantly amplified in the Sun [4]; (iii) they offer the best pre-SNO global fit of solar neutrino data [5], and (iv) a RSFP type solution, being an active–to–active conversion mechanism, has the right features to reconcile the SNO CC and SuperKamiokande results. One problem of this type of solutions was the arbitrariness in the choice of the magnetic field profile in the solar convective zone [6, 7, 8]. However, it was shown by us that this arbitrariness can be removed in a self-consistent way from magneto-hydrodynamics theory [5].

We consider the more general case of non-zero neutrino mixing were we have obtained two important results [9]: (i) we recover the resonant small-mixing solution to the solar neutrino problem found previously [5] and analyse its status in the light of the new SNO and 1258–day SK results, and (ii) we find a genuinely new non-resonant NRSFP solution in the so-called dark-side of the neutrino mixing parameter [10, 11]. Following [12] we determine the allowed solutions by considering only the total rates of the solar neutrino experiments. We find that these solutions, both the resonant spin flavour precession solution (which we call RSFP) as well as a new non-resonant one (NRSFP solution), provide excellent descriptions of the solar rates, including the recent SNO CC result. Subsequently we demonstrate how these solutions predict a substantially flat recoil energy spectrum of solar neutrinos in agreement with the observations of the Super-Kamiokande experiment [13]. Moreover, our solutions are consistent with the non-observation of electron anti–neutrinos from the sun [14, 15] in the results of the LSD experiment [16] as well as SuperKamiokande [17].

NEUTRINO EVOLUTION AND SURVIVAL/CONVERSION PROBABILITIES

Motivated by the results from reactor neutrino experiments [18] and to some extent also from atmospheric neutrinos [19] we adopt, for simplicity, a two–flavour RSFP scenario. The Majorana neutrino evolution Hamiltonian in a magnetic field in this case is well–known to be four–dimensional [1],

$$i \begin{pmatrix} \dot{v}_{eL} \\ \dot{\bar{v}}_{eR} \\ \dot{v}_{\mu L} \\ \dot{\bar{v}}_{\mu R} \end{pmatrix} = \begin{pmatrix} V_e - c_2 \delta & 0 & s_2 \delta & \mu B_+(t) \\ 0 & -V_e - c_2 \delta & -\mu B_-(t) & s_2 \delta \\ s_2 \delta & -\mu B_+(t) & V_\mu + c_2 \delta & 0 \\ \mu B_-(t) & s_2 \delta & 0 & -V_\mu + c_2 \delta \end{pmatrix} \begin{pmatrix} v_{eL} \\ \bar{v}_{eR} \\ v_{\mu L} \\ \bar{v}_{\mu R} \end{pmatrix}, \quad (1)$$

where $c_2 = \cos 2\theta$, $s_2 = \sin 2\theta$, $\delta = \Delta m^2/4E$ (assumed to be always positive) are the neutrino oscillation parameters; μ is the neutrino transition magnetic moment; $B_\pm = B_x \pm iB_y$, are the magnetic field components which are perpendicular to the neutrino trajectory; $V_e(t) = G_F \sqrt{2}(N_e(t) - N_n(t)/2)$ and $V_\mu(t) = G_F \sqrt{2}(-N_n(t)/2)$ are the neutrino vector potentials for v_{eL} and $v_{\mu L}$ in the Sun, given by the number densities of the electrons ($N_e(t)$) and neutrons ($N_n(t)$). When $\theta \to 0$ we recover the case treated in [5] while as $B \to 0$ we recover the pure oscillation case. We have solved this equation numerically and computed the survival and conversion probabilities. In our calculations we use the electron and neutron number densities from the BP00 model [20] with the magnetic field profile obtained in ref. [5] for k=6 and $R_0 = 0.6R_\odot$. We assume a transition magnetic moment of 10^{-11} Bohr magneton, consistent with experiment and a magnetic field magnitude around 80 kGauss, allowed by helioseismological observations. Finally, in order to obtain Earth matter effects we integrate numerically the evolution equation in the Earth matter using the Earth density profile given in the Preliminary Reference Earth Model (PREM) [21].

RATE FIT

In our analysis of solar neutrino data [13] we follow the techniques which have already been presented in previous papers [12, 19, 22] using the theoretical BP00 standard solar model best–fit fluxes and estimated uncertainties from ref. [20]. In addition to the solar data [13] we also use the reactor data [18] as well as the data on searches for anti-neutrinos from the sun [16].

We use the magnetic field profile obtained in ref. [5] for k=6 and $R_0 = 0.6R_\odot$. The resulting theoretical framework has therefore only 2 effective free parameters: Δm^2, $\tan^2 \theta$. The remaining parameter $\mu B_\perp$ characterizing the maximum magnitude of the magnetic field in the convective zone has been fixed at its optimum value. Since the parameter space is three-dimensional, the allowed regions for a given C.L. are defined as the set of points satisfying the condition

$$\chi^2_{SOL}(\Delta m^2, \theta, \mu B_\perp) - \chi^2_{SOL,min} \leq \Delta\chi^2(C.L., 2 \text{ d.o.f.}), \qquad (2)$$

where χ^2_{SOL} contains

$$\chi^2_{LSD}(\Delta m^2, \theta, \mu B_\perp) = \frac{\left(N^{TH}_{\bar{\nu}_e}(\Delta m^2, \theta, \mu B_\perp) - N^{EXP}_{\bar{\nu}_e}\right)^2}{\sigma^2_{LSD}} \qquad (3)$$

where $N^{EXP}_{\bar{\nu}} = -1.5$ and $\sigma_{LSD} = 22$ in order to account for the data on searches for anti-neutrinos from the sun [16]. This term plays an important role in restricting the neutrino parameters.

In our numerical calculations we use the survival/conversion probabilities of solar electron neutrino valid in the full range of Δm^2 and θ, selecting the optimum value of $\mu B_\perp$ with $B_\perp$ varying over the range from 0 to 100 kGauss.

Finally, we employ the relevant reaction cross sections and efficiencies for the all experiments used in ref. [12, 19, 22]. For the SNO case the CC cross section for deuterium was taken from [23].

Here we take into account the total rates in the chlorine, gallium, and Super-Kamiokande experiments, the SNO CC result and the anti-neutrino limit from LSD, and also the reactor neutrino data [18]. The rates from the GALLEX/GNO experiments have been averaged so as to provide a unique data point. The resulting number of degrees of freedom is therefore 4: 4 (rates) + SNO + LSD -2 (parameters: Δm^2, θ) with a fixed $\mu B_\perp$.

RESULTS AND DISCUSSION

We present in Fig. 1 the allowed solutions for the two-flavour SFP case. These include the pure two-neutrino oscillation case, as well as the conventional RSFP and the new NRSFP solution.

The contours refer to 90%, 99% CL defined with respect to the global minimum of χ^2. We find that both LMA and SMA oscillation solutions are recovered without an essential change due to the effect of the magnetic moment. The SMA solution appears (even though disfavored), but leads to an unacceptably tilted recoil energy spectrum.

An important point to notice is that this plot lacks the LOW solution as well as the characteristic region joining it through the dark side to the vacuum-type solutions [24]. In this figure we have adjusted the value of $\mu B_\perp$ to its best value (for $\mu = 10^{-11}$ Bohr magneton this corresponds to $B_\perp \sim 80$ kGauss). One sees that the relatively large μ value has important consequences. It leads in this case in the complete absence of all large mixing solutions other than the LMA solution due to the magnetic field effect. From this point of view vacuum-type solutions are *unstable* against the effect of the magnetic field. In fact the non-LMA large mixing oscillation (OSC) solutions are not re-instated even if the ^{8}B neutrino flux is left free.

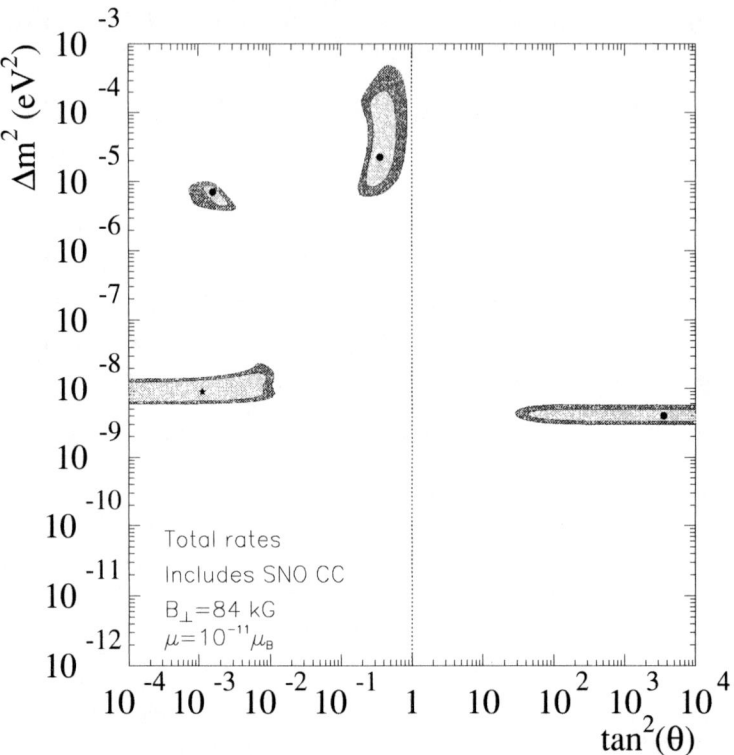

FIGURE 1. Allowed solutions to the solar neutrino rates and reactor data for $\mu_\nu = 10^{-11}\mu_B$ and $B_\perp = 84$ kGauss. The upper limit on the solar anti-neutrino flux according to LSD data is included.

A more striking feature of Fig. 1 is the appearance of two new solutions which are totally due to the effect of the magnetic field. One contains the previous resonant no-mixing solution which is recovered, after updating the solar data to the measurements from 1258 days of Super-Kamiokande data and SNO CC measurement. One sees that this RSFP solution extends up to $\tan^2\theta$ values around 10^{-2} or so. More importantly, one finds a genuinely new non-resonant (NRSFP) solution in the "dark-side" of the parameter space, for large $\tan^2\theta$ values. Note that in obtaining the shape of the RSF solutions we have made use of the data on searches for anti-neutrinos from the sun [16]. These play an important role in cutting the non-resonant RSF solution to $\tan^2\theta$ values larger than about 30.

The goodness of the fit for the various solutions in Fig. 1 is given in table 1. One notices that, of the OSC-type solutions, LMA is the best [1]. However the SFP solutions

---

[1] The first time the LMA solution was shown to be the best OSC solution was in [22] due to the details of the solar neutrino spectra measured at Super-Kamiokande. This trend is now re-inforced by the enhanced statistics. The SNO CC rate-result implies, on its own, a preference for the LMA if the BP00 boron flux

are slightly better.

TABLE 1. Best-fit points and goodness-of-fit of oscillation and spin flavour solutions to the solar neutrino problem as determined from the rates-only analysis for $\mu = 10^{-11} \mu_B$ and $B_\perp = 84$ kGauss.

Solution	Δm^2	$\tan^2(\theta)$	χ^2_{min}	g.o.f.
LMA	2.1×10^{-5}	0.34	3.99	14%
SMA	6.9×10^{-6}	1.6×10^{-3}	5.25	7%
RSF	8.9×10^{-9}	1.1×10^{-3}	2.98	22%
NRSF	4.0×10^{-9}	3.5×10^3	3.83	15%

Although we have presented here only a rates analysis, we have shown that both spin flavour precession spectra are totally consistent with the Super-Kamiokande data and, as a result, will remain as excellent solutions after the inclusion of the recoil energy spectra [9].

A full-fledged global fit of the recoil spectra for the spin flavour solutions will be presented elsewhere [24].

ACKNOWLEDGMENTS

This work has been done in collaboration with Carlos Peña-Garay, Timur Rashba, Victor Semikoz, and Jose Valle. We would like to thank Alexei Bykov, Hiroshi Nunokawa, Alexander Rez, Victor Popov and Dmitri Sokoloff for very useful discussions. This work was supported by CONACYT and SNI.

REFERENCES

1. J. Schechter and J. W. F. Valle, Phys. Rev. **D24** (1981) 1883; Erratum-ibid. **D25** (1982) 283
2. Q. R. Ahmad et al. [SNO Collaboration], nucl-ex/0106015.
3. J. Schechter and J. W. F. Valle, Phys. Rev. **D22** (1980) 2227.
4. E. K. Akhmedov, Phys. Lett. B **213** (1988) 64; C. Lim and W. J. Marciano, Phys. Rev. D **37** (1988) 1368.
5. O. G. Miranda, C. Pena-Garay, T. I. Rashba, V. B. Semikoz and J. W. F. Valle, Nucl. Phys. B **595** (2001) 360 [hep-ph/0005259].
6. E. K. Akhmedov and J. Pulido, Phys. Lett. B **485** (2000) 178 [hep-ph/0005173]; Astropart. Phys. **13** (2000) 227 [hep-ph/9907399].
7. M. M. Guzzo and H. Nunokawa, Astropart. Phys. **12** (1999) 87 [hep-ph/9810408]. H. Nunokawa and H. Minakata, Phys. Lett. B **314** (1993) 371.
8. J. Derkaoui and Y. Tayalati, Astropart. Phys. **14** (2001) 351 [hep-ph/9909512].
9. O. G. Miranda, C. Pena-Garay, T. I. Rashba, V. B. Semikoz and J. W. Valle, Phys. Lett. B **521**, 299 (2001) [arXiv:hep-ph/0108145].
10. G. L. Fogli, E. Lisi and D. Montanino, Phys. Rev. D **54** (1996) 2048 [hep-ph/9605273].

---

is assumed

11. A. de Gouvea, A. Friedland and H. Murayama, Phys. Lett. B **490** (2000) 125 [hep-ph/0002064].
12. J. N. Bahcall, M. C. Gonzalez-Garcia and C. Pena-Garay, JHEP **0108**, 014 (2001) [arXiv:hep-ph/0106258]. For recent papers on the oscillation interpretation of the solar neutrino anomaly after the SNO result see [25].
13. B.T. Cleveland et al., Astrophys. J. **496** (1998) 505; K.S. Hirata et al., Kamiokande Coll., Phys. Rev. Lett. **77** (1996) 1683; W. Hampel et al., GALLEX Coll., Phys. Lett. B **447** (1999) 127; D.N. Abdurashitov et al., SAGE Coll., Phys. Rev. Lett. **83** (1999) 4686; Phys. Rev. C **60** (1999) 055801 [astro-ph/9907113]; V. Gavrin (SAGE Collaboration), Nucl. Phys. Proc. Suppl. B **91** (2001) 36; M. Altmann et al. (GNO Collaboration), Phys. Lett. B **490** (2000) 16; E. Bellotti et al. (GNO Collaboration), Nucl. Phys. Proc. Suppl. B **91** (2001) 44; Y. Fukuda et al. (Super-Kamiokande Collaboration), Phys. Rev. Lett. **81** (1998) 1158; Erratum **81** (1998) 4279; Phys. Rev. Lett. **82** (1999) 1810; Y. Suzuki (Super-Kamiokande Collaboration), Nucl. Phys. Proc. Suppl. B **91** (2001) 29; S. Fukuda et al. (Super-Kamiokande Collaboration), Phys. Rev. Lett. **86** (2001) 5651.
14. R. Barbieri, G. Fiorentini, G. Mezzorani and M. Moretti, Phys. Lett. B **259** (1991) 119.
15. P. Vogel and J. F. Beacom, Phys. Rev. D **60** (1999) 053003 [hep-ph/9903554].
16. M. Aglietta et al., JETP Lett. **63** (1996) 791 [Pisma Zh. Eksp. Teor. Fiz. **63** (1996) 753].
17. M.B. Smy, private communication; C. Yanagisawa, Talk given at the Euroconference on Frontiers on Cosmology and Astroparticle Physics, Saint Feliu de Guixols, Spain, Proceedings Supplements, 2001, Vol. 95, ISSN 0920-5632
18. M. Apollonio et al. [CHOOZ Collaboration], Phys. Lett. B **466**, 415 (1999) [hep-ex/9907037].
19. See, for example, M. C. Gonzalez-Garcia, M. Maltoni, C. Pena-Garay and J. W. F. Valle, Phys. Rev. D **63** (2001) 033005 [hep-ph/0009350], G. L. Fogli in *Neutrino Telescopes 2001*, Venice, Italy, March 2001
20. http://www.sns.ias.edu/~jnb/SNdata/Export/BP2000; J. N. Bahcall, S. Basu and M. H. Pinsonneault, Astrophys. J. **529**, 1084 (2000); Astrophys. J. **555** (2001) 990
21. A. M. Dziewonski and D. L. Anderson, Phys. Earth Planet. Inter. **25**, 297 (1981).
22. M. C. Gonzalez-Garcia, P. C. de Holanda, C. Pena-Garay and J. W. F. Valle, Nucl. Phys. **B573** (2000) 3 [hep-ph/9906469] and references therein.
23. J.F. Beacom and S.J. Parke, Phys. Rev. D **64**, 091302 (2001) [hep-ph/0106128]; S. Nakamura, T. Sato, V. Gudkov and K. Kubodera, Phys. Rev. C **63** (2001) 034617; M. Butler, J.-W. Chen and X. Kong, Phys. Rev. C **63** (2001) 03550; I.S. Towner, Phys. Rev. C **58** (1998) 1288.
24. For a complete discussion see O. Miranda et al, in preparation.
25. A. Bandyopadhyay, S. Choubey, S. Goswami and K. Kar, Phys. Lett. B **519**, 83 (2001) [hep-ph/0106264]; G. L. Fogli, E. Lisi, D. Montanino and A. Palazzo, Phys. Rev. D **64**, 093007 (2001) [hep-ph/0106247]; P. Creminelli, G. Signorelli and A. Strumia, JHEP **0105**, 052 (2001) [hep-ph/0102234]; V. Barger, D. Marfatia and K. Whisnant, Phys. Rev. Lett. **88**, 011302 (2002) [hep-ph/0106207].

Scalar Mesons and Chiral Dynamics

Mauro Napsuciale

*Instituto de Fisica, Universidad de Guanajuato,
Lomas del Bosque 103, Facc. Lomas del Campestre
37150, Leon, Guanajuato, Mexico.*

Abstract. We discuss scalar mesons properties on the light of chiral dynamics. Considering them as the chiral partners of pseudo-scalar mesons we propose an explanation to their unusual properties based on non-trivial vacuum effects coming from the interplay between spontaneous breaking of chiral symmetry and the violation of $U_A(1)$ symmetry by instantons. Including vector mesons as external sources we work out predictions for radiative decays of vector mesons and compare some of them with recent experimental results from high luminosity Φ factories.

INTRODUCTION

The understanding of scalar excitations is a fundamental problem which we encounter in many branches of physics ranging from the dilaton in theories for gravity, the higgs in the electroweak theory, to the pairing of fermions in condensed matter. The reason is quite simple: scalar excitations have the same quantum numbers as the vacuum and their properties are strongly influenced by it. Understanding scalar excitations is at the same time, in some way, an understanding of the vacuum properties of the corresponding theory which is a particularly difficult task in the case it has an strongly coupled regime.

In this talk I will summarize our work on scalar mesons in low energy QCD. The well-established lowest lying scalar mesons are the isovector $a_0(980)$ and the isoscalar $f_0(980)$. These mesons are nearly degenerate in mass which suggests they are the scalar analogous to the $\rho(770)$ and $\omega(780)$ which would imply a $\frac{1}{\sqrt{2}}(\bar{u}u - \bar{d}d)$ and $\frac{1}{\sqrt{2}}(\bar{u}u + \bar{d}d)$ for the $a_0(980)$ and $f_0(980)$ respectively. The first problem with this identification is the experimental fact that the $f_0(980)$ strongly couples to the $\bar{K}K$ system. The second problem is the small coupling to two photons these mesons have, a problem which we will review in detail below. Before this let us mention that over the past few years compelling evidence has accumulated for the existence of a broad isoscalar structure (σ) in the very low energy region [1] which is strongly coupled to two pions and the existence of a isospinor scalar in the 800 – 900 MeV has been claimed by many authors [2], although its existence is still under debate [3].

Scalar (and pseudoscalar) mesons as $\bar{Q}Q$ states: testing the structure of hadrons with photons.

Electromagnetic decays of hadrons give valuable information on their structure for the simple reason that photons couple to charged objects, in particular to the partons constituting a hadron. In this sense the two photon decay of scalars constitute direct evidence for their quark structure. Quark model calculations for the $a_0(980) \to \gamma\gamma$ and $f_0(980) \to \gamma\gamma$ decays were summarized in [4] and a molecule structure was explored in this work which seemed to be favored by the measured branching ratios. Here we will work with the following assumptions: i) meson are composed of $\bar{Q}Q$ where Q denotes a **constituent** (colored) quark. ii)They are non-relativistic systems and we work in the zero-binding approximation ($M \approx 2m_Q$). Under these assumptions we can use either NRQCD-like calculations (singlet channels only) or we can apply the quarkonium techniques developed in [5]. The calculations lead to

$$\Gamma(^1S_0 \to \gamma\gamma) = \frac{12\alpha^2}{M_P^2} |R(0)|^2 e_Q^2$$
$$\Gamma(^3P_J \to \gamma\gamma) = \frac{N_J \alpha^2}{M_J^4} |R'(0)|^2 e_Q^2$$

where $N_0 = 432$, $N_2 = 576/5$ and $N_1 = 0$ as required by charge conjugation. Here, e_Q denotes the charge of the constituent quark (in units of e) and $R(0)$, $R'(0)$ denote the quarkonium wave funcion and its derivative evaluated at the origin respectively. For the physical process *meson* $\to \gamma\gamma$ it is necessary to consider the isospin structure which amounts to replace $e_Q^2 \to F(I)$, where $F(1) = (e_u^2 - e_d^2)/\sqrt{2}$ for a $(\bar{u}u - \bar{d}d)/\sqrt{2}$ meson and $F(0) = (e_u^2 + e_d^2)/\sqrt{2}$ for a $(\bar{u}u + \bar{d}d)/\sqrt{2}$ meson. The only unknown parameters here are the wave functions at the origin which do not allow us to predict the individual widths. However, this factor cancels out in the ratios of widths which allow us to compare with the measured ratios for mesons with the same orbital angular momentum. We show in a table the results for the ratios of different combinations of the $^3P_J \to \gamma\gamma$ decay widths $R_{th}^{M_1 M_2} \equiv \frac{\Gamma(M_1 \to \gamma\gamma)}{\Gamma(M_2 \to \gamma\gamma)} = \left(\frac{m_{M_2}}{m_{M_1}}\right)^4 \frac{F^2(I_1)}{F^2(I_2)}$. In this table a $(\bar{u}u + \bar{d}d)/\sqrt{2}$ structure has been assumed for the f_0 and f_2 mesons and a $(\bar{u}u - \bar{d}d)/\sqrt{2}$ structure for the a_0 and a_2 mesons.

$M_1 M_2$	$R_{th}^{M_1 M_2}$	$R_{exp}^{M_1 M_2}$
$a_2(1320) f_2(1270)$	0.31	0.36 ± 0.05
$a_0(980) f_0(980)$	0.35	1.62 ± 0.96
$f_0(980) f_2(1270)$	0.76	0.14 ± 0.04
$a_0(980) a_2(980)$	0.85	0.24 ± 0.09

The conclusion we extract from this table is that two photon decays of tensor mesons are consistent with the assumed composition whereas **$a_0(980) \to \gamma\gamma$ and $f_0(980) \to \gamma\gamma$ are not consistent with a $(\bar{u}u \pm \bar{d}d)/\sqrt{2}$ structure for these mesons**. Another possibility in the case of the $f_0(980)$ is a $\bar{s}s$ structure. For this case we obtain $R_{th}^{f_0 f_2} = 0.06$ vs

$R_{exp}^{f_0 f_2} = 0.14 \pm 0.04$. Thus $f_0(980) \to \gamma\gamma$ is neither well described in terms of quarkonium calculations assuming a pure $\bar{s}s$ structure. It still remain the possibility that the $f_0(980)$ be a more complicated object similar to the η or η' i.e. $f_0 = \sin\phi_s S_{ns} + \cos\phi_s S_s$ where $S_{ns} = (\bar{u}u + \bar{d}d)/\sqrt{2}$ and $S_s = \bar{s}s$. In this case $R_{th}^{f_0 f_2} = \frac{4}{15}(\frac{M_{f_2}}{M_{f_0}})^4(\sin\phi_s + \frac{\sqrt{2}}{5}\cos\phi_s)$ and agreement with the experimental ratio requires $\phi_s \approx 9°$.

What about pseudoscalars?. In this case the quarkonium state has the $J^{PC} = 0^{-+}$ quantum numbers ($^1S_0$ state in spectroscopic notation). The predicted ratio of widths in the case $\eta - \pi$, written in terms of the mixing angle in the singlet-octet basis $\theta_P = \phi_P - 54.7°$ is $R_{th}^{\eta\pi} = \frac{m_\eta^3}{m_\pi^3}(\frac{m_\pi}{m_\eta})^5 \frac{1}{\sqrt{3}}[\cos\theta_p - 2\sqrt{2}\sin\theta_p)]^2$. In this case, the results for the ratio of widths differ from those arising in naive quark model (or SU(3)) (see e.g. [6] page 117). Indeed, there is a dynamical factor $(\frac{m_\pi}{m_\eta})^5 \approx 10^{-3}$ for this ratio which considerable reduce the ratio with respect to naive quark model considerations when physical masses are used in the numerics. Reproducing experimental data require to introduce a "reference mass"(M_R) [7] whose origin is clear in our framework, $M_R \approx 2m_Q$ which represent the "constituent mass" for the quarkonium. The inability of quarkonium calculations to describe two photon decays of pseudoscalars without any additional assumption has to do with the mechanisms which deviate the values of the masses of pseudoscalars from their "constituent mass" values. It has been clear since the early days of QCD that the spectrum of pseudoscalars is strongly influenced by non-trivial effects due to the properties of the QCD vacuum. On the other hand scalars have the same quantum numbers as the vacuum and the immediate question is whether or not the scalar spectrum and decay properties are modified, from what we naively would expect, by effects due to the properties of the QCD vacuum. In the next section we will review our work on the effects of the vacuum in the structuring of the spectrum and decay properties of scalar mesons.

Chiral symmetry, vacuum effects and scalar meson properties.

We explored this topic in the framework of a Linear Sigma Model (LSM) which incorporates the most important effects in the low energy domain of QCD, namely, the spontaneous breaking of χral symmetry (SBχS) and the breakdown of the singlet axial symmetry. For the latter we use the bosonized version of the instanton induced six-quark interaction discovered by 't Hooft. The pseudoscalar and scalar matrix fields P and σ are written in terms of a specific basis spanned by seven of the standard Gell-Mann matrices, namely λ_i ($i = 1, \ldots, 7$), *and* by two unconventional matrices λ_{ns}=diag(1,1,0), and $\lambda_s = \sqrt{2}$ diag(0,0,1), respectively. We use the convention $P \equiv \frac{1}{\sqrt{2}}\lambda_i P_i$ with $i = ns, s, 1, \ldots, 7$ and similarly for the scalar field. The scalar and a pseudoscalar mesonic nonet enter in the chiral combination $M = \sigma + iP$. The lagrangian is

$$\mathcal{L} = \mathcal{L}_{sym} + \mathcal{L}_{U_A(1)} + \mathcal{L}_{SB}. \tag{1}$$

The $[U(3)_L \otimes U(3)_R]$ symmetric part is given by

$$\mathcal{L}_{sym} = \frac{1}{2}\text{tr}\left[(\partial_\mu M)(\partial^\mu M^\dagger)\right] - \frac{\mu^2}{2}X(\sigma,P) - \frac{\lambda}{4}Y(\sigma,P) - \frac{\lambda'}{4}X^2(\sigma,P), \quad (2)$$

where $X(\sigma,P) \equiv \text{tr}\left[MM^\dagger\right]$ $Y(\sigma,P) \equiv \text{tr}\left[(MM^\dagger)^2\right]$. The $U_A(1)$ symmetry breaking lagrangian is given by $\mathcal{L}_{U_A(1)} = -\beta\{\det(M) + \det(M^\dagger)\}$. Finally we have the explicit symmetry breaking term $\mathcal{L}_{SB} = \text{tr}[c\sigma] = \text{tr}\left[\frac{b_0}{\sqrt{2}}\mathcal{M}_q(M+M^\dagger)\right]$ where $c \equiv \frac{1}{\sqrt{2}}\lambda_i c_i$, is related to the quark mass matrix $\mathcal{M}_q$ by $c = \sqrt{2}b_0\mathcal{M}_q$ and has $\frac{c_{ns}}{\sqrt{2}} = \sqrt{2}\hat{m}b_0$ and $c_s = \sqrt{2}m_s b_0$ as the only non-vanishing entries. Here, b_0 is an unknown parameter with dimensions of squared mass. We work in the exact isospin limit, $\hat{m} = m_u = m_d$. The linear σ term induces σ-vacuum transitions which rearrange the vacuum. Shifting to physical fields we obtain the following masses [8, 9, 10, 11, 12] :

$$m_\pi^2 = \xi + 2\beta b + \lambda a^2, \qquad m_{a_0}^2 = \xi - 2\beta b + 3\lambda a^2$$
$$m_K^2 = \xi + 2\beta a + \lambda(a^2 - ab + b^2), \qquad m_\kappa^2 = \xi - 2\beta a + \lambda(a^2 + ab + b^2)$$
$$m_{\eta_{ns}}^2 = \underline{\xi - 2\beta b} + \lambda a^2, \qquad m_{S_{ns}}^2 = \underline{\xi + 2\beta b} + 3\lambda a^2 + 4\lambda' a^2, \quad (3)$$
$$m_{\eta_s}^2 = \xi + \lambda b^2, \qquad m_{S_s}^2 = \xi + 3\lambda b^2 + 2\lambda' b^2$$
$$m_{\eta_{s-ns}}^2 = \underline{-2\sqrt{2}\beta a}, \qquad m_{S_{s-ns}}^2 = \underline{2\sqrt{2}(\beta + \lambda' b)a}.$$

where $a = \langle\sigma_{ns}\rangle/\sqrt{2}$, $b = \langle\sigma_s\rangle$ and $m_{\eta_{s-ns}}^2$, $m_{S_{s-ns}}^2$ denote the OZI rule violating terms mixing strange and non-strange isoscalar quarkonia. Notice that the $U_A(1)$ symmetry breaking couples to the spontaneous breakdown of chiral symmetry (underlined terms in Eq.(3)) contributing to the masses of all fields (except strange fields) and this effect has the opposite sign in the scalar sector with respect to the pseudoscalar sector. In particular it is responsible for the mixing of flavor fields. We claim that this is the striking effect which explains the unusual properties of the lowest lying scalar mesons. In [8, 9] the corresponding coupling has been fixed to $\beta = -1.551 \pm 0.072$ GeV using information on the pseudoscalar spectrum as input and the members of the scalar nonet were identified as $\sigma(\approx 450)$, $f_0(980)$, $a_0(980)$, and $\kappa(\approx 900)$. The $U_A(1)$-SBχS effect has the following consequences: It pushes the pions and kaons **down** making them light (an effect driven by the quark masses since elimination of linear terms after SBχS requires $f_\pi m_\pi^2 = 2\hat{m}b_0$; $f_K m_K^2 = (\hat{m}+m_s)b_0$) and the η_{ns} **up** making it heavy (an effect only partly driven by quark masses).ii) It pushes the a_0 and κ mesons **up** making them heavy and σ_{ns} **down** making it light. As a consequence this effect simultaneously explains at the qualitative level :i) Why the pion and kaon are light, ii) The mixing of pseudoscalar and scalar mesons. iii) Why the η_{ns} is so heavy, iv) Why the sigma meson is so light, vi) The accidental degeneracy of the $a_0(980)$ and the $f_0(980)$ (and perhaps also of the κ meson) vi) The strong coupling of the $f_0(980)$ to $\bar{K}K$. The immediate question at this point is what the model predicts for the coupling of $f_0(980)$ and $a_0(980)$ to two photons.

Consistent description of $a_0(980) \to \gamma\gamma$ and $f_0(980) \to \gamma\gamma$ decays

The $a_0(980) \to \gamma\gamma$ and $f_0(980) \to \gamma\gamma$ decays have been calculated in the present framework [13]. These decays are induced by loops of charged mesons (M). The calculation of the corresponding diagrams give *finite* contributions. The analytical results depend on the SMM couplings which are related to meson masses by chiral symmetry. The amplitude for $S \to \gamma\gamma$ decay is

$$\mathcal{M}(S \to \gamma(\varepsilon,k)\gamma(q,\eta)) = \frac{i\alpha}{\pi f_K} \mathcal{A}^S (q.k\, g^{\mu\nu} - k^\mu q^\nu) \eta_\mu \varepsilon_\nu. \quad (4)$$

and the corresponding width is $\Gamma(S \to \gamma\gamma) = \frac{\alpha^2}{64\pi^3} \frac{m_S^3}{f_K^2} |\mathcal{A}^S|^2$, whereas the measured widths are [6] $\Gamma(f_0 \to \gamma\gamma)_{exp} = 0.39^{+10}_{-13}$ keV, $\Gamma(a_0 \to \gamma\gamma)_{exp} = 0.24^{+0.08}_{-0.07}$ keV (we assume $BR(a_0(980) \to \pi^0\eta) \cong 1$). From these values we extract $|\mathcal{A}^{a_0}_{exp}| = 0.34 \pm 0.05$ and $|\mathcal{A}^{f_0}_{exp}| = 0.44 \pm 0.07$.

In the model, the $a_0 \to \gamma\gamma$ decay is induced by loops of kaons and kappas. Kaon loops yield $\mathcal{A}^{a_0}_K = 0.42$. Kappa loops interfere destructively with kaon loops and results depend on the kappa mass. Using e.g. $m_\kappa = 900$ MeV we obtain $\mathcal{A}^{a_0}_{LSM} = 0.36$. The experimental result is consistent with $m_\kappa \in [800, 935]$ MeV.

In the case of the $f_0(980) \to \gamma\gamma$ decay there are contributions from loops of pions kaons and kappas. The corresponding amplitudes as calculated in the model yield $\mathcal{A}^{f_0}_M = f_K \left(\frac{g_{f_0MM}}{m^2_{f_0}} \right) N_M$ with the loop factors $N_\pi = (-1.10 + 0.48i)$, $N_K = 1.06$ and the value of N_κ depend again on the kappa mass. For $m_\kappa = 900$ MeV we obtain $N_\kappa = 0.12$. The dominant contribution again comes from kaon loops since the $f_0\pi\pi$ coupling is proportional to the $\sin\phi_s$ factor which is small, and the $f_0\kappa\kappa$ coupling is proportional to $m_f^2 - m_\kappa^2$ which is also small relative to the typical hadron energy scale of 1 GeV. Kaon loops contributions alone (using $\phi_s = -9°$) yield $\mathcal{A}^{f_0}_K = 0.46$ whereas including all contributions (using the E791 value $m_\kappa = 797$ MeV [23]) yield $|\mathcal{A}^{f_0}_{LSM}| = 0.42$ to be compared with $|\mathcal{A}^{f_0}_{exp}| = 0.44 \pm 0.06$. Reversing the argument: fixing the value of m_κ to the central value of E791, the experimental uncertainties allow for $\phi_s \in [-4°, -25°]$. In the whole we conclude that experimental data on $a_0 \to \gamma\gamma$ and $f_0 \to \gamma\gamma$ are well described by meson loops.

Complementarity of LSM and χPT in the scalar channel: $V^0 \to P^0 P^{0\prime} \gamma$ decays.

From the experimental point of view these decays are a clean place to study $P^0 - P^{0\prime}$ systems since neutral particles are involved in the final state, hence there exist no final state radiation. Branching ratios and energy spectrum were recently measured for some of these decays by the SND and CMD collaborations at Novosibirsk and improved data can be expected from DAΦNE. On the theoretical side contributions from intermediate vector mesons $V^0 \to V^{0\prime} P^0 \to P^0 P^{0\prime} \gamma$ were estimated in [14] using VMD and the corresponding results do not describe the experimental results for the BR.'s. The possibility of enhancement of these BR's due to re-scattering effects was explored in [15] using χPT with vector mesons as external fields. At leading order ($O(p^4)$) these contributions are **finite** and no ambiguities due to counter-terms exist. VMD and chiral loops contributions are collected in a table below which shows that these contributions do not account for the measured BR's either. Since the $P^0 - P^{0\prime}$ system can be in a $J^P = 0^+$ state we should expect resonant effects due to scalars manifest in these decays.

Decay	BR$_{\text{exp}} \times 10^5$	(VC)$\times 10^5$	(χPT) $\times 10^5$
$\phi \to \pi^0\pi^0\gamma$	$11.58 \pm 0.93 \pm 0.52$	1.2	5.05
$\phi \to \pi^0\eta\gamma$	$9.0 \pm 2.4 \pm 1.0$	0.54	2.95
$\phi \to K^0\bar{K}^0\gamma$		2.7×10^{-7}	1.0×10^{-3}
$\rho \to \pi^0\pi^0\gamma$	$4.2^{+2.9}_{-2.0} \pm 1.0$	1.1	0.97
$\rho \to \pi^0\eta\gamma$		4×10^{-5}	4×10^{-6}
$\omega \to \pi^0\pi^0\gamma$	7.2 ± 2.5	2.8	9×10^{-2}
$\omega \to \pi^0\eta\gamma$		1.6×10^{-2}	1.6×10^{-4}

Vector fields were introduced as external field in the model in [16, 17, 18] and contributions of intermediate scalar mesons to these processes were calculated for the most interesting processes, namely $\phi \to \pi^0\pi^0\gamma$, $\phi \to \pi^0\eta\gamma$ and $\rho, \omega \to \pi^0\pi^0\gamma$. We refer to [16, 17, 18] for details. Here we just summarize the main results. From the theoretical point of view is worth to remark that the obtained amplitudes to these processes reduce to the $O(p^4)$ χPT amplitudes obtained in [15] in the case of heavy scalars. In this sense, LSM yields results which are complementary to χPT. In addition to catch the physics of chiral loops the LSM amplitude is also able to reproduce the effects of the scalar poles at higher $P^0 P^{0\prime}$ invariant mass values. In the case of $\phi \to P^0 P^{0\prime}\gamma$ pion loops are suppressed by G-parity, hence kaon loops give the most important contribution. The decay $\phi \to \pi^0\pi^0\gamma$ is highly sensitive to the scalar mixing angle and it is not sensitive to the σ mass whenever it be light since $g_{\sigma KK} \sim m_\sigma^2 - m_K^2$. Disastrous results are obtained for $m_\sigma > 600$ MeV. The energy spectrum for this process is nicely reproduced by the calculations in the model [16]. The calculated branching ratios, including intermediate vector meson contributions is $BR(\phi \to \pi^0\pi^0\gamma)_{\text{TH}} = 1.08 \times 10^{-4}$ which is to be compared with $BR(\phi \to \pi^0\pi^0\gamma)_{\text{EXP}} = 1.08 \pm 0.17 \pm 0.09$ measured by the CMD2 coll.[19] and $BR(\phi \to \pi^0\pi^0\gamma)_{\text{EXP}} = 1.14 \pm 0.10 \pm 0.12$ obtained by the SND Coll. [20].

In the case of $\phi \to \pi^0\eta\gamma$, the only intermediate scalar is the $a_0(980)$[17]. The corresponding spectrum is also reproduced in the model. The branching ratio calculated in the model is $B(\phi \to \pi^0\eta\gamma)_{\text{LSM}} = (0.75\text{--}0.95) \times 10^{-4}$, to be compared with the values reported by the CMD2 Collaboration $B(\phi \to \pi^0\eta\gamma)_{\text{CMD2}} = (0.90 \pm 0.24 \pm 0.10) \times 10^{-4}$ [19] and the SND result $B(\phi \to \pi^0\eta\gamma)_{\text{SND}} = (0.88 \pm 0.14 \pm 0.09) \times 10^{-4}$ [21]. Improved data near the a_0 pole will be very important in the understanding of the $a_0(980)$ meson.

In the case of $\rho \to \pi^0\pi^0\gamma$ exchange of vector mesons account for $\approx 25\%$ of the measured BR, χral loops (dominated by pions) account for another $\approx 20\%$ and the calculated BR taking into account both contributions [15] is within two standard deviations from the experimental results recently reported by the SND Coll. [21] and quoted in table above. An analysis of the energy spectrum as a function of the mass and width of the σ meson shows that this process is sensitive enough to these quantities and a measurement of the spectrum can be used to extract the corresponding values. Integrating the $\pi^0\pi^0$ invariant mass spectrum and using the central values obtained by the E791 Collaboration $m_\sigma = 478^{+24}_{-23} \pm 17$ MeV and $\Gamma_\sigma = 324^{+42}_{-40} \pm 21$ MeV we obtain $BR(\rho \to \pi^0\pi^0\gamma)_{LSM} = 1.5 \times 10^{-5}$. For $m_\sigma = 478$ MeV and a narrower width $\Gamma_\sigma = 263$ MeV (as predicted by the LSM, see also [24]) we obtain $BR(\rho \to \pi^0\pi^0\gamma)_{LSM} = 2.1 \times 10^{-5}$.

Finally, in the case of $\omega \to \pi^0\pi^0\gamma$ pion loops are also suppressed by G-parity and kaon loops should give the most important contribution. However, the intermediate scalars are $f_0(980)$ and σ and the $f_0(980)$ has a large mass compared to the energy region of interest whereas σ contributions are highly suppressed since $g_{\sigma KK} \sim m_\sigma^2 - m_K^2$. Hence, whenever the mass of the sigma meson be small, scalar effects are negligible in this process, although interference with the intermediate vector meson contributions seem to close the gap between the experimental results $BR(\omega \to \pi^0\pi^0\gamma)_{EXP} = (7.2 \pm 2.5) \times 10^{-5}$ [6] and the VM contributions when we take care of including properly the $\omega - \rho$ mixing [18] which yield $BR(\omega \to \pi^0\pi^0\gamma)_{TH} = 4.5 \pm 1.1 \times 10^{-5}$ [18] (see also [25]).

Conclusions.

Summarizing, calculations for $S \to \gamma\gamma$ ($S = a_0(980), f_0(980)$) considering S as a NR-quarkonium state yield results which are not consistent with experimental data. The same is true in the case of pseudoscalars. In this case the disagreement can be traced back to the strong distortion of the spectrum from naive quarkonium expectations due to QCD vacuum effects. Since scalars have the same quantum numbers as the vacuum, highly non-trivial effects are expected in this sector due to vacuum properties. We study such effects in the framework of a phenomenological $U(3) \times U(3)$ chiral lagrangian which incorporates spontaneous χS and $U_A(1)$ symmetry breaking. In this framework there is an important effect: the coupling of the $U_A(1)$ violating interaction to the spontaneous breaking of chiral symmetry which generates mass terms ($U_A(1)$-SBχS effect). This effect simultaneously explains: the smallness of the masses of the pions and kaons (an effect driven by quark masses), the OZI-rule violating mixing of flavor fields, the accidental degeneracy of the $a_0(980)$ and $f_0(980)$, the lightness of the sigma meson and the controversial $a_0, f_0 \to \gamma\gamma$ decays. Intermediate scalar contributions to $V^0 \to P^0 P^{0'} \gamma$ are also calculated. The energy spectrum of $\phi \to \pi^0\pi^0\gamma$ is nicely described giving direct evidence for the assignment of the $f_0(980)$ and indirect evidence for a light σ. Calculations for total and partial BR's are in good agreement with measurements from SND and CMD collaborations. Energy spectrum in $\phi \to \pi^0\eta\gamma$ is well described by the chiral loops + scalar resonant effects. Improvement in the measurements near the a_0 pole are encouraged. We also calculate the $\rho \to \pi^0\pi^0\gamma$ decay. The corresponding BR is sensitive to the parameters of the σ. The branching ratio as measured by the SND Coll. is consistent with the results of the model when the values for the mass and width of this scalar as measured by the E791 Collaboration are used. Measurements of the energy spectrum are encouraged. Finally, $\omega \to \pi^0\pi^0\gamma$ is not sensitive to scalar contributions.

Acknowledgements

I wish to thank J. L Lucio, S. Rodriguez, A. Bramon, R. Escribano, M. Kirchbach and A. Wirzba for an enjoyable collaboration. This work was supported by CONCYTEG, Mexico under contract 00-16-CONCYTEG-CONACYT-075.

REFERENCES

1. See e.g. N. A. Tornqvist, "Summary talk of the conference on the sigma resonance," Kyoto, Japan, June (2000). arXiv:hep-ph/0008135. N. A. Tornqvist, "Does the light and broad sigma(500) exist?," arXiv:hep-ph/9904346 and references therein.
2. R. L. Jaffe, The Phenomenology Of (2 Quark 2 Anti-Quark) Mesons," Phys. Rev. D **15**, 267 (1977). R. L. Jaffe, Phys. Rev. D **15**, 281 (1977).
   M. D. Scadron, Phys. Rev. D **26**, 239 (1982).
   E. Van Beveren, T. A. Rijken, K. Metzger, C. Dullemond, G. Rupp and J. E. Ribeiro, Z. Phys. C **30**, 615 (1986).
   S. Ishida, M. Ishida, T. Ishida, K. Takamatsu and T. Tsuru, Prog. Theor. Phys. **98**, 621 (1997) [arXiv:hep-ph/9705437].
   D. Black, A. H. Fariborz, F. Sannino and J. Schechter, Phys. Rev. D **58**, 054012 (1998) [arXiv:hep-ph/9804273].
   J. A. Oller and E. Oset, Phys. Rev. D **60**, 074023 (1999) [arXiv:hep-ph/9809337].
3. S. N. Cherry and M. R. Pennington, Nucl. Phys. A **688**, 823 (2001) [arXiv:hep-ph/0005208]. S. Cherry and M. R. Pennington, arXiv:hep-ph/0007275.
4. T. Barnes, Phys. Lett. B **165**, 434 (1985).
5. J. H. Kuhn, J. Kaplan and E. G. Safiani, Nucl. Phys. B **157**, 125 (1979). J. G. Korner, J. H. Kuhn, M. Krammer and H. Schneider, Nucl. Phys. B **229**, 115 (1983). R. Barbieri, R. Gatto and R. Kogerler, Phys. Lett. B **60**, 183 (1976).
6. D. E. Groom *et al.* [Particle Data Group Collaboration], Eur. Phys. J. C **15**, 1 (2000).
7. C. Hayne and N. Isgur, Phys. Rev. D **25**, 1944 (1982).
   E. S. Ackleh and T. Barnes, Phys. Rev. D **45**, 232 (1992).
8. M. Napsuciale, arXiv:hep-ph/9803396.
9. M. Napsuciale, A. Wirzba and M. Kirchbach, arXiv:nucl-th/0105055, to appear in Nucl. Phys. A.
10. M. Napsuciale and S. Rodriguez, Int. J. Mod. Phys. A **16**, 3011 (2001).
11. N. A. Tornqvist, Eur. Phys. J. C **11**, 359 (1999) [arXiv:hep-ph/9905282].
12. G. 't Hooft, arXiv:hep-th/9903189.
13. J. L. Lucio Martinez and M. Napsuciale, Phys. Lett. B **454**, 365 (1999) [arXiv:hep-ph/9903234].
14. A. Bramon, A. Grau and G. Pancheri, Phys. Lett. B **283**, 416 (1992).
15. A. Bramon, A. Grau and G. Pancheri, Phys. Lett. B **289**, 97 (1992).
16. J. L. Lucio and M. Napsuciale, *Proceedings of the III International Workshop on Physics and Detectors for DaΦne*, edited by S. Bianco *et al.* (Frascati Physics Series Vol. XVI 2000), p. 591 arXiv:hep-ph/0001136.
17. A. Bramon, R. Escribano, J. L. Lucio M., M. Napsuciale and G. Pancheri, Phys. Lett. B **494**, 221 (2000) [arXiv:hep-ph/0008188].
18. A. Bramon, R. Escribano, J. L. Lucio Martinez and M. Napsuciale, Phys. Lett. B **517**, 345 (2001) [arXiv:hep-ph/0105179].
19. R. R. Akhmetshin *et al.* [CMD-2 Collaboration], Phys. Lett. B **462**, 380 (1999) [arXiv:hep-ex/9907006].
20. V. M. Aulchenko *et al.* [SND Collaboration], Phys. Lett. B **440**, 442 (1998) [arXiv:hep-ex/9807016].
21. M. N. Achasov *et al.*, JETP Lett. **71**, 355 (2000) [Pisma Zh. Eksp. Teor. Fiz. **71**, 355 (2000)].
22. E. M. Aitala *et al.* [E791 Collaboration], Phys. Rev. Lett. **86**, 770 (2001) [hep-ex/0007028].
23. C. Gobel [E791 Collaboration], arXiv:hep-ex/0110052.
24. N. Wu, arXiv:hep-ex/0104050.
25. D. Guetta and P. Singer, Phys. Rev. D **63**, 017502 (2001) [arXiv:hep-ph/0005059].

The structure of the proton as seen with the H1 detector

J. G. Contreras

Departamento de Física Aplicada, Cinvestav, Mérdia; México

Abstract. This is the text of a lecture hold during the VIII Mexican School on Particles and Fields. The first part of the lecture has been aimed to advance undergraduate students with little previous knowledge of deep inelastic scattering and it is quite elementary. The second part presents recent published results from the H1 Collaboration related to the structure of the proton.

INTRODUCTION

H1 is the name of a huge detector, composed itself by several big subdetectors, but also of the collaboration of scientists who designed, built and maintain this apparatus. The detector is located in one of the four colliding points in the accelerator HERA which is part of the laboratory DESY in the city of Hamburg in Germany.

The H1 Collaboration

The collaboration was founded in the 80's and it took several years to design, test and built the detector. It was late in 1992 that H1 took the first data and it has being doing so since then. It has undergone several relatively minor upgrades and during 2000 and 2001 it has been substantially upgraded. Now it is ready to take data again.

Some 40 institutes and 400 scientists conform the H1 Collaboration. It has published more than 100 papers in international journals and have been cited several thousand times. It also has provided topics for several hundred PhD and Diploma thesis and it is awarded plenary talks in every major meeting on High Energy Physics.

One of the main topics of investigation in H1 is the study of the structure of the proton. As it will be shown in the second part of the talk, this structure is dynamical, that is the components of the proton are not just there as the walls of a house, but are moving around like the clouds in the sky. This characteristic makes the study of this structure quite complicated, but also quite rich. H1 is, along with a few other big collaborations, an important player in the development of the field from the experimental side. For more information regarding the collaboration and the detector look up the page [1]

The structure of the matter

At this point you may be wondering: So many people, such a long time, so much money (I have not told you, but this kind of enterprises costs lots and lots of money!), such a big effort, why? Well, to study the innermost structure of matter. Oh!, ... and why, and why protons? The answer to the first question is more of a philosophical nature and best discussed in some other place, but the second one will be answered in the following.

Since a long time mankind has been struggling with two questions: What are the basic building blocks of matter? and how are they held together to form all what we see?. There have been several different answers to them along the centuries, but it is quite recently that the answers were based on experimental knowledge more than on philosophical grounds. Since then the fundamental level has been reaching smaller and smaller sizes. From molecules to atoms and then their nuclei and the protons and neutrons inside them. Nowadays it is known that protons, neutrons and a plethora of similar particles are formed by quarks and gluons, which along with the electron and its cousins (all of them named leptons) and the photon and its cousins (all of them named Gauge Bosons) form the most fundamental level of the structure of matter. But this may not be the final answer and it could well be that there will be some sort of substructure found in future experiments.

So the current answers, as embodied in the so called Standard Model, are: what are the basic building blocks of matter? Quarks and Leptons. How are they held together to form all what we see? By Gauge Bosons. There are six types of quarks: up (u) down (d), strange (s), bottom (b) and top (t); and also six kinds of leptons: electron (e), muon (μ), tau (τ) and three neutrinos (ν_e, ν_μ, ν_τ). All of the quarks and leptons also have antiparticles. There are gauge bosons for each of the known forces: photons (γ) for electromagnetism, Z^0 and $W^\pm$ for the weak force (responsible for example for the radioactive decay of some nuclei) and gluons for the strong force (also called QCD) which holds nuclei together.

The study of the structure of matter has moved thus, from molecules to the inside of protons and neutrons. The behavior of the objects there, the quarks and gluons, is described by the part of the Standard Model called QCD. QCD stands for quantum chromo dynamics, because the quarks and the gluons are quantum fields, they move inside the proton and carry a charge (similar to the electric charge) called whimsically *color*.

DIS

Now the procedure should be straight forward: take the Lagrangian of QCD, obtain the equations of motion, solve them and compare to experiments to see if this is the right theory. But there is a problem. On the one hand, free colored fields are not allowed in QCD and in the other hand it is only known how to solve the free theory of QCD or perturbative approximations of it (pQCD).

What to do then? Take an object formed by colored fields (for example the proton), go to a corner of phase space where perturbative approximation is valid (as in Deep

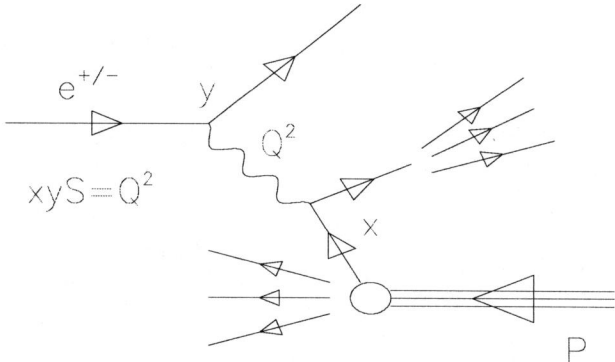

FIGURE 1. Schematic representation of a DIS process. The incoming lepton $e^{\pm}$ interacts with the hadron P through a gauge boson of virtuality Q^2 which is scattered off a quark inside the proton carrying momentum x. The total available energy in the CMS is S.

Inelastic Scattering, DIS), perform experiments and try to interpret them within pQCD (using for example HERA and H1).

The basic idea of DIS is depicted in figure 1. An elementary probe, in this case a high energy electron is the source of the light of the DIS microscope. The high energy is needed because there is a direct relation between it and the momentum of the light. And in quantum mechanics high momentum means high resolution, which is needed to see inside the target. The light in the case of HERA can be a photon, but also a Z or a $W^{\pm}$. In this case the outgoing lepton is a neutrino ν_e (this is called charged current DIS), whereas in the former case it is an electron (called neutral current DIS). The target must be formed by quasi free colored objects. In the case of HERA it is a proton moving at great velocity. Finally one needs a detector to study the result of the interaction.

The DIS process can be described using two independent variables. It is common to use Q^2, the square of the momentum of the light of the microscope and x, the fraction of the energy of the target seen by the light with resolution Q^2. One also uses the square of total energy available for the process in the center of mass system S (this is fixed by the accelerator) and the inelasticity variable y related to the others through $xyS = Q^2$. Note that $0 \leq x, y \leq 1$.

HERA

HERA is a source, provides storage and makes collisions of high energy electrons and protons for DIS. It is in a tunnel of 6.3 km of circumference and 20–30 meters below ground. It houses four experiments: HERA–B which is under construction and will study CP violation in the B system; HERMES, which studies the polarized structure of the proton; and ZEUS and H1 which are universal detectors to study ep interactions.

At HERA the protons are accelerated up to 920 GeV and the electrons to 27.5 GeV. Since the beginning of data taking HERA luminosity (a measure of the efficiency of the

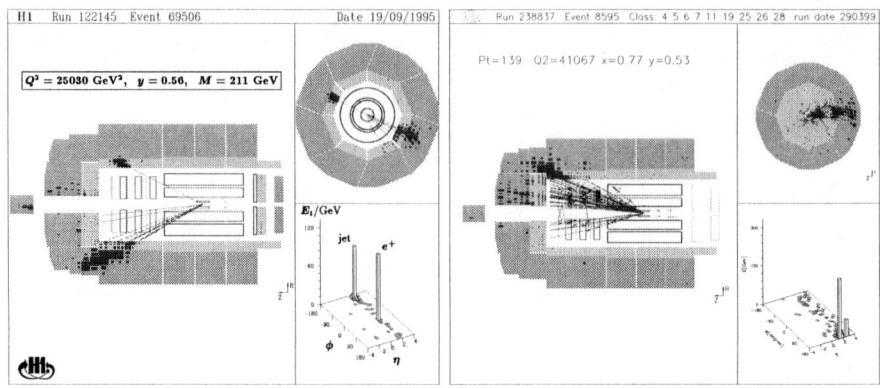

FIGURE 2. Views of NC (left) and CC (right) DIS processes in the H1 detector.

machine) has been 150 pb^{-1} for each experiment ZEUS and H1. From now on, thanks to a recent upgrade of the accelerator, there will be for each detector 150 pb^{-1} each year. This is a huge increase and, as it will be shown, opens the door to new exciting studies. This new era is called HERA II.

The main characteristic of HERA is its high energy, never before reached. This is reflected in the kinematic plane available to do experiments. In x HERA covers from 10^{-6} to almost 1, and in Q^2 from 10^{-1} Gev^2 to several times 10^4. This is an increase in several orders of magnitude with respect to previous experiments in both variables.

The H1 detector

H1 is a huge detector of some more than 10 meters in each dimension and several thousand tons of weight. It is built like an onion with layers of different subdetectors which combined allow the determination of the momentum, energy and in some cases type of particle to be measured [2].

Figure 2 shows standard DIS events as seen with the H1 detector [1, 3]. The incoming proton reaches the detector from the right of the longitudinal view and the electron enters from the left. One can distinguish the scattered electron and broken proton in the first panel and in the second one the missing transverse energy, signaling an undetected particle (assumed to be a v_e in this case).

THE STRUCTURE OF THE PROTON

The next step is to have an idea of the structure of the proton and confront it with experiments. This will be presented in steps, starting from the most general to more specific features of this structure. First the neutral current (NC) channel will be explored, then the charged current (CC) interactions and finally some comments on exclusive processes will be made.

NC DIS in pQCD

The idea obtained from pQCD for the differential cross section of NC DIS is expressed in the following formula (for an explanation of all the symbols look for example [4]):

$$\frac{d\sigma_{NC}^{e^{\pm}p}}{dxdQ^2} = \frac{2\pi\alpha^2}{xQ^4}[Y_+\tilde{F}_2(x,Q^2) \mp Y_-x\tilde{F}_3(x,Q^2) - y^2\tilde{F}_L(x,Q^2)] \quad (1)$$

here

$$\tilde{F}_2 = F_2 - v_e\frac{\kappa_W Q^2}{Q^2+M_Z^2}F_2^{\gamma Z} + (v_e+a_e)(\frac{\kappa_W Q^2}{Q^2+M_Z^2})^2 F_2^Z \quad (2)$$

and

$$x\tilde{F}_3 = -a_e\frac{\kappa_W Q^2}{Q^2+M_Z^2}xF_3^{\gamma Z} + (2v_e a_e)(\frac{\kappa_W Q^2}{Q^2+M_Z^2})^2 xF_3^Z \quad (3)$$

There are several things to notice in equation 1. The left side of it is what it is measured, the right side is the description based in a given model; in this case pQCD. There are several F_i in this formula. The goal is to extract all from experimental data. They are called structure functions and describe the content of the proton in terms of probability distribution of quarks and gluons.

Note the dependence on Q^2. remember that this is quantum mechanics, so only probabilities are measured. This means that there are statistical fluctuations in the measurements. To keep them under control it is needed to have many events, but this is more difficult for higher Q^2, due to the Q^{-4} factor, precisely where the resolution is better and the data most interesting!. So one expects a more difficult measurement at high Q^2.

Note also that for $\tilde{F}_2$ there are three terms corresponding to the interchange of a photon, F_2, of a Z boson, F_2^Z, and the interference between these two cases, $F_2^{\gamma Z}$. For $\tilde{F}_2$ there exist only the last two terms, i.e., there is no direct *only* γ contribution.

The question is how to extract the structure function from the measurements. This is done taking into account their different behavior under variations of the kinematic variables. For example F_2 has no coefficient depending on Q^{-2} nor Q^{-4} as $F_2^{\gamma Z}$, F_2^Z and all F_3 have. There is also the x coefficient to $\tilde{F}_3$ and the difference between e^+p and e^-p data which gives a handle to extract $\tilde{F}_3$. Finally there is the y^2 factor which allows access to F_L.

First let's start with the cross section (see [4] and reference therein). The first panel of figure 3 shows NCDIS data against Q^2. This is not all the data used to extract the structure functions. There are more, but due to restrictions of time and space in this talk I concentrate only in the high Q^2 data. The cross section falls seven orders of magnitude. This means that with the same apparatus one is able to measure precisely over seven

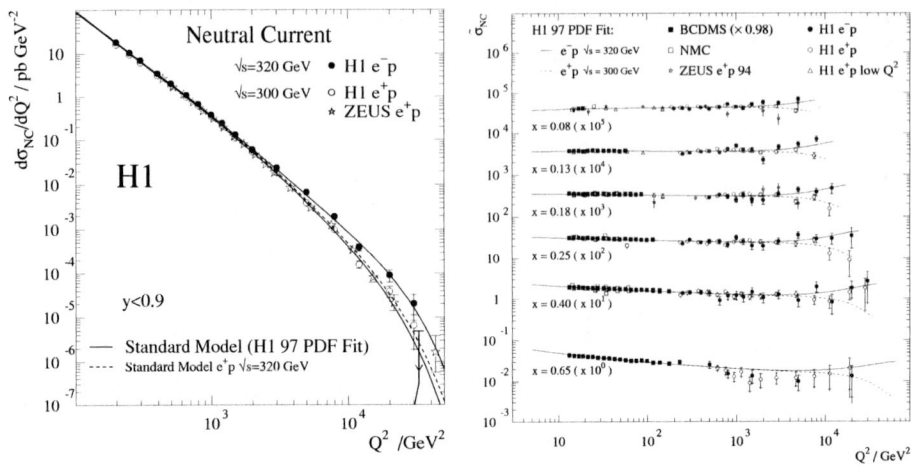

FIGURE 3. dependence of the NC DIS cross section on Q^2 (left) and on Q^2 for different fixed values of x (right).

orders of magnitude. This is no small feat! Notice that H1 and Zeus data agree and that formula 1 describes this data with its characteristic Q^{-4} drop.

The second panel of figure 3 shows the same data in more detail. Here the data is divided in strips of fixed x. In addition to the previous observations one notices the agreement also with fixed target experiments. Also here it is quite clear to see the difference at high Q^2 for electron and positron data as expected from equation 1 and leads to the interpretation of the existence of the interference and Z terms.

Figure 4 shows the structure function F_2 extracted from the data (see [4], [5], [6] and references therein). assuming the validity of equation 1. This is very precise data over several orders of magnitude in x and in Q^2. All experiments agree and the model, represented by the solid line, describes perfectly the data. Notice the change of slope with x. This is linked to the dynamics of the quarks and gluons inside the proton. They are not static, but mix, die and are born in a continuous way throughout the life of the proton. This is a fascinating subject, which unfortunately I do not have time to dwell upon.

Next, in the first panel of figure 5, the extraction of $\widehat{F}_3$ is shown for selected Q^2 values [4]. Here the ability of HERA to run with electrons and positrons is exploited using:

$$x\widehat{F}_3 = \frac{Y_+}{2Y_-}(\widehat{\sigma}_{NC}^- - \widehat{\sigma}_{NC}^+) \tag{4}$$

where $\widehat{\sigma}$ includes all the relevant kinematic factors and the high y data is excluded to be able to neglect F_L. Note that indeed one can extract this structure function, albeit still with big error bars. This will be improved rapidly using the new power of the upgraded HERA machine.

In the second panel the extracted F_L data is shown. This is new, preliminary H1 data [7]. Here one looks with special interest into the high y region where, due to the y^2 factor

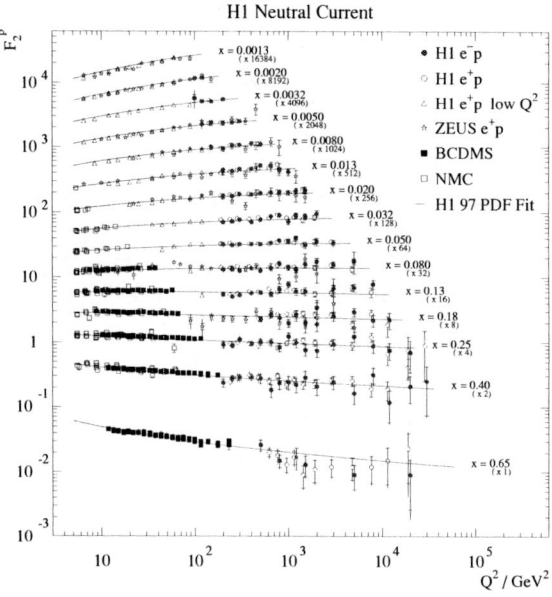

FIGURE 4. Compilation of F_2 data extracted from NC DIS processes.

in equation 1 and the fact that y is bounded by zero and one, one expects the influence of the existence of F_L to be maximal in the cross section. The extractions done with electron and positron data are consistent. Again the measurement is still dominated by statistics, so wait for HERA II!. Notice also that F_L is directly related to the gluon in the

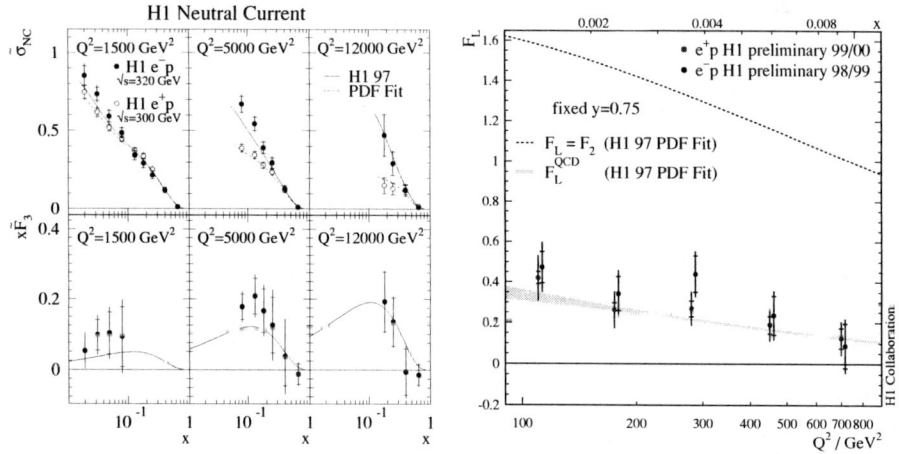

FIGURE 5. Extraction of $x\widetilde{F}_3$ (left) and F_L (right) structure functions from NC DIS data.

proton making this a quite important and exciting measurement.

CC DIS in pQCD

Up to know we have obtained information about the structure functions but not directly about the quarks and gluons in the proton. This means we still need a model of how the quarks and gluons conform each structure function. It is interesting to look at data which gives information directly on the parton level. One step in this direction was F_L which depends heavily on the gluon. Now we will look at charged current DIS which allows to tag the valence u_v and d_v distributions according to:
where

$$\tilde{\sigma}_{CC}^+ = x[(u+c) + (1-y)^2(\bar{d}+\bar{s})] \approx xu_v \tag{5}$$

and

$$\tilde{\sigma}_{CC}^- = x[(\bar{u}+\bar{c}) + (1-y)^2(d+s)] \approx (1-y)^2 xd_v \tag{6}$$

$$\frac{d\sigma_{CC}^{e^\pm p}}{dxdQ^2} = \frac{G_F^2}{2\pi x}\left(\frac{M_W^2}{Q^2+M_W^2}\right)^2 \tilde{\sigma}_{CC}^\pm \tag{7}$$

First let's look at the cross section and afterwards at the quark distributions. Figure 6 shows the CC DIS cross section as function of Q^2 for electron and positron data (see [4] and references therein). There is again agreement between both experiments, ZEUS and H1, and the model given by equation 7. The errors are very small over almost all the kinematic range. The cross section from electron data is bigger than that from positron

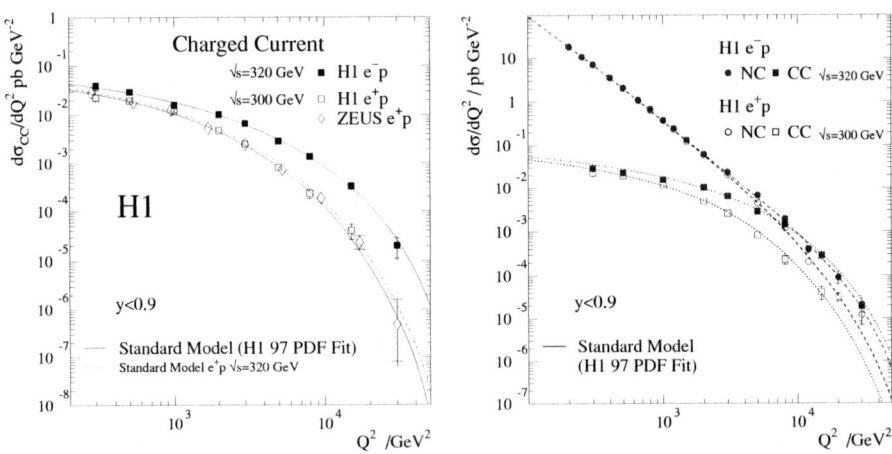

FIGURE 6. Dependence of CC DIS data on Q^2 (left) and comparison to NC DIS data (right).

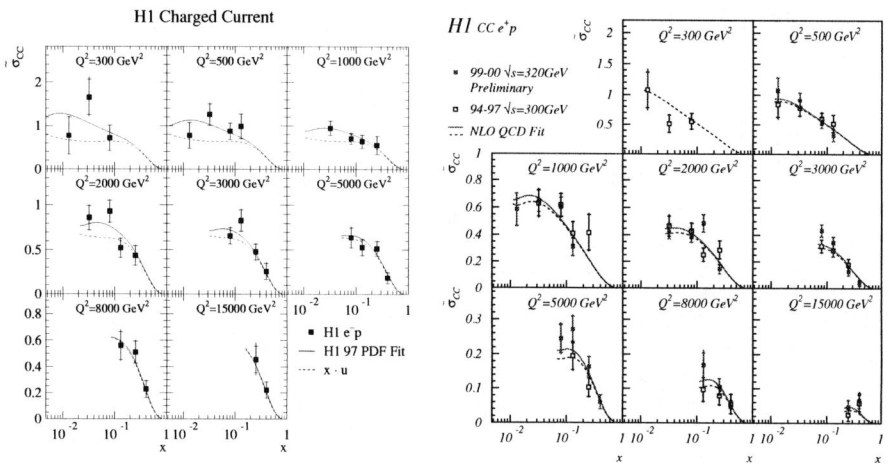

FIGURE 7. CC DIS data for e^-p (left) and e^+p (right) collisions.

data supporting the model of two u quarks of valence inside the proton for one d valence quark. In the same figure a comparison is made with the NC DIS cross section data. Remember that a cross section is a measure of the strength of the force responsible for the interaction. Notice how at small Q^2 the two cross sections are quite different, representing the fact that the electromagnetic force is quite stronger than the weak force at 'big' scales. But as Q^2 grows and the resolution improves one notice that both forces approach each other until they became only one force. This is electroweak unification. Really a plot for the textbooks!

The two panels of figure f:ccxq2 show the x dependence of the CC DIS cross section for positron and electron data separately for different fixed Q^2 values (see [4], [5] and for the H1 preliminary data in the left panel [8]). Again the model of formula 7 describes the data as shown by the solid line. Also shown, with the dashed line, is the contribution, again according to equation 7, of the d_v and u_v quarks to the cross section. One notice that specially at high x they form almost all the cross section as expected from pQCD.

CONCLUSIONS AND OUTLOOK

There are still many different ways to extract information on the structure of the proton from experiment. For example to look specifically at processes where a given quark flavor is identified, for example charm. Or processes which according to the model have to be mediated by a gluon as in two jet events. Unfortunately there is lack of time and space to discussed them all.

Here I would like to conclude with the same questions that I began: So many people, such a long time, so much money, such a big effort, why? I hope that now the answer is a little clearer: To study the innermost structure of matter, and the your reaction, kind reader, would be along the lines: Oh! ... That sounds like fun, let's try with protons!

Acknowledgements This work was done with partial support from Conacyt, México.

REFERENCES

1. http://www-h1.desy.de/h1/www/general/home/home.html
2. H1 Collab. I. Abt et al., Nucl. Inst. Meth. A **386** (1997) 310 and 348.
3. H1 Collab. C. Adloff et al., Z. Phys. C **74** (1997) 191–205.
4. H1 Collab. C. Adloff et al., Eur. Phys. J. C **19** (2001) 269–288.
5. H1 Collab. C. Adloff et al., Eur. Phys. J. C **13** (2000) 609–639.
6. H1 Collab. C. Adloff et al., Nucl. Phys. B **470** (1996) 3–38.
7. Eckstein, D., for the H1 Collaboration, *A New Measurement of the Deep Inelastic Scattering Cross Section and of F_L at Low Q^2 and Bjorken-x at HERA* to appear in the proceedings of the 9th International Workshop on Deep Inelastic Scattering, 27 April - 1 May, Bologna, Italy.
8. Dubak, A., for the H1 Collaboration, *High Q^2 NC and CC cross sections from H1* to appear in the proceedings of the 9th International Workshop on Deep Inelastic Scattering, 27 April - 1 May, Bologna, Italy.

Electroweak baryogenesis with primordial hypermagnetic fields

Alejandro Ayala[1], Jaime Besprosvany[2], Gabriel Pallares[1], Gabriella Piccinelli[3]

[1]*Instituto de Ciencias Nucleares, Universidad Nacional Autónoma de México, Apartado Postal 70-543, México Distrito Federal 04510, México.*
[2]*Instituto de Física, Universidad Nacional Autónoma de México, Apartado Postal 20-364, México Distrito Federal 01000, México.*
[3]*Centro Tecnológico Aragón Universidad Nacional Autónoma de México Avenida Rancho Seco S/N, Bosques de Aragón, Nezahualcóyotl Estado de México 57130, México.*

Abstract. Primordial magnetic fields, independently of their origin, could have had a significant influence over several physical processes that took place during the evolution of the early universe, in particular baryogenesis. Recall that for temperatures above the electroweak phase transition ($T > 100$ GeV), the symmetry of the standard model corresponded to the $U(1)_Y$ hypercharge group, instead of the $U(1)_{em}$ electromagnetic group and are therefore properly called hypermagnetic fields. In this work, we show that during a first order electroweak phase transition, the presence of hypermagnetic fields produces an axial charge segregation in the reflection and transmission of fermions off the true vacuum bubbles. We also comment on the possible consequences that these processes have for the generation of baryon number during the phase transition.

INTRODUCTION

The explanation of the observed excess of baryons over antibaryons in the universe represents one of the most outstanding problems of particle physics as applied to cosmology. A theory aimed to explain such excess has to meet the three well-known Sakharov conditions[1], namely: (1) Existence of interactions that violate baryon number; (2) C and CP violation and (3) departure from thermal equilibrium. The above conditions are met in the standard model (SM) provided the electroweak phase transition (EWPT) is of first order. This has raised the interesting possibility that the cosmological phase transition that gave rise to the mass of particles, which took place at temperatures of order 100 GeV, could also explain the generation of baryon number. Consequently, a great deal of effort has been devoted to explore this possibility [2].

It is nowadays widely accepted that the minimal SM, as such, cannot explain the observed baryon number. The reason is that the EWPT turns out to be only too weakly first order which in turn implies that any baryon asymmetry generated at the phase transition was erased by the same mechanism that produced it, *i.e.*, sphaleron induced processes [3]. Moreover, the amount of *CP* violation coming from the CKM matrix alone cannot account by itself for the observed asymmetry, given that its effect shows up in the coupling of the Higgs with fermions at a high perturbative order [4], producing a baryon to entropy ratio at least ten orders of magnitude smaller than the observed one.

Nevertheless, it has been recently pointed out that, provided a source of enough *CP* violation exists, the above scenario could significantly change in the presence of large-scale primordial magnetic fields [5, 6, 7] (see however Ref. [8]), which can be responsible for a stronger first-order EWPT. This situation is analogous to the case of a type I superconductor in which the presence of an external magnetic field modifies the order of the phase transition due to the Meissner effect. Though the nature of these fields is a subject of current research, their existence prior to the EWPT epoch cannot certainly be ruled out [9].

Independently of their origin, primordial fields could have had some influence on physical processes which occurred in the early universe, like big-bang nucleosynthesis and electroweak baryogenesis.

Recall that for temperatures above the EWPT, the $SU(2) \times U(1)_Y$ symmetry is restored and the propagating, non-screened vector modes that represent a magnetic field correspond to the $U(1)_Y$ group instead of to the $U(1)_{em}$ group, and are therefore properly called *hypermagnetic* fields.

In this paper we describe a simple model to show that the presence of such fields also provides a mechanism, working in the same manner as the existence of additional *CP* violation within the SM, to produce an axial charge segregation during the EWPT. This happens in the scattering of fermions off the true vacuum bubbles nucleated during the phase transition and is a consequence of the chiral nature of the fermion coupling to hypermagnetic fields in the symmetric phase. We use these results to look out at the possible implications of such axially asymmetric fermion reflection and transmission for baryogenesis. A more detailed discussion can be found in Ref. [10]

FERMIONS MOVING IN A BACKGROUND HYPERMAGNETIC FIELD

In a first order phase transition, the conversion from one phase to another happens through nucleation. The region separating both phases is called the wall. During the EWPT, the properties of the wall depend on the effective, finite temperature Higgs potential. Under the assumption that the wall is thin and that the phase transition happens when the energy densities of both phases are degenerate, it is possible to find a one-dimensional analytical solution for the Higgs field ϕ called the *kink*. This is given by

$$\phi(z) \sim 1 + \tanh(z/\lambda), \qquad (1)$$

where z is the coordinate along the direction of the phase change and λ is the width of the wall. When scattering is not affected by diffusion, the problem of fermion reflection and transmission through the wall can be casted in terms of solving the Dirac equation with a position dependent fermion mass, proportional to the Higgs field [11]. Let us further simplify the problem by considering the limit when the width of the wall approaches zero. In this case, the kink solution becomes a step function, $\Theta(z)$, and consequently, the expression for the particle's mass becomes

$$m(z) = m_0 \Theta(z). \qquad (2)$$

In terms of Eq. (2), we can see that $z \leq 0$ represents the region outside the bubble, that is the region in the symmetric phase where particles are massless. Conversely, for $z \geq 0$, the system is inside the bubble, that is in the broken phase and particles have acquired a finite mass m_0.

In the presence of an external magnetic field, we need to consider that fermion modes couple differently to the field in the broken and the symmetric phases. We start the analysis looking at the unbroken phase.

For $z \leq 0$, the coupling is chiral. Let

$$\Psi_R = \frac{1}{2}(1+\gamma_5)\Psi \tag{3}$$

$$\Psi_L = \frac{1}{2}(1-\gamma_5)\Psi \tag{4}$$

represent, as usual, the right and left-handed chirality modes for the spinor Ψ, respectively. Then, the equations of motion for these modes, as derived from the electroweak interaction Lagrangian, are

$$(i\slashed{\partial} - \frac{y_L}{2}g'\slashed{A})\Psi_L - m(z)\Psi_R = 0 \tag{5}$$

$$(i\slashed{\partial} - \frac{y_R}{2}g'\slashed{A})\Psi_R - m(z)\Psi_L = 0, \tag{6}$$

where $y_{R,L}$ are the right and left-handed hypercharges corresponding to the given fermion, respectively, g' the $U(1)_Y$ coupling constant and we take $A^\mu = (0, \mathbf{A})$ representing a, not as yet specified, four-vector potential having non-zero components only for its spatial part, in the rest frame of the wall.

We now turn to the corresponding equation in the symmetry-broken phase. For $z \geq 0$ the coupling of the fermion with the external field is through the electric charge e and thus, the equation of motion is simply the Dirac equation describing an electrically charged fermion in a background magnetic field, namely,

$$\{i\slashed{\partial} - eA_\mu \gamma^\mu - m(z)\}\Psi = 0. \tag{7}$$

To solve Eqs. (5), (6) and (7) with a constant magnetic field pointing along the $\hat{z}$ direction, we look for the scattering states appropriate to describe the motion of fermions in the symmetric and symmetry-broken phases. For our purposes, these are fermions incident towards and reflected from the wall in the symmetric phase. There are two types of such solutions; those coupled with y_L and those coupled with y_R. For an incident wave coupled with y_L (y_R), the fact that the differential equation mixes up the solutions means that the reflected wave will also include a component coupled with y_R (y_L). In analogy, the solution to Eq. (7) is found by looking for the scattering states appropriate for the description of transmitted waves. The solutions are explicitly constructed in Ref. [10] to where we refer the reader. Here, we use such solutions to construct the transmission and reflection probabilities.

REFLECTION AND TRANSMISSION PROBABILITIES

To quantify the asymmetry, we need to compute the corresponding reflection and transmission coefficients. These are built from the reflected, transmitted and incident currents of each type. Recall that for a given spinor wave function Ψ, the current normal to the wall is given by

$$J = \Psi^\dagger \gamma^0 \gamma^3 \Psi. \tag{8}$$

The reflection and transmission coefficients, R and T, are given as the ratios of the reflected and transmitted currents, to the incident one, respectively, projected along a unit vector normal to the wall.

The probabilities for finding a left or a right-handed particle in the symmetric phase after reflection, PR_L, PR_R are given, respectively by

$$PR_L = R_{L \to L} + R_{R \to L} \tag{9}$$

$$PR_R = R_{L \to R} + R_{R \to R}, \tag{10}$$

whereas the probabilities for finding a left or a right-handed particle in the symmetry broken phase after transmission, PT_L, PT_R are given, respectively by

$$PT_L = T_{L \to L} + T_{R \to L} \tag{11}$$

$$PT_R = T_{L \to R} + T_{R \to R}. \tag{12}$$

Figure 1 shows the probabilities PR_L and PR_R as a function of the magnetic field parametrized as $B = bT^2$ for a temperature $T = 100$ GeV, a fixed $E = 184$ GeV and for a fermion taken as the top quark with a mass $m_0 = 175$ GeV, $y_R = 4/3$, $y_L = 1/3$ and for a value of $g' = 0.344$, as appropriate for the EWPT epoch. Notice that when $b \to 0$, these probabilities approach each other and that the difference grows with increasing field strength. Also, in order to be able to safely neglect the contribution from the negative energy solutions, we are bound to consider not too large values of the parameter b. For the purposes of this work, we take a maximum value of $b = 1$ which for the values of T and m_0 considered, amounts for a maximum fraction of the magnetic energy to the particle's rest mass of order $\sqrt{eB}/m_0 \sim 0.3$.

Figure 2 shows the reflection and transmission probabilities as a function of the particle's energy E. Figure 2a shows the probabilities PR_L and PT_L and Fig. 2b the probabilities PR_R and PT_R for $b = 1$. As before, the parameters considered correspond to a top quark. Since the solutions are computed assuming that the transmitted waves are not exponentially damped, the energy has to be taken such that $E \geq \sqrt{m_0^2 + eB}$. It can be numerically checked that $PR_L + PT_L = PR_R + PT_R = 1$ to within a maximum deviation of one part in one thousand. The fact that these probabilities add up to one is equivalent to current conservation.

CONCLUSIONS AND OUTLOOK

In the symmetric phase, fermions couple chirally to the magnetic field, which receives the name of *hypermagnetic*, given that it belongs to the $U(1)_Y$ group. We have shown

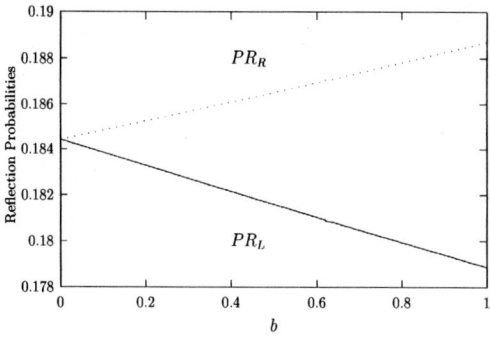

FIGURE 1. Probabilities PR_L and PR_R as a function of the magnetic field parametrized as $B = bT^2$ for $T = 100$ GeV, $E = 184$ GeV and a top quark with a mass $m_0 = 175$ GeV, $y_R = 4/3$, $y_L = 1/3$. The value for the $U(1)_Y$ coupling constant is taken as $g' = 0.344$, corresponding to the EWPT epoch.

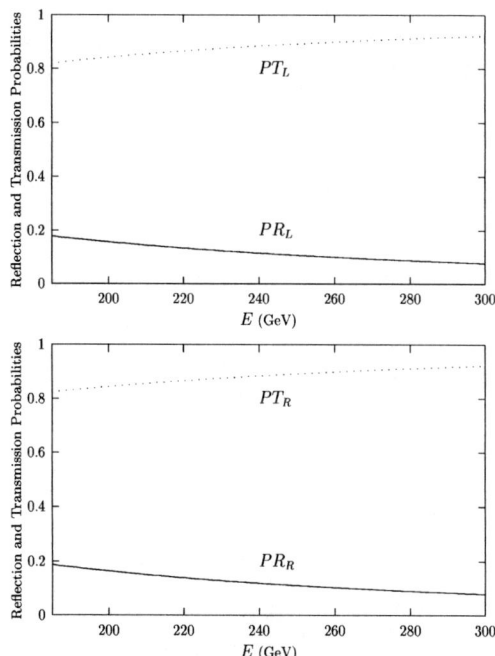

FIGURE 2. Reflection and transmission probabilities as a function of the particle's energy E. Figure 2a (upper panel) shows the probabilities PR_L and PT_L. Figure 2b (lower panel) shows the probabilities PR_R and PT_R. In both cases, the strength of the magnetic field is taken with $b = 1$ and $T = 100$ GeV. Also $m_0 = 175$ GeV, $y_R = 4/3$, $y_L = 1/3$, as corresponds to a top quark.

that the chiral nature of this coupling implies that it is possible to build an axial asymmetry during the scattering of fermions off the wall. We have computed reflection and transmission probabilities showing explicitly that they differ for left and right-handed incident particles from the symmetric phase.

Recall that under the very general assumptions of CPT invariance, together with conservation of unitarity, which are satisfied in the present analysis, the total axial asymmetry (which includes contributions both from particles and antiparticles) is quantified in terms of the particle (axial) asymmetry.

This asymmetry, built on either side of the wall, is dissociated from non-conserving baryon number processes and can subsequently be converted to baryon number in the unbroken phase where sphaleron induced transitions are taking place with a large rate. This mechanism receives the name of *non-local baryogenesis* [12, 13, 14, 15] and, in the absence of the external field, it can only be realized in extensions of the SM where a source of CP violation is introduced *ad hoc* into a complex, space-dependent phase of the Higgs field during the development of the EWPT [16].

Due to the sphaleron dipole moment, another consequence of the existence of an external magnetic field is the lowering of the barrier between topologically inequivalent vacuua [17]. This effect acts in such a way that any baryon asymmetry generated by the building of an axial charge during the asymmetric reflection of fermions into the unbroken phase, in the presence of a magnetic field, stands little chance of surviving in the broken phase. Nonetheless, if such primordial fields indeed existed during the EWPT epoch and the phase transition was first order, as is the case, for instance, in minimal extensions of the SM, the mechanism advocated in this work has to be considered as acting in the same manner as a source of CP violation that can have important consequences for the generation of a baryon number.

ACKNOWLEDGMENTS

Support for this work has been received in part by DGAPA-UNAM under grants number IN118600 and IN108001 and by CONACyT under grant number ICM 35792-E.

REFERENCES

1. A. D. Sakharov, Pis'ma Zh. Eksp. Teor. Fiz. **5**, 32 (1967) [JETP Lett. **5**, 24 (1967)].
2. For a recent review on the subject see M. Trodden, Rev. Mod. Phys. **71**, 1463 (1999).
3. K. Kajantie, M. Laine, K. Rummukainen and M. Shaposnikov, Nucl. Phys. B **466**, 189 (1996).
4. M. Dine, "Baryogenesis: Electroweak and otherwise", Proceedings of the 1994 TASI summer school *CP violation and the limits of the standard model*, Ed. by J.F. Donoghue, p.p. 507-548.
5. M. Giovannini and M. E. Shaposhnikov, Phys. Rev. D **57**, 2186 (1998).
6. P. Elmfors, K. Enqvist and K. Kainulainen, Phys. Lett. B **440**, 269 (1998).
7. M. Giovannini, Phys. Rev. D **61**, 063004 (2000).
8. V. Skalozub and V. Demichik, "Can baryogenesis survive in the standard model due to strong hypermagnetic field?", **hep-ph/9909550**.
9. For recent reviews on the origin, evolution and some cosmological consequences of primordial magnetic fields see: K. Enqvist, Int. J. Mod. Phys. **D7**, 331 (1998); R. Maartens, "Cosmological magnetic fields", International Conference on Gravitation and Cosmology, India, Jan. 2000, Pramana **55**, 575 (2000) and references therein; D. Grasso and H.R. Rubinstein Phys. Rep. **348**, 163 (2001).
10. A. Ayala, J. Besprosvany, G. Pallares and G. Piccinelli, Phys. Rev. D **64**, 123529 (2001).
11. A. Ayala, J. Jalilian-Marian, L. McLerran and A. P. Vischer, Phys. Rev. D **49**, 5559 (1994).
12. A. E. Nelson, D. B. Kaplan and A. G. Cohen, Nucl. Phys. B **373**, 453 (1992).

13. M. Dine, O. Lechtenfield, B. Sakita, W. Fischel and J. Polchinski, Nucl. Phys. B **342**, 381 (1990).
14. A. G. Cohen, D. B. Kaplan and A. E. Nelson, Phys. Lett. B **263**, 86 (1991).
15. M. Joyce, T. Prokopec and N. Turok, Phys. Lett. **B338**, 269 (1994).
16. E. Torrente-Lujan, Phys. Rev. D **60**, 085003 (1999).
17. D. Comelli, D. Grasso, M. Pietroni and A. Riotto, Phys. Lett. **B458**, 304 (1999).

Rare Kaon Decays

Peter S. Cooper

Fermi National Accelerator Laboratory, PO. Box 500 M.S. 122, Bavatia, Illinois 60510 USA

Abstract. There has been great recent progress in measurements of rare and ultra-rare kaon decays, particularly those involving flavor changing neutral currents. I review here those recent results and the prospects for future measurements.

INTRODUCTION

The subject of rare and ultra-rare kaons decays is very active at this time with 8 experiments from 4 labs reporting new results recently and 6 new experiments recently approved. The physics topics under study include flavor changing neutral currents (FCNC), measurements of direct and indirect CP violation and searches for lepton flavor violation (LFV). These measurements are all characterized by very high sensitivities, studying modes with branching ratios in the $10^{-7} - 10^{-12}$ range with single event sensitivities approaching 10^{-12}. This is a region of sensitivity beyond the reach of charm and B experiment, at least for now. The development of beams with fluxes well above $1 MHz$ of kaon decays enable these experiments. In this paper I will cover some of the important new results and the goals for the new experiments.

FLAVOR CHANGING NEUTRAL CURRENTS

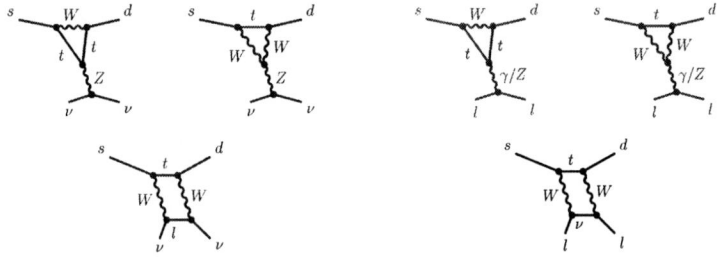

FIGURE 1. FCNC Feynman diagrams.

The flavor changing neutral currents in leptonic kaon decay, shown in Fig. 1, always involve at least two electroweak currents and an up type quark loop. These short distance amplitudes tend to be dominated by top quarks in the loops, which give direct access to $\lambda_t = V_{ts}^* V_{td} = A\lambda^5(1-\rho-i\eta)$ the complex parameter of the CKM matrix, V_{td}

which controls CP violation in the Standard Model. Decay modes where long distance amplitudes are small or absent become laboratories to directly measure λ_t.

As an example of the state of the art in ultra-rare decays the E871 experiment at BNL has published the rate of the rarest decay mode every observed, Br[$K_L \to e^+e^-$] = $(8.7^{+5.7}_{-4.7}) \times 10^{-12}$ [1] with 4 clean events and a 6200 events signal for the relatively common mode Br[$K_L \to \mu^+\mu^-$] = $(7.18 \pm 0.17) \times 10^{-9}$ [2]. These signals are shown in Fig. 2. These dilepton modes are dominated by long distance $K_L \to \gamma^*\gamma^*$ diagrams. The unitary bound from these long distance diagrams saturates the $K_L \to \mu^+\mu^-$ rate observed leaving little room for short distance effects like V_{td}. The main goal of this experiment was the search for the LFV mode $K_L \to \mu^\pm e^\mp$ where E871 set a limit of Br[$K_L \to \mu^\pm e^\mp$] < 4.7×10^{-12} [3]

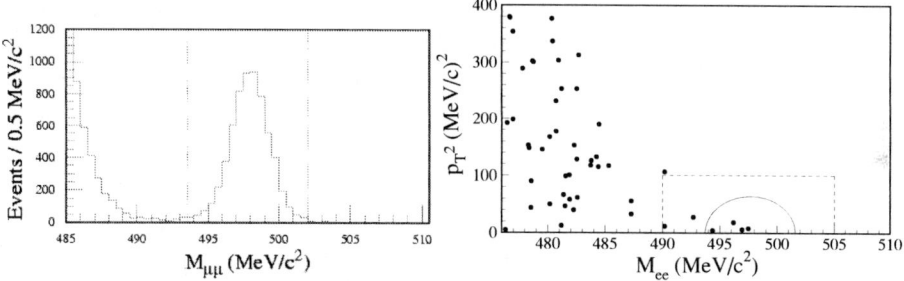

FIGURE 2. E871 signals for a) $K_L \to \mu^+\mu^-$ b) $K_L \to e^+e^-$.

KTeV at Fermilab has searched for $K_L \to \pi^0 l^+ l^-$ and set a limit of Br[$K_L \to \pi^0 \mu^+\mu^-$] < 3.8×10^{-10} [4] and Br[$K_L \to \pi^0 e^+ e^-$] < 5.1×10^{-10} [5] in both cases observing 2 events expecting about 1 background in their signal box. The Standard Model prediction for the direct CP violation contribution to $K_L \to \pi^0 e^+ e^-$ is $(4.3 \pm 2.1) \times 10^{-12}$. These measurements are getting close to the interesting region. There are also CP conserving and indirect CP violating contributions to these modes. V_{td} must be extracted once these modes are reliably measured.

NA48 at CERN has undertaken a program to measure $K_S \to \pi^0 e^+ e^-$ to help with this extraction. In a two day test run they set a limit of Br[$K_S \to \pi^0 e^+ e^-$] < 1.4×10^{-7} [6]. The Standard Model prediction for $K_S \to \pi^0 e^+ e^-$ is $\sim 5 \times 10^{-9}$. The measurement of this mode is a major goal of the approved NA48/1 [7] run. They expect a ×50 improvement with 120 days of running to achieve a sensitivity of $\sim 3 \times 10^{-9}$, in the Standard Model range.

There have been 3 measurements of Br[$K^+ \to \pi^+ \mu^+ \mu^-$]: $(5.0 \pm 0.4 \pm 0.9) \times 10^{-8}$ by BNL E787 [8], $(9.22 \pm 0.60 \pm 0.49) \times 10^{-8}$ by BNL E865 [9], and $(9.8 \pm 1.0 \pm 0.5) \times 10^{-8}$ by Fermilab HyperCP [10]. This mode is long distance dominated but of considerable value in constraining chiral perturbation theory models.

The purely leptonic channels appear to be completely dominated by the long distance two photon diagram. The only other observable is the final state lepton polarization which is unlikely to contain important physics given the saturated unitarity bound. The semi-leptonic $K \to \pi l^+ l^-$ measurements are approaching the Standard Model predictions where these 3 body modes should be observed. NA48/1 will help measure the CP conserving amplitudes. The theoretical problem of disentangling the long and short distance contributions look formidable (to me). None of the new experiments plan to measure these modes in K_L^0. Perhaps the experiments planned for the JHF will go after these important measurements.

$$K \to \pi \nu \bar{\nu}$$

The $K \to \pi \nu \bar{\nu}$ decays are where the precious metals live in FCNC kaon decays. The $K_L \to \pi^0 \nu \bar{\nu}$ decay is the *golden mode* which directly measures η, the imaginary part of V_{td} with negligible theoretical uncertainty; $\text{Br}[K_L \to \pi^0 \nu \bar{\nu}] = 1.8 \times 10^{-11} A^4 X(x_t)^2 \eta^2$. The challenge is that the Standard Model prediction is $\text{Br}[K_L \to \pi^0 \nu \bar{\nu}] = (2.6 \pm 1.2) \times 10^{-11}$ where the uncertainty comes from the present knowledge of the top quark mass and V_{cb}. The $K^+ \to \pi^+ \nu \bar{\nu}$ decay is the *silver mode*. It's rate is proportional to $|V_{td}|^2$ but with a significant correction for loops with a charmed quark; $\text{Br}[K^+ \to \pi^+ \nu \bar{\nu}] = 4.11 \times 10^{-11} A^4 X(x_t)^2 [(\rho - \rho_0)^2 + \eta^2]$. This correction depends on the poorly known charmed quark mass which gives rise to a theoretical uncertainty of $\sim 7\%$ in the latest NLO QCD analysis. The Standard Model prediction today is: $\text{Br}[K_L \to \pi^0 \nu \bar{\nu}] = (7.5 \pm 2.9) \times 10^{-11}$ [11]. This mode is *the* best place to directly measure $|\lambda_t| = |V_{ts}^* V_{td}|$.

On the experimental side there has been much activity and recent progress in searching for these ultra-rare decay modes. The KTeV experiment has reported two upper limits for $\text{Br}[K_L \to \pi^0 \nu \bar{\nu}]$; 1.6×10^{-6} [12] in a measurement where the only thing observed are 2 photons from a high P_t π^0 decay and 5.9×10^{-7} [13] where the π^0 Dalitz decay is observed. The former measurement is background limited, the latter is essentially background free. These limits are still 4-5 orders of magnitude above the Standard Model prediction.

The BNL E787 experiment has reported a major milestone in the progress toward measuring these modes with the report of a clean 2 event signal for $K^+ \to \pi^+ \nu \bar{\nu}$ [14, 15]. The second of these events and their final sensitivity plot are shown in Fig. 3. The experiment is done with stopped kaons where the entire $K^+ \to \pi^+ \to \mu^+ \to e^+$ decay chain is observed with precision measurements of π^+ energy, momentum and range.

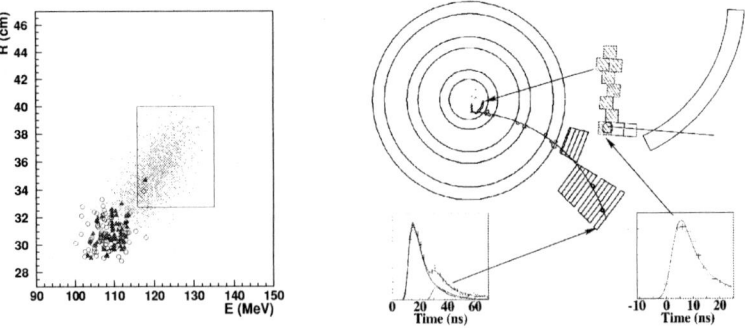

FIGURE 3. E787 - a) signal for $K^+ \to \pi^+ \nu \bar{\nu}$ b) the new event.

The branching ratio reported by E787 is Br[$K^+ \to \pi^+ \nu \bar{\nu}$] = $\left(1.58^{+1.80}_{-0.81}\right) \times 10^{-10}$ [15], about twice the Standard Model prediction but well within the uncertainty of a two event signal. The probability that both of these events are background is only $\sim 0.02\%$.

There are 4 recently approved new efforts in the $K \to \pi \nu \bar{\nu}$ sector. In $K^+ \to \pi^+ \nu \bar{\nu}$ the BNL E787 collaboration was approved for another major run with an upgraded detector, as E949 [16]. This experiment, which I and several of my Fermilab and IHEP colleagues have joined, has a goal of a 5-10 events signal for $K^+ \to \pi^+ \nu \bar{\nu}$. Our data taking run resumes in February 2002. As the next step in $K^+ \to \pi^+ \nu \bar{\nu}$ my colleagues and I, including several BNL members of E949, have been approved for a new Fermilab experiment, CKM (E921) [17], to measure the $K^+ \to \pi^+ \nu \bar{\nu}$ branching ratio with 100 events and less than 10% background using a decay in flight technique in a new Superconducting RF separated K^+ beam at 22 GeV/c. The CKM apparatus and expected signal are shown in Fig. 4.

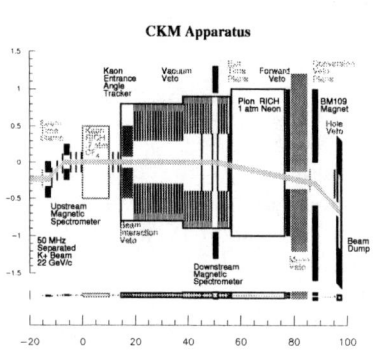 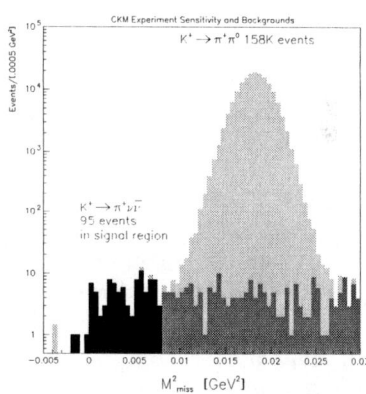

FIGURE 4. CKM - a) Apparatus and b) signal for $K^+ \to \pi^+ \nu \bar{\nu}$.

There are two new experiments attempting to observe and measure the rate of $K_L \to \pi^0 \nu \bar{\nu}$. E391a [18] at KEK has a goal of reaching the 1 event level with a hermetic photon calorimeter as a π^0 spectrometer and photon veto. They plan to start data taking in fall 2003. KOPIO [19] is a major new effort at BNL based on an innovative beam technique with a low momentum neutral beam (800 MeV/c) and time-of-flight to measure the K_L^0 momentum. Their goal is a 50 event signal with $S/N > 2$. Both these experiments require outstanding performance from their photon detector systems.

Successful measurements from these experiments, together with B sector results, will allow a direct test of the Standard Model hypothesis that the only source of CP violation is from η, the imaginary part of Vtd. These measurements are all theoretically robust enough that a disagreement among them would require sources of CP violation beyond the Standard Model.

There are four such observables; two in the kaon system and 2 in the B systems. The sensitives expected from these measurements are shown in Fig. 5. on the unitarity triangle plane. If these four measurements are not all consistent with a single point for the apex of the unitarity triangle then the Standard Model description of the source of CP violation is, at least, incomplete.

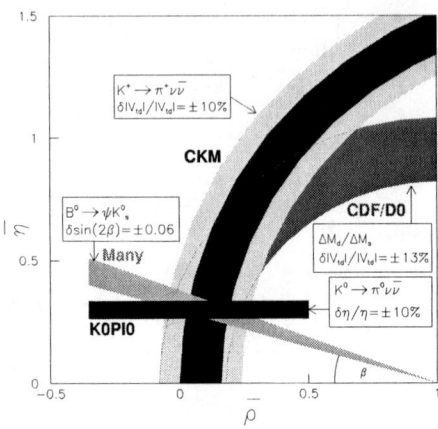

FIGURE 5. Unitary triangle with expected sensitivities from theoretically clean measurements.

CP/T VIOLATION

The increased sensitivity of recent experiments have now permitted the measurement of new CP or T violating observables in K_L^0 decays. The radiative decay $K_L \to \pi^+\pi^-\gamma$ which has both CP conserving direct photon emission (DE) and CP violating internal bremstralung (IB) amplitudes. KTeV has measured both the rate, Br[$K_L \to \pi^+\pi^-\gamma, E_\gamma^* > 20 MeV$] $= 4.28 \times 10^{-5}$ and the relative fraction $DE/(DE+IB) = 0.683 \pm 0.011$ [20] which is a CP violating effect in the interference of the DE and IB amplitudes. In the $K_L \to \pi^+\pi^-e^+e^-$ decay this same interference manifests itself as an asymmetric azimuthal angular correlation between the $\pi^+\pi^+$ and e^+e^- decay planes. KTeV has measured both the rate; Br[$K_L \to \pi^+\pi^-e^+e^-$] $= (3.6 \pm 0.6 \pm 0.4) \times 10^{-7}$ [21] and CP (and T) violating asymmetry; $0.136 \pm 0.025 \pm 0.012$ [22]. This is the largest CP violating effect ever seen. In a preliminary result [23], NA48 has reported a measurement of this same asymmetry in $K_S \to \pi^+\pi^-e^+e^-$ where there is no CP violation. They report Br[$K_S \to \pi^+\pi^-e^+e^-$] $= (4.3 \pm 0.2 \pm 0.3) \times 10^{-5}$ and A $= -0.002 \pm 0.034 \pm 0.012$, consistent with zero as expected. KTeV has also just recently reported [24] a new measurement of the charge asymmetry in $K_L \to \pi^\pm e^\mp \nu$ decay. Their new value, $\delta_L = (3322 \pm 58 \pm 46) \times 10^{-6}$ is in good agreement with previous measurements and 2.4 times more precise than the best prior result.

At KEK E246 is re-measuring the transverse μ^+ polarization in $K^+ \to \pi^0 \mu^+ \nu$ decay. This is a T odd observable. They report $P_T = -0.0033 \pm 0.0037 \pm 0.0009$ [25] and expect to achieve a final uncertainty of $\sigma(P_T) = 0.0030$ when all their data is analyzed. E246 has also made new measurements of the form factors in $K^+ \to \pi^0 e^+ \nu$ decay. They report [26] results consistent with zero for both the scaler [$|f_S/f_+| = -0.002 \pm 0.026 \pm 0.014$] and tensor [$|f_T/f_+| = -0.01 \pm 0.14 \pm 0.09$] form-factors contradicting the 3 σ non-zero effects in the current PDG averages.

Direct CP violation is now well established in the neutral kaon system with the recent results from KTeV and NA48. The connections between ε and ε'/ε and Standard Model parameters are still seriously limited by theoretical uncertainties. The $K_L \to \pi^+\pi^-e^+e^-$ asymmetry measurement is one of the most beautiful results I've seen in years. This asymmetry and the charge asymmetry in $K^+ \to \pi^0 e^+ \nu$ are due to indirect CP violation, like ε, so they suffer similar theoretical difficulties. The studies of Kaon semi-leptonic decays are beginning to achieve the levels of sensitivity and precision to the details of the electro-weak interaction we normally associate only with muon decay.

LEPTON FLAVOR VIOLATION

The search for lepton flavor violation (LFV) has been a topic of enduring interest. With the recent evidence for neutrino oscillations indicating that lepton flavor is not conserved this topic has become even hotter. In the kaon system LFV has usually been parameterized by positing a new flavor changing neutral current decaying to $\mu^\pm e^\mp$ which has full Fermi weak couplings and is suppressed only by its large mass in the propagator. By dimensional analysis a branching ratio limit of 10^{-12} corresponds to a lower mass limit on this LFV current of $\sim 1000 M_W$ or $\sim 82 TeV$.

In the kaon system this has lead to new searches for the LFV modes; $K_L \to \pi^0 \mu^\pm e^\mp$ $K^+ \to \pi^+ \mu^+ e^-$, and $K_L \to \mu^\pm e^\mp$. BNL E871 has reported Br[$K_L \to \mu^\pm e^\mp$] < 4.7×10^{-12} [3] in a background free measurement, as mentioned above. KTeV has a preliminary upper limit for $K_L \to \pi^0 \mu^\pm e^\mp$; Br[$K_L \to \pi^0 \mu^\pm e^\mp$] < 4.4×10^{-10} [27] with 2 events observed. BNL E865 has reported their final limit on $K^+ \to \pi^+ \mu^+ e^-$; Br[$K^+ \to \pi^+ \mu^+ e^-$] < 2.8×10^{-11} [28]. BNL E865 also has a new limit on "neutrino-less double μ decay": Br[$K^+ \to \mu^+ \mu^+ \pi^-$] < 3.0×10^{-9} [29].

Lepton flavor violation is a topic with so little theoretical guidance that there are no clear "goal posts". If the only source of LFV is mixing in the neutrino sector then the LFV effects induced in the kaon sector are unobservably small ($\sim 10^{-25}$). Technicolor, Leptoquark and compositeness models might apply, each with their own parameters and predictions for observable effects in the quark, charged lepton and neutrino sectors. This field has been driven by better experimental ideas. The present state of the art in the kaon sector has been established by BNL E871, BNL E865 and KTeV. In the immediate future the new searches will be back in the charged lepton sector with the search for $\mu^+ Z \to e^+ Z$ in the new MECO experiment at BNL and new searches for $\mu^+ \to e^+ \gamma$ and $\mu^+ \to e^+ e^+ e^-$ at PSI. Further progress in the kaon sector awaits some new experimental ideas.

SUMMARY

The kaon sector continues to be a gold mine 55 years after the discovery of the kaon. I've reviewed here 23 recent results from 8 different experiments (BNL787, BNL865, BNL871, NA48, KTeV/FNAL799, KTeV/FNAL832, HyperCP/FNAL871 and KEK246) from 4 laboratories on 3 continents. This is far from from a complete review of recent work. Good progress is being made on understanding the sources of CP violation with FCNC and other decays. The searches for lepton flavor violation have achieved substantially improved limits; sufficient to mute the discussions of whole classes of formerly fashionable models.

The next round of experiments are underway with a new experiment or two at each of the labs. The main theme of these measurements is a direct confrontation of the Standard Model's description of the source of matter - anti-matter asymmetry in nature. We're not done learning from the kaon system yet.

ACKNOWLEDGEMENTS

Takeshi Komatsubara's *XXI Physics in Collision* talk was kindly made available to me and liberally used in preparing this talk.

REFERENCES

1. D. Ambrose *et al.* [BNL E871 Collaboration], Phys. Rev. Lett. **81**, 4309 (1998)
2. D. Ambrose *et al.* [E871 Collaboration], Phys. Rev. Lett. **84**, 1389 (2000).
3. D. Ambrose *et al.* [BNL Collaboration], Phys. Rev. Lett. **81**, 5734 (1998) [arXiv:hep-ex/9811038].
4. A. Alavi-Harati *et al.* [KTEV Collaboration], Phys. Rev. Lett. **84**, 5279 (2000)
5. A. Alavi-Harati *et al.* [KTeV Collaboration], Phys. Rev. Lett. **86**, 397 (2001)
6. A. Lai *et al.* [NA48 Collaboration], Phys. Lett. B **514**, 253 (2001).
7. [CERN Experiments NA-48-1/2], http://www1.cern.ch/NA48/Welcome.html
8. S. C. Adler *et al.* [E787 Collaboration], Phys. Rev. Lett. **79**, 4756 (1997) [arXiv:hep-ex/9708012].
9. H. Ma *et al.*, Phys. Rev. Lett. **84**, 2580 (2000) [arXiv:hep-ex/9910047].
10. H. K. Park *et al.* [HyperCP Collaboration], arXiv:hep-ex/0110033.
11. A. J. Buras, arXiv:hep-ph/0101336.
12. J. Adams *et al.* [KTeV Collaboration], Phys. Lett. B **447**, 240 (1999) [arXiv:hep-ex/9806007].
13. A. Alavi-Harati *et al.* [The E799-II/KTeV Collaboration], Phys. Rev. D **61**, 072006 (2000)
14. S. C. Adler *et al.* [E787 Collaboration], Phys. Rev. Lett. **79**, 2204 (1997) [arXiv:hep-ex/9708031].
15. S. Adler *et al.* [E787 Collaboration], Phys. Rev. Lett. **88** (2002) 041803 [arXiv:hep-ex/0111091].
16. [Brookhaven E949 Experiment], http://www.phy.bnl.gov/e949
17. [Fermilab CKM Experiment], http://www.fnal.gov/projects/ckm/Welcome.html
18. [KEK E391a Experiment], http://psux1.kek.jp/ e391
19. [Brookhaven KOPIO Experiment], http://pubweb.bnl.gov/users/e926/www/
20. A. Alavi-Harati *et al.* [The KTeV Collaboration], Phys. Rev. Lett. **86**, 761 (2001)
21. J. Adams *et al.* [The KTeV Collaboration], Phys. Rev. Lett. **80**, 4123 (1998).
22. A. Alav-Harati *et al.* [KTeV Collaboration], Phys. Rev. Lett. **84**, 408 (2000)
23. R. Sacco [NA48 Collaboration], Acta Phys. Polon. B **32**, 1969 (2001).
24. A. Alavi-Harati [KTeV Collaboration], arXiv:hep-ex/0202016.
25. M. Abe *et al.* [KEK-E246 Collaboration], Phys. Rev. Lett. **83**, 4253 (1999).
26. S. Shimizu *et al.* [KEK-E246 Collaboration], Phys. Lett. B **495**, 33 (2000).
27. B. Cox [KTeV Collaboration], Nucl. Phys. Proc. Suppl. **99B**, 96 (2001).
28. R. Appel *et al.*, Phys. Rev. Lett. **85**, 2450 (2000) [arXiv:hep-ex/0005016].
29. R. Appel *et al.*, Phys. Rev. Lett. **85**, 2877 (2000) [arXiv:hep-ex/0006003].

A Survey of Charm Hadroproduction Results

James S. Russ

Physics Department, Carnegie Mellon University, Pittsburgh, PA 15213 USA

Abstract. We review the original goals of charm hadroproduction experiments at fixed target energies, initially intended to explore perturbative QCD at the charm mass scale. High-statistics studies with π^- beams suggest strong non-perturbative effects at large x_F. Recent results from proton and Σ^- beams show further systematics of non-perturbative behavior. This review summarizes the systematics of these effects as developed for different charm hadrons and different beam particles.

INTRODUCTION

Cross section measurements are a fundamental test of the calculational basis for a theory of the strong interactions. Students at this workshop will recognize that Fermi's Golden Rule for the rate of a potential-mediated interaction in non-relativistic quantum mechanics tells us:

$$\Gamma = \frac{2\pi}{\hbar} \mid S_{fi} \mid^2 \times [\text{density of states}]$$

The quantity Γ is a scattering <u>rate</u> (number of scatters per sec) per unit scatterer. Therefore, we can relate it to the scattering <u>cross section</u> by dividing by the incident flux F (number of incident particles per area per time): $\sigma = \Gamma/F$

When we convert to relativistic language, the matrix element of the potential becomes the 4-dimensional S-matrix element, and the 4-dimensional density of states expression includes factors relevant to the spins of the particles involved. Therefore, we see that a cross section measurement is aimed at measuring a scattering matrix element - something which Strong-Interaction Theory can (in principle) calculate.

With this goal in mind, charm hadroproduction studies in fixed target experiments have been carried out for the past 2 decades. Having two heavy flavor species (c,b) to compare allows us in principle to test the flavor-independence of hadroproduction predicted by QCD. However, thorny theoretical issues have been recognized by many authors. [1] Since the actual observables are color-singlet final state hadrons, there may be significant non-perturbative hadronization effects: (a) color-drag effects between either the c or $\bar{c}$ quark and the outgoing colored fragment from the beam or target hadron; and (b) other hadronization effects, including rescattering of the separating charmed and anti-charmed hadrons because of the limited p_T range of fixed target experiments.

The consensus is that the original goal of testing theory is unattainable. Instead, the intent of the latest round of charm hadroproduction experiments is to systematize the non-perturbative factors at work by comparing production of several charm species at different energies by several different beam hadrons. This report summarizes the current understanding of non-perturbative features of the data on charm hadroproduction. I should note that the strength of non-perturbative physics in charm photoproduction at fixed target energies is much weaker than in hadroproduction. The photoproduction data are predominantly diffractive, i.e., are dominated by quark pair production, rather than through the partonic interactions of the photon. [2] Such a distinction from fixed target hadroproduction suggests that many of the non-perturbative features of hadroproduction stem from the the effects listed above. We shall try to systematize what is known.

GENERAL THEORETICAL FEATURES OF CHARM HADROPRODUCTION

The conventional picture of Heavy Quark production assumes factorization, i.e., a separation between the quarks produced in the hard scattering process and the hadron fragments that are left in the beam and target hadrons. Factorization allows one to use parton ideas from Deep Inelastic Scattering to analyze the quark production process at an unphysical high mass scale where the calculation is understood, then evolve the distributions to the m_Q-scale. The NLO parton calculation includes single gluon emission effects in the production process and was first analyzed by Nason, Dawson and Ellis. [3, 4] To use the NLO picture one chooses: (a) the renormalization scale for the parton-level process; (b) the factorization scale for the parton distributions (generally the same as (a)); and (c) the effective quark constituent mass. One adds parton transverse momentum k_T, attributable to confinement, by hand. The NLO calculation leaves the LO prediction for single-quark distributions in x_F or p_T unchanged, except in scale. NLO effects mainly modify $Q\bar{Q}$ pair distributions, chiefly the azimuthal angle distribution between Q and $\bar{Q}$.

The NLO calculation leaves us with quark-level distributions. Observables are usually related to hadrons, and non-perturbative fragmentation mechanisms have to be introduced. As a consequence of the heavy quark Q's fragmenting into a hadron, distributions in x_F and p_T for charm hadrons are expected to be softer than those for the quarks themselves. NLO calculations are presented as computational packages with adjustable scale parameters and use various parton distribution functions. [1] Typically hadronization is treated by applying the empirical Peterson function developed for e^+e^- production to the quark-level distributions.

As shown in Fig. 1 the NLO top calculation with logarithmic resummation gives good (5%) agreement with varying scales, indicating that higher-order effects do not change the cross section significantly. Theory and data agree well within the experimental errors. The b-quark calculation, on the other hand, shows significant scale sensitivity (30% level) even at the b-quark mass. Moreover, the data lie significantly above even the most optimistic theory calculation. The b-quark data do NOT agree with NLO calculation. The theory works well for a clearly-perturbative problem, top production. It already shows signs of strain at the b-quark scale.

Charm quark production is much more problematic for NLO calculations. m_c is not large compared to Λ_{QCD}. Adding NLO diagrams to the LO theory increased the calculated cross section by a factor of 3-5, so NNLO may be important for charm. No

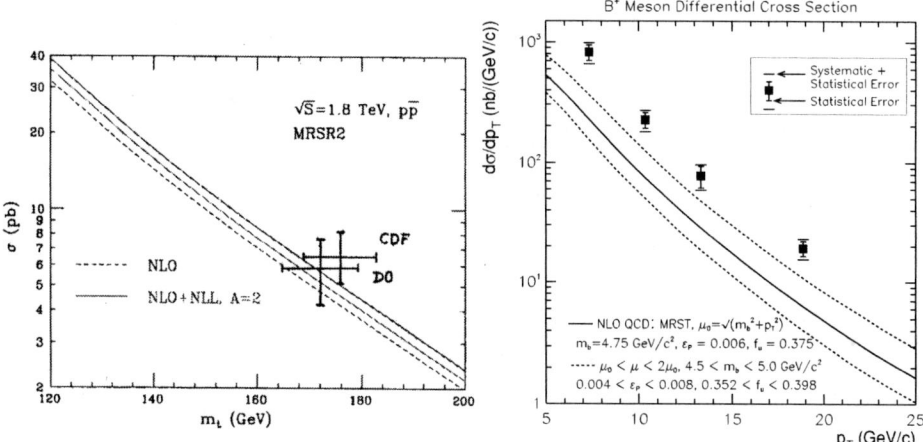

FIGURE 1. left: NLO top cross section and data; right: NLO b cross section and CDF data.

one has tried to do such a complicated calculation. The present theory scale variation in the total charm hadroproduction cross section calculation, seen at left in Fig. 2, is huge - a factor of 10. Experimental data do fall within that range, suggesting that the quark-level computation has some connection to the data. However, total cross section data do not depend on hadronization effects. We consider next what happens for single-charm-hadron distributions

FEATURES OF CHARM HADROPRODUCTION DATA

As an illustration of the general features of the data, consider the Fermilab E791 (500 GeV/c π^-) x_F distribution for D^0 mesons [+ c.c.] covering the central and forward production regions shown at right in Fig. 2. [5]

One sees that the overall x_F shape matches the quark-level calculation. Normalization here is to the data; calculation falls well below data and has a wide range of scale dependence. There is some experimental suggestion of a slower decrease at large x_F than the NLO calculation. The hadronized calculation (Petersen fragmentation) is softer than the data. These results are typical of most hadroproduction data.

Particle-Antiparticle Asymmetries in Charm Hadroproduction

Both LO and NLO pictures of charm hadroproduction predict very small asymmetries in any single-particle distribution comparison of particle and antiparticle. Therefore, it was a major surprise when the E791 high-statistics data confirmed earlier suggestions of significant asymmetry between D^+ and D^- production, as shown in the left panel of Fig. 3. [6] However, the right panel shows that interpretation is not simple. $D^{*\pm}$

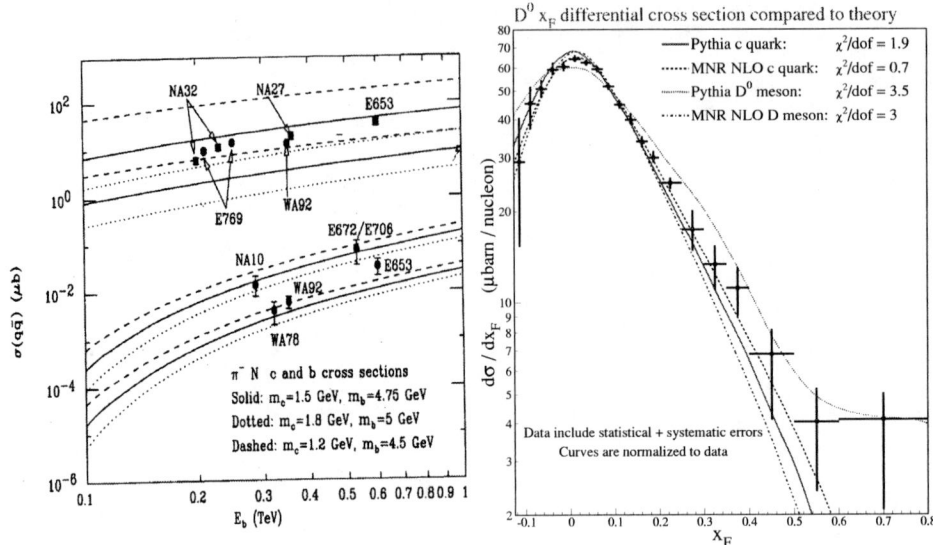

FIGURE 2. left: charm hadroproduction total cross sections; right: E791 D^0 + c.c. x_F distribution (500 GeV/c π^-)

distributions, ostensibly having the same valence quark effects as the $D^\pm$, show NO strong asymmetry, as we will discuss later. [7]

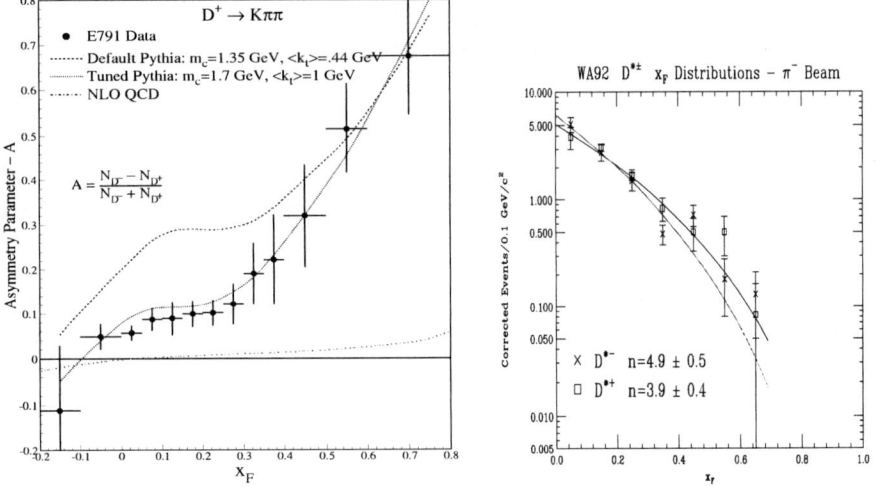

FIGURE 3. (left) E791(500 GeV/c π^-) $D^\pm$ production asymmetry; (right) WA92(350 GeV/c π^- beam) $D^{*\pm}$ asymmetry

Because these data cannot be explained by perturbative QCD, the questions raised focus on other physics: (a) why do some but not all charm hadrons show large asymmetry?

(b) is there a beam-hadron dependence of the effect? and (c) why is it x_F dependent? These questions can be framed within two possible pictures of large non-perturbative effects: the Quark-Gluon-String Model (QGSM) originally included in the PYTHIA Monte Carlo [8] and the Intrinsic Charm Model (ICM) [9].

The QGSM recognizes that the low- p_T environment of fixed target production may permit color-field interactions between the c- or $\bar{c}$ -quark and the fragments of the beam hadron to produce a colorless hadron. This color-drag effect will *increase* the x_F of the charm hadron when it occurs, compared to the perturbative value. Such effects are usually termed *leading* behavior and occur when the outgoing charm hadron shares at least one valence quark in common with the beam hadron. When the charm quark is well-separated from the beam fragments (at large p_T), the effect should be smaller.

The ICM, having rather different kinematics, came from QED analyses by Brodsky and co-workers. In the ICM the valence quark distribution of the beam hadron includes an intrinsic c- $\bar{c}$ component with some probability. Therefore, in soft collisions the beam hadron may dissociate, leaving the c- or $\bar{c}$ quark carrying a large fraction of the beam momentum. This leads to an excess cross section at low p_T, large x_F.

The strong asymmetries seen in hadroproduction have shifted experimental emphasis from testing pQCD to systematizing non-perturbative effects. Experimental studies focus on comparing the x_F distributions of particle and antiparticle. In the following sections we shall see what the data for fixed target charm production by these three beam hadrons say about the leading systematics.

Comparison of $D^\pm$ and $D^{*\pm}$ Meson Production

The D* mesons are the S=1 hyperfine excitations of the D mesons and have the same valence quark content. Any leading behaviour should be similar for the two systems, yet we saw in Fig. 3 that the $D^{*\pm}$ behavior is clearly quite different from that of the $D^\pm$ family for a π^- beam. What about production by different beam hadrons? SELEX has PRELIMINARY data on D* and $D^\pm$ production by Σ^-, shown in Fig. 4.

These results are remarkable for the steep x_F fall of the *leading* D^{*-} events. Even though the integral yield for $x_F \geq 0.15$ favors the leading hadron, the asymmetry changes sign for $x_F \geq .3$. This is quite different from the π^- production case.

For $D^\pm$ production, shown at right, we get a different kind of surprise. There is no shape distinction between the leading D^- and the non-leading D^+, only an x_F-independent normalization preference for leading particle production.

What should one conclude? The valence d-quark structure is different in the π^- and Σ^-, but the impact of having a single quark or diquark seems to be quite different for the S=0 and S=1 mesons. Adding proton data (single d quark) from future SELEX analysis may be instructive.

$D_S^{\pm}$ and D^0 Meson Production

Consider next D_S production. For the π^- beam there is no valence s-quark. E791 reports a very small asymmetry with no particular x_F structure. [10] In contrast, the Σ^- beam has an s quark that makes the D_S^- leading. Preliminary SELEX Σ^- data are shown at left in Fig. 5. Strong leading behavior is clearly seen, unlike the nominally-comparable case of the D^-. Thus, D_S production is consistent with color-drag effects for both π^- and Σ^- beams.

The remaining meson to be considered is the D^0. The production distributions here are contaminated by feed-down from the dominant decays of $D^{*\pm,0}$ mesons to final states with D^0s. SELEX has made the first effort to look at the x_F dependence for D^0 and $\overline{D}^0$ separately, shown at right in Fig. 5. Feeddown from the steep D^{*-} distribution for the Σ^- beam can be seen in these data as a sharpening of the $\overline{D}^0$ distribution for $x_F \leq 0.3$. The D^{*+} x_F shape, on the other hand, is very similar to what we see here for D^0. If one fits both D^0 and $\overline{D}^0$ distributions for $x_F \geq 0.3$ to $(1-x_F)^n$, the n-values for the two particles are very similar, as shown in Fig. 5. The curve for $\overline{D}^0$ in the figure includes a contribution with a fixed $(1-x_F)^{11}$ behavior to account for the D^{*-} contribution at small x_F. The overall conclusion is that Σ^- production shows no leading behavior for the D^0 system, consistent with expectation. The upcoming π and proton beam results from SELEX, where leading effects are expected, will be important to see if the D^0 system is predicted correctly by non-perturbative effects.

Λ_c Production

Finally we turn our attention to charm baryon production. E791 has a good-statistics measurement of the asymmetry in the central production region showing approximately

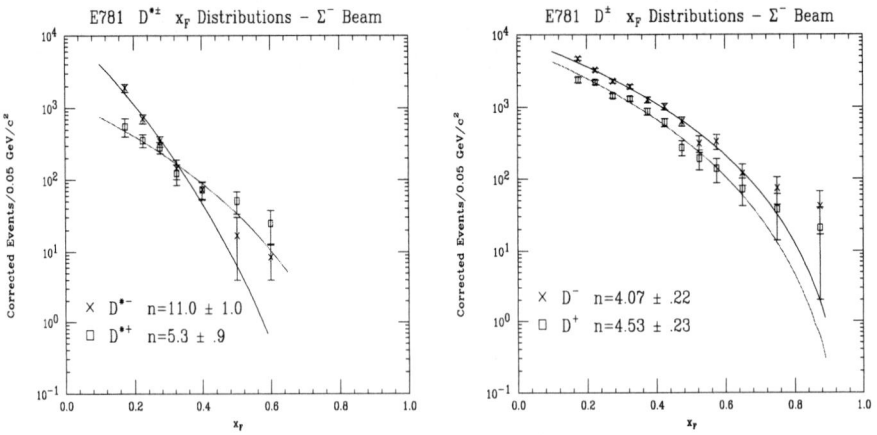

FIGURE 4. SELEX (600 GeV Σ^- beam) x_F distributions for (left) D^{*-} and D^{*+}; (right) D^- and D^+

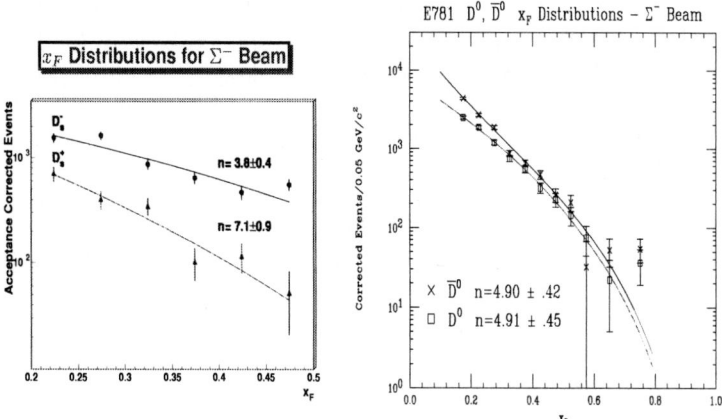

FIGURE 5. SELEX (600 GeV Σ^- beam) x_F distributions for (left) $D_S^\pm$; (right) D^0 and $\overline{D}^0$

equal production of Λ_c^+ and $\overline{\Lambda_c}^-$, consistent with non-perturbative expectations since the pion can contribute a valence quark or antiquark. [11]

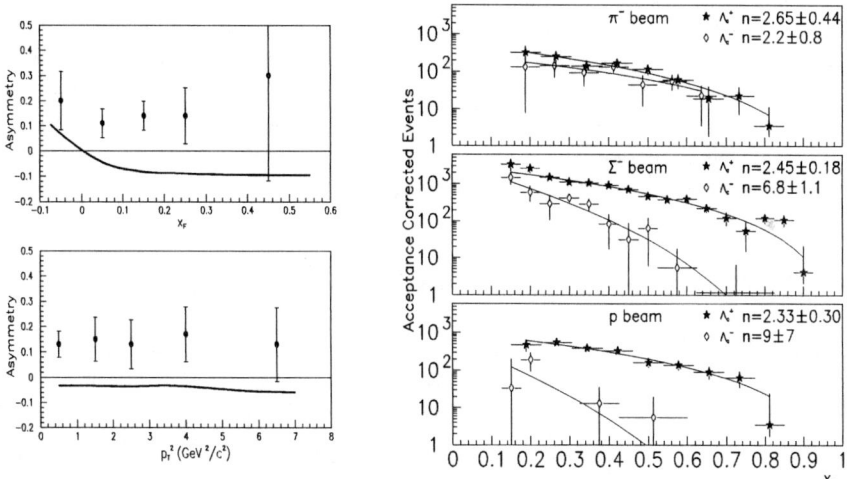

FIGURE 6. (left)E791 (500 GeV π^-) Λ_c^+ asymmetry vs x_F; (right) SELEX (600 GeV Σ^-) Λ_c^+ and $\overline{\Lambda_c}^-$ x_F distributions

SELEX presents x_F distributions from all 3 beam hadrons for Λ_c^+ and $\overline{\Lambda_c}^-$ separately, as seen in the right panel of Fig. 6. For a *baryon* beam, antibaryon production is strongly suppressed, and the suppression grows stronger at large x_F. These findings again agree with leading behavior. For Λ_c^+ production both baryon beams have asymmetry parameter $A \sim 1$ for $x_F \geq 0.5$.

The Λ_c^+ p_T distribution from the Σ^- beam is shown at left in Fig. 7. The break in slope at $p_T^2 \sim 2 \mathrm{GeV}/c^2$ has been previously seen in D^0 data (E791) and D-meson combined data (WA92) from pion beams. This common behavior suggests that it is a feature of c-quark production and is independent of non-perturbative issues like hadronization or color-drag effects. SELEX has looked at the x_F distributions in the diffractive regime ($p_T \leq 1$ GeV/c^2) and in the power-law regime ($p_T^2 \geq 2$ GeV/c^2). These two distributions are shown at right in Fig. 7.

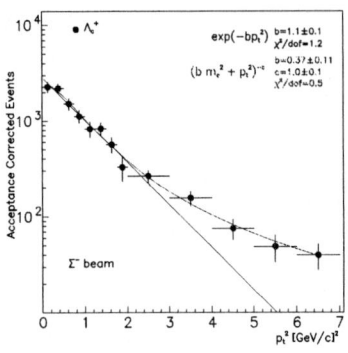 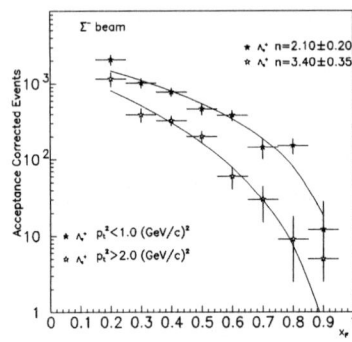

FIGURE 7. (left) SELEX (600 GeV Σ^-) Λ_c^+ p_T^2 distribution ; (right) x_F distributions for $p_T^2 \leq 1 (GeV/c)^2$ and $p_T^2 \geq 2 (GeV/c)^2$

The x_F distribution falls faster for the large p_T data than does the one at low p_T. This is consistent with QGSM effects. The c-quark at large p_T is more isolated from the beam fragments and experiences less color drag than one at low p_T. SELEX has also looked for ICM-type effects in these data, viz., a sharpening of the p_T distribution to favor small p_T in the large x_F sample. No such effect is observed.

PRODUCTION SUMMARY AND STATUS

Not all production channels for the 3 beam hadrons have been discussed here. Continuing SELEX analysis will fill in the table. Results can be expected within the coming year. We see the following pattern in the currently-available data:

1. D_S, D^0, and Λ_c^+ production by π^- and Σ^- beams show leading effects where predicted.
2. D^+ and D^{*+} + c.c. production by π^- and Σ^- beams show strong non-perturbative effects, but they are not at all consistent with model predictions.

The surprises in item 2 come from production differences in S=0 and S=1 mesons of the same valence quark content. The other charm hadrons follow the non-perturbative model rather well. Can we understand this?

There are several Russian groups who have attempted to deal with all charm hadroproduction data in a self-consistent fashion. They treat non-perturbative effects by using different charm quark fragmentation descriptions along with including color-drag effects invoking detailed valence quark configurations in the parton distribution functions. There are many parameters in the models, but Likhoded and coworkers [12] and Piskounova and coworkers [13] use e^+e^- and HERA ep data as well as hadroproduction data to fit parameters and predict the high-energy results. It will be a real challenge to the models to confront the complete set of high energy data that will be available soon. An important question is whether the information gleaned from the model fits can be related to a more fundamental picture of the non-perturbative interactions in charm hadroproduction.

We end this section by recalling the original reason to study heavy quark (HQ) production: testing flavor-independence in HQ processes. Are charm cross sections at a perturbative scale the same as b cross sections? What is the right way to make the comparison? The collider data from Tevatron Run II may offer an interesting new look into these issues if one compares charm and beauty production at m_T scales far above m_b.

AND NOW SOMETHING COMPLETELY DIFFERENT

Soon after charmonium states were discovered, theoretical calculation of ccq baryon states were done and mass spectra predicted using a Coulomb-like potential. Further refinements were added, but the absence of experimental stimulus has stymied progress. The current situation is summarized in Ref. [14].

From the experimental viewpoint, one expects that a ccq baryon will have the cc pair in a symmetric spin state, like the J/ψ(2S) at 3.69 GeV/c². The lowest ccq mass according to Richard's summary is expected to be in the range 3.63-3.75 GeV/c². Decay modes have not been explored. The likely products are a charm baryon plus meson, with a lifetime comparable to or shorter than the Λ_c^+ lifetime of 200 fs (half the c-quark lifetime).

SELEX has searched for the decay $ccu^{++} \to \Lambda_c^+ + (K^-\pi^+\pi^+)$. The $(K^-\pi^+\pi^+)$ vertex is formed topologically, with no particle ID requirements. This 3-prong vertex must be distinct both from the primary vertex (vertex separation significance of 1-4 σ) and the Λ_c^+ decay vertex (similar requirement). The results are shown in the left panel of Fig. 8. One sees an interesting bump (3.3 σ) at a mass of 3.79 GeV/c². The resolution of the gaussian fit to the data is consistent with experimental simulation. The lifetime appears to be very short, but not zero.

Background evaluations are shown, along with the signal channel, in the right panel of Fig. 8. We switched mass assignments of the 3 mesons, requiring $(K^+\pi^-\pi^+)$ (top right). We looked at overall neutral systems which *cannot* be (ccq) systems. Fig. 8 shows these neutral states with the meson system corresponding to $(K^-\pi^+\pi^-)$ (lower left) or $(K^+\pi^-\pi^-)$ (lower right). Only one bin in the neutral system shows any significant fluctuation up from the nominal background shape.

These data suggest a possible ccu^{++} state at 3.79 GeV/c², but the statistics are limited. The yield of this state, if real, into this mode is about 1% of the yield of $\Lambda_c^+ \to pK^-\pi^+$. Moinester quotes an estimate by Bjorken that a ccu^{++} state would have about a 5% branching ratio into our mode, i.e., comparable to the Λ_c branching ratio. [15]. SELEX

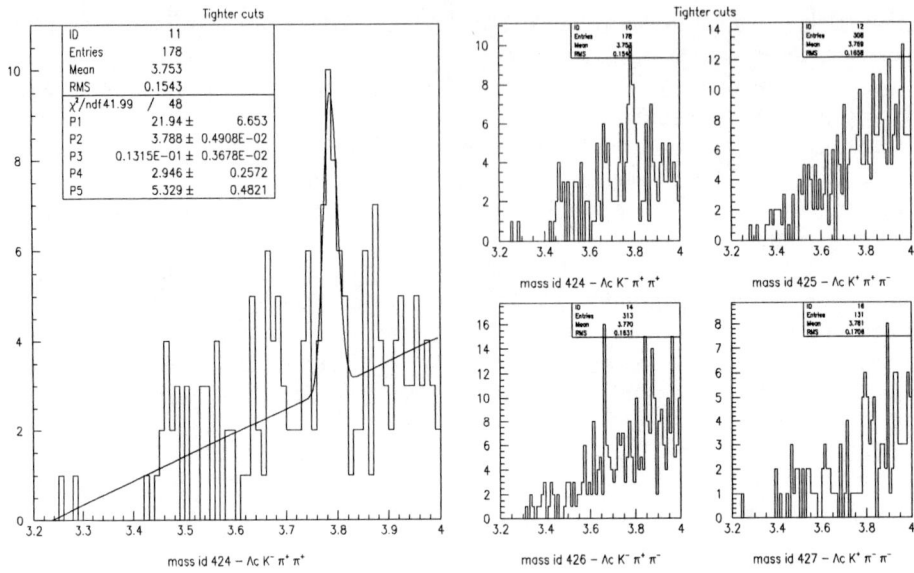

FIGURE 8. (left) SELEX ccu^{++} search ; (right) mass distributions for ccu^{++} signal and background channels as discussed in text

will pursue a search for the isospin partner ccd^{+} state. It would be very exciting to see a confirmation of such an effect from, e.g., FOCUS, which has a very different production environment and different background.

REFERENCES

1. Stefano Frixione, Michelangelo L. Mangano, Paolo Nason, and Giovanni Ridolfi. *Adv. Ser. Direct. High Energy Phys.*, 15:609–706, 1998.
2. P. L. Frabetti *et al. Phys. Lett.*, B308:193–200, 1993.
3. P. Nason, S. Dawson, and R. K. Ellis. *Nucl. Phys.*, B327:49–92, 1989.
4. P. Nason, S. Dawson, and R. K. Ellis. *Nucl. Phys.*, B335:260, 1990.
5. E. M. Aitala *et al. Phys. Lett.*, B462:225–236, 1999.
6. E. M. Aitala *et al. Phys. Lett.*, B371:157–162, 1996.
7. M. I. Adamovich *et al. Nucl. Phys.*, B547:3–14, 1999.
8. T. Sjostrand. *Comput. Phys. Commun.*, 82:74–90, 1994.
9. R. Vogt, S. Brodsky, and P. Hoyer. *Nucl. Phys.*, B383:643–684, 1992.
10. E. M. Aitala *et al. Phys. Lett.*, B411:230–236, 1997.
11. E. M. Aitala *et al. Phys. Lett.*, B495:42–48, 2000.
12. A. K. Likhoded and S. R. Slabospitsky. 2000.
13. O. I. Piskounova. *Nucl. Phys. Proc. Suppl.*, 93:144–147, 2001.
14. S. Fleck and J. M. Richard. *Prog. Theo. Phys.*, 82:760–774, 1989.
15. M. A. Moinester. *Z. Phys.*, A355:349–362, 1996.

Latest Oscillation Results from SNO

R.G. Van de Water, for the SNO collaboration

Los Alamos National Laboratory, P-25 Group, MS H846, Los Alamos, NM, 87545
(vdwater@lanl.gov)

Abstract. The Sudbury Neutrino Observatory has been successfully taking quality neutrino data for over two years now. This paper reports on analysis of the solar charge current and elastic scattering neutrino fluxes from the first year of pure D_2O running. Combined with the SuperK elastic scattering flux, there is strong evidence that solar neutrinos are oscillating, and that predictions of the standard solar model are in good agreement with data.

INTRODUCTION

The solar neutrino deficit problem has been with us for over thirty years now [1], [2], [3], [4], [5], [6]. The small observed flux compared to theoretical solar model calculations is hard to reconcile with standard astrophysical sources [7]. Coupled with energy dependencies as indicated from multiple independent experiments, the preferred solution is vacuum or matter enhanced neutrino oscillations.

The Sudbury Neutrino Observatory (SNO) was designed to solve this long standing mystery by measuring the disappearance of electron neutrinos (ν_e) and the the subsequence reappearance of other flavor neutrinos (ν_μ or ν_τ). This would be compelling evidence that neutrinos do indeed oscillate. We present here the first solar neutrino results from SNO, which have also been described in detail elsewhere [8], [9].

The long story of the solar neutrino problem stems from the fact that experiments to detect the whimsical neutrino are extremely difficult. The realization that we are now coming to a resolution of the problem, and possible hints of physics beyond the standard model, is a tribute to the tenacity and intelligence of neutrino experimentalists.

THE SNO DETECTOR

The Sudbury Neutrino Observatory is a 1 kiloton heavy water, 9,500 phototube, real-time, state of the art second generation water Cerenkov detector designed primarily to study solar neutrinos. It is situated 2 km deep underground in an active mine at the INCO Creighton #9 shaft, near Sudbury, Canada. The SNO detector has been taking high quality pure D_2O production neutrino data for nearly two years. Water assays and many source calibration runs have been performed to study the detector cleanliness, response and systematics. It has met or exceeded all design goals and expectations.

The use of heavy water permits the unique detection of neutrinos though the three main reactions,

$$(ES) \quad \nu_e + e^- \rightarrow \nu_e + e^-$$
$$(CC) \quad \nu_e + d \rightarrow e^- + p + p \quad (1)$$
$$(NC) \quad \nu_x + d \rightarrow \nu_x + n + p$$

where ν_x refers to any active flavor of neutrino. Both the elastic scattering (ES) and charge current (CC) reactions are detected via the recoil/emitted electron that subsequently produces Cerenkov light. In the case of the neutral current (NC) reaction, the liberated neutron captures on deuterium, producing a 6.25 MeV gamma ray that subsequently Compton scatters and produces Cerenkov photons. An important fact to remember is that the ES reaction actually contains a small component of ν_μ and ν_τ.

The ES reaction is strongly pointed towards the sun, while CC has a $\sim 1 - 1/3 cos\theta_{sun}$ distribution, and the NC angular distribution is flat with respect to the sun. As well, the radial and energy distribution have different behaviors, which can be used to separate out the various reactions. This will be discussed in more detail below.

For more detector details, technical specifications, and description of calibration sources mentioned in this paper, the reader is referred to [10], [11].

RAW DATA ANALYSIS

The goal of data analysis is to separate neutrino events from the various types of backgrounds, and then count the relative sizes of the three signals in Equation 1. The data set comprises data taken from Nov 2, 1999 to January 15, 2000, with pure D_2O only. The total live days was 240.95 days, with about one third of that data kept blind, and was only opened after all cut selections and systematic studies completed.

The analysis energy threshold is taken to be 6.75 MeV, which is well above the 2 MeV hardware threshold. The reason for this is to remove backgrounds from neutrons (6.25 MeV gamma energy), and low energy radioactive decays from Uranium and Thorium. In this preliminary analysis, we are only focusing on the CC and ES reactions, thus the NC (i.e. neutrons) is considered a background.

Another major source of backgrounds are instrumental effects in the detector, such as flashing PMTs, electronic pickup, light from static discharges, etc. Data cleaning cuts were developed using only low level detector quantities such as tube time and charge, hit patterns, vetos, etc. These cuts were applied to all the data and removed most of the instrumental backgrounds, as demonstrated in Figure 1. Calibration sources were then used to measure the sacrifice of these cuts, which was estimated to be less than 0.5%. Residual contamination levels were estimated to be less than 1%.

Events that passed this first stage then had their vertex position and direction of the particle fitted using the time calibrated PMTs. The fit algorithm used a time and angle PDF which were determined from calibration sources and Monte Carlo. The fit vertex resolution is found to be 16 cm, and the angular resolution is 26.7^o.

After the position fit, an energy estimate was assigned to the event that assumed the hypothesis that it was generated by Cerenkov light from a single electron and based on

Instrumental Background Cuts

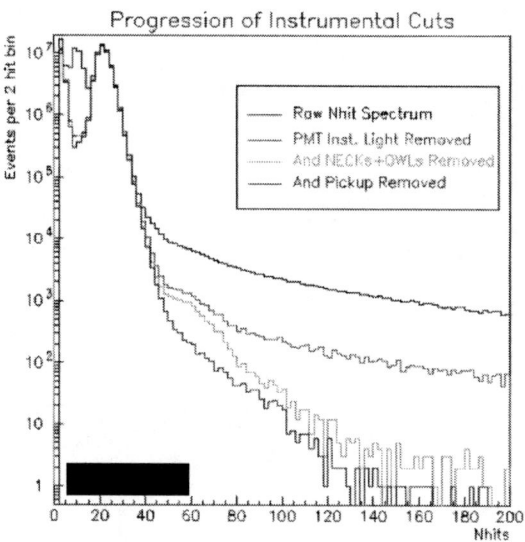

FIGURE 1. The effects of progressive instrumental data cleaning cuts applied to the data.

the number of hits. The estimate was also corrected for detector conditions such as the number of online and fully functional PMT channels, as well as optical effects. Figure 2 show the detector response to various calibration sources and the response simulated from the detector Monte Carlo. The excellent fit demonstrates an understanding of our energy response withing the D_2O in the solar neutrino energy range. An overall systematic error on the energy scale has been estimated at 1.4%.

With the event fitted, and the energy determined, two more higher level cuts were performed to ensure that the events were consistent with Cerenkov radiation. We define Cerenkov light with two orthogonal cuts, one which tests the narrowness of the timing distribution (prompt light), and the other the angular distribution of hit PMTs. Figure 3 shows these distributions for instrumental backgrounds, calibration source, and neutrino data. We require the final selected data to reside in the box shown in Figure 3, and should ensure we have removed all contamination from instrumental backgrounds, while keeping total signal loss at the 1% level.

The final cuts are kinematical, which require events to be within a fiducial volume of 550 cm, and have a kinematic energy of $T > 6.75$ MeV. The fiducial volume cut minimizes contamination from the acrylic vessel (at 600 cm), light water (from 600 to 890 cm) and PMT backgrounds (at 890 cm), while the energy threshold reduces radioactive backgrounds and neutron events. Monte Carlo simulation of the backgrounds from

SNO Energy Calibrations

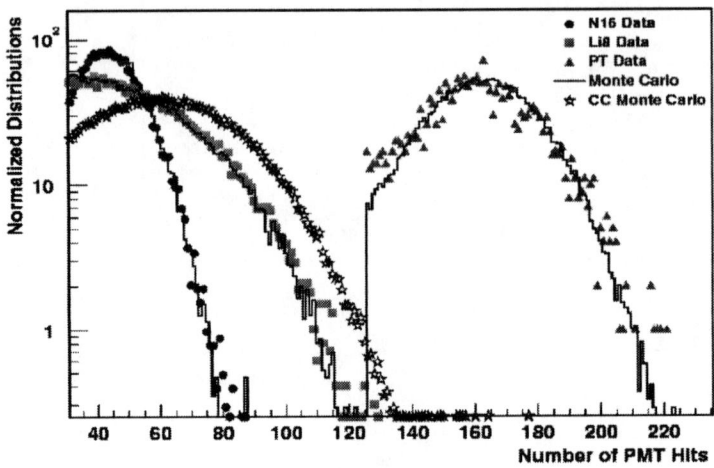

FIGURE 2. Detector kinetic energy response to various calibration sources. Also shown is the expected undistorted CC spectrum. In the region of interest, the energy response of the detector is well quantified by the Monte Carlo, and corresponds to about 9 hits/MeV.

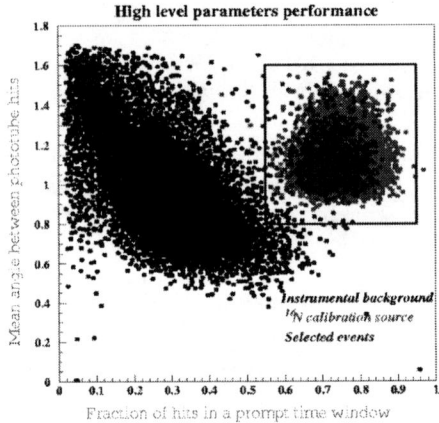

FIGURE 3. Cerenkov box cuts defined by average angle between hit PMTs, and ratio of prompt to total light. The box region denotes the accepted signal region.

TABLE 1. Breakdown of data processing and selection steps.

Analysis step	No. of events
Total event triggers	355 320 964
Neutrino data triggers	143 756 178
$N_{hit} \geq 30$	6 372 899
Instrumental background	1 842 491
Muon followers	1 809 979
Cerenkov box cuts	923 717
550 cm fiducial volume cut	17 884
6.75 MeV threshold cut	1 169
Total events remaining	1 169

Uranium and Thorium decay, with concentration input from water assay measurements, shows that the contamination above the energy threshold cut is minimal. Table 1 summarizes the event numbers as a function of processing or selections applied. The final total of 1169 events represents a clean sample of solar neutrinos.

RESULTS

Now that we have a clean sample of solar neutrinos in hand, the next step is to fit the data to the known angular ($cos\theta_{sun}$), radial (R), and energy (E) distributions. These signal probability density functions (PDF) were generated with the detailed detector Monte Carlo, which are shown in Figure 4. As well, the energy spectrum was derived from an undistorted ^8B spectrum [12].

Figure 5, 6 and 7 shows the resulting fits to the neutrino sample for the three SNO reactions, i.e. ES, CC, and neutrons. Remember, not all neutrons are necessarily from solar NC reactions, with some contribution from U/Th decays. The fits to the data are good, and are consistent with the hypothesis that the events are from solar neutrinos. As well, no statistical difference was found between the open and blind data. A maximum likelihood fit using these distributions yields 975 ± 39.7 CC events, 106.1 ± 15.2 ES events, and 87.5 ± 24.7 neutron events.

Correcting for detector acceptance and cross section normalization [13], we find the following fluxes,

$$\phi_{SNO}^{ES} = 2.39 \pm 0.34(stat)_{-0.14}^{+0.16}(sys) \times 10^6 cm^{-2} s^{-1} \quad (2)$$

and,

$$\phi_{SNO}^{CC} = 1.75 \pm 0.07(stat)_{-0.11}^{+0.12}(sys) \pm 0.05(theory) \times 10^6 cm^{-2} s^{-1} \quad (3)$$

A breakdown of the systematic errors can be found in [8], and the theoretical error is from the calculated CC cross section uncertainty [13].

From simple electroweak physics, the ES reaction is sensitive to all three flavors of neutrinos approximately as follows,

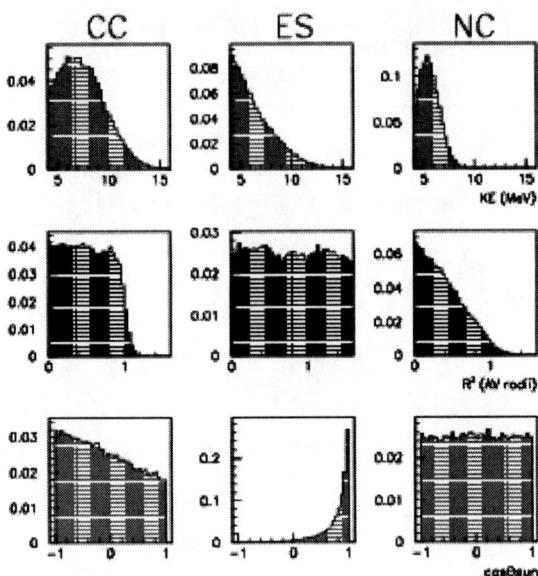

FIGURE 4. Monte Carlo generated energy (top), radial (middle), and directional (bottom) distributions used to fit the data for each of the three reaction types.

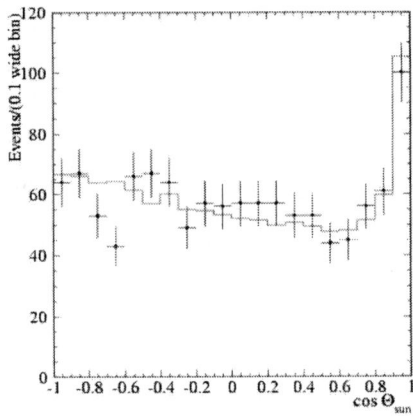

FIGURE 5. Reconstructed direction with respect to the Sun. The error bars are the data, and the line the signal Monte Carlo. The peak at $cos\theta \sim 1$ is the expect ES, while gradually increasing slope out to -1 is from CC.

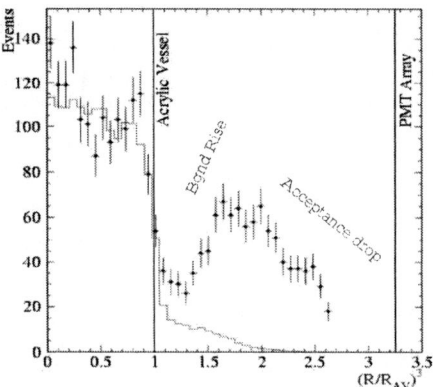

FIGURE 6. The reconstructed radial distribution (normalized to the acrylic vessel). The error bars are the data, and the line the signal Monte Carlo. The sudden drop at $R \sim 1$ is from the CC flux suppressed in the light water region.

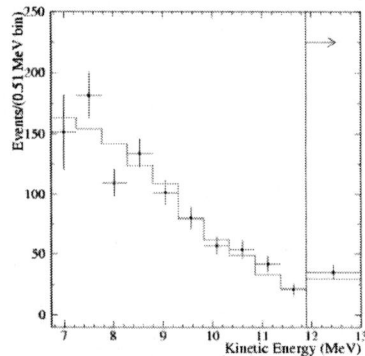

FIGURE 7. The reconstructed neutrino energy. The error bars are the data, and the line the signal Monte Carlo. The fit is consistent with a ^8B spectrum.

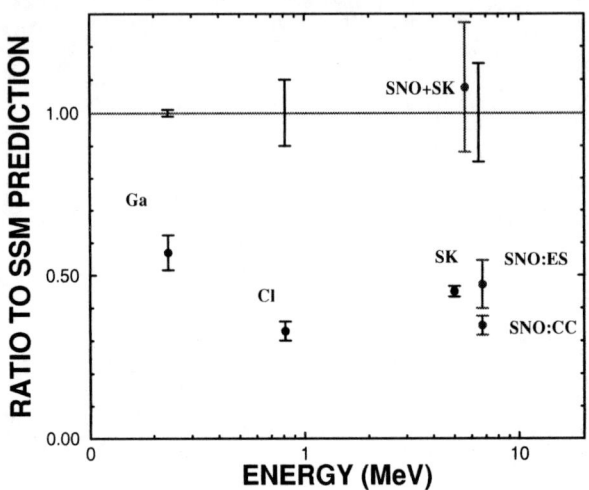

FIGURE 8. Summary of solar neutrino rate measurements relative to the standard solar model (BPB01). Also shown is the total flux derived from the SNO+SuperK difference.

$$\phi^{ES}(\nu_x) \approx \phi(\nu_e) + 1/7\phi(\nu_\mu) + 1/7\phi(\nu_\tau) \qquad (4)$$

while the CC reaction is only sensitive to electron neutrinos,

$$\phi^{CC}(\nu_e) = \phi(\nu_e) \qquad (5)$$

Thus, a difference between the two measured fluxes (assuming identical effective thresholds), is sensitive to the appearance of muon and tau neutrinos from the sun, which is evidence of oscillations.

The difference in the measured fluxes in equation 2 and 3, yield only a 1.6σ effect. This low significance is driven by the limited statistics in the ES reaction. Fortunately, a precise ES flux measurement, with nearly identical threshold, exists from SuperK [5],

$$\phi^{ES}_{SuperK} = 2.32 \pm 0.03(stat)^{+0.08}_{-0.07}(sys) \times 10^6 cm^{-2}s^{-1} \qquad (6)$$

This yields a more significant difference $\phi^{ES} - \phi^{CC} = 0.57 \pm 0.17 \times 10^6 cm^{-2}s^{-1}$, or 3.3$\sigma$, which is clear indication that muon and tau neutrinos are appearing from oscillations. Releasing the ^8B spectrum constraint, yield similar results (though with larger errors), suggesting that the solar ^8B spectrum is essentially undistorted.

Figure 8 summarizes the situation for all published solar neutrino experiments. The 3.3σ difference between SuperK and SNO can be seen, as well as the large suppression relative to the standard solar model [14]. Also shown in the figure, and given below, is the derived total neutrino flux from the sun assuming the difference between SNO and SuperK flux measurements are from ν_μ and ν_τ neutrinos,

$$\phi(\nu_x) = 5.44 \pm 0.99 \times 10^6 cm^{-2} s^{-1} \qquad (7)$$

As can be seen, the agreement is good, within the large ~20% errors.

FUTURE PROSPECTS

SNO has measured the CC and ES neutrino flux from the sun, and comparison with SuperK ES flux indicates that neutrinos are undergoing flavor oscillation. Future measurements from SNO will include the NC flux, day-night asymmetry, and CC spectrum. These will help to pin down exactly the details of the oscillation solution, e.g. LMA, LOW, SMA, Vacuum. As well, salt has been added to the detector in May of 2001, which increases the sensitivity of the NC reaction. This will make for a more precise measurement of the NC flux.

REFERENCES

1. B.T. Cleveland et al. Astrophys. J. **496**, 505 (1998).
2. Y. Fukuda et al. Phys. Rev. Lett. **77**, 1683 (1996).
3. J.N. Abdurashitov et al. Phys. Rev. C **60**, 055801 (1999).
4. W. Hampel et al. Phys. Lett. B **447**, 127 (1999).
5. S. Fukuda et al. Phys. Rev. Lett. **86**, 5651 (2001).
6. M. Altmann et al. Phys. Lett. B **490**, 16 (2000).
7. N. Hata, S. Bludman, P. Langacker, Phys. Rev. D **49**, 3622 (1994).
8. Q.R. Ahmad et al. Phys. Rev. Lett. **87**, 071301 (2001).
9. J.R. Klein, preprint hep-ex 0111040.
10. The SNO collaboration, Nucl. Instr. and Meth. **A449**, 172 (2000)
11. R.G. Van de Water, Second Tropical Workshop, San Jaun, Puerto Rico, AIP volume 540, (2000).
12. C.E. Ortiz et al. Phys. Rev. Lett. **85**, 2909 (2000).
13. M. Butler, J.W. Chen, X. Kong, Phys. Rev. C **63**, 035501 (2001).
14. J.N. Bahcall, M.H. Pinsonneault, S. Basu, (BPB01), Astro-phys. J. **555**, 990 (2001). The reference ^8B neutrino flux is $5.05 \times 10^6 cm^{-2} s^{-1}$.

Candidates for non–baryonic dark matter

Nicolao Fornengo

Dipartimento di Fisica Teorica, Università di Torino and INFN, Sezione di Torino
via P. Giuria 1, I–10125 Torino, Italy

Abstract. A review is given of the present status of non–baryonic dark matter candidates in supersymmetry and of the possibility to probe them through direct and indirect detection.

EVIDENCE OF DARKNESS

The presence of large amounts of non–luminous components in the Universe has been identified along the years by different means and on different scales: on the galactic scale, the flatness of the rotational curves of many galaxies indicates a dark component which is presumably distributed as a halo around the galaxies; clusters points toward a sizeable contribution of unseen matter distributed between galaxies; more recently, on cosmological scales, the combination of the results on high–redshift supernovae and on the anisotropies of the cosmic microwave background radiation is pointing toward a flat Universe whose energy density is dominated by a dark vacuum component (cosmological constant, quintessence) together with a sizeable dark component of matter. In terms of the density parameter Ω, the current view can be summarized as follows [1]: the total amount of matter/energy of the Universe is $\Omega_{\rm tot} = 1.02^{+0.06}_{-0.05}$, and this is composed of a matter component $\Omega_{\rm M} = 0.31^{+0.13}_{-0.12}$ and a vacuum–energy component $\Omega_\Lambda = 0.71 \pm 0.11$. The clear indication of the latest data is therefore that the Universe is strongly dominated by dark (and unknown) components. In fact the numbers above cannot be reconciled with a Universe made only of standard components: primordial nucleosynthesis tells us that baryons can contribute only at the level of $\Omega_{\rm b} = 0.037 \pm 0.11$, while luminous matter is known to provide only a contribution of order $\Omega_{\rm lum} \sim 0.003$. We are therefore facing the presence of at least *three dark components* in the Universe: dark baryons, dark matter and dark energy. The existence of both dark (relativistic or non–relativistic) exotic matter and dark energy asks for extension of the standard model of fundamental interactions, since no known particle or field can explain either of these components. In this review, we will discuss the efforts to explain the amount of dark matter in the Universe, which we summarize as:

$$0.05 \lesssim \Omega_{\rm M} h^2 \lesssim 0.3 \tag{1}$$

and the studies related to the searches for dark matter particles. For an updated and extensive list of references on all the topics covered in this paper and for a more general discussion of particle candidates to non–baryonic dark matter, we refer to Ref. [2].

SUPERSYMMETRY AND DARK MATTER: THE NEUTRALINO

The existence of a relic particle in supersymmetric theories arises from the conservation of a symmetry, R–parity, which prevents the lightest of all the superpartners from decaying. The nature and the properties of this particle depend on the way supersymmetry is broken. The neutralino can be the dark matter candidate in models where supersymmetry is broken through gravity– (or anomaly–) mediated mechanisms. The actual implementation of a specific susy scheme depends on a number of assumptions on the structure of the model and on the relations among its parameters. This induces a large variability of the phenomenology of neutralino dark matter.

The simplest and most direct implementation of supersymmetry is represented by the *minimal supergravity* (mSUGRA) scheme, where, in addition to requiring gauge coupling constant unification at the GUT scale, also all the mass parameters in the susy breaking sector are universal at the same GUT scale. The low–energy sector of the model is obtained by evolving all the parameters through renormalization group equations (RGE): this process also induces the breaking of the electroweak (EW) symmetry in a radiative way (rEWSB). This model is very predictive, since it relies only on four free parameters, but at the same time it has a very constrained phenomenology at low–energy. It also appears to be quite sensitive to some standard model parameters, like the mass of the top and bottom quarks (m_t and m_b) and the strong coupling constant α_s.

A more relaxed implementation of this susy scheme is offered by *non–universal supergravity* (nuSUGRA), where some of the unification conditions at the GUT scale are relaxed: non–universality has been studied in the Higgs, in the sfermion and in the gaugino sectors.

Specific patterns of non–universality may be originated through mechanisms which involve effects of extra–dimensions, like in *D–brane* and *string models*. These models also may be very predictive, with very few free parameters, but the relations among them is different from what is postulated in mSUGRA models.

It has also been realized that unification conditions, both for gauge couplings and/or mass parameters, may occur at scales which are different from the standard GUT scale at about 10^{16} GeV. This unification scale may be lower than the usual GUT scale (*intermediate unification scale models*), and this induces a modification of neutralino phenomenology at low energy.

A different approach is offered by the *low–energy supersymmetric model* (effMSSM), defined directly at the EW scale, which is where the phenomenology of neutralino dark matter is actually studied. Also in this case we have to make assumptions in order to reduce the number of free parameters to a manageable number. These assumptions must be mild enough not to represent an arbitrary over–constraint on the model, and all the relevant parameters at the EW scale must be represented. It is possible in this case to work with six or seven free parameters.

Other models which have been discussed in the literature in connection with neutralino dark matter are *dilaton domination* models, models with *CP–violation* and *anomaly mediated models*.

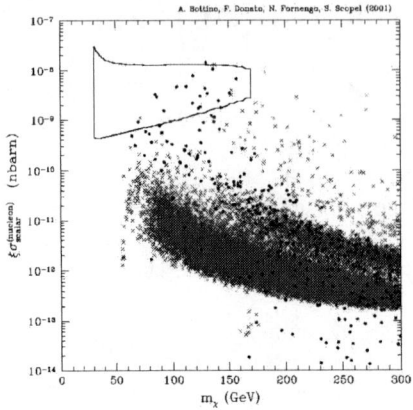 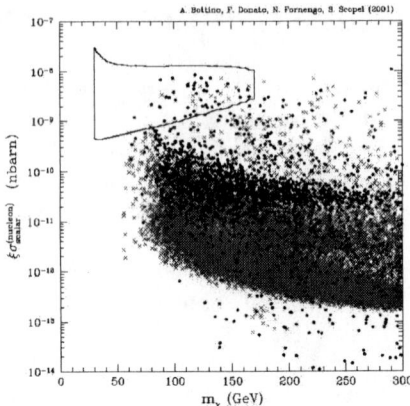

FIGURE 1. Relevant direct detection cross section times the fractional amount of neutralino dark matter in a mSUGRA (left) and nuSUGRA (right) scheme, with m_t, m_b and α_s varied inside their 2–σ allowed intervals. The (red) crosses refer to dominant neutralino dark matter, while the (blue) dots refer to sub–dominant neutralinos ($\Omega_\chi h^2 < 0.05$).

DIRECT DETECTION OF RELIC NEUTRALINOS

Direct detection relies on the scattering of dark matter particles off the nuclei of a low–background detector. This method is sensitive to the local properties of the neutralinos in the halo, *i.e.* its local abundance ρ_χ and its local velocity distribution, and depends on the neutralino–nucleus scattering cross section, which is usually dominated by the coherent interaction. The detection rate is proportional to the product $[\rho_\chi \times \sigma_{\text{scalar}}^{(\text{nucleon})}]$ for any given velocity distribution [4]. The current sensitivity of direct detection experiments on the neutralino–*nucleon* cross section is: $4 \cdot 10^{-9}$nbarn $\lesssim \xi\sigma_{\text{scalar}}^{(\text{nucleon})} \lesssim$ few $\cdot 10^{-8}$nbarn for neutralino masses in the range: 30 GeV $\lesssim m_\chi \lesssim$ 200 GeV [4]. These ranges consider the uncertainties in the local value of the dark matter density: $\rho_l = (0.2 \div 0.7)$ GeV cm^{-3}, and some uncertainties in the halo models [4]. The quantity ξ measures the fraction of local dark matter to be ascribed to the neutralino: $\rho_\chi = \xi\rho_l$ with $\xi \leq 1$ [3]. This experimental sensitivity of direct detection is shown in fig. 1 as a closed region. The contour which is reported in these figures is actually the DAMA/NaI region which is obtained when the annual modulation effect observed by the DAMA Collaboration is interpreted as due to a dark matter particle [8]. Fig. 1 shows the comparison between the experimental results and the theoretical calculations in the mSUGRA (left) and nuSUGRA (right) schemes. The quantity ξ, which determines whether the neutralino is a dominant or sub–dominant component, is calculated according to its relic abundance as: $\xi = \min(1, \Omega_\chi h^2/0.05)$ [3].

The question whether current direct detection sensitivities are probing dominant or

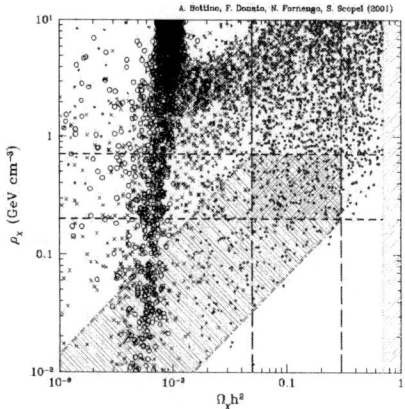

FIGURE 2. Neutralino local density ρ_χ derived by requiring that $[\rho_\chi \sigma_{\text{scalar}}^{(\text{nucleon})}]$ is at the level of the current experimental sensitivity in direct detection, plotted vs. the neutralino relic abundance $\Omega_\chi h^2$ in the effMSSM.

subdominant relic neutralinos may be answered in terms of the plot shown if fig.2, which translates directly in terms of astrophysical and cosmological quantities the direct detection results (see, for instance Ref. [4]). By considering the current interval of sensitivities on the quantity $[\rho_\chi \times \sigma_{\text{scalar}}^{(\text{nucleon})}]$, the calculation of $\sigma_{\text{scalar}}^{(\text{nucleon})}$ allows us to determine the values of ρ_χ which are required for each susy configuration in order to provide a detectable signal. Fig.2 shows the calculated values of ρ_χ vs. the neutralino relic abundance, for the effMSSM. We see that a fraction of susy models overlap with the region of main cosmological and astrophysical interest: $0.05 \lesssim \Omega_\chi h^2 \lesssim 0.3$ and $0.2 \leq \rho_\chi/(\text{GeVcm}^{-3}) \leq 0.7$. For points in this region, the neutralino is the dominant component of dark matter both in the Universe and at the galactic level. For points which fall inside the band delimited by the slant dot–dashed lines, the neutralino would provide only a fraction of the cold dark matter both at the level of local density and at the level of the average Ω, a situation which would be possible if the neutralino is not the unique cold dark matter particle component. On the other hand, configurations above the upper dot–dashed line and below the upper horizontal solid line would imply the somewhat more ulikely situation of a stronger clustering of neutralinos in our halo as compared to their average distribution in the Universe. Finally, configurations above the upper horizontal line are incompatible with the upper limit on the local density of dark matter in our Galaxy.

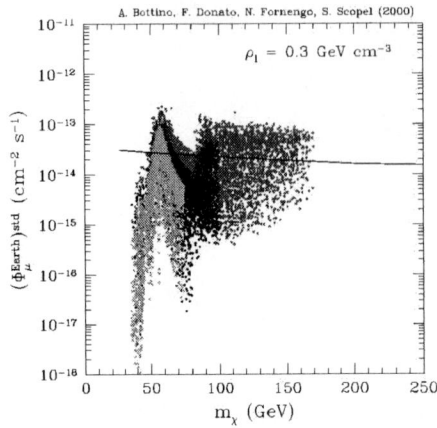
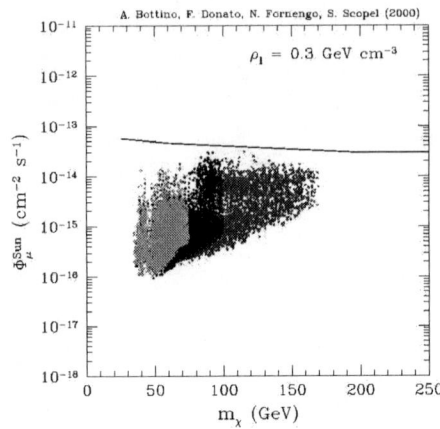

FIGURE 3. Flux of upgoing muons from neutralino annihilation inside the Earth (left) and Sun (right) in the effMSSM and for susy configurations currently explored by direct detection [6]. The horizontal line is the upper bound from MACRO.

INDIRECT DETECTION OF RELIC NEUTRALINO

Indirect detection relies on the possibility to identify signals which are originated by neutralino self–annihilations.

The first type of signal, due to neutralino annihilation taking place in celestial bodies (Earth or Sun) where the neutralinos have been gravitationally captured and accumulated, is a neutrino flux, detected in a neutrino telescope as a flux of *up–going muons*. Since the process relevant for accumulation of neutralino is capture, which relies on neutralino scattering off the nuclei of the Earth and the Sun, this detection technique is sensitive only to local properties of the Galaxy, like direct detection. In fig. 3 we show the flux of upgoing muons calculated in the effMSSM for the signal coming from the Earth and the Sun[6]. The susy configurations which are shown are only the ones which are currently at the level of being explored by direct detection and the upper limit on the upgoing muon flux from MACRO is shown as a solid line. We can see that indirect detection at neutrino telescopes is partially competitive with direct detection, although a large fraction of configurations which are explored by direct detection require more sensitive neutrino telescopes in order to be probed.

Neutralino annihilations can take place also in the galactic halo and the signals can consist in: a *diffuse neutrino flux* ; a *diffuse gamma–ray flux* or a *gamma–ray line*; exotic components in cosmic rays: *positrons*, *antiprotons* and *antideuterium*. In this case, global properties of the halo are relevant, and therefore the matter distribution of neutralinos is an important quantity. In particular, the overdensities which would be present if the halo were clumpy would have the effect of largely enhancing the

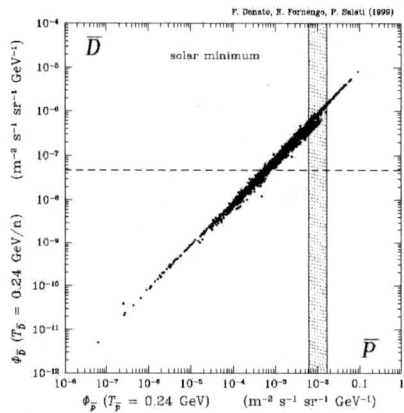

FIGURE 4. Flux of antideuterium vs. the flux of antiprotons from neutralino annihilation in the effMSSM for a smooth halo [7]. The vertical band is the measurement of BESS95+97. The horizontal line denotes the reaching capability of AMS on a 3–year flight on board of the space station

predicted signals. Neutrino and gamma fluxes require clumpiness to reach detectable levels. Also the recent anomaly in the HEAT data [9] on positrons may be explained by neutralino annihilations if overdensities are present in the galactic halo [10]. On the contrary, antiprotons and antideuterium may be detectable also in a smooth halo. In fig. 4 we show the correlation between the antiproton and the antideuterium flux calculated in the effMSSM for a smooth halo [7]. The vertical band represents the BESS 95+97 measurement of the antiproton flux. At present, there are no measurement of an antideuterium component in cosmic rays. The horizontal line in fig. 4 shows the reaching capability of the AMS detector in a 3–year flight on board of the space station. Recently, a detector designed specifically to look for antideuterium in cosmic rays has been proposed [11].

CONCLUSIONS

Two are the main issues in particle dark matter studies: *i)* to explain the observed amount of dark matter in the Universe ($0.05 \lesssim \Omega_M h^2 \lesssim 0.3$) by finding suitable particle candidates; *ii)* to detect a relic particle. For both of them there appear to be good prospects of success.

As for the candidates, there are many proposed particles which could act as dark matter. Some of these candidates turn out to be quite natural, like *e.g.* the massive neutrino, the axion or the neutralino [2]. Almost all of the proposed candidates can play the role of the dominant dark matter component, although for some of them a non–

standard cosmology is required. An important remark is that, from the particle physics point of view, dark matter may naturally be multi–component. A multi–component dark matter scenario offers opportunity for interesting phenomenology not only to the dominant candidate, which would explain the cosmological observation on the Ω_M parameter, but also to the sub–dominant candidates, since usually these are the ones which are easier to detect. The detection of a particle which is a relic from the early Universe would be a very important and exciting result.

As for detection, perspectives are good, both for direct and for indirect detection techniques, especially for the most interesting and studied candidate, the neutralino. The possibility to have detectable rates for neutralinos depends on the specific susy model which is considered, and quite generally it appears simpler to detect a relic neutralino which is a sub–dominant dark matter component. Nevertheless, there are many susy schemes where relic neutralinos can provide enough cosmological abundance to explain the observed amount of dark matter, and at the same time they can have detection rates large enough to be accessible to direct, and also to some indirect, detection methods. The positive indication of annual modulation in the detection rate of the DAMA/NaI Collaboration, which is at the moment, the most compelling indication for a particle dark matter signal, may be interpreted as originated from relic neutralinos and explained in a number of realization of supersymmetry. It is worth noticing that the presence of a signal from dark matter, like the annual modulation effect or signals which could hopefully come in future experiments, can be very important not only for astrophysics and cosmology but also for particle physics, since the need to explain the effect can help in deriving properties of particle physics models and possibly discriminate among different realizations, for instance of supersymmetry.

ACKNOWLEDGMENTS

This work was partially supported by the Research Grants of MURST within the *Astroparticle Physics Project*. I also wish to thank Conacyt Mexico for partial support. Special thanks to the Organizers for all what they have done in organizing such a wonderful workshop. Finally I wish to warmly thank Omar Miranda for his kind invitation to Zacatecas and CINVESTAV–IPN.

REFERENCES

1. P. de Bernardis et al, astro-ph/0105296.
2. N. Fornengo, invited review talk at TAUP 2001, LNGS, hep-ph/0201156.
3. A. Bottino, F. Donato, N. Fornengo and S. Scopel, hep-ph/0105233.
4. A. Bottino, F. Donato, N. Fornengo and S. Scopel, Phys. Rev. D63 (2001) 125003 [hep-ph/0010203].
5. A. Bottino, N. Fornengo and S. Scopel, Nucl. Phys. B608 (2001) 461 [hep-ph/0012377].
6. A. Bottino, F. Donato, N. Fornengo and S. Scopel, Phys. Rev. D62 (2000) 056006 [hep-ph/0001309].
7. F. Donato, N. Fornengo and P. Salati, Phys. Rev. D62 (2000) 043003 [hep-ph/9904481].
8. R. Bernabei *et al.* (DAMA Collab.), Phys. Lett. B480 (2000) 23; Eur. Phys. J. C18 (2000) 283.
9. S. W. Barwick *et al.*, Astrophys. J. 482 (1997) L191; S. Coutu *et al.*, Proc. of 27th ICRC (2001).
10. E.A. Baltz *et al.*, astro-ph/0109318; G.L. Kane *et al.*, hep-ph/0108138.
11. K. Mori *et al.*, astro-ph/0109463.

Results from the Mainz Neutrino Mass Experiment and Perspectives of the next Generation Experiment KATRIN

J. Bonn

Joh. Gutenberg University, Mainz, Germany
email: jochen.bonn@uni-mainz.de

Abstract. The most direct way to get information on neutrino masses is the investigation of weak decays and tritium is the best candidate due to its high specific activity and low endpoint energy. The relevance of this approach is based on the following arguments:

- Neutrinos are not massless as we have convincing evidence for neutrino oscillations from experiments detecting atmospheric and solar neutrinos. The signals are interpreted as differences in the squares of neutrino masses from which a lower bound for the neutrino masses can be derived.

- As these differences are rather small one can set limits for all neutrino masses from the lowest upper bound which is set by the Mainz group from tritium β decay:
  $m_\nu^2 = -1.6 \pm 2.5 \pm 2.1 \text{ eV}^2/c^4$ corresponding to an upper limit of $m_\nu < 2.2 \text{ eV}/c^2$ (95 % C.L.). Hence the sum of the three neutrino masses is less than about $7 \text{ eV}/c^2$.

The results are of interest for cosmology as neutrinos are part of the hot dark matter. The minimum total mass deduced from oscillation experimentes is about equal to the sum of the stellar masses. The present upper bound from the direct mass mesurements is about equal to the limit given by stucture formation.

The result of the two presently working experimentes in Mainz and Troitsk and their systematic uncertainties will be presented.

These experiment have almost reached their sensitivity limits. A next generation experiment KATRIN will be set up at Forschungszentrum Karlsruhe. Its aim is to push the sensitivity limit to $m_\nu < 0.3 \text{ eV}/c^2$ (90 % C.L.). This would fully cover the cosmologically relevant parameter space and set the neutrino mass scale to discriminate between models with degenerated and hierachical neutrino mass scenarios.

INTRODUCTION

With the recent evidences for neutrino oscillation and consequently for non-zero neutrino masses from atmospheric [1] and solar neutrino [2] experiments the question of the neutrino mass scale has become very important, as it has strong consequences for particle physics as well as for astrophysics and cosmology. The investigation of the tritium β spectrum near its endpoint measures the mean square of the mass of the "electron neutrino" $m_{\nu_e}^2 = \sum_j |U_{ej}|^2 m_j^2$ with neutrino mixing matrix U and neutrino mass eigenstates m_j; it is the most sensitive of the so-called direct methods and provides information complementary to the searches for neutrinoless double β decay and oscillation experiments [3]. Tritium β decay is the ideal method to distinguish between models with hier-

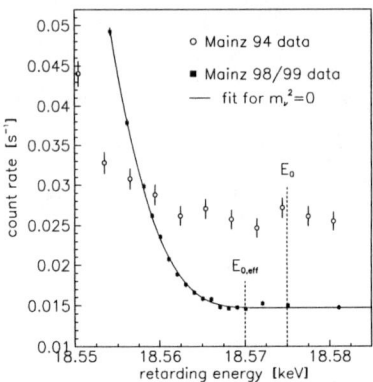

FIGURE 1. Averaged count rate of the combined 1998 and 1999 data Q3–Q8 (points) with fit (line) in comparison with previous Mainz data from 1994 [12] as function of the retarding energy near the endpoint E_0, and effective endpoint $E_{0,eff}$. The latter is shifted about 5 eV below E_0 due to the finite width of the filter of 4.8 eV and a mean rovibrational excitation energy of 1.7 eV of the daughter molecule $(T\,^3He)^+$.

archical and those with degenerate neutrino masses. Furthermore, neutrino masses up to about 1 eV/c^2 are especially interesting for cosmology because of their contribution to the missing hot dark matter in the universe. A mass of about 1 eV/c^2 is fitting models of structure formation [4].

The tritium β decay experiments running at Mainz and Troitsk [5, 6] are currently reaching their sensitivity limit of 2–3 eV. Both experiments use spectrometers, which are called MAC-E-Filter (Magnetic Adiabatic Collimation followed by a retarding Electrostatic Filter) [7, 8]. A future experiment based on this principle, the KArlsruhe TRItium Neutrino experiment (KATRIN) is proposed to improve the neutrino mass sensitivity by one order of magnitude into the sub-eV region. This will allow to check the cosmological relevance of neutrino masses and to distinguish between hierarchical and degenerate neutrino mass scenarios.

THE MAINZ NEUTRINO MASS EXPERIMENT

The Mainz experiment uses as β source a solid source of molecular tritium, quench condensed on a graphite substrate (HOPG), which is atomic flat on a microscopic scale. The Mainz experiment has been recently upgraded [5, 9, 10] to improve the sensitivity to m_ν down to a limit of 2 eV/c^2 and to check the anomalous excess in the spectrum close to the endpoint reported by the Troitsk group [15, 6]. This upgrade allows to use films of a typical thickness of about 480 Å, which corresponds to about 140 monolayers. The source diameter is 17 mm giving a luminosity which is equal to the Troitsk one.

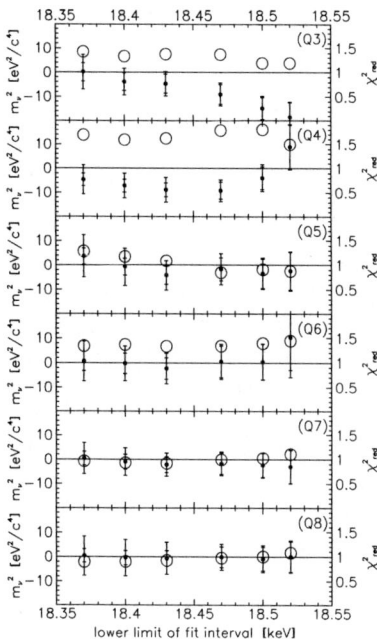

FIGURE 2. Fit results on m_ν^2 (left scale, filled circles) for the different runs Q3–Q8 with statistical (inner bars) and total uncertainties (outer bars) in dependence on the lower limit of the fit interval (upper limit: 18.66 keV, well above E_0). The corresponding values of $\chi^2_{red} = \chi^2/d.o.f.$ of the fits (open circles) can be read from the right scale.

The 6 runs of 1998 and 1999

With the improved setup 6 runs (labelled Q3-Q8) of 7 month measurement time in total have been taken in 1998 and 1999. Additional studies on quench condensed T_2 films clarified their energy loss function [11], their self-charging [9, 10], and their dewetting as a function of temperature [13, 14].

Fig. 1 shows the integral β spectrum as function of the filter potential for the last 15 eV below the endpoint E_0= 18575 eV plus the background region above E_0. Plotted is the count rate averaged over the runs Q3 to Q8; data obtained in 1994 [12] are shown for comparison. The improved spectrometer yields a signal to background ratio 10 times better than before and much better statistics has been obtained meanwhile. A fit with m_ν^2 fixed to zero perfectly fits the new data set over these last 15 eV of the β spectrum. This limits any persistent spectral anomaly in this range to an amplitude below 10^{-3}/s (as against a total flux of 10^8/s entering the spectrometer). A spectral anomaly, like the fluctuating anomaly reported by the Troitsk group [15, 6], on the other hand, reaches amplitudes up to 10^{-2}/s.

Figure 2 shows the fit results on m_ν^2 with statistical and total uncertainties for the 6 different runs Q3 to Q8 as function of the lower energy limit of the data interval used for the analysis. The monotonous trend towards negative values of m_ν^2 for larger fit intervals as it was observed for the Mainz 1991 and 1994 data [12] has vanished. This shows that the dewetting of the T_2 film from the graphite substrate [13, 14] indeed was the reason for this behavior. Now this effect is safely suppressed at the much lower temperature of the T_2 film. However, the first two data sets of 1998 (Q3,Q4) still do not fulfill the requirement of being stable with respect to the variation of the lower limit of the fit range and being compatible with the physical allowed range of $m_\nu^2 \geq 0$. But deviations were limited to -10 eV^2/c^4 corresponding to a 2 σ level. The origin of this small residual spectral distortion has not yet been completely clarified, but is certainly connected to a minute instrumental effect depending on running conditions. This statement is corroborated by the fact that the later runs Q5 to Q8 showed absolutely "clean" spectra (see Fig. 2). This improvement was probably due to a lowering and a stabilization of background rate achieved by applying rf pulses to one of the electrodes of the electrostatic filter in the 2 s measuring pauses during which the filter potential was changed every 20 s. This way some trapped charges were swept out of the spectrometer, apparently. Trapped particles cause background events through secondary reactions with the residual gas.

Taking the clean data sets Q5–Q8 we use the data down to 70 eV below E_0. For this interval the combined statistical and systematic uncertainty attains a minimum. The result for m_ν^2 is
$$m_\nu^2 = -1.6 \pm 2.5 \pm 2.1 \; eV^2/c^4$$
which corresponds to an upper limit of
$$m_\nu \leq 2.2 \; eV/c^2 \quad (95 \% \; C.L.)$$

The 2 runs of 2000

From October to the end of the year 2000 the Mainz group took data. No significant Troitsk-like anomaly was found in particular not in those two subsets of data which were obtained in parallel to data taking at Troitsk (6.12-13.12 and 22.12 - 28.12). But most (not all) of the data subsets suffered again from small residual perturbations which resulted in negative m_ν^2 fit results in the range of -10 eV^2/c^4. We ascribe the reappearance of slight spectral irregularities in our data to a somewhat less favorable status of the spectrometer. Contrary to earlier runs we have not baked it beforehand this time. All our observations leave little doubt that the residual spectral anomalies discussed here are of instrumental origin and connected to the particular electric and magnetic design of MAC-E-Filters, and, moreover, that they can be overcome by paying utmost attention to UHV and surface conditions as well as to the electromagnetic design. [1]

---

[1] This statement is supported by the fact that – after having prepared every part of the whole setup in the best way of which we know – our new measurement of October 2001 results in very clean spectra like in late 1998 and 1999.

CONCLUSION AND OUTLOOK

The upgraded Mainz experiment has obtained very clean data sets in late 1998 and 1999, which are fully compatible with an ordinary β spectrum at neutrino mass zero. The intensive studies of the systematics of quench condensed tritium films allowed to avoid the tritium film roughening and to characterize the film properties with small uncertainties. From the result on m_ν^2 an upper limit on the neutrino mass of 2.2 eV/c^2 is deduced, which is practically the sensitivity limit of the experiment.

But this sensitivity is not enough to clarify the open questions concerning the neutrino mass scale and the cosmological role of neutrinos mentioned above. An improvement by another order of magnitude is needed. Currently the only method to reach a sub-eV sensitivity on the neutrino mass is again the investigation of tritium β decay: The use of β emitters with even lower endpoint energy like Rhenium in combination with cryogenic bolometers is a very important new approach with high potential, but it is still in the development phase much behind present tritium β decay experiments. As mentioned above the very important search for neutrinoless double β decay gives complementary results and cannot replace the direct searches especially in the case of degenerated neutrino masses [3]. Therefore, the most promising and reliable way to reach this sub-eV sensitivity is to utilize the very successful MAC-E-Filter investigating the very upper end of the tritium β spectrum.

To clearly reach the sub-eV sensitivity a significantly better energy resolution (1 eV instead of 5 eV at Mainz) and a much higher count rate (the region of interest scales with the neutrino mass) is needed, both requiring a significantly larger spectrometer. In a previous paper [17] the possibility of a spectrometer based on the same MAC-E filter principle but 5 times larger (in linear dimensions) than the present one was investigated. By an additional time-of-flight analysis the spectrometer can transform from an integrating high pass filter into a narrow band filter (MAC-E-TOF mode), an interesting feature for the investigation of systematics [17].

Now a strong collaboration with institutions from Fulda/Germany, Karlsruhe/Germany, Mainz/Germany, Prague/Chech Republic, Seattle/USA and Troitsk/Russia has been formed to carry out the KArlsruhe TRItium Neutrino experiment KATRIN at the German national laboratory Forschungszentrum Karlsruhe. This idea, the motivation and implications for physics was discussed at an international workshop at Bad Liebenzell/Germany in January 2001 [18] resulting in a strong support by the international community. Meanwhile a letter of intent is available [19]. KATRIN will build up a large tritium β spectrometer with 7 m diameter and 20 m length following a pre spectrometer which rejects already β electrons except the ones with energies very close to the endpoint. As tritium source a molecular gaseous tritium source will be used, like in the Troitsk experiment but having a factor of 40 times higher count rate. A quench condensed molecular tritium source like in Mainz is aimed for additionally as an alternative source. First simulations with quite conservative assumptions show that this experiment should reach an uncertainty on the electron neutrino mass squared Δm_ν^2 of 0.08 eV^2/c^4. Of course, we are hoping to find a non-zero neutrino mass, but if not this would correspond to an upper limit of 0.35 eV/c^2 at 90 % C.L. First funding for KATRIN has been given by the German Bundesministerium für Bildung und Forschung to built a pre-spectrometer and to check new ideas to improve the experiment. A full proposal will be ready within next year, the real construction phase should start in 2003, first data are expected for 2006.

ACKNOWLEDGEMENTS

The work of the Mainz group was supported by the Deutsche Forschungsgemeinschaft under contract Ot33/13 and by the German Bundesministerium für Bildung und Forschung under contract 06MZ866I/5.

REFERENCES

1. Y. Fukuda, *et al.*, Phys. Rev. Lett. **85** (2000) 3999
2. Q.R. Ahmad, *et al.*, nucl-ex/0106015
3. Y. Farzan *et al.*, hep-ph/0105105
4. M. Tegmark *et al.*, Phys. Rev. **D63** (2001) 043007
5. Ch. Weinheimer *et al.*, Phys. Lett. **B460** (1999) 219
6. V.M. Lobashev, *et al*, Phys. Lett. **B460** (1999) 227
7. A. Picard *et al.*, Nucl. Inst. Meth. **B63** (1992) 345
8. V.M. Lobashev *al.*, Nucl. Inst. Meth. **A240** (1985) 305
9. H. Barth *et al.*, Prog. Part. Nucl. Phys. **40** (1998) 353
10. B. Bornschein, PhD thesis, Mainz university, 2000
11. V.N. Aseev *et al.*, Eur. Phys. J. **D10** (2000) 39
12. H. Backe *et al.*, Proc. of Neutrino 96, Helsinki, World Scientific/Singapure
13. L. Fleischmann *et al.*, J. Low Temp Phys. **119** (2000) 615
14. L. Fleischmann *et al.*, Eur. Phys. J. **B16** (2000) 521
15. A.I. Belesev *et al.*, Phys. Lett. **B350** (1995) 263
16. J. Bonn *et al.*, Nucl. Phys. B (Proc. Suppl.) **91** (2001) 273
17. J. Bonn *et al.*, Nucl. Inst. and Meth. **A421** (1999) 256
18. International Workshop on "Neutrino Masses in the sub-eV Range" Bad Liebenzell, Germany, January 18–21, 2001, http://www-ik1.fzk.de/tritium/liebenzell/
19. A. Osipowicz *et al.*, hep-ex/0109033

Two Phases of Neutrino Physics

José Wudka

Physics Department, University of California Riverside CA 92521-0413, USA

Abstract. In this talk I will review some of the leading effects of gravity in the quantum evolution of various particles, concentrating on the effects on neutrinos (both Majorana and Dirac).

INTRODUCTION

Neutrinos due to their weak interactions are well suited for providing information about the properties of otherwise inaccessible regions, such as the core of active galactic nuclei. Being copiously produced in such regions, they then travel mostly unaffected by their environment. Upon detection their observed properties such as the energy spectrum, flux and flavor distribution can be used to obtain information about the creation region. Nonetheless there are some interactions that can modify these properties. In this talk I will discuss some such effects produced by the interactions of neutrinos with the gravitational and electromagnetic fields.

NEUTRINOS IN GRAVITY

I will separate the effects produced by the interactions of neutrinos with a gravitational field into two types. Strong local effects near a deep gravitational well and weak extended effects. I will discuss each of these separately.

Strong gravitational effects

I will assume that the neutrinos move semiclassically along a geodesic of affine parameter ℓ and use standard techniques to extract the effective Hamiltonian describing the evolution of a neutrino wave-packet as a function of ℓ. The Dirac equation for neutrinos in a gravitational field is [1] $[ie_a^\mu \gamma^a(\partial_\mu + \omega_\mu) - m + \rlap{/}{J} P_L]\psi = 0$ where e_μ^a denote the tetrads, J_L the weak current and $\omega_\mu = \frac{1}{8}[\gamma_a, \gamma_b] e^{va} e^b_{v;\mu}$ [1]. In the semiclassical expansion

---

[1] Greek and Latin indices denote the global and local Lorentz coordinates respectively. A semicolon denotes a covariant derivative, a comma the usual derivative. Greek (Latin) indices are raised and lowered using the global (flat) metric $g_{\mu\nu}$ (η_{ab}).

the neutrino wave function takes the form

$$\psi = e^{iS}\chi \tag{1}$$

where χ denotes a slowly-varying amplitude and S the classical action satisfying the usual Hamilton-Jacobi equation [2] $g^{\mu\nu}\partial_\mu S\,\partial_\nu S = 0$ (neglecting mass effects). The solution gives the expression for the geodesic $x(\ell)$. In finding the semiclassical corrections to the purely classical case $\chi = $ constant, it is convenient to construct an adapted coordinate system obtained by using a set of 3 vectors v_A^μ, $A = 1,2,3$ satisfying the geodesic deviation equation [3]. I then choose coordinates [5] $\{\ell, \xi^A\}$ such that a point x in the neighborhood of the geodesic can be written $x = x^\mu(\ell) + v_A^\mu(\ell)\xi^A$.

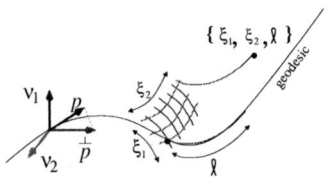

In terms of S the classical momentum is given by [2] $p_\mu = -\partial_\mu S$ and it proves useful to define the time-like projection of p, namely [6]

$$\overset{\perp}{p}{}^\mu = \left(\delta^\mu_\sigma - v_{A\sigma}(N^{-1})^{AB}v_B^\mu\right)p^\sigma; \qquad N_{AB} = v_A^\mu v_B^\nu g_{\mu\nu} \tag{2}$$

I will be interested in the evolution of the amplitude χ which determines the probability of neutrino oscillations. Assuming that the wave packet remains localized I expand χ in a double power series, in ξ and $1/pR$, where p denotes the momentum of the neutrinos and R the scale of the gravitational interactions. When this is substituted into the Dirac equation yields the following effective Hamiltonian (see [6] for details),

$$\tilde{H}_{\text{eff}} = \frac{1}{2}m^2 - p\cdot(J_W - J_G)P_L \qquad J_G^a = \frac{1}{4}\varepsilon^{abcd}\lambda_{fcd}\left(\eta_b^f + 2p^f\overset{\perp}{p}_b/\overset{\perp}{p}{}^2\right) \tag{3}$$

where $\lambda_{fcd} = (e_{f\mu,\nu} - e_{f\nu,\mu})e_c^\mu e_d^\nu$. Denoting by u^a the speed of the matter in the local Lorentz frame and N_p, N_n the proton and neutron number densities, the weak current for electron-type neutrinos equals $J^a_{W;\nu_e} = (G_F/\sqrt{2})(2N_p - N_n)u^a$ while for muon-type neutrinos $J^a_{W;\nu_\mu} = -(G_F/\sqrt{2})N_n u^a$. The final expressions, including a dipole-moment coupling to the electromagnetic field are [6]

$$i\frac{d}{d\ell}\begin{pmatrix}\nu_{eL}\\ \nu_{\mu L}\\ \nu_{eR}\\ \nu_{\mu R}\end{pmatrix} = H_{\text{eff}}\begin{pmatrix}\nu_{eL}\\ \nu_{\mu L}\\ \nu_{eR}\\ \nu_{\mu R}\end{pmatrix} \tag{4}$$

$$H_{\text{eff}} = \begin{pmatrix} -p \cdot J_{\text{eff}}^{\nu_e} + \frac{1}{2}\Delta m_{12}^2 \sin^2\vartheta & \frac{1}{4}\Delta m_{12}^2 \sin 2\vartheta & E\mu_{ee}\Omega^* & E\mu_{e\mu}\Omega^* \\ \frac{1}{4}\Delta m_{12}^2 \sin 2\vartheta & -p \cdot J_{\text{eff}}^{\nu_\mu} + \frac{1}{2}\Delta m_{12}^2 \cos^2\vartheta & E\mu_{\mu e}\Omega^* & E\mu_{\mu\mu}\Omega^* \\ E\mu_{ee}\Omega & E\mu_{e\mu}\Omega & 0 & 0 \\ E\mu_{\mu e}\Omega & E\mu_{\mu\mu}\Omega & 0 & \frac{1}{2}\Delta m_{12}^2 \end{pmatrix}$$
(5)

where ϑ is the neutrino mixing angle, E the neutrino (conserved) energy and $\Omega = (B^1 - E^2) + i(E^1 + B^2)$ in terms of the electromagnetic field in the local frame [2] coupled through the dipole moments μ_{ij}. Note that the mass terms in H_{eff} are simply $m^2/2$.

For Majorana neutrinos [7],

$$i\frac{d}{d\ell} \begin{pmatrix} \nu_e \\ \nu_\mu \\ \bar{\nu}_e \\ \bar{\nu}_\mu \end{pmatrix} = H_{\text{eff}}^{\text{Maj}} \begin{pmatrix} \nu_e \\ \nu_\mu \\ \bar{\nu}_e \\ \bar{\nu}_\mu \end{pmatrix}$$
(6)

$$H_{\text{eff}}^{\text{Maj}} = \begin{pmatrix} p \cdot J_{\text{eff}} & \frac{1}{4}\Delta m^2 \sin 2\vartheta & 0 & E\mu_t\Omega^* \\ \frac{1}{4}\Delta m^2 \sin 2\vartheta & p \cdot J_{\text{eff}} + \frac{1}{2}\Delta m^2 \cos 2\vartheta & -E\mu_t\Omega^* & 0 \\ 0 & E\mu_t\Omega & -p \cdot J_{\text{eff}} & \frac{1}{4}\Delta m^2 \sin 2\vartheta \\ E\mu_t\Omega & 0 & \frac{1}{4}\Delta m^2 \sin 2\vartheta & -p \cdot J_{\text{eff}} + \frac{1}{2}\Delta m^2 \cos 2\vartheta \end{pmatrix}$$
(7)

I will investigate some of the effects that can be derived from the above expressions for those interesting cases: (i) resonances and spin precession in strong gravitational environments, and (ii) the evolution in weak gravity for long distances.

Resonances and spin precession

There effective Hamiltonian for Dirac neutrinos implies the possible occurrence of two types of resonances. (i) L↔L: the MSW resonances which remain unaffected by the gravitational interactions (assuming the principle of equivalence is not violated). (ii) L↔R: when $EJ_{\text{eff}} \sim \Delta m^2$ and do depend on the presence of gravitational fields.

For Majorana neutrinos I'll concentrate on the $\nu_\mu \leftrightarrow \bar{\nu}_e$ transitions. Using a standard procedure [8] the persistence and transition probabilities for the ν_μ are

$$1 - P(\nu_\mu \to \bar{\nu}_e) = P(\nu_\mu \to \nu_\mu) = \frac{1}{2} + \cos 2\vartheta \left(\frac{1}{2} - P_{LZ}\right) \cos(2\vartheta_G)$$
(8)

where the gravitational mixing angle ϑ_G is given by

$$\tan 2\vartheta_G = E\mu_t B/[(\Delta m^2 \cos 2\vartheta)/4 + p \cdot J_G]\big|_{\text{production}}$$
(9)

---

[2] The third direction is defined by the neutrino momentum.

and the Landau-Zener probability equals

$$P_{LZ} = e^{-2\pi^2\beta^2\Lambda}, \quad \beta = E\mu_t B\,|_{\text{resonance}}, \quad \Lambda^{-1} = |d(p\cdot J_{\text{eff}})/d\ell|_{\text{resonance}} \quad (10)$$

For typical AGN models $J_W \lesssim 10^{-29}$ eV^{-1} while $J_G \sim 1/R$ (R denotes the scale of the metric). Near the horizon R is of the order of the gravitational radius so that $J_G \gg J_W$ in the vicinity of an AGN. Resonances then occur for phenomenologically acceptable values of the parameters: e.g., a black hole with $M \sim 10^8 M_\odot$, $\Lambda \sim r_g$, $\Omega \sim 10^4$G, and when $\Delta m^2 \sim 10^{-5}$ eV^2, resonances will occur provided $E \sim 10$TeV, $\mu > 3 \times 10^{-15} \mu_B$.

A competing effect is that of coherent precession generated by the off-diagonal terms in H_{eff}. These effects are important provided $\mu > |R\Omega|^{-1}\left(\Delta m^2 R/E + 1\right) = \mu_{\min}^{\text{prec}}$. Note that $\mu_{\min}^{\text{prec}}/\mu_{\min}^{\text{res}} = \sqrt{R\Delta m^2/E} + 1/\sqrt{R\Delta m^2/E} \geq 2$

Neutrinos near a Kerr black hole

Denoting the gravitational radius by r_g the resonance condition is $f = (r_g/E)|p \cdot J_G| = O(1)$ which is realized in a wide region in the vicinity of the horizon:

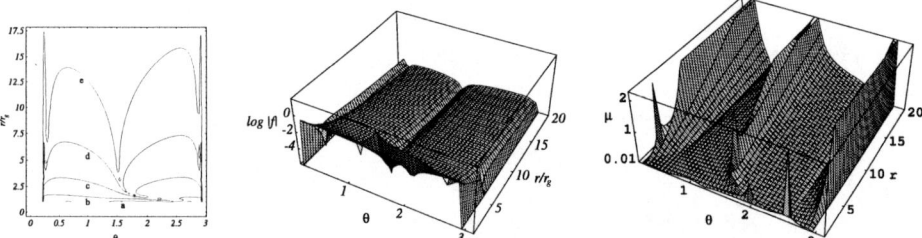

(the lines $a,\ldots,e$ correspond to $f = 10^0,\ldots,10^{-4}$). The last graph gives $\mu_{\min}$ for Dirac neutrinos in units of $10^{-13}\mu_B$ and for $E = 1$TeV, $\delta m^2 = 10^{-6}$eV^2,

For Majorana neutrinos [7], for $\mu_t = 10^{-17}\mu_B$ at a distance of $r = 6r_g$, $\theta = \pi/2$,

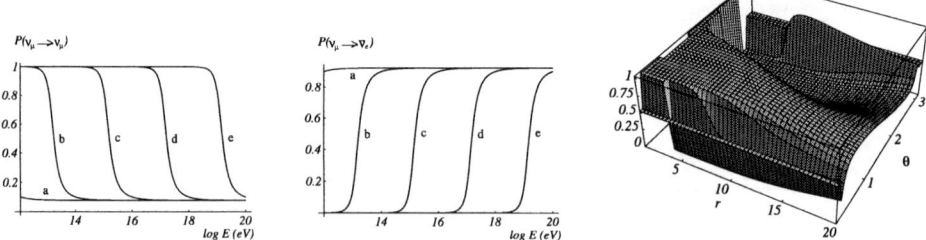

(where $a,\ldots,e$ correspond to $\Delta m^2 = 10^{-10},\ldots 10^{-2}$eV^2) and in the last plot I chose $\mu_t = 10^{-17}\mu_B$, $E = 1$PeV, $\Delta m^2 = 10^{-8}$eV^2. The transition probability will be large provided $(\pi\nu)^2 > 2\delta$; $f > \max.\{\delta,\nu\}$ where $\delta = r_g|\Delta m^2|/(4E)$, $\nu = |r_g\mu_t B|$

The gravitational resonances will spread an initial flux over all the neutrino flavors. For example if the initial flavor is mainly ν_μ with flux $F \sim$

$(F_0/E^2)\exp(-20E/E_{\max})$, $F_0 \sim 2.5 \times 10^{-12}$ erg/(s cm^2) [9], the observed ν_τ flux is given by the plots below [7].

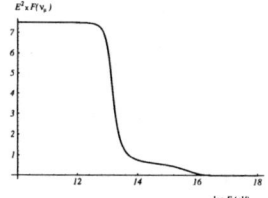 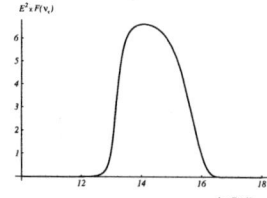

WEAK EFFECTS OVER LONG DISTANCES

From (5,7) it follows that the leading mass effects for neutrinos is simply given by the term $m^2/2$ in the (effective) Hamiltonian [3]. This result holds for all particles provided the semiclassical approximation is valid. To see this [10] note that for $m \neq 0$ the action in (1) obeys $\partial_\mu S \partial_\nu S = m^2$. Assuming m is small I look for solutions that are power series in m^2, namely $S = S_0 + m^2 S_1 + \cdots$. When substituted into the equation for S this yields $p^\mu \partial_\mu S_1 = dS_1/d\ell = -1/2$ so that $S_1 = -m^2 \ell/2$.

Though the quantum phase S_1 *per se* is not an observable, phase *differences* are. I will consider the case where two different mass eigenstate wave packets, of masses m_1 and m_2 are observed in a region B [11]. For this to be possible the amplitudes for the corresponding wave packets will be large simultaneously at B, which can be achieved in two ways. The first has both wave packets are created in the same regions A but at slightly different (local) times. In this case the phase difference is simply $\Delta S = \frac{1}{2}\Delta m^2(\ell_B - \ell_A)$ where $\Delta m^2 = m_1^2 - m_2^2$. The second possibility [12] is for the particles to start from different regions, A, A', simultaneously. In this case the phase difference has an additional contribution, $\Delta S = \Delta S_0 + \frac{1}{2}\Delta m^2(\ell_B - \ell_A)$. Note that $\Delta S_0 = O(\Delta m^2)$. These two possibilities are illustrated in the figures below

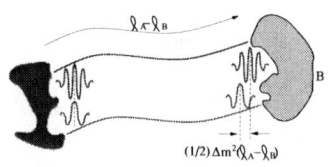 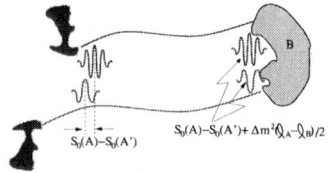

---

[3] The Hamiltonian has units of mass^2 since it generates translations in the affine parameter ℓ which has units of length^2. In a static space times geodesics are characterized (in part) by a constant energy E, in this case the usual definitions are recovered by defining a new affine parameter $\ell \to E\ell$ so that $H \to m^2/(2E) + \cdots$.

Distances

In order to determine the phase difference in terms of observable parameters, the expression of the affine parameter in terms of the distance from A to B is needed. The explicit expressions will of course depend on the choice of distance, which in curved space is not unique. Below I will consider two possibilities. One possible definition of distance is given by the time it takes light to travel from A to B [2],

$$d_{A \to B} = E \int_A^B \frac{d\ell}{\sqrt{g_{00}}} \qquad (11)$$

(E is the particle energy and I assumed a static space-time with metric $g_{\mu\nu}$). The second possibility is obtained by measuring the perimeter of a large circle with the source of the gravitational field at the center (assuming this is a localized object), the distance is then the perimeter divided by 2π. These two definitions coincide only in flat space.

Kerr metric example

Using the standard expression for the Kerr space-time [2] the corresponding expressions for the m^2 term in the phase are, in terms of r and d

$$\begin{aligned} S_1 &= -\frac{m^2}{2E}\sqrt{r^2 - (L/E)^2} + \frac{m^2[(K-L^2) + 2ELa]}{4E^3\sqrt{r^2 - (L/E)^2}} - \frac{m^2 r_g(r - \sqrt{r^2 - (L/E)^2})}{4E\sqrt{r^2 - (L/E)^2}} + \cdots \\ &= -\frac{m^2}{2E}\left[\ell - \frac{r_g}{2}\mathrm{arcsinh}(E\ell/L) + \cdots\right] \end{aligned} \qquad (12)$$

where the first terms on the right-hand side give the flat space contributions. Note that when expressed in terms of r the non-flat contribution is proportional to an angular momentum, either of the particle (L, K) or of the black hole a [13]. For radial geodesics in the field of a non-rotating black hole these gravitational effects vanish so that there are no effects proportional to m^2 in this case [12]. This is *not* the case when the phase is written in terms of d, even in the zero angular momentum limit a non-trivial term $\propto m^2 \ln d$ remains [14]. It follows that the conclusions of [14, 12] are not contradictory, but are based on a different definition of the gravitational effects: the complete phase is a global scalar and unambiguous; the segregation into flat and gravitational contributions is convention-dependent. These results are reproduced in a variety of metrics[10].

Observability

A different issue is the observability of the above phase difference. In order to measure any gravitational effects in the phase we should be able to segregate the flat-space contributions. This, however, necessitates a very accurate measure of the distance to the source.

For definiteness I will consider a black hole of $10^9 M_\odot$. If we measure distance using r then $L \sim ER$, where R denoting the scale of the metric at the creation point. Then the mass terms in the phase are $\sim m^2(r+R)/E$. The contribution $\propto R$ denotes the leading gravitational effects, and it is clear that we need to measure r to a precision better than R in order to measure (or even detect) them. This however, proves to be an unrealistic constraint: since $R \sim 10^9$km$= 0.0001$ light years, and $r \gg 100$ light years this requires a precision of $\ll 0.0001\%$ in r, which is unrealistic.

Similarly, if d is chosen as the flat-space distance then the mass terms in the phase become $\sim (m^2/E)[d+0.5 r_g \ln(d/r_g)]$. Assuming that the distance is known, say, to 1% (which is very optimistic) then $d \lesssim 280 r_g \sim 0.03$ light years. So, even though there is a logarithmic enhancement when using d, the gravitational-mass effects can be measured only near the source (needless to say, if we had a super-massive black hole so close to the solar system, measuring this phase would be a minor concern).

CONCLUSIONS

Gravity can induce spin-flavor transitions among neutrinos (and other particles as well) provided the off-diagonal terms in the Hamiltonian $\sim \mu B$ are large enough and/or for large enough E. In generic AGN scenarios these effects will evenly populate all neutrino species for magnetic moments well below the most stringent astrophysical bounds.

Non-resonant effects are also of interest. These again generate oscillations between the neutrino species, yet they are always subdominant to the flat-space contributions and are very difficult to disentangle them in realistic scenarios.

REFERENCES

1. B. S. DeWitt, *Dynamical theory of groups and fields* (Gordon and Breach; New York; 1965).
2. L .D. Landau, E. M. Lifshitz, *The classical theory of fields* (Pergamon; Oxford, New York; 1987).
3. S. Weinberg, *Gravitation and cosmology: principles and applications of the general theory of relativity* (Wiley; New York; 1972)
4. A. Messiah, *Quantum mechanics* (North-Holland; Amsterdam; 1961-62)
5. B. Sakita, R. Tsani, in: *Rationale of beings: recent developments in particle, nuclear, and general physics*, Festschrift in honor of Gyo Takeda; edited by K. Ishikawa *et al.* (World Scientific; Singapore, Philadelphia; 1986).
6. D. Piriz, M. Roy and J. Wudka, Phys. Rev. D **54**, 1587 (1996) [arXiv:hep-ph/9604403].
7. M. Roy and J. Wudka, Phys. Rev. D **56**, 2403 (1997) [arXiv:hep-ph/9703362].
8. S. J. Parke, Phys. Rev. Lett. **57**, 1275 (1986)
9. A. P. Szabo, R. J. Protheroe, in proceedings of the workshop on *High Energy Neutrino Astrophysics*, edited by V. I. Stenger *et al.* (World Scientific; singapore; 1992). R. J. Protheroe, T. Stanev, *ibid.*
10. J. Wudka, Phys. Rev. D **64**, 065009 (2001) [arXiv:gr-qc/0010077].
11. Y. Grossman and H. J. Lipkin, Phys. Rev. D **55**, 2760 (1997) [arXiv:hep-ph/9607201].
12. T. Bhattacharya, S. Habib and E. Mottola, arXiv:gr-qc/9605074; Phys. Rev. D **59**, 067301 (1999).
13. Y. Kojima, Mod. Phys. Lett. A **11**, 2965 (1996) [arXiv:gr-qc/9612044].
14. D. V. Ahluwalia and C. Burgard, Gen. Rel. Grav. **28**, 1161 (1996) [gr-qc/9603008]

Neutrino oscillations in dense neutrino media

S. Pastor

Max-Planck-Institut für Physik, Föhringer Ring 6, 80805 München, Germany

Abstract. In the early universe or in some regions of supernovae, the neutrino refractive index is dominated by the neutrinos themselves. Previous studies found numerically that these self-interactions have the effect of coupling different neutrino modes in such a way as to synchronize the flavor oscillations which otherwise would depend on the energy of a given mode. We provide a simple explanation for this baffling phenomenon in analogy to a system of magnetic dipoles which are coupled by their self-interactions to form one large magnetic dipole which then precesses coherently in a weak external magnetic field. In this picture the synchronized neutrino oscillations are perfectly analogous to the weak-field Zeeman effect in atoms.

INTRODUCTION

In a two-flavor system of mixed neutrinos, the flavor content of a given state oscillates with the frequency $\Delta m^2/2p$ where $\Delta m^2 = m_2^2 - m_1^2$ is the neutrino mass-squared difference and p is the momentum. Therefore, if a neutrino ensemble encompasses many modes with many different momenta, these modes develop growing relative phases so that the overall flavor content of the ensemble quickly decoheres. This trivial effect is illustrated in Fig. 1 (dotted line) for an ensemble of neutrinos (no anti-neutrinos) with a thermal momentum distribution at temperature T. We took the vacuum mixing angle to be $\sin 2\theta = 0.8$ and all neutrinos were originally in a pure ν_e state. In our example the momentum distribution is very broad so that the flavor decoherence takes place within about one oscillation period.

This behavior changes dramatically when the neutrinos feel a significant weak-interaction potential caused by the presence of background neutrinos. We express the strength of the neutrino-neutrino potential in terms of the parameter $\kappa \equiv 2\sqrt{2}G_F n_\nu p_0/\Delta m^2$, where $p_0 = \langle p^{-1}\rangle^{-1}$. When the neutrino-neutrino potential is comparable or much larger than a typical $\Delta m^2/2p$, corresponding to $\kappa = O(1)$ or larger, the modes get locked to each other—the entire ensemble oscillates with a common frequency ω_{synch} which corresponds to a certain average of $\Delta m^2/2p$ (Fig. 1). This stunning effect was first discovered in numerical studies of early-universe neutrino oscillations [1] and then elaborated and applied in a large series of papers [2]-[9].

The mode synchronization effect illustrated in Fig. 1 is a strictly nonlinear effect caused by the neutrino-neutrino self-interactions and as such seems difficult to understand. We have developed a very simple and physically transparent theory of this effect, taking full advantage of the equivalence of our problem with the spin-precession of a magnetic dipole in magnetic fields [10]. As far as we can tell, our picture perfectly accounts for all of the relevant oscillation phenomena discussed in the literature.

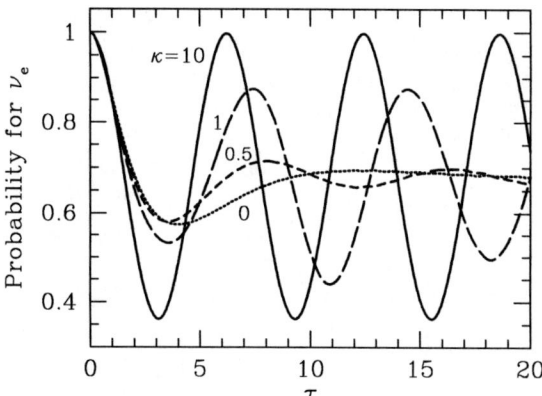

FIGURE 1. Total ν_e survival probability as a function of time, where $\tau \equiv (\Delta m^2/2p_0)t$ and $p_0 = \langle p^{-1}\rangle^{-1} \simeq 2.2T$ for a Fermi-Dirac distribution of momenta. The curves are for different values κ of the neutrino self-coupling as indicated where $\kappa = 0$ corresponds to vacuum oscillations.

EXPLANATION OF SYNCHRONIZED OSCILLATIONS

Our starting point is the evolution equation for neutrino oscillations in vacuum,

$$\frac{d\mathbf{P}}{dt} = \dot{\mathbf{P}} = \frac{\Delta m^2}{2p}\mathbf{B}\times\mathbf{P} = \mathbf{B}\times\mathbf{M}\,, \tag{1}$$

in the well-known spin-precession picture [11]. Here, $\mathbf{P}$ is the polarization vector in flavor space of a mode with momentum p. In the usual way, the z-component of $\mathbf{P}$ gives us the probability for finding the neutrino, say, in the electron flavor state by virtue of $\text{prob}(\nu_e) = \frac{1}{2}(1+P_z)$. The unit vector $\mathbf{B} = (\sin 2\theta, 0, -\cos 2\theta)$ with mixing angle θ is an effective "magnetic field" around which $\mathbf{P}$ precesses. Therefore, $\mathbf{P}$ and $\mathbf{M}$ play the role of an angular momentum vector and its associated magnetic dipole moment, respectively, while $\Delta m^2/2p$ gives us the proportionality between $\mathbf{M}$ and $\mathbf{P}$ and thus plays the role of the "gyromagnetic ratio" for a given mode, determining the rate of precession.

We will frequently consider the polarization vector for an entire ensemble of neutrinos, $\mathbf{J} \equiv \sum_{j=1}^{N_\nu} \mathbf{P}_j$, i.e. we consider a large volume $\mathcal{V}$ filled homogeneously by N_ν neutrinos with an isotropic distribution of the momenta so that it suffices to specify $p_j = |\mathbf{p}_j|$ for a given mode. There is no closed equation of motion for $\mathbf{J}$ because the individual modes oscillate with different frequencies, but its projection on $\mathbf{B}$ is evidently conserved. On the other hand, the fast precession of the individual $\mathbf{P}_j$ around $\mathbf{J}$ average the transverse components of the individual modes to zero so that the asymptotic value $\mathbf{J}_\infty = (\mathbf{B}\cdot\mathbf{J})\mathbf{B}$ obtains. For maximum mixing, and if initially all neutrinos were in a flavor eigenstate, $\mathbf{J}_\infty = 0$, corresponding to an incoherent mixture of both flavors.

A background medium consisting, say, of electrons modifies the "magnetic field." Assuming our two-flavor system involves electron neutrinos the substitution is

$$\frac{\Delta m^2}{2p}\mathbf{B} \to \frac{\Delta m^2}{2p}\mathbf{B} + \sqrt{2}G_F n_e \hat{\mathbf{z}} \tag{2}$$

with n_e the electron number density. Therefore, the precession is no longer around a common direction for all modes. If we started with a situation of maximum mixing, then the medium reduces the effective mixing angle for all modes, and a very dense medium suppresses flavor oscillations entirely.

This is very different if we consider a neutrino ensemble so dense that the neutrinos themselves produce a significant refractive index. In that case the medium's contribution to the refractive index is not along the flavor direction (the z-axis), but rather along the direction of **J**. Put another way, neutrinos produce an "off-diagonal refractive index" [12] because a given background neutrino may be a coherent superposition of flavor states. The equations of motion for a single mode j and for **J** now read [13]

$$\dot{\mathbf{P}}_j = \frac{\Delta m^2}{2p_j} \mathbf{B} \times \mathbf{P}_j + \frac{\sqrt{2}G_F}{\mathcal{V}} \mathbf{J} \times \mathbf{P}_j, \quad \longrightarrow \quad \dot{\mathbf{J}} = \sum_{j=1}^{N_v} \mathbf{P}_j = \mathbf{B} \times \sum_{j=1}^{N_v} \frac{\Delta m^2}{2p_j} \mathbf{P}_j. \qquad (3)$$

where the term proportional to G_F represents the self-interactions. The first derivative $\dot{\mathbf{J}}$ of the ensemble's polarization vector is not affected by the self-interactions. Still, the evolution of **J** is changed because the evolution of the individual modes is affected. However, the vacuum oscillations are not obviously suppressed even in a dense gas of neutrinos, in contrast to a standard background medium.

Our case of active-active neutrino oscillations is very different from the active-sterile case [13, 14]. Sterile neutrinos do not produce a weak potential so that there is no off-diagonal refractive index. The neutrino contribution to the "effective magnetic field" is along the z-direction, i.e. not proportional to **J**. Therefore, the self-interaction term is somewhat more similar to the effect of an external background medium, although the oscillation equations, of course, remain non-linear.

It is now easy to demonstrate that Eq. (3) implies a synchronized precession of all modes around **B** if the neutrino density is sufficiently large. To this end we first imagine the vacuum oscillation term to be absent, i.e. we consider the equation of motion

$$\dot{\mathbf{P}}_j = \frac{\sqrt{2}G_F}{\mathcal{V}} \mathbf{J} \times \mathbf{P}_j. \qquad (4)$$

Every individual mode precesses around the direction of **J**. Of course, if all neutrinos were initially prepared in a specific flavor state, then all $\mathbf{P}_j$ as well as **J** are aligned along the z-direction, and no precession takes place (perfectly coherent state). Likewise, if the individual $\mathbf{P}_j$ point in random directions so that $\mathbf{J} = 0$, again there are no precessions (perfectly incoherent flavor mixture). We consider the general case where the $\mathbf{P}_j$ initially point in many different directions, but do not add to zero.

Next we switch on the vacuum term from Eq. (1), i.e. a weak external "magnetic field". With "weak" we mean that for a typical mode the precession frequency around **J** is much larger than the one around **B**. This implies that the evolution of a given mode remains dominated by **J**. The fast precession around **J** implies that the projection of $\mathbf{P}_j$ on **J** is conserved while the transverse component averages to zero on a time scale relevant for the slow precession around **B**. If **J** moves slowly, then the individual modes will follow **J**. Put another way, the individual modes are coupled to each other by their strong "internal magnetic fields," forming a compound system with one large magnetic moment. It is this compound object which precesses around **B**. Of course, if the external field is much larger than the internal ones (dilute neutrino gas), then the modes will decouple and precess individually around the external field with their separate vacuum oscillation frequencies.

Our picture is perfectly analogous to the Zeeman effect in atoms. An atomic state is characterized by its spin angular momentum **S**, its orbital angular momentum **L**, and the total angular momentum $\mathbf{J} = \mathbf{L} + \mathbf{S}$. In a weak external magnetic field the spin-orbit coupling caused by the internal magnetic fields (Russell-Saunders coupling) remains intact, the external field is only a perturbation. In this case it is the total angular momentum **J** which precesses, i.e. which determines the atomic level splittings caused by the external B-field. On the other hand, if the external field is much stronger than the internal one, then **L** and **S** decouple and precess independently around the external field: the atomic levels are determined by the separate quantization of **L** and **S** along the external B-field (Paschen-Back effect).

Granting that in the neutrino case the polarization vectors of the individual modes are locked in the sense that **J** indeed forms one large "angular momentum," the evolution of the compound system is governed by the equation

$$\dot{\mathbf{J}} = \omega_{\text{synch}} \mathbf{B} \times \mathbf{J}. \tag{5}$$

It remains to determine the value of ω_{synch} which plays the role of the gyromagnetic ratio for the compound system. Of the individual modes, the external field "sees" only the projection along **J** because the transverse components average to zero. Therefore, if $\hat{\mathbf{J}}$ is a unit vector in the direction of **J**, the contribution of mode j to the total magnetic moment **M** is $(\Delta m^2/2p_j)\hat{\mathbf{J}} \cdot \mathbf{P}_j$ so that

$$\mathbf{M} = \hat{\mathbf{J}} \sum_{j=1}^{N_v} \frac{\Delta m^2}{2p_j} \hat{\mathbf{J}} \cdot \mathbf{P}_j \quad \longrightarrow \quad \omega_{\text{synch}} = \frac{|\mathbf{M}|}{|\mathbf{J}|} = \frac{1}{|\mathbf{J}|} \sum_{j=1}^{N_v} \frac{\Delta m^2}{2p_j} \hat{\mathbf{J}} \cdot \mathbf{P}_j. \tag{6}$$

In particular, if all modes started aligned (coherent flavor state as in the case shown in Fig. 1) then $|\mathbf{J}| = N_v$ and $\hat{\mathbf{J}} \cdot \mathbf{P}_j = 1$ so that $\omega_{\text{synch}} = \Delta m^2/2\langle p^{-1}\rangle$ (see also [5]).

If an external medium is present, the uncoupled modes precess around different **B**-field vectors rather than a common direction. When the modes are coupled by self-interactions, it is still the one **J** which precesses around one common **B** field which is a suitable average of the individual $\mathbf{B}_j$ which can be easily worked out.

Returning to the simpler case of a common **B** for all modes, the calculation of ω_{synch} amounts to determining the Landé-factor for **J** in the atomic analogy. This is the problem of calculating the magnetic moment of a system if the angular momentum is the sum of individual components which have magnetic moments with different gyromagnetic ratios. In this case the vector sums of the angular momenta and of the magnetic moments are not co-linear. In atomic physics, the spin angular momentum produces twice the magnetic moment of the orbital angular momentum, hence the complication.

We stress that, contrary to the previous literature, our analysis shows that there is nothing special about the initially aligned state, even though this state may be most motivated by the neutrino applications when all of them start in one flavor state. Any initial configuration of **J** precesses as one vector. Again, in atomic physics **L** and **S** can be combined in different ways to form one **J**. For example, a p-state with $L = 1$ and $S = 1/2$ can combine to a $J = 3/2$ or a $J = 1/2$ state; there is nothing special about the behavior of the "aligned" state ($J = 3/2$). Moreover, even though the initial **J** precesses as a compound system, the individual modes do not oscillate in unison as suggested by the previous literature, except for the special case of perfect initial alignment and infinitely strong self-coupling. In the general case the motion of every $\mathbf{P}_j$ is a fast precession around **J**, superimposed on a slow precession of **J** around **B**. The compound motion of $\mathbf{P}_j$ is generally rather complicated and different for every mode.

Our treatment predicts that any set of initial $\mathbf{P}_j$ leads to synchronized oscillations in the sense of a precession of the initial **J** with a single frequency ω_{synch}. The length of the initial-state **J** is conserved while the oscillation frequency depends on details of the initial distribution of polarization vectors. The system does not prefer one particular synchronized state over another. For example, the completely incoherent state will stay that way, and will not spontaneously align the $\mathbf{P}_j$ to form a coherent state, in agreement with the stability analysis of [9]. In our treatment it is obvious that any initial configuration is "neutral" with regard to stability considerations. Put another way, the nonlinear aspect of the neutrino system manifests itself in the coupling of the individual $\mathbf{P}_j$ to each other to form a compound **J** which acts as one large angular momentum. Beyond this, the system is easily understood in terms of a linear equation of motion.

Our analysis is also entirely independent of the number of neutrinos or modes. For example, if there are only two modes nothing in our analysis changes. In fact, the atomic example was one consisting of two coupled magnetic moments, the orbital and spin terms. In the previous literature the synchronized oscillations were called a "collective effect" or a "soliton-like" solution, suggesting that the presence of many degrees of freedom was crucial. In contrast, the full effect is present in a system consisting of only two degrees of freedom. We have checked with a simple numerical code that this is indeed the case for a variety of two- and three-mode examples.

There are exceptions to our general statements which will be of interest in the next section. Evidently we can construct "pathological" cases where several angular momenta $\mathbf{P}_j$ add up to a vanishing or arbitrarily small **J** while the magnetic moments do not, leading to a gyromagnetic ratio which can be constructed to become arbitrarily large. In this case ω_{synch} will no longer represent some typical $\Delta m^2/2p$, but can be constructed to become arbitrarily large. In this case the motion of **J** is not necessarily slow compared to the internal precessions of the individual $\mathbf{P}_j$ so that our treatment is no longer adequate. However, the perfectly incoherent state of an ensemble of many modes will not be pathological in this sense because the random distribution of $\mathbf{P}_j$ will ensure that both $|\mathbf{J}| \sim |\mathbf{M}| \sim 1/\sqrt{N_\nu}$. In the limit $N_\nu \to \infty$ both **J** and **M** vanish simultaneously.

NEUTRINOS PLUS ANTINEUTRINOS

As a next step we may extend our analysis to the case where neutrinos and antineutrinos are simultaneously present, an inevitable situation in a realistic system such as the early universe unless the neutrino chemical potentials are extremely large. To first order in G_F

the equations of motion are [13]

$$\dot{\mathbf{P}}_j = +\frac{\Delta m^2}{2p_j}\mathbf{B}\times\mathbf{P}_j + \frac{\sqrt{2}G_F}{\mathcal{V}}(\mathbf{J}-\bar{\mathbf{J}})\times\mathbf{P}_j\,,\quad \dot{\bar{\mathbf{P}}}_k = -\frac{\Delta m^2}{2p_k}\mathbf{B}\times\bar{\mathbf{P}}_k + \frac{\sqrt{2}G_F}{\mathcal{V}}(\mathbf{J}-\bar{\mathbf{J}})\times\bar{\mathbf{P}}_k, \tag{7}$$

where overbarred quantities refer to anti-neutrinos. In particular, the total polarization vectors are $\mathbf{J} = \sum_{j=1}^{N_\nu}\mathbf{P}_j$ and $\bar{\mathbf{J}} = \sum_{k=1}^{N_{\bar\nu}}\bar{\mathbf{P}}_k$. The definition of the polarization vector for anti-neutrinos is "reversed" in the sense that in vacuum it precesses in the opposite direction of neutrinos [13]. This corresponds to fermions with a true magnetic moment which have opposite gyromagnetic ratios for particles and anti-particles.

It is obvious that a system consisting of neutrinos and anti-neutrinos behaves the same way as one consisting of neutrinos only, except that the role of the total angular momentum is now played by $\mathbf{I} = \mathbf{J} - \bar{\mathbf{J}}$. The anti-particles appear as normal modes of the system, except that they sport negative gyromagnetic ratios. It is $\mathbf{I} = \mathbf{J} - \bar{\mathbf{J}}$ which precesses slowly around the external field, while all $\mathbf{P}_j$ and $\bar{\mathbf{P}}_k$ remain pinned to $\mathbf{I}$. The corresponding evolution equation for the compound system and the the total gyromagnetic ratio are

$$\dot{\mathbf{I}} = \omega_{\text{synch}}\mathbf{B}\times\mathbf{I},\quad \omega_{\text{synch}} = \frac{1}{|\mathbf{I}|}\left(\sum_{j=1}^{N_\nu}\frac{\Delta m^2}{2p_j}\hat{\mathbf{I}}\cdot\mathbf{P}_j + \sum_{k=1}^{N_{\bar\nu}}\frac{\Delta m^2}{2p_k}\hat{\mathbf{I}}\cdot\bar{\mathbf{P}}_k\right) \tag{8}$$

We have checked with a numerical code several situations with a thermal population of neutrinos and antineutrinos, and the results were always as expected.

In the absence of a chemical potential we have $N_\nu = N_{\bar\nu}$. If all neutrinos start in a given flavor state we have $\mathbf{I} = \mathbf{J} - \bar{\mathbf{J}} = 0$. However, we are now in the "pathological" situation that a vanishing or very small $\mathbf{I}$ is associated with a large magnetic moment because particles and anti-particles enter with exactly opposite magnetic moments. To illustrate this case we consider only one mode of particles and anti-particles so that the equations of motion are

$$\dot{\mathbf{P}} = +\omega\mathbf{B}\times\mathbf{P} + (\mathbf{P}-\bar{\mathbf{P}})\times\mathbf{P}\,,\quad \dot{\bar{\mathbf{P}}} = -\omega\mathbf{B}\times\bar{\mathbf{P}} + (\mathbf{P}-\bar{\mathbf{P}})\times\bar{\mathbf{P}}\,, \tag{9}$$

$$\dot{\mathbf{P}} - \dot{\bar{\mathbf{P}}} = \omega\mathbf{B}\times(\mathbf{P}+\bar{\mathbf{P}})\,, \tag{10}$$

with a suitable ω. Evidently $\mathbf{I} = \mathbf{P} - \bar{\mathbf{P}}$ is not conserved if the two polarization vectors start aligned so that at first $\mathbf{I} = 0$. The effect of the external field is to drive $\mathbf{P}$ and $\bar{\mathbf{P}}$ apart, creating a net $\mathbf{I}$ which is orthogonal to $\mathbf{B}$. In the case of large mixing both $\mathbf{P}$ and $\bar{\mathbf{P}}$ will oscillate much faster than they would in vacuum, yet convert to the other flavor.

In the case of small mixing the result strongly depends on the sign of ω, i.e. that of Δm^2. For $\omega > 0$, the directions of the vectors $\mathbf{P}$, $\bar{\mathbf{P}}$ and $\mathbf{B}$ almost coincide. The role of the strong neutrino self-potential term is just to increase the oscillation frequency, while the amplitude of P_z is the same as in the vacuum case. For an inverted mass hierarchy ($\Delta m^2 < 0$) and small mixing angle, $\mathbf{B}$ is close to the z-axis, but it is almost opposite to the initial directions of $\mathbf{P}$ and $\bar{\mathbf{P}}$. A small seed of $|\mathbf{I}| \neq 0$ is enough to drive $\mathbf{P}$ and $\bar{\mathbf{P}}$ to the opposite direction from their initial orientation, i.e. one can achieve complete flavor conversion (see fig. 3 in [10]). This resonance has a different origin from the Mikheev-Smirnov-Wolfenstein mechanism because it comes from the neutrino self-potential term.

CONCLUSIONS

In summary, we have provided a simple and physical explanation of the synchronized oscillations observed in dense neutrino ensembles. The effect is perfectly analogous to the coupling of several angular momenta, for example spin and orbital angular momentum in an atom, to form one large compound angular momentum with one large associated magnetic dipole moment which precesses as one object in a weak external field. The nonlinear nature of neutrinos oscillating in a background of neutrinos is thus reduced to a very simple and well-known coupling effect of magnetic moments to each other.

The analysis of neutrino oscillations in the early Universe in the presence of neutrino asymmetries constitutes an application of our work. Preliminary results show that, for the parameters favored by recent experimental data on solar and atmospheric neutrinos, active-active oscillations would be effective before the big-bang nucleosynthesis (BBN) epoch, equilibrating the different neutrino flavors. Thus the BBN limit on the v_e degeneracy parameter applies to all flavors, which leads to stringent limits on a putative extra cosmic radiation contribution from degenerate neutrinos [15].

ACKNOWLEDGMENTS

The author thanks Conacyt Mexico for partial support and G.G. Raffelt and D.V. Semikoz for an enjoyable collaboration. He was supported by a Marie Curie fellowship of the European Commission under contract HPMFCT-2000-00445 and, in Munich, partly supported by the Deutsche Forschungsgemeinschaft under grant No. SFB 375 and the ESF network Neutrino Astrophysics.

REFERENCES

1. S. Samuel, *Phys. Rev. D* **48**, 1462 (1993).
2. V. A. Kostelecký and S. Samuel, *Phys. Rev. D* **49**, 1740 (1994).
3. V. A. Kostelecký, J. Pantaleone and S. Samuel, *Phys. Lett. B* **315**, 46 (1993).
4. V. A. Kostelecký and S. Samuel, *Phys. Lett. B* **318**, 127 (1993).
5. V. A. Kostelecký and S. Samuel, *Phys. Rev. D* **52**, 621 (1995).
6. V. A. Kostelecký and S. Samuel, *Phys. Rev. D* **52**, 3184 (1995).
7. S. Samuel, *Phys. Rev. D* **53**, 5382 (1996).
8. V. A. Kostelecký and S. Samuel, *Phys. Lett. B* **385**, 159 (1996).
9. J. Pantaleone, *Phys. Rev. D* **58**, 073002 (1998).
10. S. Pastor, G. G. Raffelt and D. V. Semikoz, hep-ph/0109035.
11. L. Stodolsky, *Phys. Rev. D* **36**, 2273 (1987).
12. J. Pantaleone, *Phys. Lett. B* **287**, 128 (1992).
13. G. Sigl and G. Raffelt, *Nucl. Phys. B* **406**, 423 (1993).
14. B. H. McKellar and M. J. Thomson, *Phys. Rev. D* **49**, 2710 (1994).
15. A.D. Dolgov et al., in preparation.

Chiral Quark Models

H. J. Weber

Institute of Nuclear and Particle Physics, University of Virginia, Charlottesville, VA 22904, USA

Abstract. Chiral quark models are briefly sketched. Then the main question is addressed: Are chiral quark models consistent with the proton spin data? The proton spin fractions of the nonrelativistic quark model (NQM) are improved by chiral field theory. Such chiral models agree with the proton spin data only because the chiral quark-Goldstone boson couplings are taken to be pure spinflip which is consistent with chiral perturbation theory. However, for constituent quarks non-spinflip chiral transitions are not negligible, so that chiral quark models disagree with the proton spin data.

INTRODUCTION

For massless up (u), down (d) and strange (s) quarks, quantum chromodynamics (QCD) is invariant under global chiral rotations of the quark fields $q_{R,L} \to e^{i\gamma_5 \lambda_f \alpha} q_{R,L}$, $\bar{q}_{R,L} \to \bar{q}_{R,L} e^{i\gamma_5 \lambda_f \alpha}$ with $q_{R,L} = \frac{1}{2}(1 \pm \gamma_5)q$: QCD has $SU(3)_L \times SU(3)_R$ chiral symmetry involving independent rotations of left- and right-handed quark fields. Large mass splittings of chiral partners, that is, states of the same spin but opposite parity, such as $m_{S11} - m_N = 1535 - 939$ MeV, $m_{a_1} - m_\rho = 1260 - 770$ MeV, show that chiral symmetry is broken to such an extent that the small current quark masses cannot credibly explain. Non-zero matrix elements such as $\langle 0|A^\mu(0)|\pi(p)\rangle = if_\pi p^\mu$ suggest that chiral symmetry $SU(3)_L \times SU(3)_R \to SU(3)_V$ is broken spontaneously.

The first and simplest model of a spontaneous chiral symmetry breakdown is the linear σ-model [1] with the mesonic Lagrange density $\mathcal{L}_\sigma$ displaying a Mexican-hat-type potential $\lambda \left(\sigma^2 + \vec{\pi}^2 - f_\pi^2\right)^2$ for the massless σ-meson and pions. Its degenerate ground states lie in the rim of the hat at the non-zero vacuum expectation value $\langle \sigma \rangle_0$ of the σ-meson. Massless current quarks become massive, $m_q \sim \langle \sigma \rangle_0$, in the broken symmetry phase where one of the vacuum states is chosen spontaneously. The rim is flat in the pion directions, so that pions retain their zero mass; they are Goldstone bosons (GBs).

The MIT bag model realizes color confinement in a spherical cavity, but violates chiral symmetry at the surface. When a massless quark is reflected by the surface, it changes helicity and thus chirality because its momentum changes but not its spin projection. However, with pions on the surface the quark helicity can be chirally rotated along with the momentum. This is the seminal idea of chiral bag models [2] which spawned many other models. Minimal linear-σ Lagrange densities are

$$\mathcal{L}_{CBM} = \theta_V \left(\bar{q}i\gamma \cdot \partial q - B\right) - \delta_S \left[g \, \bar{q}(\sigma + i\vec{\tau} \cdot \vec{\pi}\gamma_5)q\right] + \Phi_R \mathcal{L}_\sigma, \tag{1}$$

where the step function θ_V is 1 inside the bag and vanishes outside and $\Phi_R = 1 - \theta_V$, ("little" CBM [3]), $\Phi = 1$, ("cloudy" CBM [4]), $\Phi = \theta_V$, ("Yale" CBM [5]). That is, in the "little" bag the mesonic Lagrangian $\mathcal{L}_\sigma$ acts only outside the bag, in the

Yale version only inside, and in the cloudy bag everywhere. The main ingredients of all chiral bag models are MIT bag quark model wave functions and hedgehog pion wave functions. Most CBMs badly violate Lorentz invariance. Typical nonlinear CBM Lagrange densities are, to lowest order in the pion field,

$$\mathcal{L}_{nl} = \theta_V(\bar{q} i \gamma \cdot \partial q - B) + \Phi_R \left\{ \frac{1}{2} (\partial_\mu \vec{\pi})^2 - \frac{1}{2} \delta_S \bar{q} e^{i \vec{\tau} \cdot \vec{\pi} \gamma_5 / f_\pi} q + \ldots \right\}. \qquad (2)$$

Nambu–Jona-Lasinio models [6] are patterned after superconductivity with a four-point fermion interaction $g_\pi[(\bar{q}q)^2 + (\bar{q}i\gamma_5\vec{\tau}q)^2]$ in the chiral invariant quark Lagrangian involving an isodoublet color triplet quark field in its simplest version. Color confinement is not implemented. The effective meson field theory involves quark loops and Goldstone bosons, but no kinetic energies for the mesons. The gap equation for the quarks requires an ultraviolet cutoff, the scale introduced by the spontaneous chiral symmetry breakdown for sufficiently large coupling g_π.

Chiral quark-soliton models [7] are based on the instanton-fluid model of the vacuum. These models are closer to QCD and the large N_c limit than NJL models and still rapidly developing, lacking only color confinement. After bosonization the effective chiral Lagrangian resembles that of ref. [8].

A general feature of chiral quark models is that they involve Goldstone bosons as fundamental degrees of freedom. Because pseudoscalar mesons are bound quark-antiquark states in QCD, a double counting problem arises; it is addressed in ref. [8]. The general purpose of chiral quark models is to estimate pionic and other mesonic corrections to hadronic properties derived from valence quarks.

Now we turn to the main question: Do chiral quark models explain the proton spin?

SPIN FRACTIONS FROM THE NQM AND DIS

The NQM spin-flavor wave function of the proton implies the well known quark fractions $u_\uparrow = 5/3$, $u_\downarrow = 1/3$, $d_\uparrow = 1/3$, $d_\downarrow = 2/3$, $s_\uparrow = 0$, $s_\downarrow = 0$, and the spin fractions $\Delta u = u_\uparrow - u_\downarrow = 4/3$, $\Delta d = d_\uparrow - d_\downarrow = -1/3$, $\Delta s = 0$, which are to be compared, somewhat naively, with the spin data from deep inelastic lepton scattering (DIS)

SMC [9] at 5 GeV^2: $\Delta u = 0.82 \pm 0.06$, $\Delta d = -0.44 \pm 0.04$, $\Delta s = -0.10 \pm 0.06$.

Now the spin data really measure $\Delta q = q_\uparrow - q_\downarrow + \bar{q}_\uparrow - \bar{q}_\downarrow$, so that one has to subtract the antiquark contribution $\Delta q_v = \Delta q - \Delta \bar{q}$ before comparing with valence quark spin fractions from the NQM. This is a small effect, because the $\Delta \bar{q}$ are small $\Delta u_v = 0.77 \pm 0.10 \pm 0.08$, $\Delta d_v = -0.52 \pm 0.14 \pm 0.09$, but the errors are significantly increased. The axial charge of the nucleon $a_3 = g_A^{(3)} = \Delta u - \Delta d = \mathcal{F} + \mathcal{D} = (G_A/G_V)_{n \to p} = 5/3$ is obtained from the weak decay of the neutron to be compared with the data [10] $a_3 = 1.2670 \pm 0.0035$. The octet axial charge is defined as $a_8 = \Delta u + \Delta d - 2\Delta s = 1$. The ratio of axial charges is the $SU(3)_F$ symmetric value, so that the discrepancy between the NQM value $a_3/a_8 = 5/3$ and the data $a_3/a_8 = 2.18 \pm 0.12$ suggests $SU(3)_F$ breaking. The proton spin $\Delta\Sigma = \Delta u + \Delta d + \Delta s = 2J_z = 1$ differs from the data–known as the spin problem–$\Delta\Sigma = 0.28 \pm 0.17$ at 5 GeV^2, and for all proton data: $\Delta\Sigma = 0.37 \pm 0.11$.

The spin data are measured in deep inelastic scattering (DIS) lepton scattering. In the Bjorken limit the data scale depending only on the longitudinal quark momentum fraction x and not separately on momentum transfer and energy transfer. The unpolarized DIS cross section becomes a sum of elastic lepton-quark differential cross sections weighted by the probability $q(x)$ of a quark with momentum fraction x in the proton called the quark distribution function. This is the parton model. The spin fractions defined by the forward axial vector current matrix elements are related to the polarized nucleon structure functions in leading order $\int_0^1 dx g_1^p(x) = \frac{1}{2}\sum_q e_q^2 \Delta q$, but with QCD corrections the polarized structure functions depend on the momentum scale Q^2 to which all data are evolved. However, in the chiral limit of massless quarks the triplet and octet axial currents are conserved just like the weak vector currents, so that these axial charges are scale independent. Thus, apart from the small strange quark spin fraction, a qualitative comparison of the NQM spin fractions with the spin data is not unreasonable after all. Typically DIS analyses only parameterize initial polarized structure functions at some low momentum scale in a generic form $\Delta q(x, Q^2) = A_f x^{\alpha_f}(1-x)^{\beta_f}\left(1 + \gamma_f x^{\delta_f}\right)$, but do not explain them. A dynamical description of the spin reduction is a motivation of the chiral field theory approach to estimate spin fractions that we discuss next.

QUARK SPIN FRACTIONS FROM CHIRAL DYNAMICS

Chiral field theory applies at scales from Λ_{QCD} up to the chiral symmetry restoration scale $\sim \Lambda_\chi = 4\pi f_\pi \sim 1.17$ GeV, where $f_\pi = 0.093$ GeV is the pion decay constant. If the chiral symmetry breakdown is based on $SU(3)_L \times SU(3)_R$, then the effective interaction between quarks and the octet of Goldstone boson (GB) fields Φ_f involves the axial vector coupling $\mathcal{L}_{int} = -\frac{g_A}{2f_\pi}\sum_{f=1}^8 \bar{q}\partial_\mu \gamma^\mu \gamma_5 (1+\varepsilon\lambda_8)\lambda_f \Phi_f q$ that is well known from soft-pion physics. Incorporating $SU(3)_F$ breaking[11, 12] in the λ_8 direction involving the parameter ε is suggested by the large strange quark mass compared to m_u, m_d and is motivated by the discrepancy between a_3/a_8 and the data. Despite the nonperturbative nature of the chiral symmetry breakdown, the interaction between quarks and Goldstone bosons is small enough for a perturbative expansion to apply. It induces the chiral quark transitions $q_\downarrow \to q_\uparrow + GB$, $q_\uparrow \to q_\downarrow + GB$ with probabilities that can be directly read off $\mathcal{L}_{int}$ so that $u_\uparrow \to \pi^0 + u_\downarrow$ occurs with probability $\frac{a}{2}\left(1 + \frac{\varepsilon}{\sqrt{3}}\right)^2$, etc., where a is the chiral strength parameter. Chiral field theory dissolves a dynamical or constituent quark into a current quark and a cloud of virtual Goldstone bosons

$$|U\rangle = \sqrt{Z}|u_0\rangle + \alpha_{\pi/U}|d\pi^+\rangle + \frac{1}{2}\alpha_{\pi/U}|u\pi^0\rangle + \alpha_{K/U}|sK^+\rangle + ... \qquad (3)$$

The valence quark fractions of the NQM are reduced by the chiral fluctuation probability P and chiral fluctuations to the quark-GB states of Eq. 3 add terms proportional to a. Therefore, after one chiral fluctuation the quark spin fractions become

$$\Delta u = \frac{4}{3}(1-P) - \frac{5}{9}a\left(1+\frac{\varepsilon}{\sqrt{3}}\right)^2, \quad \Delta d = -\frac{1}{3}(1-P) - \frac{10}{9}a\left(1+\frac{\varepsilon}{\sqrt{3}}\right)^2,$$

$$\Delta s = -a\left(1 - \frac{2\varepsilon}{\sqrt{3}}\right)^2. \tag{4}$$

Table 1 Quark Spin Fractions of the Proton

	Data at 5 GeV^2 SMC	NQM	SU(3)$_F$ a=0.21 $\varepsilon = 0$	no SU(3)$_F$ a=0.12 $\varepsilon = 0.2$	η' a=0.12 $\varepsilon = 0.2$ $\zeta = -0.3$
Δu	0.82±0.06	4/3	0.47	0.82	0.81
Δd	-0.44±0.06	-1/3	-0.38	-0.39	-0.39
Δs	-0.10±0.06	0	-0.21	-0.07	-0.07
$\Delta\Sigma$	0.28±0.17 0.37±0.11	1	-0.12	0.36	0.35
a_3/a_8	2.18±0.12	5/3	2.76	2.12	2.13
$g_A^{(3)}$	1.2670±0.0035	5/3	1.12	1.22	1.21
$\mathcal{F}/\mathcal{D}$	0.575±0.016	2/3	2/3	0.58	0.58
I_G	0.235±0.026	1/3	0.236	0.27	0.25

Except for Δs, the spin fractions are reduced more for $\varepsilon \neq 0$ than the $SU(3)$ symmetric case, and these changes all go in the right direction compared with the spin data. As can be seen from the $\varepsilon = 0$ column in Table 1, the SU(3) symmetric chiral spin fraction model [13] explains the spin fractions of the proton qualitatively, except for the ratio of axial charges $a_3/a_8 = 5/3$ and the weak axial vector coupling constant of the nucleon, $g_A^{(3)} = a_3$. The effects of SU(3) breaking[11] ($\varepsilon \neq 0$) lead to a remarkable further improvement of the spin and quark sea fractions in comparison with the data (see the first $\varepsilon = 0.2$ column in Table 1). The last column in Table 1 shows that the changes caused by including the η' [14] are well within the error bars of the spin fraction data. These remarkably successful spin fraction estimates are in sharp contrast to [16]

$$\Delta u = 0.86, \quad \Delta d = -0.29, \quad \Delta s = -0.006, \quad a_3 = 1.15, \quad \Delta\Sigma = \Delta u + \Delta d + \Delta s = 0.56,$$

from polarized valence quark distributions dressed by a pion cloud and other GBs in a nonrelativistic chiral constituent quark model. From this fairly large value $\Delta\Sigma = 0.56$ the authors conclude: The nucleon spin problem cannot be solved by Goldstone dressing of constituent quarks alone. To resolve this puzzle let us look into polarized DIS initial quark distributions (at a low momentum scale) dressed by GBs from which the spin fractions can be obtained by integrating over Bjorken x. Chiral fluctuations occur with probability densities $f(u_\uparrow \to \pi^+ + d_\downarrow)$,... and corresponding ones for the other quarks and initial helicity. After integrating over transverse momentum in the infinite momentum frame, the polarized ($-$ sign) and unpolarized ($+$ sign) chiral splitting functions are obtained, $P_{GB/q}^\pm(x) = \int d^2k_\perp f_{q \to q'GB}^\pm(x, \vec{k}_\perp)$. The splitting of quarks into a Goldstone boson and a recoil quark corresponds to a factorization of DIS structure functions that leads to a convolution of valence quark distributions (q^0) with splitting functions $P^\pm \otimes q^0 = \int_x^1 \frac{dx_1}{x_1} P^\pm_{q \to q'GB}(\frac{x}{x_1}) q^0(x_1)$. Via P-wave quark-$GB_m$ couplings spin is diluted into orbital angular momentum of sea quarks. The unpolarized splitting functions P^+ determine the (spinflip plus non-spinflip) probability for finding a Goldstone boson of mass m_{GB} carrying the longitudinal momentum fraction x_{GB} of the parent quark's momentum and a recoil quark q' with momentum fraction x' for each fluctuation. Thus, chiral fluctuations in lowest order of perturbation theory contribute convolution integrals $\sum_{q,m} P_m P^+_{u_\uparrow \to q_\downarrow m} \otimes u^0_\uparrow + \sum_{q,m} P_m P^+_{d_\uparrow \to q_\downarrow m} \otimes d^0_\uparrow$, etc. to the quark distribution $q_\downarrow$, etc. Chiral fluctuations also cause a reduction of the valence quark probabilities $(1 - P_q) q^0_{\uparrow,\downarrow}$, where P_q are the total fluctuation probabilities. On the microscopic level, this reduction corresponds to that displayed in Eq. 4. In (non-renormalizable) chiral field theory with cutoff Λ_χ of ref. [13], the unpolarized chiral splitting function takes the form

$$P^+_{q \to q' + GB}(x_{GB}) = \frac{g_A^2}{f_\pi^2} \frac{x_{GB}}{32\pi^2} (m_q + m_{q'})^2 \int_{-\Lambda_\chi^2}^{t_{min}} dt \frac{(m_q - m_{q'})^2 - t}{(t - m_{GB}^2)^2}, \qquad (5)$$

where $t = k^2 = -[(k_\perp)^2 + x_{GB}[(m'_q)^2 - x'(m_q)^2]]/x'$ is the square of the Goldstone boson four-momentum. The polarized splitting function is obtained using that it contains the difference of non-flip and helicity-flip probabilities. Since quarks are on their mass shell in the light front dynamics used here, the axialvector quark-Goldstone boson interaction is equivalent to γ_5 coupling. Except for an overall factor, the relevant unpolarized chiral transition probability is proportional to

$$-\frac{1}{2} tr[(\gamma \cdot p' + m'_q) \gamma_5 (\gamma \cdot p + m_q) \gamma_5] = 2(pp' - m_q m'_q) = (m_q - m'_q)^2 - k^2, \qquad (6)$$

where $2pp' = m'^2_q + m^2_q - k^2$. Eq. 6 is the numerator in Eq. 5 which can be written as

$$\frac{1}{x'}[(k_\perp)^2 + [m'_q - x' m_q]^2], \qquad (7)$$

and has the following physical interpretation. The axialvector quark-GB coupling $\gamma_\mu \gamma_5 k^\mu$ in $\mathcal{L}_{int}$ involves the spin raising and lowering operators $\sigma_1 \pm i\sigma_2$ in a scalar product with the transverse momentum $\vec{k}_\perp$ of the recoil quark, which suggests that the $k_\perp^2$ term in P^+ of Eq. 7 represents the helicity-flip probability of the chiral fluctuation, while the longitudinal and time components, $\gamma_\pm \gamma_5 k^\pm$, induce the non-spinflip probability, which depends on the quark masses. This can be seen from the helicity non-flip probability

$$|\bar{u}'_\uparrow \gamma_5 u_\uparrow|^2 = |\bar{u}'_\downarrow \gamma_5 u_\downarrow|^2 \sim (m'_q - x' m_q)^2, \quad x' = 1 - x_{GB}, \qquad (8)$$

using light cone spinors and suppressing the spinor normalizations. Thus Eq. 8 identifies the mass term in Eq. 7 as the helicity non-flip chiral transition. Similarly, the helicity-flip probability is obtained from

$$|\bar{u}'_\downarrow \gamma_5 u_\uparrow|^2 \sim (p'_\perp)^2 + x'^2 (p_\perp)^2 - x' p'_\perp \cdot p_\perp \qquad (9)$$

which, in frames where $\vec{p}_\perp = 0$, reduces to $(\vec{k}_\perp)^2$, i.e. the net helicity flip probability generated by the chiral splitting process. In the nonrelativistic limit, where $|\vec{p}'_\perp| \ll m'_q$, $|\vec{p}_\perp| \ll m_q$, non-spinflip dominates over spinflip, while spinflip dominates at high momentum. The polarized splitting function P^- therefore has the same quark mass dependence as P^+, but involves the helicity flip probability with the opposite sign, i.e. has

$$\frac{1}{x'}[-(\vec{k}_\perp)^2 + (m'_q - x'm_q)^2] \tag{10}$$

in its numerator. Only for massless quarks there are no non-spinflip chiral transitions, so that $P^- = -P^+$ holds which is characteristic of pure spinflip chiral transitions. Let us now return to the polarized quark distributions and their lowest moments, the spin fractions.

Upon generalizing the chiral spin fraction formalism of [11, 14] to the polarized quark distributions, the polarized quark distributions are

$$q_\uparrow(x) = (1 - P_q)q_\uparrow^0(x) + \sum_{m,q'} p_m P^+_{q' \to qm} \otimes q_\downarrow'^0 + \ldots \tag{11}$$

which obviously are based on pure spinflip chiral transitions. The corresponding result holds for the other quark helicity. The ellipses in Eq. 11 denote double convolution terms with $q_\uparrow^0$ from a Goldstone boson m that cancel in $q_\uparrow - q_\downarrow$. The opposite quark helicity on the rhs of Eq. 11 implies the **negative** sign of all chiral contributions to the spin distributions

$$\Delta q(x) = (1 - P_q)\Delta q^0(x) - \sum_{m,q'} p_m P^+_{q' \to qm} \otimes \Delta q'^0. \tag{12}$$

This result is common to all recent successful chiral models of spin fractions. Let us emphasize that the general reduction of the valence quark spin fractions Δq^0 by chiral fluctuations in lowest order in Eq. 12 is the crucial property responsible for the remarkable success of chiral field theory for the proton spin fractions. Eq. 12 can be compared to the corresponding one from DIS involving the polarized splitting functions which has the same form as Eq. 12, except for the replacement of $-P^+$ by the corresponding polarized splitting function P^-. A comparison with Eqs. 7,10 shows that this approximation, $P^- \approx -P^+$, corresponds to neglecting the non-spinflip probability, which is valid only if the quark mass term in the numerator (i.e. Eq. 10 of the splitting functions is negligible compared to the tranverse momentum scale. This is not the case for constituent quarks [16].

When these $\Delta q(x)$ of Eq. 12 are integrated over Bjorken x, the lowest moments reproduce precisely the structure of the results for the spin fractions [11] (cf. Table 1) because moments factorize. Therefore integrating

$$\Delta u(x) = (1 - P_u)\Delta u^0 + \int_x^1 \frac{dy}{y} \left[\Delta u^0(y) \left(\frac{1}{2} P^-_{u \to u\pi^0}(\frac{x}{y}) + \frac{1}{6} P^-_{u \to u\eta} \right) + \Delta d^0 P^-_{d \to u\pi^-} \right], \tag{13}$$

and similar results for $\Delta d(x)$, $\Delta s(x)$, over Bjorken x and using $P^- \approx -P^+$ reproduces Eq. 4, where the chiral strength parameter a and the $SU(3)_F$ breaking parameter ε are now related to the unpolarized splitting functions by

$$a(1+\frac{\varepsilon}{\sqrt{3}})^2 = \int_0^1 dx P^+_{u \to d\pi^+}, \quad a(1-\frac{2\varepsilon}{\sqrt{3}})^2 = \int_0^1 dx P^+_{u \to sK^+}.$$

CONCLUSIONS

Because non-spinflip chiral transitions are important for constituent quarks, chiral quark models disagree with the proton spin data. If chiral field theory of current quarks is used as in chiral perturbation theory to improve the spin fractions of the NQM, the results agree with the proton spin data [11, 12, 13, 14, 15], explaining the success of these chiral models.

It is interesting to note that proton spin data are successfully described by chiral models based on an instanton fluid in the vacuum, where the chiral instanton-quark interaction is also pure spinflip [17]. Moreover, the dynamical quark mass $m_q(p^2)$ in such models may go to zero for $p^2 \to 0$ [18].

REFERENCES

1. M. Gell-Mann and M. Lévy, Nuovo Cim. **16**, 705 (1960).
2. A. Chodos and C. B. THorn, Phys. Rev. **D12**, 2733 (1975); T. Inoue and T. Maskawa, Prog. Theor. Phys. **54**, 1833 (1975).
3. G. E. Brown and M. Rho, Phys. Lett. **82B**, 177 (1979).
4. G. A. Miller, A. W. Thomas, S. Théberge, Phys. Lett. **91B**, 192 (1980).
5. A. Chodos and H. Nadeau, Phys. Rev. **D33**, 1450 (1986).
6. S. P. Klevansky, Rev. Mod. Phys. **64**, 649 (1992).
7. H. Weigel, hep-ph/0110236; M. Praszalowicz, H.-Ch. Kim and K. Goeke, hep-ph/0110135.
8. A. Manohar and H. Georgi, Nucl. Phys. **B234**, 189 (1984).
9. D. Adams *et al.* (SMC), Phys. Rev. **D56**, 5530 (1997).
10. Particle Data Group, D. E. Groom *et al.*, Eur. Phys. J. **C15**, 1 (2000).
11. H. J. Weber, X. Song, and M. Kirchbach, Mod. Phys. Lett. **A12**, 729 (1997).
12. X. Song, J. S. McCarthy, and H. J. Weber, Phys. Rev. **D55**, 2624 (1997).
13. E. J. Eichten, I. Hinchcliffe and C. Quigg, Phys. Rev. **D45**, 2269 (1992).
14. T. P. Cheng and L.-F. Li, Phys. Rev. **D57**, 344 (1998); Phys. Rev. Lett. **74**, 2872 (1995).
15. J. Linde, T. Ohlsson and H. Snellman, Phys. Rev. **D57**, 452 (1998).
16. K. Suzuki and W. Weise, Nucl. Phys. **A634**, 141 (1998).
17. A. E. Dorokhov, N. I. Kochelev, and Yu. A. Zubov, Int. J. Mod. Phys. **A8**, 603 (1993).
18. A. E. Dorokhov and W. Broniowski, hep-ph/0110056.

In-Medium Chiral Perturbation Theory

Ulf-G. Meißner*, José A. Oller† and Andreas Wirzba*

*Institut für Kernphysik (Theorie), Forschungszentrum Jülich, D-52425 Jülich, Germany
†Departamento de Física, Universidad de Murcia, E-30071 Murcia, Spain

Abstract. We report about a systematic study (see U.-G. Meißner, J.A. Oller and A. Wirzba, nucl-th/0109026) of the properties of pions and external sources in nuclear matter. An explicit expression of the generating functional of two-flavor low-energy QCD with external sources in the presence of non-vanishing nucleon densities has been derived recently (see J.A. Oller, hep-ph/0101204). The corresponding in-medium chiral perturbation theory (CHPT) lagrangian is non-covariant as well as non-local and formulated in terms of pions and external sources only. Within this approach the (chiral) power counting rules for the calculation of in-medium pion properties are derived in the so-called standard and in the non-standard scenario where the residual nucleon energies are of the order of the pion mass or of the order of the nucleonic kinetic energy, respectively. Also the pertinent scales for the range of applicability of these two perturbative expansions are established. In the standard scenario, we report on a systematic analysis (in contrast to the mean-field approach where nucleon correlations are neglected) of the pionic n-point in-medium Green functions up to and including next-to-leading order. These include the in-medium contributions to the quark condensates, pion propagators, pion masses, and pion couplings to the axial-vector and pseudoscalar current for symmetric and non-symmetric nuclear matter. Finally, we discuss the in-medium pion-pion scattering where the non-standard counting scenario holds.

INTRODUCTION

The three pion states, π^+, π^- and π^0, play a special role in nuclear and particle physics. This follows from the fact that for light up and down quarks, quantum chromodynamics (QCD) possesses an approximate chiral symmetry. In other words, it is a good first approximation to the theory to consider the light quarks as massless. This symmetry is not present in the ground-state or the particle spectrum. Instead, it is believed that this symmetry is spontaneously broken down to its vectorial subgroup, $SU(2)_L \times SU(2)_R \to SU(2)_{L+R}$, with the appearance of three (Pseudo-)Goldstone bosons which can be identified with the three pion states, $\pi^\pm, \pi^0$. The unique order parameter signaling this symmetry violation is the finiteness of the square of the weak pion decay constant in the chiral limit, $f^2 \neq 0$. In fact, $f \simeq f_\pi = 92.4\,\text{MeV}$. Another order parameter, often considered, is the quark condensate in the vacuum, $\langle 0|\bar{q}q|0\rangle$, where $|0\rangle$ denotes the highly complicated vacuum. However, it should be noted that, in principle, chiral symmetry could be broken even if $\langle 0|\bar{q}q|0\rangle \approx 0$, as long as $f^2 \neq 0$. In addition, the chiral symmetry is also explicitly broken because the current up and down quarks have a small mass (small compared to a typical hadronic scale of 1 GeV). Due to Goldstone's theorem, the interactions of the pions with themselves or matter fields must vanish as three-momentum and energy go to zero in the chiral limit. This in turn allows for a systematic treatment of such processes in the framework of an effective field theory (chiral perturbation theory, henceforth CHPT[1, 2]), in terms of a simultaneous expansion in external momenta and mass terms.

It is also believed that with increasing temperature and/or density, the chiral symmetry of QCD is restored. While lattice studies indicate that the critical temperature is $T_c \simeq$ 150 MeV, much less is known about the critical density. Note that lattice QCD applied to finite chemical potential μ (density ρ) is only in its infancy (for a recent method to tackle this problem see [3] and references therein).

Therefore, the study of pion properties, calculated at finite temperature and/or density, is a useful tool to gain an understanding how these transitions are approached and to improve our knowledge of QCD at finite density. The recent measurements of deeply bound pionic states in heavy nuclei performed at GSI [4, 5] can be interpreted in terms of a pion mass shift due to the high nuclear density in the center of such nuclei [6, 7].

In this talk we will report about a *systematic* study of the properties of pions and external sources in nuclear matter [8, 9]. Nuclear matter is a system of an infinite number of protons and neutrons. For typical nuclear densities, the interactions of the constituents of nuclear matter are strong, such that one has to deal with phenomena in the non-perturbative regime of QCD. Thus, approximations are unavoidable. However, one needs to be able to control these, as in the methodology of effective field theories. Consequently, the aim of an in-medium QCD effective field theory is to create a scheme which allows to estimate the errors when the expansion is truncated at some order. As mentioned above, the low energy effective field theory of QCD is CHPT. Indeed CHPT allows not only to tackle processes involving pions, but as well to consider nucleons (baryons). These massive states are included as matter fields chirally coupled to pions and external sources. There are many articles [10, 11, 12, 13] in the literature where CHPT lagrangians, at most bilinear in the nucleon fields, are applied to the nuclear case in the following way: The bilinears $\overline{N}DN$ (with D a generic differential operator including the coupling to pions and external fields) are replaced by the non-relativistic mean-field approach $\rho_p \mathrm{tr} D_{11} + \rho_n \mathrm{tr} D_{22}$, with ρ_p (ρ_n) the proton (neutron) density, where the symbol tr refers to the trace over spinor indices and the subscripts run in flavor space. Proceeding in this way one keeps track of the information contained in the vacuum CHPT lagrangians, but the chiral counting in the medium is lost, since nucleon correlations, due to the baryon propagators, are not considered. In fact, such contributions can be of the same or even of lower chiral order as those terms kept in the mean-field approach. As shown in Ref. [9], they are the dominant contributions when the energy flowing through the baryon propagator is of the order of a nucleonic kinetic energy. Note that the approach of Ref. [9] not only encompasses, but also transcends the so-called low–density theorems as formulated in Refs. [14].

IN-MEDIUM GENERATING FUNCTIONAL

In order to go beyond the mean field approach we followed Ref. [2] and considered the CHPT lagrangian supplemented by external fields [9]. According to Ref. [15], a general chiral lagrangian can be expanded in an increasing number of baryon fields, $\mathcal{L} = \mathcal{L}_{\pi\pi} + \mathcal{L}_{\overline{\psi}\psi} + \mathcal{L}_{\overline{\psi}\psi\overline{\psi}\psi} + \ldots$, where ψ denotes the nucleon field. As in the mean-field approaches, the multi-nucleon local interactions will be left out here. So only the perturbative expansions of the chiral lagrangians $\mathcal{L}_{\pi\pi}$ and $\mathcal{L}_{\bar{\psi}\psi}$ (the latter is bilinear in the nucleon field) in powers of the external four-momenta (derivatives) of the pions, the external three-momenta of the nucleons and the quark masses will be considered. For this purpose, we have made use of the results from Ref. [8], where, after inserting the nuclear matter state as ground state at asymptotic times and integrating out the baryonic fields in the path integral representation, the *in-medium* SU(2)×SU(2) generating functional is derived as a path integral over the usual chiral quaternion field $U \equiv uu \equiv \exp(i\phi/f)$ where $\phi = \begin{pmatrix} \pi^0 & \pi^+/\sqrt{2} \\ \pi^-/\sqrt{2} & -\pi^0 \end{pmatrix}$:

$$\begin{aligned}
e^{iZ[v,a,s,p]} &= \int [dU] \exp\Big\{ i\int dx\,\mathcal{L}_{\pi\pi} + \int \frac{d\vec{p}}{(2\pi)^3 2E(p)} \int dx\,dy\, e^{ip(x-y)} \\
&\quad \times \mathrm{Tr}\Big[-iA\left(I_4 - D_0^{-1}A\right)^{-1}\Big|_{(x,y)} (\slashed{p} + m_N) n(p)\Big] \\
&\quad -\frac{1}{2}\int \frac{d\vec{p}}{(2\pi)^3 2E(p)} \int \frac{d\vec{q}}{(2\pi)^3 2E(q)} \int dx\,dx'\,dy\,dy'\, e^{ip(x-y)} e^{-iq(x'-y')} \\
&\quad \times \mathrm{Tr}\Big[-iA\left(I_4 - D_0^{-1}A\right)^{-1}\Big|_{(x,x')}(\slashed{q}+m_N) n(q) \\
&\quad \times (-iA)\left(I_4 - D_0^{-1}A\right)^{-1}\Big|_{(y',y)}(\slashed{p}+m_N) n(p)\Big] + \cdots \Big\} \\
&\equiv \int [dU] \exp\Big\{ i\int dx\,\widetilde{\mathcal{L}}_{\pi\pi}[U;v,a,s,p]\Big\}.
\end{aligned} \quad (1)$$

Here, the operator A is defined as the difference between the full and free Dirac operators, i.e. $\mathcal{L}_{\bar{\psi}\psi} \equiv \bar{\psi}(x) D(x) \psi(x) \equiv \bar{\psi}(x)(D_0(x) - A(x))\psi(x)$ with $D_0 = i\gamma^\mu \partial_\mu - m_N$, while the diagonal flavor matrix $n(p) = \mathrm{diag}[\theta(k_F^{(p)} - |\vec{p}|), \theta(k_F^{(n)} - |\vec{p}|)]$ parametrizes the upper cutoff for the three-momentum integrations in terms of the proton and neutron Fermi momenta, respectively. The operator I_4 is the unit operator in 4 dimensions, while $E(p) = \sqrt{\vec{p}^2 + m_N^2}$ is the on-shell energy of the nucleon of mass m_N.

As a result the *in-medium* CHPT lagrangian $\widetilde{\mathcal{L}}_{\pi\pi}$, in terms of *only* pions and external (vector v, axial-vector a, scalar s, and pseudoscalar p) sources is given. Thus the problem reduces to that of vacuum CHPT, except for the important difference that the resulting lagrangian is *non-covariant* as well as *non-local* (for a general analysis of the structure of non-relativistic, but local effective field theories see [16]). Especially, it contains the non-local *vacuum* vertex $\Gamma \equiv -iA(I_4 - D_0^{-1}A)^{-1}$ that generates a geometric series in terms of the local interaction operator A and the free Dirac propagator D_0^{-1}. The vertex-operator A is itself subject to a chiral power expansion and contains both pion legs and external sources. The interaction operators of first and second order read

$$A^{(1)} = -i\gamma^\mu \Gamma_\mu - ig_A^0 \gamma^\mu \gamma_5 \Delta_\mu \quad \text{with} \quad 2\begin{Bmatrix} \Gamma_\mu \\ \Delta_\mu \end{Bmatrix} \equiv [u^\dagger, \partial_\mu u]_\mp - iu^\dagger(v_\mu + a_\mu)u \pm iu(v_\mu - a_\mu)u^\dagger,$$

$$A^{(2)} = -c_1 \langle \chi_+ \rangle - \frac{c_2}{8} \langle \Delta_\mu \Delta_\nu \rangle \frac{D^\mu}{m_N} \frac{D^\nu}{m_N} + \frac{c_3}{8} \langle \Delta_\mu \Delta^\mu \rangle - \frac{c_4}{16} \gamma^\mu \gamma^\nu [\Delta_\mu, \Delta_\nu] - c_5 \left(\chi_+ - \tfrac{1}{2} \langle \chi_+ \rangle \right)$$

$$+ \cdots \quad \text{with} \quad \chi_+ \equiv u^\dagger \chi u^\dagger + u \chi^\dagger u \quad \text{and} \quad \chi \equiv -2 \frac{\langle \bar{q} q \rangle}{f^2} (s + ip).$$

Here $\langle \cdots \rangle$ is the trace in the flavor space and $D_\mu \equiv \partial_\mu + \Gamma_\mu$. The c_i's are finite $O(p^2)$ low-energy constants, see Ref. [17, 18] for more details. Note that $A^{(1)}$ corresponds to the S-wave Weinberg-Tomozawa term and the derivative P-wave pion-nucleon coupling, while the sigma-term (proportional to c_1) and the so-called range terms (proportional to c_2 and c_3) are included in $A^{(2)}$. Finally, the generalized *in-medium* vertices, which still may be linked to each other by pion legs from $\mathcal{L}_{\pi\pi}$, consist of several non-local vacuum vertices Γ connected through the exchange of on-shell Fermi-sea states (see the traces in Equation (1)) with three-momenta smaller than $k_F^{(p)}(k_F^{(n)})$. The lowest order *in-medium* vertices are obtained when only $A^{(1)}$ operators are used; the next-to-leading order is obtained when one $A^{(2)}$ operator is included. In Ref. [9] the pertinent Feynman rules for the computation of connected in-medium graphs in momentum space can be found.

RESULTS

We have systematically studied in [9] several low–energy QCD Green functions up to next-to-leading order, $O(p^5)$, when the standard counting holds, or, by working out the leading in-medium contributions, in the non-standard case. The novel results that we have obtained there can be summarized as follows:

1. In contrast to previous works, which apply the mean-field approach or many–body calculations, the in-medium chiral counting has been worked out including contributions from baryon propagators. The choice of the counting scheme depends on the energy flowing into the nucleon lines. This induces a separate consideration of the standard and the non–standard case, respectively, as summarized in Table 1. In the former case, the chiral expansion of pion properties in the medium starts with terms at $O(p^4)$, and the next–to–leading order corrections appear at $O(p^5)$, quite different to the in–vacuum power counting which starts at $O(p^2)$ with next-to-leading corrections of $O(p^4)$.

2. We have also established the relevant scales of the problem (under restriction to the $\mathcal{L}_{\pi\pi}$ and $\mathcal{L}_{\pi N}$ lagrangians of Ref. [8]). In the vacuum, the pertinent scale at which the chiral expansion breaks down is $\Lambda_\chi \simeq 1$ GeV $\sim 4\pi f_\pi$. In the medium, one has two new scales. These are $\sqrt{6\pi} f_\pi \simeq 0.7$ GeV and $6\pi^2 f_\pi^2 / 2 m_N \simeq 0.27$ GeV, in case that the standard or the non-standard counting rules apply, see Table 1. In the case of P-wave interactions, these scales have to be reduced by factors $1/g_A$ and $1/g_A^2$, respectively.

3. We have studied the quark condensates and re-derived, from the effective field theory point of view, known results in symmetric nuclear matter (see the derivation of Ref. [19] via the Hellmann-Feynman theorem and also Ref. [20]), and have further extended them to the non-symmetric case:

$$\langle \Omega | \bar{u} u | \Omega \rangle = \langle \bar{u} u \rangle_{\text{vac}} \left[1 - \frac{2\sigma}{f^2 M_\pi^2} \hat{\rho} + \frac{4 c_5}{f^2} \bar{\rho} \right], \quad \langle \Omega | \bar{d} d | \Omega \rangle = \langle \bar{d} d \rangle_{\text{vac}} \left[1 - \frac{2\sigma}{f^2 M_\pi^2} \hat{\rho} - \frac{4 c_5}{f^2} \bar{\rho} \right].$$

TABLE 1. Chiral counting and pertinent breakdown scales for in–medium CHPT for the standard and the non–standard case. Here, "energy flow" means the energy flowing into the nucleon lines mediated by pions or external sources. The integer d_j counts the chiral dimension of a local vacuum vertex, while the integer $\delta_{\rho j} = 4 + \sum_{i=1}^{V}(d_i - 1) \geq 4$ counts the chiral dimension of a non-local in-medium vertex with V local vertices. E_x is the number of external legs, and κ stands for the sum $\kappa = I_B - I_B^* + \sum_{m \geq 3}(m-2)p_m + \sum_{m \geq 2}(m-1)n_m \geq 0$, where I_B and I_B^* are the total number and the $O(p^{-2})$-restricted number of internal baryon lines, respectively. Finally p_m is the number of local vacuum vertices with m legs from the pion lagrangian $\mathcal{L}_{\pi\pi}$, while n_m counts the number of A-type vacuum vertices with m attached legs.

	Standard counting	Non-standard counting
energy flow	$Q^0 \sim M_\pi \sim O(p)$	$Q^0 \sim Q^2/2m_N \sim O(p^2)$
nucleon propagator	$D_0^{-1} \sim O(p^{-1})$	$D_0^{-1} \sim O(p^{-2})$
counting index ν	$2 + 2L_\pi + \sum_{i=1}^{V_\pi}(d_i - 2) + \sum_{i=1}^{V_\rho}(\delta_{\rho i} - 2)$	$4 - E_x + \sum_{i=1}^{V_\pi}(d_i - 2) + \sum_{i=1}^{V_\rho}(\delta_{\rho i} - 3) + \kappa$
breakdown scale Λ	$\sqrt{6\pi} f_\pi \simeq 0.7\,\text{GeV}$	$6\pi^2 f_\pi^2 / 2m_N \simeq 0.27\,\text{GeV}$

Here $|\Omega\rangle$ is the nuclear matter background, while $\hat{\rho} \equiv \frac{1}{2}(\rho_p + \rho_n)$ and $\bar{\rho} \equiv \frac{1}{2}(\rho_p - \rho_n)$ are defined as the isospin symmetric and asymmetric combinations of the proton and neutron density, ρ_p and ρ_n respectively. The pion-nucleon sigma term [21, 22] reads $\sigma = -4c_1 M_\pi^2 \approx 44 \pm 8 \pm 7\,\text{MeV}$, where M_π is the vacuum pion mass. The isospin breaking contribution is parametrized by the low-energy constant $c_5 = -(0.09 \pm 0.01)\,\text{GeV}^{-1}$ [22] which is very suppressed as compared to $2c_1 = 2(-0.81 \pm 0.12)\,\text{GeV}^{-1}$ [23]. Hence the presence of the terms proportional to c_5 does not alter sizably the values of the quark condensates at finite density, studied in Refs. [12, 19, 20] for the symmetric nuclear matter case. In fact, at normal nuclear matter density $\rho = \rho_0$, there is a 35 % reduction of the condensates. This is compatible with a partial restoration of chiral symmetry.

4. We have considered the propagation of pions in the medium and obtained the (inverse) pion propagator up to $O(p^5)$. The spectral relations between the energy ω and the three-momentum $\vec{q}$ of on-shell in-medium pions are given by

$$\pi^0: \quad \omega^2 - M_{\pi^0}^2\left(1 + c_1 \frac{8\hat{\rho}}{f^2}\right) + \frac{4\hat{\rho}}{f^2}\omega^2\left(c_2 + c_3 - \frac{g_A^2}{8m_N}\right) - \vec{q}^{\,2}\left(1 + \frac{4\hat{\rho}}{f^2}c_3 - \frac{g_A^2 \hat{\rho}}{m_N f^2}\right)$$
$$- \frac{g_A^2 \hat{\rho}}{2f^2 m_N} \frac{(\vec{q}^{\,2})^2}{\omega^2} - \widehat{M}_\pi^2 \frac{4\bar{\rho}}{f^2} \frac{m_u - m_d}{m_u + m_d} c_5 = 0, \quad (2)$$

$$\pi^\pm: \quad \omega^2 - M_{\pi^\pm}^2\left(1 + c_1 \frac{8\hat{\rho}}{f^2}\right) + \frac{4\hat{\rho}}{f^2}\omega^2\left(c_2 + c_3 - \frac{g_A^2}{8m_N}\right) - \vec{q}^{\,2}\left(1 + \frac{4\hat{\rho}}{f^2}c_3 - \frac{g_A^2 \hat{\rho}}{m_N f^2}\right)$$
$$- \frac{g_A^2 \hat{\rho}}{2f^2 m_N} \frac{(\vec{q}^{\,2})^2}{\omega^2} \pm \frac{g_A^2 \bar{\rho}}{f^2 \omega}\vec{q}^{\,2} \mp \frac{\bar{\rho}\omega}{f^2} = 0. \quad (3)$$

Here $\widehat{M}_\pi^2$ is the lowest order CHPT pion mass, and $M_{\pi^0}, M_{\pi^\pm}$ are the vacuum pion masses at $O(p^4)$. All of them can be taken to coincide with the physical masses since differences will be $O(p^6)$. For symmetric matter ($\bar{\rho} = 0$) and in the chiral limit ($m_q = 0$) we can extract the following dispersion law from the previous expressions:

$$\omega^2 = \vec{q}^{\,2}\left(1 - \frac{4\hat{\rho}}{f^2}c_2\right). \quad (4)$$

This implies that the in-medium pion velocity $\tilde{v} = \frac{d\omega}{d|\vec{q}|} = 1 - \frac{2\hat{\rho}}{f^2}c_2$ is less than the vacuum speed of light ($c = 1$) for the empirical value of the low-energy constant $c_2 = (3.2 \pm 0.25)\,\text{GeV}^{-1}$ of Ref. [17, 18]. Using furthermore the value $c_3 = (-4.70 \pm 1.16)\,\text{GeV}^{-1}$ from Ref. [23], we have established from Equation (3) that chiral symmetry can account for the observed shift of the mass of the negative pion in deeply bound pionic states in ^{207}Pb. Our numerical result $\Delta M_{\pi^-} = 18 \pm 5\,\text{MeV}$ is compatible with the experimental number, $\Delta M_{\pi^-} = 23 - 27\,\text{MeV}$ [4, 5] within errors. The main contribution to this shift, around 16 MeV, results from the division of the Weinberg term by the wave-function renormalization for in-medium pions in symmetric nuclear matter at threshold (see below), which for nuclear saturation density is ~ 0.5.

5. The wave function renormalizations of the in-medium pion fields, $\langle\Omega|\pi^\alpha(x)|\pi^\beta(p)\rangle \equiv Z_\alpha(\vec{q}^2)^{-1/2}\delta_{\alpha\beta}e^{-ipx}$, established from general principles via the equal-time commutation relations, read

$$Z_{\pi^\alpha}(\vec{q}^2) = 1 + \frac{4\hat{\rho}}{f^2}\left(c_2 + c_3 - \frac{g_A^2}{8m_N}\frac{M_\pi^2}{\omega_0(\vec{q}^2)^2}\right) \mp \left\{\frac{\bar{\rho}}{2f^2\omega_0(\vec{q}^2)} - \frac{g_A^2\bar{\rho}\vec{q}^2}{2f^2\omega_0(\vec{q}^2)^3}\right\}\delta_{\alpha\pm},$$

where $\omega_0(\vec{q}^2) = \sqrt{\vec{q}^2 + M_\pi^2}$ and where the last two terms contribute only to the case of charged pions, $\pi^\alpha = \pi^\pm$, and in asymmetric nuclear matter.

6. The in-medium generating functional also allows the study of the coupling of pions with axial-vector and pseudoscalar sources [9]. In particular, there is the splitting of the temporal and space-like component of the in-medium pion decays constant (see also Refs. [24, 12, 13]) which in the isospin-limit ($\bar{\rho} = m_u - m_d = 0$) and at threshold read

$$f_t = f_\pi\left\{1 + \frac{2\hat{\rho}}{f^2}\left(c_2 + c_3 - \frac{g_A^2}{8m_N}\right)\right\} = f_\pi\left\{1 - \frac{\rho}{\rho_0}(0.26 \pm 0.04)\right\}, \quad (5)$$

$$f_s = f_\pi\left\{1 - \frac{2\hat{\rho}}{f^2}\left(c_2 - c_3 + \frac{g_A^2}{8m_N}\right)\right\} = f_\pi\left\{1 - \frac{\rho}{\rho_0}(1.23 \pm 0.07)\right\}. \quad (6)$$

The ratio $f_s/f_t = 1 - \frac{4\hat{\rho}}{f^2}c_2 = \tilde{v}^2 < 1$ is consistent with the discussion about the in-medium pion velocity. Moreover, we have shown that in-medium corrections up to $O(p^5)$ do not spoil the validity of the Gell-Mann-Oakes-Renner relation, see also Ref. [12]. Note that there is a decrease with increasing density for both, the quark condensates and the temporal component of the pion decay constant f_t. The two effects seem to indicate a partial chiral symmetry restoration with increasing density. However, a systematic study of the in-medium order parameters still has to be performed. A drastic quenching with density has also been obtained for the spatial component of the pion decay constant f_s. To $O(p^5)$, the QCD Ward identities relating both the temporal and spatial components of the axial-vector currents with the pseudoscalar ones and quark masses hold.

7. Finally, $\pi\pi$ scattering has been studied up to $O(p^3)$ since in this case the non-standard counting occurs. This implies that the in-medium corrections start at lower orders than in the standard case, namely already at $O(p^3)$, see Table 1. In addition the scale, below which the perturbative expansion is applicable, decreases. As a result the in-medium corrections increase very rapidly with density. Already at $k_F \simeq 200$ MeV, or at a density of just $\sim 0.4\rho_0$, they are of the same size as the lowest order CHPT results.

OUTLOOK

There still exist challenges for the future, e.g. the inclusion of multi-nucleon contact interactions remains an open problem. These interactions are enhanced because of the largeness of the S-wave scattering lengths related to the presence of shallow NN bound states. Furthermore, the simultaneous calculation of pion loops necessary for determination of the full $O(p^6)$ contributions has still to be performed. Finally, the possibility of some non-perturbative scheme that allows for the recovery of the scale $\sqrt{6}\pi f_\pi$, even in the case of the non-standard counting or when the multi-nucleon local interactions are included, should be pursued.

REFERENCES

1. S. Weinberg, *Physica* **A96**, 327 (1979).
2. J. Gasser and H. Leutwyler, *Ann. Phys. (N.Y.)* **158**, 142 (1984).
3. Z. Fodor and S. D. Katz, *Nucl. Phys. Proc. Suppl.* **106**, 441 (2002).
4. H. Gilg et al., *Phys. Rev.* **C62**, 025201 (2000).
5. K. Itahasi et al., *Phys. Rev.* **C62**, 025202 (2000).
6. T.-S. Park, H. Jung and D.-P. Min, nucl-th/0101064.
7. N. Kaiser and W. Weise, *Phys. Lett.* **B512**, 283 (2001).
8. J. A. Oller, *Phys. Rev.* **C65**, 025204 (2002), hep-ph/0101204.
9. U.-G. Meißner, J. A. Oller and A. Wirzba, nucl-th/0109026.
10. D. B. Kaplan and A. E. Nelson, *Phys. Lett.* **B175**, 57 (1986).
11. G. E. Brown, K. Kubodera, M. Rho and V. Thorsson, *Phys. Lett.* **B291**, 355 (1992);
    T. Muto, R. Tamagaki and T. Tatsumi, *Prog. Theor. Phys. Suppl.* **112**, 159 (1993);
    V. Thorsson, M. Prakash and J. M. Lattimer, *Nucl. Phys.* **A572**, 693 (1994); (E) **A574**, 851 (1994);
    G. E. Brown, C. Lee, M. Rho and V. Thorsson, *Nucl. Phys.* **A567**, 937 (1994);
    V. Thorsson, M. Prakash and J. M. Lattimer, *Nucl. Phys.* **A572**, 693 (1994);
    C. Lee, G. E. Brown, D. Min and M. Rho, *Nucl. Phys.* **A585**, 401 (1995).
12. V. Thorsson and A. Wirzba, *Nucl. Phys.* **A589**, 633 (1995);
    A. Wirzba and V. Thorsson, "In-medium effective chiral lagrangians and the pion mass in nuclear matter", in *Hirschegg '95: Dynamical Properties of Hadrons in Nuclear Matter*, edited by H. Feldmeier and W. Nörenberg, GSI-print, Darmstadt, 1995, pp. 31-43, hep-ph/9502314.
13. M. Kirchbach and A. Wirzba, *Nucl. Phys.* **A604**, 395 (1996);
    M. Kirchbach and A. Wirzba, *Nucl. Phys.* **A616**, 648 (1997).
14. E. G. Drukarev and E. M. Levin, *Nucl. Phys.* **A511**, 697 (1988);
    T. D. Cohen, R. J. Furnstahl and D. K. Griegel, *Phys. Rev. Lett.* **67**, 961 (1991);
    M. C. Birse, *J. Phys.* **G 20**, 1537 (1994).
15. S. Weinberg, *Phys. Lett.* **B251**, 288 (1990); *Nucl. Phys.* **B363**, 3 (1991).
16. H. Leutwyler, *Phys. Rev.* **D49**, 3033 (1994).
17. N. Fettes, U.-G. Meißner and S. Steininger, *Nucl. Phys.* **A640**, 199 (1998).
18. N. Fettes and U.-G. Meißner, *Nucl. Phys.* **A676**, 311 (2000).
19. T.D. Cohen, R.J. Furnstahl and D.K. Griegel, *Phys. Rev.* **C45** (1992).
20. E.G. Drukarev and E.M. Levin, *Prog. Part. Nucl. Phys.* **27**, 77 (1991).
21. J. Gasser, M.E. Sainio and A. Svarc, *Nucl. Phys.* **B307**, 779 (1988).
22. V. Bernard, N. Kaiser and U.-G. Meißner, *Nucl. Phys.* **A615**, 483 (1997).
23. P. Büttiker and U.-G. Meißner, *Nucl. Phys.* **A668**, 97 (2000).
24. M. Kirchbach and D. O. Riska, *Nucl. Phys.* **A578**, 511 (1994).

Chiral Symmetry and the Medium Modification of Hadrons

J. Wambach

*Institut für Kernphysik, Technische Universität Darmstadt, Schloßgartenstr. 9,
D-64289 Darmstadt, Germany*

Abstract. The medium modification of mesons in dense hadronic matter is discussed with a focus on the relationship to the chiral structure of the non-perturbative QCD vacuum.

INTRODUCTION

One of the fundamental questions in strong interaction physics concerns the mass generation of hadrons containing light quarks. In contrast to the heavy-quark sector where the mass is set at the electroweak scale, light-quark hadrons receive their mass through the non-perturbative structure of the QCD vacuum itself through the condensation of quark and gluons.

When hadronic matter is subjected to extreme conditions in density and temperature such as in the interior of neutron stars or in central relativistic heavy-ion collisions, the QCD vacuum will be altered, eventually leading to the liberation of the elementary constituents in a new state of matter, the 'quark gluon plasma' (QGP). Such a restructuring of the vacuum must be accompanied by significant changes in the spectral properties of light hadrons, commonly called 'medium modifications'. Here mesons are of particular importance, since they constitute the 'elementary' $\bar{q}q$ excitations of the vacuum and hence serve as experimental probes of the underlying vacuum changes. For the structure of light mesons chiral symmetry and its spontaneous break-down plays a decisive role as is well established from lattice QCD simulations [1]. The physical mechanism for spontaneous symmetry breaking and mass generation is provided by instantons which largely saturate the euclidian gauge-field action [2].

These induce effective quark-(anti)quark interactions which are strong enough to cause a BCS-like transition [3] to a $\bar{q}q$ condensate. As a result a mass gap appears, the 'constituent quark' mass of $\sim$ 300-500 MeV. Indeed the phenomenon of spontaneous mass generation is observed on the lattice [5] as seen in Fig. 1 where the euclidian quark mass is displayed in the Landau gauge. One also notices that chiral symmetry breaking is a long-distance phenomenon occurring only at small momenta. It has been demonstrated that this picture also yields an excellent description of hadronic correlation functions [2] and hence forms the basis for a variety of approaches to deal with medium modifications. An obvious starting point is NJL-type models [3] where the instanton-induced interaction is assumed to be point-like. They are very successful in describing mesonic spectral functions, especially when treated beyond the mean-field approximation [4].

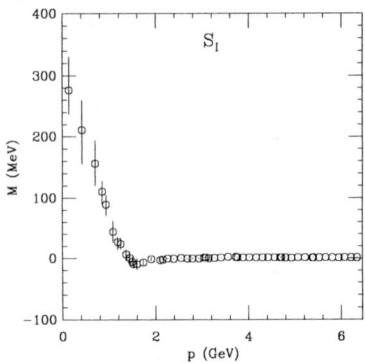

FIGURE 1. The quark mass function, $M(p^2)$, from lattice QCD simulations [5].

Such calculations are very demanding but finally lead to results in accordance with 'effective theories' (such as the vector dominance model (VDM) [6]) in which hadronic fields feature as the elementary degrees of freedom. It is more 'economical' to start with those from the outset, especially when dealing with 'precursor' phenomena of the QGP transition. This strategy will be followed in the present contribution.

CHIRAL EFFECTIVE THEORY AND CORRELATORS

The spontaneous breaking of chiral symmetry has two important consequences. One is the appearance of massless Goldstone bosons and the other the absence of parity doublets in the hadron spectrum ($m_\pi \neq m_{f_0}, m_\rho \neq m_{a_1}$ etc). For the present discussion the second feature will be more relevant. Chiral symmetry does, however, more than just predict the existence of Goldstone bosons. It also prescribes and severely restricts their mutual interactions as well as those with other hadrons. This forms the basis for constructing 'chiral effective theories'.

The formal starting point are the pertinent quark currents

$$J_i(x) = \bar{q}(x)\Gamma_i q(x), \Gamma_i = 1, \gamma_5, \gamma_\mu, .. \tag{1}$$

which are identified with *elementary* hadronic fields $\phi_i(x)$. One then writes down the most general effective Lagrangian, consistent with the underlying symmetries and anomaly structure of QCD. Relevant examples are the linear- or nonlinear sigma model, gauged sigma models, etc.

The physical information about hadronic spectral properties and their in-medium modification is contained in the current-current correlation functions $D_i(x) \propto \langle J_i(x)J_i(0)\rangle$. In terms of the elementary hadronic fields and for matter in thermal and chemical equilibrium they are given by the (retarded) momentum-space Green's function

$$D_i(\omega, \vec{q}) = i\int d^4x\, e^{iqx}\theta(x_0)\langle\langle[\phi_i(x),\phi_i(0)]\rangle\rangle \tag{2}$$

where the average is taken in the grand canonical ensemble. All information about the physical excitations of the medium is contained in the spectral function

$$A_i(\omega, \vec{q}) = -\frac{1}{\pi} \text{Im} D_i(\omega, \vec{q}). \tag{3}$$

In the first step the parameters of the chiral Lagrangian are adjusted to reproduce the vacuum properties of the hadronic correlators in question (mass, width..) as well as possible. Here either perturbative, or in some cases, non-perturbative methods [7] are used. Once the vacuum model is fixed, the medium modifications can be predicted with a good degree of accuracy.

MESONS IN THE HADRONIC MEDIUM

Chiral Condensate evolution

With increasing density (baryo-chemical potential μ_q) and temperature T the quark condensate will diminish and eventually vanish, thus restoring chiral symmetry. For hadronic matter in equilibrium the QCD partition function is given by

$$Z_{QCD}(V, T, \mu_q) = \text{Tr} e^{-(H_{QCD} - \mu_q N_q)/T}, \tag{4}$$

where N_q is the quark number operator and μ_q the quark chemical potential. The in-medium quark condensate $\langle\langle \bar{q}q \rangle\rangle$ can be directly inferred from the free energy density in the thermodynamic limit

$$\Omega_{QCD}(T, \mu_q) = -\lim_{V \to \infty} \frac{T}{V} \ln Z_{QCD}(V, T, \mu_q) \quad \text{as} \quad \langle\langle \bar{q}q \rangle\rangle = \frac{\partial \Omega_{QCD}(T, \mu_q)}{\partial m_q^\circ}, \tag{5}$$

where m_q° denotes the current quark mass.

In the broken phase, the obvious first step is to approximate the free energy density by an ideal gas of hadrons. The resulting condensate ratio then becomes

$$\frac{\langle\langle \bar{q}q \rangle\rangle}{\langle \bar{q}q \rangle} = 1 - \sum_h \frac{\Sigma_h \rho_h^s(T, \mu_q)}{f_\pi^2 m_\pi^2}; \quad \Sigma_h = m_q^\circ \frac{\partial m_h}{\partial m_h^\circ} \tag{6}$$

where ρ_h^s denotes the scalar density of hadrons and m_h their vacuum mass. At low temperature and small μ_q in which the hadron gas is dominated by thermally excited pions and a free Fermi gas of nucleons, Eq. (6) leads to the model-independent leading-order result

$$\frac{\langle\langle \bar{q}q \rangle\rangle}{\langle \bar{q}q \rangle} = 1 - \frac{T^2}{8f_\pi^2} - 0.3 \frac{\rho}{\rho_0} \ldots, \tag{7}$$

where $\rho_0 = 0.16/\text{fm}^3$ is the saturation density of symmetric nuclear matter. Thus the mere presence of an ideal gas of hadrons already alters the vacuum and leads to a decrease of the condensate, without changing the vacuum properties of the hadrons!

Obviously medium-modifications and the corresponding non-trivial change of the QCD vacuum have to involve hadronic interactions. They become increasingly important as the matter grows hotter and denser, i.e. as the point of chiral restoration is approached. The theoretical description involves more and more degrees of freedom, which severely limits the description in terms of elementary hadronic fields near the phase transition.

Fluctuations of the chiral condensate

Being dependent on the renormalization scale, the chiral condensate is not observable. However, the fluctuations $\langle\langle(\bar{q}q)^2\rangle\rangle - \langle\langle\bar{q}q\rangle\rangle^2$ are. They correspond to the scalar (chiral) susceptibility and are given as the second derivative of the free energy density

$$\chi_+ = \langle\langle(\bar{q}q)^2\rangle\rangle - \langle\langle\bar{q}q\rangle\rangle^2 = \frac{\partial^2 \Omega_{QCD}(T,\mu_q)}{\partial^2 m_q^\circ} \tag{8}$$

w.r.t. to the current quark mass. To make contact with the hadronic spectrum we consider the bilinear scalar quark current $J_+ = \bar{q}q$. From the corresponding correlator

$$D_+(\omega,\vec{q}) = i \int d^4x\, e^{iqx} \theta(x_0) \langle\langle[(J_+(x),J_+(0)]\rangle\rangle \tag{9}$$

the scalar susceptibility is obtained as the 'polarizability sum rule' for the scalar spectral function $A_+(\omega,\vec{q})$:

$$\chi_+ = \int_0^\infty d\omega A_+(\omega,\vec{q}=0) , \tag{10}$$

and hence relates to the properties of the scalar f_0 meson in the medium.

From direct lattice simulations it is established that QCD with two flavors and vanishing μ_q exhibits a second-order phase transition in the chiral limit [8]. The transition at $T = 0$ but finite μ_q is not as well understood, since it cannot be simulated on the lattice, due to complex fermion determinants. Model calculations indicate however that it may be first order at several times ρ_0. As a second- or weak first-order phase transition is approached the fluctuations of the order parameter become large. A measure for the growth are the appropriate susceptibilities which, in a second-order transition, diverge with critical exponents that are determined by universal behavior. The approach to criticality in the case of two-flavor QCD is clearly seen on the lattice [9].

Following the suggestion in ref. [10] that two-flavor QCD lies in the same universality class as the $O(4)$ Heisenberg model we can use the linear sigma model to asses the chiral fluctuations in the hadronic medium. In this case the scalar correlator D_+ in Eq. (9) is identified with the in-medium σ-propagator:

$$D_+ = D_\sigma(\omega,\vec{q}) = i \int d^4x\, e^{iqx} \theta(x_0) \langle\langle[\sigma(x),\sigma(0)]\rangle\rangle . \tag{11}$$

Applying a low-density expansion for the in-medium condensate according to Eq. (7)

$$\langle\langle\sigma\rangle\rangle \equiv \langle\sigma\rangle\Phi(\rho); \quad \Phi(\rho) = 1 - \alpha\frac{\rho}{\rho_0} \tag{12}$$

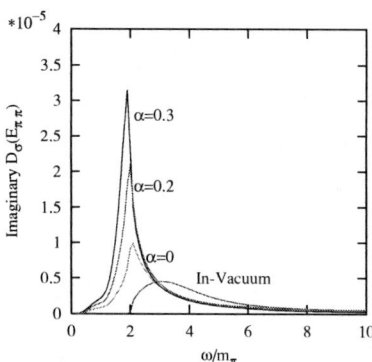

FIGURE 2. The imaginary part of $D_\sigma(\omega,0)$ [11] at nuclear saturation density for various values of α, the parameter that controls the linear density dependence of $\langle\langle\sigma\rangle\rangle$ (Eq. 7).

together with pionic loop corrections the results [11] depicted in Fig. 2 predict a dramatic near-threshold enhancement of the σ spectral function already at ρ_0 (corresponding to $\Phi(\rho) \sim 0.7 - 0.8$). It should be noted that this effect is also found in the non-linear realization of chiral symmetry [12] and is therefore generic.

What are the experimental signatures? Since the σ meson strongly couples to two-pion states an obvious experiment is the production of two $J = I = 0$ pions near threshold in nuclei. Such experiments have been conducted by the CHAOS collaboration at TRIUMF using an incident π^+ beam on various nuclear targets [13] identifying charged pions in the final state. A second experiment is by the Crystal Ball (CB) collaboration at BNL with an incident π^- beam [14] detecting a π^0 pair in the final state through coincident 4γ decay. For the invariant mass distribution of the produced pion pair both measurements consistently observe a significant reshaping as function of mass number as indicated in Fig 3. Here the composite ratio

$$C_{\pi\pi}^A = \frac{\sigma^A(M_{\pi\pi})}{\sigma_T^A} / \frac{\sigma^N(M_{\pi\pi})}{\sigma_T^N} \quad (13)$$

is displayed, where $\sigma^A(M_{\pi\pi})$ ($\sigma^N(M_{\pi\pi})$) denotes the invariant mass distribution in the nucleus (nucleon), while σ_T^A (σ_T^N) is the corresponding total cross section for the $\pi 2\pi$ process. The various curves in the right panel of Fig. 3 summarize the theoretical predictions. The full [17] and dotted [18] lines only include contributions from p-wave modifications of the pion propagator while the dashed-dotted line [19] only considers the dropping of the σ mass without dressing the pion loop. Finally, the dashed line [11] includes both effects and gives a reasonable description of the data.

The ρ meson and Dileptons

The ρ meson is featured as a prominent resonance in the e^+e^- annihilation cross section and therefore plays a central role for dilepton production in heavy-ion collisions

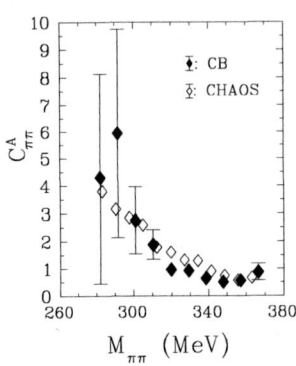
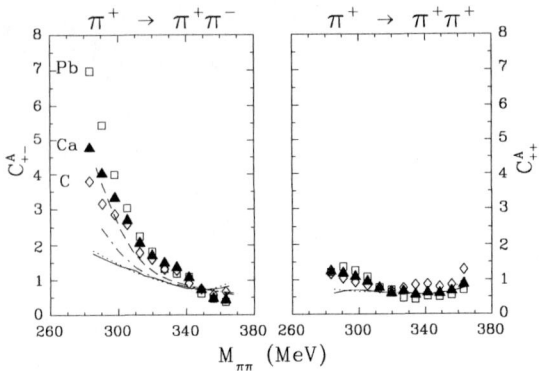

FIGURE 3. The composite ratio $C_{\pi\pi}^A$ for various nuclear targets [15, 16].

at invariant masses below 1GeV. In general, the production rate is given in terms of the in-medium electromagnetic current-current correlation function as

$$\frac{dR_{l^+l^-}}{d^4q} = f(q_0)L^{\mu\nu}\mathrm{Im}D_{\mu\nu}^{\mathrm{elm}}; \quad D_{\mu\nu}^{\mathrm{elm}}(\omega,\vec{q}) = i\int d^4x\, e^{iqx}\theta(x_0)\langle\langle[J_\mu^{\mathrm{elm}}(x),J_\nu^{\mathrm{elm}}(0)]\rangle\rangle \quad (14)$$

where $L^{\mu\nu}$ denotes the lepton tensor and $f(q_0)$ is a thermal Bose factor. Invoking vector dominance, the hadron tensor $D_{\mu\nu}^{\mathrm{elm}}$ directly relates to the in-medium properties of the ρ meson and can be evaluated in the VDM. Two important medium effects can be identified. The first is the modification of the intermediate two-pion state to which the ρ meson strongly couples. Here proper care has to be taken to ensure gauge invariance. The second is a direct coupling to baryonic resonances, most prominently the $N^*(1550)$ resonance. The inclusion of both effect leads to a large broadening of the spectral function at high density and temperature and results in a dramatic reshaping of the dilepton rate as indicated in the left panel of Fig. 4. Once the local rate is space-time evolved through a realistic fire ball expansion until thermal freeze out, and detector acceptances and background rates from Dalitz decays are properly accounted for, the resulting rates compare favorably with the measurements of the CERES collaboration at the CERN SpS [22] (right panel of Fig. 4). Also the transverse spectra are well reproduced.

The connection to the restoration of chiral symmetry is not apparent, however. This requires a simultaneous evaluation of both the vector- and axialvector correlator. In the vacuum and in the chiral limit both are related by two 'Weinberg sum rules' [23]:

$$\int_0^\infty ds\,(\rho_V^\circ(s) - \rho_A^\circ(s)) = f_\pi^2; \quad \int_0^\infty ds\,s\,(\rho_V^\circ(s) - \rho_A^\circ(s)) = 0 \quad (15)$$

where the first directly links the vacuum spectral functions to f_π, the order parameter of chiral symmetry restoration. Similar sum rules also hold in the hadronic medium [24] and serve as an important constraint of models that intend to properly implement chiral symmetry in the correlators. One such model has recently been proposed in ref. [25].

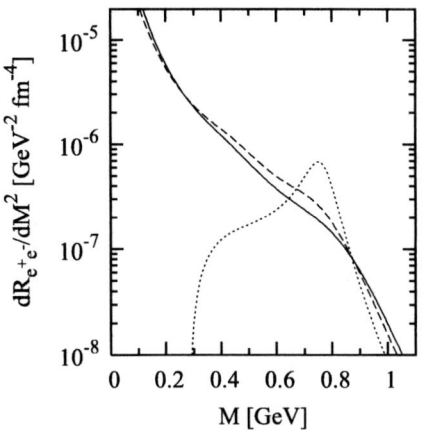 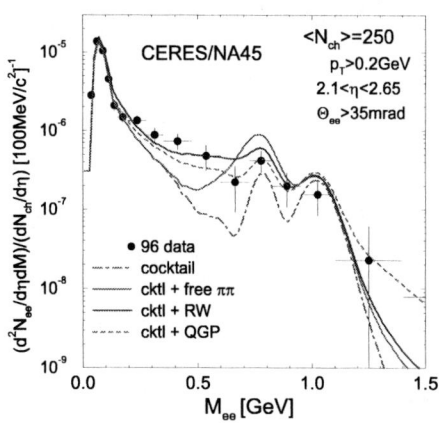

FIGURE 4. Dilepton rates: the left panel displays the theoretical prediction at $T=150$ MeV and $\mu_B = 3\mu_q = 452$ MeV [20] while the right panel shows a comparison [21] with the measured rates of the CERES collaboration [22].

The starting point is the linear sigma model (a non-linear realization of chiral symmetry could also be chosen). Combining the ρ- and a_1 fields as chiral partners

$$Y^\mu = \vec{\rho}^\mu \cdot \vec{T} + \vec{a}_1^\mu \cdot \vec{T}_5 \tag{16}$$

one then writes down the most general Lagrangian consistent with *global* chiral symmetry to a given order in the derivative coupling. This introduces a set of bare parameters to be determined from the vacuum phenomenology, in particular the measured vector- and axialvector spectral functions. In the one-loop approximation one can achieve an excellent description of the pion electromagnetic form factor, p-wave $\pi\pi$ phase shifts, e^+e^- cross sections and τ-decay [25]. The extension to finite temperature is straightforward and results in the spectral distributions displayed in Fig. 5 for the chiral limit. At high temperatures a significant reshaping of the strength distribution is observed, especially at low energy. Nonetheless the ρ and a_1 peak remain present, even in the vicinity of the phase boundary. The parameters of the model Lagrangian can be easily adjusted so that, at tree level, the ρ-meson mass is proportional to the chiral condensate $\langle\langle\sigma\rangle\rangle$. This is the scenario of 'Brown-Rho scaling' [27]. To one-loop order, however, it can be proven analytically that the pole mass of the ρ meson remains unchanged to order T^2 and only receives contributions of order T^4 and higher hence invalidating the 'Brown-Rho scenario'.

SUMMARY

Based on general considerations of spontaneous breaking of chiral symmetry and its restoration in hot and dense hadronic matter I have discussed the in-medium properties

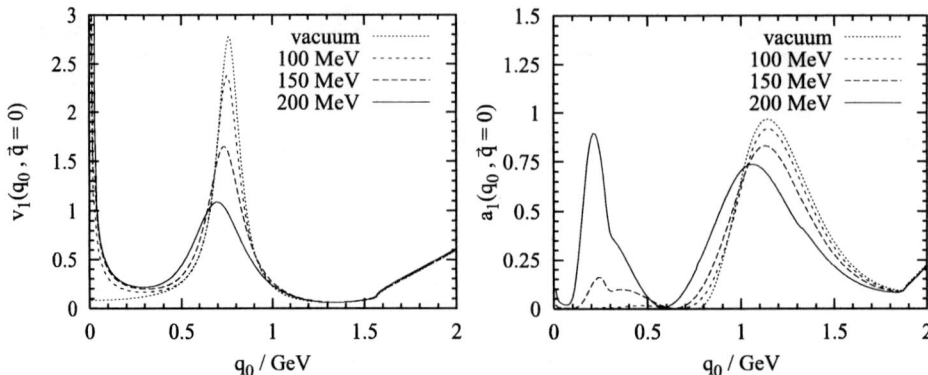

FIGURE 5. The temperature dependence of the vector spectral function (left panel) and the axialvector spectral function (right panel) in the chiral limit [26].

of σ- and ρ mesons. The former relate to the increased fluctuations of the chiral order parameter which are accessible through measurements of s-wave isoscalar two-pion correlations in nuclei. Interesting near-threshold enhancements in the invariant-mass distribution are observed in $\pi 2\pi$ reactions, which can be interpreted as a signal of the partial restoration of chiral symmetry. The ρ meson and its medium modification plays an important role in the dilepton production in relativistic heavy-ion collisions. Realistic calculations of the spectral function in a hot and dense fireball indicate a significant broadening due to collisions with baryons and mesons and explain the observed production rates. The relation to chiral symmetry is not obvious, however. To address this issue one has to consider the vector- and axialvector correlators simultaneously. Invoking global chiral symmetry in constructing a chiral Lagrangian from elementary σ, π, ρ and a_1 fields it can be shown that, in the chiral limit, there is no direct relationship between the pole mass of the ρ meson and the chiral condensate.

ACKNOWLEDGMENTS

I wish to thank my collaborators Z. Aouissat, M. Buballa, C. Isselhorst, M. Oertel, P. Schuck, R. Rapp and M. Urban. This work has been supported by the BMBF and GSI.

REFERENCES

1. J. W. Negele, Nucl. Phys. Proc. Suppl. 73 (1999) 92.
2. T. Schäfer, E. V. Shuryak, Rev. Mod. Phys. 70 (1998) 323.
3. Y. Nambu and G. Jona-Lasinio, Phys. Rev. 122 (1961) 345.
4. M. Oertel, M. Buballa, J. Wambach, Nucl. Phys. A676 (2000) 247;
   M. Oertel, M. Buballa, J. Wambach, Phys. Atom. Nucl. 64 (2001) 698.
5. J. Skullerud, D.B. Leinweber and A.G. Williams, Phy. Rev. D64 (2001) 074508.
6. J. J. Sakurai, Ann. Phys. 11 (1960) 1.

7. Z. Aouissat, P. Schuck, J. Wambach, Nucl. Phys. A618 (1997) 402.
8. F. Karsch, hep-ph/0103314.
9. F. Karsch, Nucl. Phys. A590 (1995) 367.
10. R. Pisarski, F. Wilczek, Phys. Rev. D29 (1984) 338.
11. Z. Aouissat, G. Chanfray, P. Schuck, J. Wambach, Phys. Rev. C61 (2000) 12202.
12. D. Jido, T. Hatsuda, T. Kunihiro, hep-ph/0008076.
13. F. Bonutti et al., Phys. Rev. Lett. 77 (1996) 603.
14. A. B. Starostin et al., Phys. Rev. Lett. 85 (2000) 5539.
15. P. Camerini et al., Phys. Rev. C64 (2001) 067601.
16. reprinted from F. Bonutti et al., Nucl. Phys. A677 (2000) 213 with permission from Elsevier Science.
17. M. J. Vicente Vacas, E. Oset, Phys. Rev. C60 (1999) 64621.
18. R. Rapp et al., Phys. Rev. C59 (1999) R1237.
19. T. Hatsuda, T. Kunihiro, H. Shimizu, Phys. Rev. Lett. 82 (1999) 2840.
20. M. Urban, M. Buballa, J. Wambach, Nucl. Phys. A673 (2000) 357.
21. R. Rapp, J. Wambach, Adv. Nucl. Phys. 25 (2000) 1.
22. G. Agakichiev et al., Phys. Rev. Lett. 75 (1995) 1272.
23. S. Weinberg, Phys. Rev. Lett. 18 (1967) 507.
24. J. Kapusta, E. V. Shuryak, Phys. Rev. D49 (1994) 4694.
25. M. Urban, M. Buballa, J. Wambach, Nucl. Phys. A697 (2002) 338.
26. M. Urban, M. Buballa, J. Wambach, nucl-th/0110005, Phys. Rev. Lett. to be publ.
27. G. E. Brown, M. Rho, Phys. Rev. Lett. 66 (1991) 2720.

Topics on heavy baryon chiral perturbation theory in the large N_c limit

Rubén Flores-Mendieta

Instituto de Física, Universidad Autónoma de San Luis Potosí, Alvaro Obregón 64, San Luis Potosí, S.L.P. 78000, Mexico

Abstract. We compute nonanalytical pion-loop corrections to baryon masses in a combined expansion in chiral symmetry breaking and $1/N_c$, where N_c is the number of colors. Specifically, we compute flavor-**27** baryon mass splittings at leading order in chiral perturbation theory. Our results, at the physical value $N_c = 3$, are compared with the expressions obtained in heavy baryon chiral perturbation theory with no $1/N_c$ expansion.

INTRODUCTION

Chiral perturbation theory has been a useful tool in the understanding of low-energy QCD hadron dynamics. Its application to baryons through a new formulation of the low-energy chiral effective Lagrangian in which the baryons appear as heavy static fields, was first introduced in Refs. [1, 2]. The chiral Lagrangian thus obtained was used to compute the leading nonanalytic in m_s corrections to baryon axial currents [1, 2, 3], masses, and non-leptonic decays [1, 2], to name but a few.

Similarly, the $1/N_c$ expansion has proved to be useful in the analysis of the spin-flavor structure of baryons in QCD [4, 5, 6, 7, 8]. Evidence for the predictions of the $1/N_c$ expansion for baryons has been found in the analysis of masses [7, 8, 9], magnetic moments [7, 8, 10], and axial and vector currents [8, 11, 12].

The next natural step has been to combine both approaches so that baryon matrix elements are obtained in a simultaneous chiral and $1/N_c$ expansion [7, 5, 13, 14, 15]. The resulting approach, referred to as large-N_c chiral perturbation theory, has proven to have a significant predictive power in the analysis of the baryon mass spectra [13, 15, 16], magnetic moments [10] and axial current [16].

Here we present an explicit calculation of flavor-**27** mass splittings of the octet and decuplet, which are calculable and nonanalytic in the quark masses and baryon hyperfine mass splittings at leading order in chiral perturbation theory. This analysis illustrates how to implement the procedure.

BARYONS IN CHIRAL PERTURBATION THEORY

The heavy baryon chiral Lagrangian can be constructed in terms of the pion field Π, the baryon octet field B_v, and the baryon decuplet field T^μ_{abc}, where the Π and B_v fields are

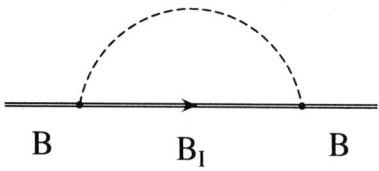

FIGURE 1.

represented by 3×3 matrices and T^μ_{abc} is a Rarita-Schwinger field which contains both spin-1/2 and spin-3/2 pieces and obeys the constraint $\gamma^\mu T_\mu = 0$ [1, 2].

The coupling of the pseudoscalar pion field with the baryon matter fields occurs through the vector and axial vector combinations $V^\mu = (1/2)(\xi \partial^\mu \xi^\dagger + \xi^\dagger \partial^\mu \xi)$ and $A^\mu = (i/2)(\xi \partial^\mu \xi^\dagger - \xi^\dagger \partial^\mu \xi)$, where $\xi = e^{i\Pi/f}$, $\Sigma = \xi^2 = e^{2i\Pi/f}$, and $f \approx 93$ MeV is the pion decay constant. Further considerations can be found in Refs. [1, 2].

The most general Lagrangian at lowest order is

$$\mathcal{L}_{\text{baryon}} = i \operatorname{Tr} \bar{B}_v (v \cdot \mathcal{D}) B_v - i \bar{T}^\mu_v (v \cdot \mathcal{D}) T_{v\mu} + \Delta \bar{T}^\mu_v T_{v\mu} + 2D \operatorname{Tr} \bar{B}_v S^\mu_v \{A_\mu, B_v\} \\ + 2F \operatorname{Tr} \bar{B}_v S^\mu_v [A_\mu, B_v] + C (\bar{T}^\mu_v A_\mu B_v + \bar{B}_v A_\mu T^\mu_v) + 2\mathcal{H} \bar{T}^\mu_v S^\nu_v A_\nu T_{v\mu}, \quad (1)$$

where D, F, C, and $\mathcal{H}$ are the baryon-pion couplings and $\Delta = m_T - m_B$ is the decuplet-octet mass difference. $\mathcal{L}_{\text{baryon}}$ describes massless pion fields interacting with degenerate SU(3) multiplets of baryons.

Three nonanalytic terms for the baryon masses are calculable, namely, the ones which vary as $m_s^{3/2}$, $m_s^2 \ln m_s$, and $(\Delta) m_s \ln m_s$. These contributions result from one-loop diagrams and have been computed in Ref. [17]. Here we analyze the leading nonanalytic contributions arising from the Feynman diagram displayed in Fig. 1. This diagram involves π, K, and η emission and reabsorption. The most general form of this contribution can be written as

$$-\delta M_i = I_1(\pi, K, \eta; \Delta) m_{i,1} + I_8(\pi, K, \eta; \Delta) m_{i,8} + I_{27}(\pi, K, \eta; \Delta) m_{i,27}. \quad (2)$$

$I_j(\pi, K, \eta; \Delta)$ are flavor lineal combinations of the integral over the loop, $F(m_\Pi, \Delta)$ [14]. Here m_Π is the pion mass and $\Pi = \pi, K, \eta$. Specifically,

$$I_1(\pi, K, \eta; \Delta) = \frac{1}{8} [3F(\pi, \Delta) + 4F(K, \Delta) + F(\eta, \Delta)], \quad (3)$$

$$I_8(\pi, K, \eta; \Delta) = \frac{2\sqrt{3}}{5} \left(\frac{3}{2} F(\pi, \Delta) - F(K, \Delta) - \frac{1}{2} F(\eta, \Delta) \right), \quad (4)$$

$$I_{27}(\pi, K, \eta; \Delta) = \frac{1}{3} F(\pi, \Delta) - \frac{4}{3} F(K, \Delta) + F(\eta, \Delta) \quad (5)$$

with

$$m_{i,1} = \bar{\lambda}^\pi_i + \bar{\lambda}^K_i + \bar{\lambda}^\eta_i, \qquad m_{i,8} = \frac{1}{\sqrt{3}} \left(\bar{\lambda}^\pi_i - \frac{1}{2} \bar{\lambda}^K_i - \bar{\lambda}^\eta_i \right) \\ m_{i,27} = \frac{3}{40} \left(\bar{\lambda}^\pi_i - 3 \bar{\lambda}^K_i + 9 \bar{\lambda}^\eta_i \right). \qquad (6)$$

The $\bar{\lambda}_i^\Pi$ coefficients for the octet baryons read

$$\bar{\lambda}_N^\pi = \frac{9}{4}(F+D)^2 + 2C^2, \qquad \bar{\lambda}_\Sigma^\pi = 6F^2 + D^2 + \frac{1}{3}C^2,$$

$$\bar{\lambda}_N^K = \frac{1}{2}(9F^2 - 6FD + 5D^2 + C^2), \qquad \bar{\lambda}_\Sigma^K = 3(F^2 + D^2) + \frac{5}{3}C^2,$$

$$\bar{\lambda}_N^\eta = \frac{1}{4}(3F - D)^2, \qquad \bar{\lambda}_\Sigma^\eta = D^2 + \frac{1}{2}C^2,$$

$$\bar{\lambda}_\Xi^\pi = \frac{9}{4}(F-D)^2 + \frac{1}{2}C^2, \qquad \bar{\lambda}_\Lambda^\pi = 3D^2 + \frac{3}{2}C^2, \qquad (7)$$

$$\bar{\lambda}_\Xi^K = \frac{1}{2}(9F^2 + 6FD + 5D^2 + 3C^2), \qquad \bar{\lambda}_\Lambda^K = 9F^2 + D^2 + C^2,$$

$$\bar{\lambda}_\Xi^\eta = \frac{1}{4}(3F + D)^2 + \frac{1}{2}C^2, \qquad \bar{\lambda}_\Lambda^\eta = D^2.$$

whereas for the decuplet baryons one has

$$\bar{\lambda}_\Delta^\pi = \frac{25}{36}\mathcal{H}^2 + \frac{1}{2}C^2, \qquad \bar{\lambda}_{\Xi^*}^\pi = \frac{5}{36}\mathcal{H}^2 + \frac{1}{4}C^2,$$

$$\bar{\lambda}_\Delta^K = \frac{5}{18}\mathcal{H}^2 + \frac{1}{2}C^2, \qquad \bar{\lambda}_{\Xi^*}^K = \frac{5}{6}\mathcal{H}^2 + \frac{1}{2}C^2,$$

$$\bar{\lambda}_\Delta^\eta = \frac{5}{36}\mathcal{H}^2, \qquad \bar{\lambda}_{\Xi^*}^\eta = \frac{5}{36}\mathcal{H}^2 + \frac{1}{4}C^2,$$

$$\bar{\lambda}_{\Sigma^*}^\pi = \frac{10}{27}\mathcal{H}^2 + \frac{5}{12}C^2, \qquad \bar{\lambda}_{\Omega^-}^\pi = \frac{10}{27}\mathcal{H}^2, \qquad (8)$$

$$\bar{\lambda}_{\Sigma^*}^K = \frac{20}{27}\mathcal{H}^2 + \frac{1}{3}C^2, \qquad \bar{\lambda}_{\Omega^-}^K = \frac{5}{9}\mathcal{H}^2 + C^2,$$

$$\bar{\lambda}_{\Sigma^*}^\eta = \frac{1}{4}C^2, \qquad \bar{\lambda}_{\Omega^-}^\eta = \frac{5}{9}\mathcal{H}^2 + C^2.$$

Equation (2) can be used to analyze the two flavor-**27** combinations of baryon masses, namely, the Gell-Mann–Okubo combination for octet baryons and the equal spacing rule combination for decuplet baryons, which are respectively

$$\frac{3}{4}\Lambda + \frac{1}{4}\Sigma - \frac{1}{2}(N + \Xi), \qquad -\frac{4}{7}\Delta + \frac{5}{7}\Sigma^* + \frac{2}{7}\Xi^* - \frac{3}{7}\Omega, \qquad (9)$$

where particle labels denote the corresponding masses. These relations are a direct consequence of the fact that SU(3) symmetry breaking is purely octet [1, 2].

The terms proportional to $m_{i,1}$ and $m_{i,8}$ in Eq. (2) have no effect on the above mass relations. The **27** piece produces the corrections

$$\left[-\frac{3}{4}(D^2 - 3F^2)\bar{I}(0) + \frac{1}{8}C^2\bar{I}(\Delta)\right], \qquad \left[\frac{5}{18}\mathcal{H}^2\bar{I}(0) - \frac{1}{4}C^2\bar{I}(-\Delta)\right] \qquad (10)$$

to the Gell-Mann–Okubo mass relation and to the equal spacing rule, respectively, where $\bar{I}(\Delta)$ is an abbreviation for $I_{27}(\pi, K, \eta; \Delta)/N_c$ [14]. The integral $I_{27}(\pi, K, \eta; \Delta)$, which is around 4 MeV, is highly suppressed relative to its singlet and octet counterparts. It explains, however, the small violation to the Gell-Mann–Okubo mass relation.

BARYONS IN LARGE N_C CHIRAL PERTURBATION THEORY

The derivation of the $1/N_c$ baryon chiral Lagrangian is given in Ref. [14]. Recent developments on the large-N_c formalism applied to baryons can be found in excellent reviews [18, 19]. One can also refer to Jenkins' contribution to this proceedings.

The $1/N_c$ baryon chiral Lagrangian can be expressed as [14]

$$\mathcal{L}_{\text{baryon}} = i\mathcal{D}^0 - M_{\text{hyperfine}} + \text{Tr}\left(\mathcal{A}^i \lambda^a\right) A^{ia} + \text{Tr}\left(\mathcal{A}^i \frac{2I}{\sqrt{6}}\right) A^i + \ldots \tag{11}$$

Each term in Eq. (11) involves a baryon operator which can be given in terms of polynomials in the spin-flavor generators J^i, T^a, and G^{ia} [8]. For instance, at the physical value $N_c = 3$, the baryon axial current A^{ia} is expressed as

$$A^{ia} = a_1 G^{ia} + b_2 \frac{1}{N_c} J^i T^a + b_3 \frac{1}{N_c^2} \mathcal{D}_3^{ia} + c_3 \frac{1}{N_c^2} O_3^{ia}, \tag{12}$$

where the operators $\mathcal{D}_3^{ia}$ and O_3^{ia} are defined in [8].

We now compute again the leading nonanalytic corrections to the baryon masses but now within the combined formalism in $1/N_c$ and chiral corrections. The computation is complicated by the presence of the hyperfine and quark mass splittings. In the chiral limit $m_i \to 0$ the baryon propagator is diagonal in spin and can be written as [14]

$$\frac{i\mathcal{P}_j}{k^0 - \Delta_j}, \tag{13}$$

where $\mathcal{P}_j$ is a spin projection operator for spin $J = j$. For $N_c = 3$ one has [14]

$$\mathcal{P}_{\frac{1}{2}} = -\frac{1}{3}\left(J^2 - \frac{15}{4}\right), \quad \mathcal{P}_{\frac{3}{2}} = \frac{1}{3}\left(J^2 - \frac{3}{4}\right). \tag{14}$$

On the other hand, Δ_j in Eq. (13) stands for the difference of the hyperfine mass splitting for spin $J = j$ and the external baryon.

The diagram in Fig. 1, given by the product of a baryon operator times the pion flavor tensor, can be expressed as

$$\frac{1}{N_c}\sum_j \left(A^{ia}\mathcal{P}_j A^{ib}\right) \Pi^{ab}(\Delta_j), \tag{15}$$

where Π^{ab} is the symmetric tensor defined as

$$\Pi^{ab}(\Delta) = I_1(\pi, K, \eta; \Delta)\delta^{ab} + I_8(\pi, K, \eta; \Delta) d^{ab8}$$
$$+ I_{27}(\pi, K, \eta, \Delta)\left(\delta^{a8}\delta^{b8} - \frac{1}{8}\delta^{ab} - \frac{3}{5}d^{ab8}d^{888}\right). \tag{16}$$

For $N_c = 3$, Eq. (15) reduces to

$$\frac{1}{N_c}\left[A^{ia}\mathcal{P}_{\frac{1}{2}}A^{ib} \Pi^{ab}(\Delta_{\frac{1}{2}}) + A^{ia}\mathcal{P}_{\frac{3}{2}}A^{ib} \Pi^{ab}(\Delta_{\frac{3}{2}})\right]. \tag{17}$$

The evaluation of Eq. (17) involves the computation of the baryon operators $A^{ia}A^{ib}\Pi^{ab}$ and $A^{ia}J^2A^{ib}\Pi^{ab}$. One can follow the approach implemented by Jenkins [14] to perform the operator reduction of the spin operators involved in the latter expressions by using spin projection operators. However, we will use a simplified version of such analysis. After a long but otherwise standard calculation we find

$$A^{ia}A^{ia} = \frac{3}{16}N_c(N_c+6)a_1^2 + \frac{3}{4}\left(1+\frac{6}{N_c}\right)a_1c_3 + \left[-\frac{5}{12}a_1^2 + \frac{2}{3}\left(1+\frac{3}{N_c}\right)a_1b_2\right.$$
$$+ \left(\frac{1}{2}+\frac{3}{N_c}+\frac{4}{N_c^2}\right)a_1b_3 + \frac{1}{12}\left(1+\frac{6}{N_c}\right)b_2^2 + \left(\frac{1}{2}+\frac{3}{N_c}-\frac{9}{N_c^2}\right)a_1c_3\right]J^2$$
$$+ \frac{1}{N_c^2}\left[\frac{2}{3}a_1b_3 + b_2^2 - 2a_1c_3 + \frac{8}{3}\left(1+\frac{3}{N_c}\right)b_2b_3\right]J^4 + O\left(\frac{1}{N_c^4}\right), \quad (18)$$

$$d^{ab8}A^{ia}A^{ib} = \left[\frac{3}{8}(N_c+3)a_1^2 + \frac{3}{2N_c}\left(1+\frac{3}{N_c}\right)a_1c_3\right]T^8 + \left[-\frac{7}{12}a_1^2 + \frac{3}{N_c^2}a_1b_3\right.$$
$$+ \frac{1}{6}\left(1+\frac{3}{N_c}\right)a_1b_2 - \frac{9}{2N_c^2}a_1c_3\right]\{J^r,G^{r8}\} + \frac{1}{N_c}\left[\frac{1}{6}a_1b_2\right.$$
$$+ \frac{1}{2}\left(1+\frac{3}{N_c}\right)\left(a_1b_3 - \frac{1}{6}b_2^2 + a_1c_3\right)\right]\{J^2,T^8\} + \frac{1}{N_c^2}\left[-\frac{1}{3}a_1b_3 + \frac{1}{2}b_2^2 - a_1c_3\right.$$
$$+ \left(\frac{1}{3}+\frac{1}{N_c}\right)b_2b_3\right]\{J^2,\{J^r,G^{r8}\}\} + \frac{2}{3N_c^3}b_2b_3\{J^4,T^8\} + O\left(\frac{1}{N_c^4}\right), \quad (19)$$

$$A^{i8}A^{i8} = \frac{1}{2}a_1^2\{G^{i8},G^{i8}\} + \frac{1}{2N_c}a_1b_2\{T^8,\{J^i,G^{i8}\}\} + \frac{1}{N_c^2}a_1b_3\{\{J^r,G^{r8}\},\{J^i,G^{i8}\}\}$$
$$+ \frac{1}{N_c^2}a_1c_3\left[\{J^2,\{G^{i8},G^{i8}\}\} + [G^{i8},[J^2,G^{i8}]] - \frac{1}{2}\{\{J^r,G^{r8}\},\{J^i,G^{i8}\}\}\right]$$
$$+ \frac{1}{4N_c^2}b_2^2\{J^2,\{T^8,T^8\}\} + \frac{1}{N_c^3}b_2b_3\{J^2,\{T^8,\{J^i,G^{i8}\}\}\} + O\left(\frac{1}{N_c^4}\right). \quad (20)$$

Explicit expression for the T^8 and G^{i8} operators can be found in Ref. [8]. Similarly,

$$A^{ia}J^2A^{ia} = \frac{3}{8}N_c(N_c+6)a_1^2 + \left(\frac{3}{2}+\frac{9}{N_c}\right)a_1c_3 + \left[\left(\frac{3}{16}N_c(N_c+6)-\frac{7}{2}\right)a_1^2\right.$$
$$+ \left(\frac{13}{4}\left(1+\frac{6}{N_c}\right)-\frac{18}{N_c^2}\right)a_1c_3\right]J^2 + \left[-\frac{5}{12}a_1^2 + \frac{2}{3}\left(1+\frac{3}{N_c}\right)a_1b_2\right.$$
$$+ \left(\frac{1}{2}+\frac{1}{N_c}\right)\left(1+\frac{4}{N_c}\right)a_1b_3 + \frac{1}{12}\left(1+\frac{6}{N_c}\right)b_2^2 + \left(\frac{1}{2}+\frac{3}{N_c}-\frac{27}{N_c^2}\right)a_1c_3\right]J^4$$
$$+ \frac{1}{N_c^2}\left[\frac{2}{3}a_1b_3 + b_2^2 - 2a_1c_3 + \frac{8}{3}\left(1+\frac{3}{N_c}\right)b_2b_3\right]J^6 + O\left(\frac{1}{N_c^4}\right), \quad (21)$$

$$d^{ab8}A^{ia}J^2A^{ib} = \left[\frac{3}{4}(N_c+3)a_1^2 + \frac{3}{N_c}\left(1+\frac{3}{N_c}\right)a_1c_3\right]T^8 - \left[2a_1^2 + \frac{9}{N_c^2}a_1c_3\right]\{J^r,G^{r8}\}$$

$$+ \left[\frac{3}{16}(N_c+3)a_1^2 + \frac{13}{4}\left(\frac{1}{N_c}+\frac{3}{N_c^2}\right)a_1c_3 \right] \{J^2, T^8\} + \frac{2}{3N_c^3}\{J^6, T^8\}$$

$$+ \left[-\frac{7}{24}a_1^2 + \frac{1}{12}\left(1+\frac{3}{N_c}\right)b_2 + \frac{1}{2N_c^2}a_1b_3 - \frac{33}{4N_c^2}a_1c_3 \right] \{J^2, \{J^r, G^{r8}\}\}$$

$$+ \frac{1}{N_c^2}\left[-\frac{1}{3}a_1b_3 + \frac{1}{2}b_2^2 - a_1c_3 + \frac{1}{3}\left(1+\frac{3}{N_c}\right)b_2b_3 \right] \{J^4, \{J^r, G^{r8}\}\}$$

$$+ \frac{1}{N_c} \left[\frac{1}{6}a_1b_2 + \frac{1}{2}\left(1+\frac{3}{N_c}\right)\left(a_1b_3 - \frac{1}{6}b_2^2 + a_1c_3\right) \right] \{J^4, T^8\} + O\left(\frac{1}{N_c^4}\right), \quad (22)$$

$$A^{i8}J^2A^{i8} = \frac{1}{4}a_1^2\left(\{J^2, \{G^{i8}, G^{i8}\}\} + 2[G^{i8},[J^2, G^{i8}]]\right) + \frac{1}{4N_c}a_1b_2\{J^2, \{T^8, \{J^i, G^{i8}\}\}\}$$

$$+ \frac{1}{2N_c^2}a_1b_3\{J^2, \{\{J^r, G^{r8}\}, \{J^i, G^{i8}\}\}\} + \frac{1}{8N_c^2}b_2^2\{J^2, \{J^2, \{T^8, T^8\}\}\}$$

$$+ \frac{1}{2N_c^2}a_1c_3 \Big[\{J^2, \{J^2, \{G^{i8}, G^{i8}\}\}\} + \{J^2,[G^{i8},[J^2, G^{i8}]]\} + 2[G^{i8},[J^2, \{J^2, G^{i8}\}]]$$

$$- \frac{1}{2}\{J^2, \{\{J^r, G^{r8}\}, \{J^i, G^{i8}\}\}\} \Big] + \frac{1}{2N_c^3}b_2b_3\{J^2, \{J^2, \{T^8, \{J^i, G^{i8}\}\}\}\}. \quad (23)$$

The full evaluation of the baryon operators leads to

$$\frac{3}{4}\Lambda + \frac{1}{4}\Sigma - \frac{1}{2}(N+\Xi) = \frac{1}{N_c}\left[\left(\frac{1}{16}a_1^2 + \frac{3}{4}\frac{1}{N_c}a_1b_2 + \frac{9}{16}\frac{1}{N_c^2}b_2^2 + \frac{3}{8}\frac{1}{N_c^2}a_1b_3\right.\right.$$
$$\left.\left. + \frac{9}{4}\frac{1}{N_c^3}b_2b_3\right)I(0) + \left(\frac{1}{8}a_1^2 + \frac{9}{8}\frac{1}{N_c^2}a_1c_3\right)I(\Delta) + O\left(\frac{1}{N_c^4}\right)\right], \quad (24)$$

for the Gell-Mann Okubo mass formula and

$$-\frac{4}{7}\Delta + \frac{5}{7}\Sigma^* + \frac{2}{7}\Xi^* - \frac{3}{7}\Omega = \frac{1}{N_c}\left[\left(\frac{5}{8}a_1^2 + \frac{15}{4}\frac{1}{N_c}a_1b_2 + \frac{45}{8}\frac{1}{N_c^2}b_2^2 + \frac{75}{4}\frac{1}{N_c^2}a_1b_3\right.\right.$$
$$\left.\left. + \frac{225}{4}\frac{1}{N_c^3}b_2b_3\right)I(0) - \left(\frac{1}{4}a_1^2 + \frac{9}{4}\frac{1}{N_c^2}a_1c_3\right)I(-\Delta) + O\left(\frac{1}{N_c^4}\right)\right], \quad (25)$$

for the equal spacing rule. The above equations, at the physical value $N_c = 3$, can be straightforwardly compared with the analogous expressions obtained in heavy baryon chiral perturbation theory Eqs. (10) by using the identifications (26).

$$\begin{aligned} D &= \frac{1}{2}a_1 + \frac{1}{6}b_3, & C &= -a_1 - \frac{1}{2}c_3, \\ F &= \frac{1}{3}a_1 + \frac{1}{6}b_2 + \frac{1}{9}b_3, & \mathcal{H} &= -\frac{3}{2}a_1 - \frac{3}{2}b_2 - \frac{5}{2}b_3, \end{aligned} \quad (26)$$

Both expressions agree.

CONCLUSIONS

We have exemplified how to compute corrections to baryon masses in a calculational scheme that simultaneously exhibits both the m_q and $1/N_c$ expansions. Flavor-**27** baryon mass splittings at leading order in chiral perturbation theory have been computed in detail to illustrate the procedure. Furthermore, our results for three colors have been compared with the ones obtained in heavy baryon chiral perturbation theory with no $1/N_c$ corrections and an agreement has been found term by term in the series.

Some other applications of the combined approach to baryon properties will be presented elsewhere.

ACKNOWLEDGMENTS

The author is grateful to Consejo Nacional de Ciencia y Tecnología (Mexico) for partial support.

REFERENCES

1. Jenkins, E., and Manohar, A.V., *Phys. Lett. B* **255**, 558-62 (1991).
2. Jenkins, E., and Manohar, A.V., *Phys. Lett. B* **259**, 353-58 (1991).
3. Borasoy, B., *Phys. Rev. D* **59**, 054021 (1999).
4. Carone, C.D., Georgi, H., and Osofsky, S.T., *Phys. Lett. B* **322**, 227-32 (1994).
5. Luty, M.A., and March-Russell, J., *Nucl. Phys. B* **426**, 71-93 (1994).
6. Dashen, R., and Manohar, A.V., *Phys. Lett. B* **315**, 425-30 (1993); Phys. Lett. B 315, 438-40 (1993).
7. Dashen, R.F., Jenkins, E., and Manohar, A.V., *Phys. Rev. D* **49**, 4713-38 (1994).
8. Dashen, R.F., Jenkins, E., and Manohar, A.V., *Phys. Rev. D* **51**, 3697-3727 (1995).
9. Jenkins E., and Lebed, R., *Phys. Rev. D* **52**, 282-94 (1995).
10. Luty, M.A., March-Russell, J., and White, M., *Phys. Rev. D* **51**, 2332 (1995).
11. Dai, J., Dashen, R.F., Jenkins, E., and Manohar, A.V., *Phys. Rev. D* **53**, 273 (1996).
12. Flores-Mendieta, R., Jenkins, E., and Manohar, A.V., *Phys. Rev. D* **58**, 094028 (1998).
13. Luty, M.A., *Phys. Rev. D* **51**, 2322-31 (1995).
14. Jenkins, E., *Phys. Rev. D* **53**, 2625-44 (1996).
15. Bedaque, P.F., and Luty, M.A., *Phys. Rev* **D** 54, 2317-27 (1996).
16. Oh, Y., and Weise, W., *Eur. Phys. J.* **A4** 363-80 (1999).
17. Jenkins, E., *Nucl. Phys. B* **368**, 190-203 (1992).
18. Manohar, A.V., "Large N QCD," in *Les Houches Session LXVIII, Probing the Standard Model of Particle Interactions*, edited by F. David and R. Gupta, Amsterdam, Elsevier, 1998.
19. Jenkins, E., *Annu. Rev. Nucl. Part. Sci.* **48**, 81-119 (1998).

Science and Technology of the TESLA Electron-Positron Linear Collider

Albrecht Wagner

University of Hamburg and DESY

Abstract. Recent analyses of the long term future of particles physics in Asia, Europe, and the U.S.A. have led to the consensus that the next major facility to be built to unravel the secrets of the micro-cosmos is an electron-positron linear collider in the energy range of 500 to 1000 GeV. This collider should be constructed in an as timely fashion as possible to overlap with the Large Hadron Collider, under construction at CERN. Here, the scientific potential and the technological aspects of the TESLA projects, a superconducting collider with an integrated X-ray laser laboratory, are summarised.

UNDERSTANDING THE ORIGIN OF MASS AND THE GRAND UNIFICATION OF FORCES

Elementary particle physics has the ambitious goal to explain the innermost building blocks of matter and the fundamental forces acting between them. Symmetry principles determine the ordering of the building blocks and the nature of their interactions. The masses of particles and the exact strength of the forces played a key role in the evolution of the universe from the Big Bang to its present appearance in terms of galaxies, stars, black holes, chemical elements and biological systems. Discoveries in particle physics thus go to the very core of our existence.

Particle physics has made enormous progress in the past thirty years by pushing back the frontiers of accelerators, experiments and theory. Today we know that matter is composed of few fundamental building blocks, called quarks and leptons. A concise theoretical framework for the forces between these constituents has been developed which is based on the theoretical principle of gauge invariance. By applying this principle it has become possible to unify the seemingly disparate electromagnetic and weak interactions into a single electroweak interaction, and to develop a quantum field theory of the strong interaction, called Quantum Chromodynamics (QCD). The forces are mediated by the photon, the W and Z bosons and the gluons, the particles or quanta of the corresponding fields.

However, such a theory of matter particles and force fields suffers from a serious deficiency. The underlying gauge principle requires at first sight all field quanta to be mass-less. This is in striking contrast to the large masses of the W and Z bosons, which are 80 – 90 times heavier than the proton. At present the only compelling way to give the particles a mass while preserving the gauge principle is the so- called Higgs mechanism. The basic idea is that the a priori mass-less particles acquire "effective masses" by interaction with a background medium, the Higgs field. The idea of mass generation by the Higgs mechanism leads to a Higgs field which spreads out in all space. An analogous mechanism is in fact responsible for the attenuation of a magnetic field inside a superconductor. Associated with the Higgs field is a new observable particle, the Higgs particle, whose analogue in superconductivity is the Cooper pair.

The matter particles, the force fields of the electroweak theory and QCD, and the Higgs mechanism are the basis of the so-called "Standard Model" of particle physics, an extremely successful theory which has been tested and validated with high precision in a broad range of experiments. Only the Higgs particle has so far escaped observation.

Therefore, one of the most pressing challenges of particle physics is to establish the Higgs mechanism, to find the Higgs particle and to study its properties, or to find an alternative explanation of the masses of particles. With the help of the Heisenberg uncertainty principle of quantum mechanics one can get a first glimpse of the Higgs particle, even if it is too heavy to be produced directly. This principle allows a heavy particle to appear as a "virtual" particle for a tiny fraction of time and thereby to influence the measurements. Precision measurements made at electron-positron colliders have confirmed this effect with two very important and striking results: First, the mass of the top quark has been determined in experiments at colliders which did not have enough energy to produce it directly, before it was observed in proton-antiproton collisions at that mass. This was convincing proof that the Standard Model is correct even at the level of effects produced by virtual particles, i.e., at the quantum level. Second, the upper limit for the mass of the Higgs particle has been determined to be less than 200 GeV in the Standard Model. Recently, events observed at the highest energy of the Large Electron Positron Collider (LEP) at CERN in Geneva have given a tantalising hint that the Higgs particle might indeed have a mass around 115 GeV. In order to produce the Higgs particle directly and in particular, to study its properties precisely, we require more energy, i.e. accelerators of higher energy than those available today. The Higgs particle is expected to play a central role in the experimental program at TESLA.

Discovering the Higgs particle and establishing the Higgs mechanism would not, however, close the book of particle physics because the Standard Model is based on too many assumptions and leaves too many facts unexplained. General arguments clearly point to the existence of an even more fundamental theory. Supersymmetry is the favoured idea underlying such an extension of the Standard Model. It leads to a consistent and calculable theory in which the Higgs mechanism can be accommodated in a natural way. Most importantly, supersymmetry provides a framework for the unification of the electromagnetic, weak and strong forces at large energies. It is deeply related to gravity, the fourth of the fundamental forces. Supersymmetry predicts several Higgs particles. The lightest Higgs particle should have a mass below 200 GeV, or even below 135 GeV in some specific models. In supersymmetric theories for every fundamental particle we know today another related and as yet undiscovered particle should exist. The lightest supersymmetric particle most likely is stable. Many of these supersymmetric particles may have masses such that they can be produced and studied in detail at TESLA.

A particularly tantalising challenge of fundamental physics is posed by gravity, the interaction responsible for the large-scale structure of the universe. Gravity can not be incorporated in the Standard Model because gravity can not be formulated consistently as a quantum field theory. Great efforts are devoted to formulating a theory in which gravity is unified with the weak, electromagnetic and strong forces. A goal of such a fundamental theory will be to predict the masses and properties of all particles based on a few fundamental principles. It will synthesise quantum physics and the theory of relativity, thus unifying the physical laws of the microcosm and the macrocosm. Recent theories, known as super-string theories, suggest that at very high energies, as they existed shortly after the Big Bang, all four forces between particles are united into a single force. The discovery of supersymmetry and the precision measurements of the properties of supersymmetric particles could provide a glimpse of the underlying fundamental theory.

Probing matter at its smallest dimensions thus leads us to a better understanding of the laws governing the cosmos. The theories of particle physics describe matter also under extreme conditions, as realised during the earliest moments of the universe, when everything was very hot and dense. Collisions of particles in accelerators allow us to recreate in the laboratory what happened immediately after the Big Bang, 15 billion years ago, when matter in the form of quarks and leptons was created from energy. Nature has passed through this stage on the way to its current state. If we succeed in determining these laws of nature we will get clearer insights into the current state of development of the universe.

Astronomical evidence strongly suggests that more than 90% of the mass in the universe is invisible and of a nature totally different from the matter from which stars, planets and humans are made of. The nature of this so-called dark matter is completely unknown. Supersymmetric particles might be the explanation. In future large accelerators we expect to find such particles, if supersymmetry is indeed realised in nature. Thus, in developing more powerful accelerators, better experiments and theories, particle physicists contribute, together with astronomers and astrophysicists, to the understanding of the origin, evolution and destiny of the universe and the nature of space and time.

The large accelerators in operation today are the electron-proton collider HERA at DESY and the proton-antiproton collider Tevatron at Fermilab near Chicago. The next milestone on the road of particle physics is set by the large proton- proton collider (LHC), which is being built at CERN and which is scheduled to be completed in 2006. Many new discoveries are expected to come from the experiments there.

However, our previous experience clearly shows that a proton collider alone is not sufficient to adequately explore the subatomic world. It must be complemented with a high energy collider for electrons and positrons. A telling example from the past is the heavy Z boson, which was discovered at a proton-antiproton collider, while its detailed properties have only been measured with high precision at electron- positron colliders. These measurements were crucial for establishing the Standard Model. In particular, they allowed an indirect determination of the mass of the top quark and are responsible for our present constraint on the Higgs mass. Another illustrative example is the discovery of the carrier of the strong force, the gluon, which was not seen at a proton collider, where the strong force dominates, but was discovered at the electron-positron collider PETRA at DESY.

The complementarity of proton and lepton colliders is due to the different nature and properties of the particles which collide in the two types of accelerator. Electrons and positrons have no internal structure. Being fundamental particles they carry the full beam energy and interact through weak and electromagnetic forces, which can be calculated precisely. The conditions under which the collision takes place are defined very well, so that we can predict precisely what to expect after the collision. One can therefore determine the properties of new particles, such as mass, lifetime, spin and quantum numbers, unambiguously and with high precision. On the other hand, it is easier to accelerate protons to very high energies than it is to accelerate electrons. Protons are, however, complicated objects, composed of quarks, antiquarks and gluons. The detailed collision process can not be well controlled or selected, the effective energy of the colliding fundamental particles is usually well below the total energy of the two protons, and the rate of unwanted collision processes is very high. For these reasons TESLA complements the LHC and will provide important new insights.

The electron-positron linear collider TESLA, in its baseline design, reaches a centre-of-mass energy of 500 GeV, five times higher than the first linear collider SLC at Stanford and 2.5 times higher than LEP at CERN. At the same time the luminosity of TESLA, a measure for the event rate which a collider can deliver, is about 1000 times higher than that of LEP at 200 GeV. Both, energy and luminosity are prerequisites for new discoveries. In a second phase, the energy range of TESLA can be extended to about 800 GeV without increasing the length of the machine.

The energy range and luminosity of TESLA will make possible precise measurements of masses, lifetimes, and interactions of particles. These measurements will be needed to understand the mechanism responsible for the generation of mass. If the world is supersymmetric, with matter and forces united in one theory, TESLA will be uniquely positioned to explore these new particles. The great experimental precision characteristic for electron-positron colliders can be exploited to probe for physics in an energy range well beyond the reach of the collider. The substantial understanding gained during the studies of the physics potential of linear colliders provides us with a firm prediction: With TESLA we expect unique and crucial new insights into the laws of particle physics.

REVIEW OF THE ACCELERATOR TECHNOLOGY

Accelerators have become a key tool to study the microcosm. Their development started about eighty years ago, and has since been boosted by many new ideas and technologies which extended the attainable energy by a factor of around 10 every decade, and thus the capacity to resolve ever smaller objects and to create heavier particles. This development of accelerators has led to important applications in other fields of science including medicine. Especially electron storage rings used for the production of X-ray beams of unprecedented brilliance provide key tools for modern research. Today they may be considered the most important spin-offs of particle physics driven accelerator research.

The most powerful accelerator concept to reach high energies is that of colliders, in which particles are made to collide head-on. Among existing facilities electron-positron colliders have played a very special role. In the collision electrons and their anti-particles, the positrons, annihilate each other and the resulting energy is converted back into new particles, whose properties are measured in the detectors surrounding the interaction point. Since electrons and positrons are elementary point- like particles, they are the precision tools of particle physics, providing accurate knowledge of the reactions under study. These features were essential for many discoveries made with electron-positron colliders, including new quarks and leptons and the particle mediating the strong interaction, the gluon, and for precision tests of the Standard Model.

Except for the Stanford Linear Collider (SLC), electron-positron colliders have so far been built as storage rings, the largest being the Large Electron Positron collider (LEP) at CERN, with a 27 km circumference and a maximum energy per beam of just over 100 GeV. This concept, however, is not suitable for reaching even higher energies, as electrons radiate electromagnetic energy when forced on a circular path. The related energy loss increases by a factor 16 when doubling the particle energy.

Therefore the only way to reach electron energies substantially above 100 GeV is by accelerating them on a straight line. This leads directly to the concept of a linear collider. In this concept electrons and positrons are accelerated in opposite directions in two linear accelerators and made to collide in the middle of a detector. Each linear accelerator consists mainly of a large series of electromagnetic radiofrequency resonators (cavities), which efficiently generate the required electric fields to accelerate the electrons and positrons.

Over the past decades, several groups world-wide have been pursuing different linear collider design concepts. Already in 1971 a group at the Institute of Nuclear Physics in Novosibirsk started detailed design work for a linear collider of several hundred GeV, addressing many of the relevant problems. Several years later, groups at CERN, at the Stanford Linear Accelerator Center (SLAC) in California, and the Japanese National Laboratory for High Energy Physics (KEK) in Tsukuba began work on linear collider designs. The feasibility of the concept has been demonstrated by the successful operation of the Stanford Linear Collider. All these concepts were based on normal-conducting copper cavities.

A major challenge for all linear collider concepts is to obtain a large collision rate (luminosity) of electrons and positrons at the interaction point. This requires very small spot sizes of the beams at the collision point and high beam powers.

The TESLA approach differs from the other designs by the choice of superconducting accelerating structures as its basic technology. The TESLA linear collider based on superconducting accelerating structures is ideally suited to meet the requirements needed for a large collision rate, namely very small beam sizes and high beam power. The advantage of superconducting technology, combined with the high effciency to convert electrical energy to beam energy, has been acknowledged from the very beginning of the research and development on linear colliders, but the technology was considered to be considerably more expensive than conventional technologies.

By a focused development program, started in 1992, the international TESLA collaboration in co-operation with industry succeeded in developing superconducting microwave cavity structures which are capable of generating an accelerating voltage per meter, called gradient, five times larger than before 1990. In addition a reduction of the cost per meter of accelerator by a factor of four was achieved. Together, these achievements provide the basis for a realistic superconducting linear collider with all its advantages.

A prototype superconducting linear accelerator was built as part of the TESLA Test Facility in order to gain long term operating experience. To date it has been operated successfully for more than 10000 hours.

The development of a powerful linear accelerator for particle physics has also created the ideal accelerator for a light source with completely new properties: An X-ray free-electron Laser (XFEL) producing X-rays with true laser properties, as first pro-posed for the Stanford Linear Accelerator. The laser light is generated when electrons travel through a special magnet structure, after having been accelerated in a linear accelerator. The TESLA linear accelerator based on superconducting cavities is ideally suited to provide electron beams of the necessary quality. Using the TESLA Test Facility accelerator laser light was generated for the very first time in the wavelength range from 180 nm to 80 nm with an X-ray free-electron Laser. This was a first proof of principle that such an X-ray laser can be built and has stimulated intense activities in the field of XFELs world-wide.

Summarising the work of the past decade the following milestones in accelerator technology and development have been reached:

Cavities exceeding an accelerating gradient of 25 MV/m are being produced routinely by industry, thus fulfilling the needs for a 500 GeV collider. Using a new surface treatment, gradients of greater than 40 MV/m have been reached in single cell cavities, giving access to energies of 800 GeV. The superconducting linear accelerator of the TESLA Test Facility has been operated for more than 10000 hours. The free-electron Laser principle has been demonstrated at wavelengths of 80 – 180 nm. Other technologically challenging components needed for the accelerator, like high-power klystrons, have successfully been developed, built and operated at the TESLA Test Facility.

These successes provide the firm basis for this technical proposal.

CONCLUSION

The technology to build TESLA is available today and the technological readiness of the project has been shown through extensive R&D work and the operation of TESLA Test Facility. The detailed studies of the scientific aims and potential of TESLA have demonstrated its unique research possibilities for particle physics, for condensed matter physics, chemistry and material science, and for structural biology. In this way TESLA satisfies the criteria for new large endeavours in science: they should be unique, open completely new research possibilities and should carry the promise to advance our knowledge of nature in many branches of science.

REFERENCES

1. The TESLA Technical Design Report,
   `http://tesla.desy.de/new_pages/TDR_CD/start.html`

CDF at the Tevatron Collider in Run 2

R. D. Erbacher

Fermilab, Batavia, IL, 60510-0500, USA

Abstract.
Run 2 of the Tevatron began in early 2001 after extensive upgrades to both the machine and the CDF and DØ detectors. For CDF, new tracking detectors, increased muon coverage, state-of-the-art front end electronics, pipelined triggering, and a complete overhaul of the DAQ have made it a very powerful tool to explore physics of all kinds. The status of CDF in Run 2 is presented, along with a first glimpse of CDF data.

INTRODUCTION

Run 1 of the Tevatron Collider at Fermilab lasted from 1992 to 1996, during which time $\sim$110 pb^{-1} of data were collected at an energy of 1.8 TeV. The Tevatron Run 2 began in the Spring of 2001, and for the next half decade (or until the LHC at CERN begins operation) the Tevatron will remain the high energy frontier in particle physics, with collisions at $\sqrt{s} \sim 2$ TeV. For Run 2, improvements to the Tevatron were made with the goal of achieving instantaneous luminosities on the order of 10^{32}cm^{-2}s^{-1}. To take advantage of the expected collision rates, CDF underwent a major overhaul, with upgrades to almost every system. CDF is now poised to exploit the approximate 150-fold increase in integrated luminosity for Run 2 in almost all areas of study.

RUN 2 TEVATRON COLLIDER UPGRADES

In the last five years, a new 150 GeV "Main Injector" synchrotron was built to boost Tevatron capabilities. The new injector is both faster in producing antiprotons and more efficient in transferring protons and antiprotons into the Tevatron ring. The upgrade is expected to provide a factor of five improvement in instantaneous luminosity. In addition, a new antiproton recycler has been built, in which antiprotons are re-cooled and re-used rather than dumped at the end of a store. The recycler is currently being commissioned and is expected to provide another factor of two in instantaneous luminosity. With these improvements, and the planned decrease in bunch spacing from 396 ns to 132 ns for Run 2b, the Tevatron hopes to eventually achieve luminosities $> 2 \times 10^{32}$ cm^{-2}s^{-1}. Although the accelerator complex has had a slower start than expected so far, it is hoped that as the machine performance is understood, the Tevatron can achieve a very high growth curve, and perhaps collect more than the projected 15 fb^{-1}.

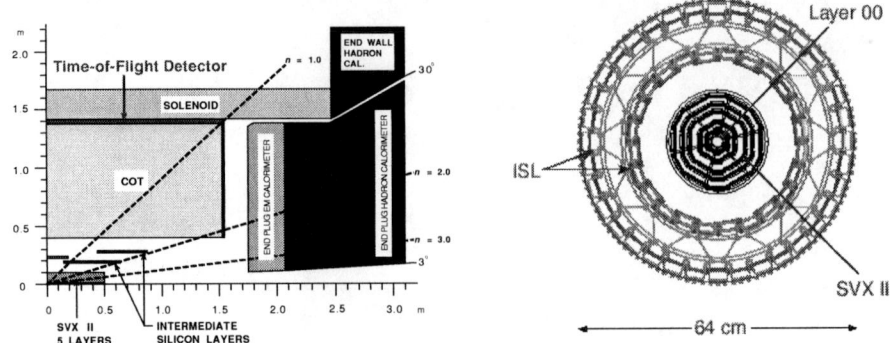

FIGURE 1. *(left)* Cutaway view of a quadrant of the Run 2 CDF detector. *(right)* Cross-sectional schematic of the CDF Run 2 silicon detector system, showing Layer 00, SVX II, and the ISL.

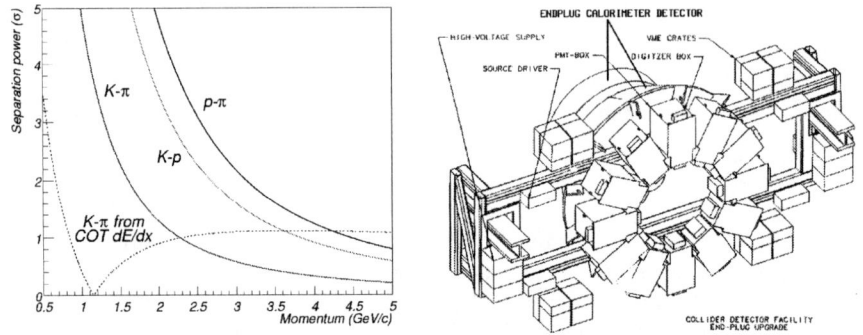

FIGURE 2. *(left)* Expected time-of-flight detector particle ID resolution versus momentum for low momentum events, compared to K-π separation from COT dE/dx. The measurements are complementary. *(right)* End-view diagram of the endplug upgrade and support structure.

CDF DETECTOR IMPROVEMENTS

The CDF detector is almost entirely new for Run 2. While the solenoid, the central calorimeters, and portions of the muon detectors remain the same as for Run 1, the rest of the detector systems were replaced or upgraded. Figure 1 shows a cutaway view of a quadrant of the Run 2 detector.

CDF will enjoy an entirely new tracking system, consisting of a three-component silicon detector package, a new wire chamber, and a time-of-flight detector. A cross-sectional layout of the silicon system is shown in Figure 1. The five-layer SVX II detector extends coverage in z and ϕ, allows 3-D vertexing with stereo channels on double-sided silicon, and enables Level 2 triggering on displaced vertices. The Intermediate Silicon Layers (ISL) is the largest double-sided silicon detector ever made. It extends b-tagging to $|\eta| = 2$, and helps to link tracks in the drift chamber to the SVX. The Layer 00 (L00) detector, which sits on the beryllium beam pipe inside of SVX, was added to the

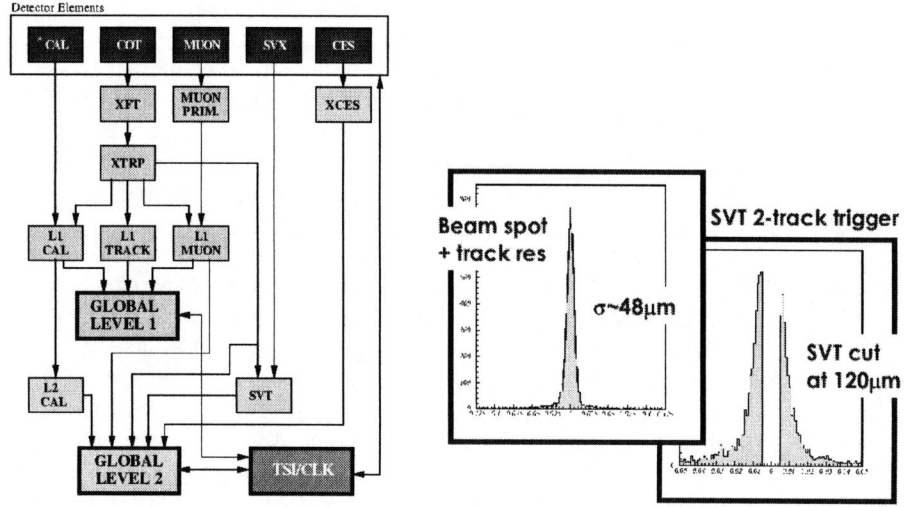

FIGURE 3. *(left)* Trigger pathways from the detectors through Level 2. *(right)* Secondary Vertex Trigger (SVT) data, showing a beam spot plus track resolution of σ ~48μm, and a potential impact parameter cut at 120μm.

silicon system to improve the impact parameter resolution and hence increase b-tagging efficiency; it may be used in the level 2 SVT trigger for b-tagging. L00 could also extend the CDF lifetime, as it will likely outlast SVX II as radiation exposure accumulates. Tracking with the silicon system is going well, with signal-to-noise levels and alignment tolerances meeting design specifications.

The new CDF drift chamber, the Central Outer Tracker (COT), consists of 30,240 sense wires: ~five times that of the drift chamber in Run 1. There are 96 wire planes total, and eight superlayers, half of which are 3° stereo. In place of field wires, the COT employs Au-plated Mylar cathode sheets to enclose "supercells", each containing 12 sense and potential wires. This design provides uniform drift over the 0.88 cm cells. An Ar:Eth:CF$_4$ gas mixture allows 100 ns drift times, well within the bunch crossing window. Although there is some variation with luminosity, the expected momentum resolution for events with transverse momenta p_T >2 GeV is $\delta p_T/p_T^2 \approx 0.33\%$ (GeV/c)$^{-1}$ [1].

The Time-of-Flight detector (TOF) for particle identification was added to the Run 2 upgrade to complement the dE/dx measurement from the COT. The detector consists of 216 scintillator bars mounted longitudinally on the inside of the solenoid, and read out on both ends by fine-mesh phototubes. The TOF is expected to obtain 100 ps resolution, allowing 2σ separation between K/π for p < 1.6 GeV, K/p for p < 2.7 GeV, and p/π for p < 3.2 GeV. Figure 2 shows the separation power for the TOF compared to the COT. The ability to identify low momentum kaons will significantly improve flavor-tagging in B decays– important for the measurement of *sin*2β and the observation of B$_s$-mixing for example. [2]

Calorimetry in the central (low η) portion of CDF has not changed from Run 1 aside from the front end read-out electronics and triggering. The endplug calorimeters, however, were replaced entirely. Figure 2 shows a diagram of the new endplug and support structure. The new scintillating tile detectors are faster and have a better sampling fraction than the gas calorimeters used in Run 1. They also provide more complete coverage, with the 10° forward gap gone and the 30° gap greatly reduced. The design is such that CDF now has the same calorimeter technology over the full solid angle to $|\eta| = 3.6$. Consequently, identical readout electronics and triggering are used for this entire region. The new custom-built VME readout boards hold single-channel daughter cards, which feature the QIE (Charge Integrating and Encoding) ASIC [1] [3], developed originally at Fermilab for the SSC, and used initially on the KTeV experiment. This chip together with a 10-bit ADC provides 18 bits of dynamic range per channel. The readout boards also have Xilinx FPGAs which contain a fully pipelined 42-clock-cycle event buffer for the Level 1 trigger decision. The upgraded calorimeter and readout systems were the earliest systems ready for commissioning data, and have been the most stable.

The muon detector upgrades for Run 2 focused on preserving the existing Run 1 detectors and increasing the muon coverage in η and ϕ. Additional coverage in the $|\eta| > 1$ region was added: new IMU detectors will allow triggering on muons in some regions out to $|\eta| = 1.5$ and identification out to $|\eta| = 2$. Since the $1.4\,\mu s$ drift times in the muon chambers are greater than the bunch spacing, scintillators are used to tag muon events. Due to the higher rates expected in Run 2, some central counters from Run 1 were removed and shielding was added to reduce occupancies in the muon detectors.

The DAQ and trigger system was completely overhauled for Run 2. Figure 3 shows the the Run 2 trigger paths. The pipelined Level 1 trigger and buffered Level 2 design gives CDF a nearly deadtimeless trigger. At Level 1, calorimeter energies, tracking p_T and ϕ (with the help of the new eXtremely Fast Tracker (XFT)), and muon stubs are combined to form objects, such as electrons and photons, on which to trigger at a <50 kHz accept rate. Level 2 uses custom Alpha processors and four event buffers to first refine Level 1 information, and then to combine it with silicon tracks and shower position information, forming a Level 2 decision at 300 Hz output rate. The silicon information is acquired at Level 2 via the SVT (Secondary Vertex Trigger), which finds displaced tracks and provides an impact parameter cut at Level 2. The SVT greatly enhances the efficiency for collecting B events. Figure 3 shows SVT data, with resolution of the beam spot plus tracks of $\sigma \sim 48\,\mu m$. Also shown is an impact parameter cut at $120\,\mu m$ for triggering on B mesons. The Level 3 trigger consists of a farm of Linux PCs running fast versions of the offline code to make more sophisticated selections on events. The upgraded DAQ model is similar to the successful Run 1b system. The design logging rate of 20 MB/sec was met long before the commissioning run, and recently was achieved with 8 different data streams. Finally, data from a new Cerenkov Luminosity Counter (CLC) [4], designed for precise measurement even at the highest luminosities, is now both part of the data stream and sent from CDF to the Tevatron for monitoring purposes.

PREPARING FOR PHYSICS: FIRST DATA

CDF had the opportunity to commission the upgraded detector during a short run in Fall 2000, and much experience was gained with the new detector and DAQ systems.

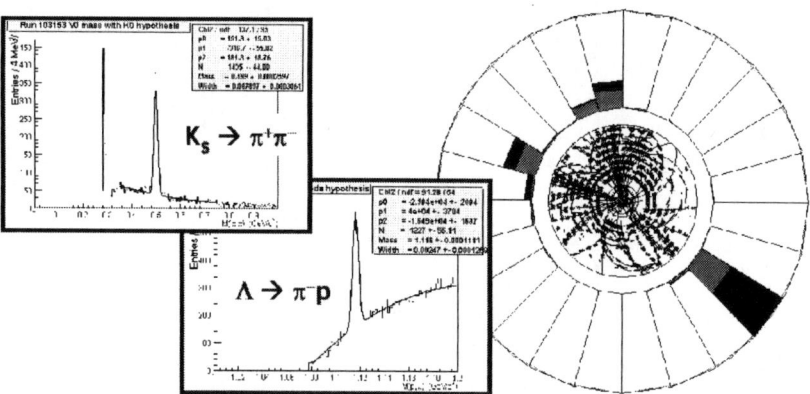

FIGURE 4. K_s and Λ peaks, found during the CDF commissioning run in Fall 2000. Silicon tracking, COT tracking, and calorimetry are shown in the event display on the right.

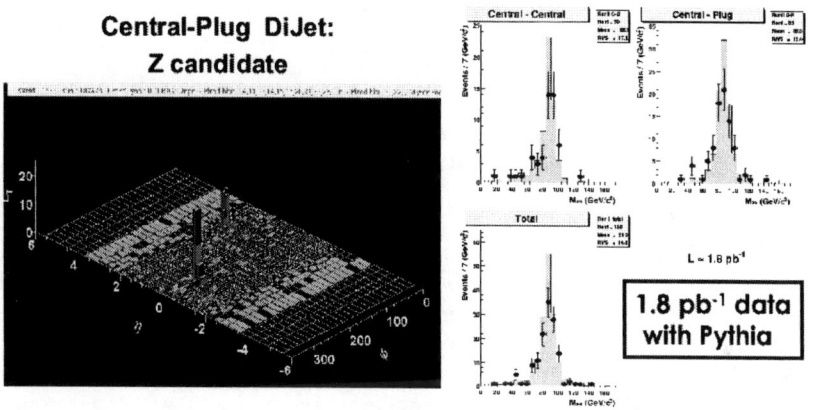

FIGURE 5. *(left)* η-ϕ map of the calorimeter showing two lepton jets from a Z candidate. *(right)* Z event distributions for leptons in the central calorimeter, in the endplug, and for one in each of these. 1.8 pb^{-1} was used and Pythia distributions are superimposed.

Figure 4 shows a three-jet event containing tracks in the silicon, the COT, and energy in the calorimeter. Also shown are K_s and Λ peaks as reconstructed from early data. One can see that CDF was functioning well from the start. Figure 5 shows a sample of Z events using 1.8 pb^{-1} with Pythia events superimposed. Such samples are being used to calibrate the detectors and to understand efficiencies and detector behavior. Some samples will already be used in upcoming CDF physics analyses as well. Figure 6 contains a selection of Run 2 physics: reconstructed charged B mesons, K°'s, and J/ψ's. While studies of the new detector's behavior and performance are ongoing, CDF is well on its way to many interesting physics results.

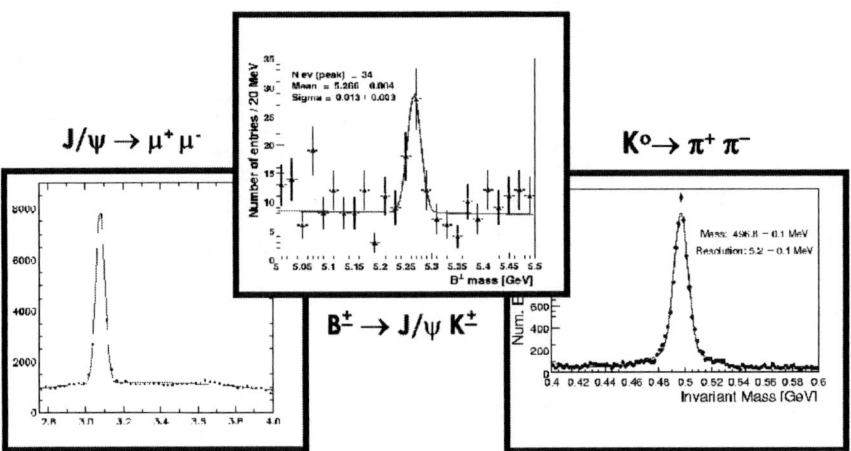

FIGURE 6. First Run 2 samples of $J/\psi \to \mu^+\mu^-$ *(left)*, $B^\pm \to J/\psi K^\pm$ *(center)*, and $K^\circ \to \pi^+\pi^-$ *(right)*, demonstrating that the tracking system is in good shape, and that we are well on our way to first physics analyses.

PHYSICS GOALS FOR RUN 2

With successful upgrades to the entire detector, trigger, and offline reconstruction package, CDF is poised to lead the world with insights into the high energy frontier for the remainder of the decade. Run 2 of the Tevatron will provide both the CDF and DØ collaborations with exciting results in top physics, the electroweak sector, B physics, QCD, and physics beyond the Standard Model. One of the major goals for CDF is elucidation of the properties of the newly-discovered top quark. We can look forward to the careful studies of top mass (an important electroweak parameter), top cross-sections, single top production, Drell-Yan, $M_t - M_{\bar{t}}$, spin correlations, anomalous couplings, QCD tests, form factors, and possibly new physics. Using the $t\bar{t}$ dilepton channel, the lepton+jets channel, and the all hadronic decay channel, CDF can expect to obtain a top mass measurement to $\delta M_t \sim 3$ GeV/c^2 with the first 2 fb^{-1}. CDF will also make high-precision electroweak measurements. With 2(10) fb^{-1}, CDF and DØ combined will obtain the world's most precise measurement of the mass of the W, to $\delta M_W = 30(20)$ MeV/c^2. Combined knowledge of M_{top} and M_W will set even more stringent limits on the Standard Model Higgs. As shown in Figure 7, with current measurements, the Standard Model Higgs is already close to being ruled out, making Run 2 top and electroweak studies extremely important. For more on Higgs and SUSY-Higgs prospects in Run 2, see the contribution by J. Conway in these proceedings [5].

The Tevatron will serve as a B-Factory for many years, as all species of B mesons are produced, and B-production rate is very high: data collection is limited only by offline bandwidth. The addition of silicon Layer 00 and the TOF detector will greatly improve our B capabilities. CDF can look forward to strong measurements of the CKM parameters: goals include CP violation in $B^\circ \to J/\psi K_s^\circ$ and $sin2\beta$ to better than ± 0.08, a significant measurement of the CP asymmetry in $B^\circ \to \pi^+\pi^-$, determination of $|V_{td}/V_{ts}|$

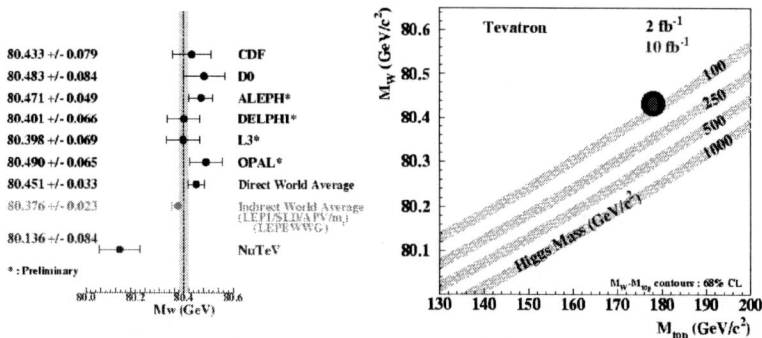

FIGURE 7. *(left)* Status from 2001 of world data on precision W mass measurements. CDF and DØ combined will have the most accurate result, with $\delta M_W < 20$ MeV/c^2 by the end of Run 2. *(right)* Prediction for indirect electroweak constraints on Standard Model Higgs (for different Higgs mass regimes) with knowledge of M_W and M_{top} for two (dark circle) or ten (light circle) fb^{-1}.

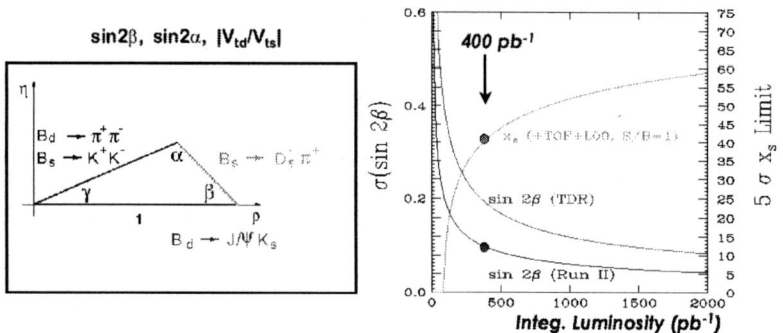

FIGURE 8. *(left)* The CKM triangle along with various decays necessary to measure some of the parameters of it. *(right)* Physics reach for $\sigma(\sin 2\beta)$ and for the B$_s$ mixing parameter x_s with increasing luminosity for CDF, both with the L00 and TOF detector additions (blue and green curves) and without these (red curve).

to 20% over the full Standard Model range, and a rich program of rare B decays, among other things. Figure 8 shows that with only 400 pb^{-1} we can place a 5σ limit on the B$_s$ mixing parameter out to $x_s = 40$, and we can achieve $\sigma(\sin 2\beta)$ of 0.1.

REFERENCES

1. CDF II Detector - Technical Design Report - Fermilab-Pub-96/390-E
2. CDF Collaboration, *A Time-of-Flight System for CDF*, CDF2573 (May 1994)
3. CDF Collaboration, *Results of the QIE6 Test with Calorimeter Phototubes*, CDF4388 (Nov. 1997)
4. CDF Collaboration, *The Cerenkov Luminosity Counter*, NIM A461:540(2001), FNALPUB-01-031-E
5. J. Conway, *Search for the Higgs at the Tevatron in Run 2*, Proceedings of the VIII Mexican Workshop on Particles and Fields (held Nov. 2001).

Search for the Higgs at the Tevatron in Run 2

J. S. Conway

Dept. of Physics and Astronomy, Rutgers University, Piscataway, New Jersey, USA

Abstract. In Run 2, with an upgraded accelerator and experiments, the Tevatron experiments are poised to extend the search for the Higgs boson of the Standard Model to higher masses. With sufficient integrated luminosity, coupled with improved b jet tagging capability and $b\bar{b}$ mass reconstruction techniques, the CDF and DØ experiments are sensitive to Higgs prooduction in several channels, including at high masses (140 – 190 GeV). The Standard Model search can be used to explore the Higgs sector of minimal supersymmetry also, and in that model the production of Higgses in association with b quark can be greatly enhanced.

INTRODUCTION

The search of the origin of electroweak symmetry breaking is the central question in high energy physics today. The most recent fits to the world's combined electroweak data[1] favor the existence of a Standard-Model-like Higgs with mass in the range 100-200 GeV. The lower limit on the Higgs mass from the LEP2 experiments is 113.4 GeV; the data from all four experiments show a 2-sigma excess at a Higgs mass of about 115 GeV.

The Tevatron experiments have the opportunity, in the years before the LHC turns on, to search for the Higgs both in the Standard Model (SM) and in supersymmetry, using a variety of search channels discussed here.

STANDARD MODEL HIGGS

Events with a SM scalar Higgs can be produced at the Tevatron in several ways. The most copious production mode is gluon-gluon fusion via a heavy quark loop, giving a single Higgs produced. The Higgs can also be produced in association with a W or Z boson via its couplings to the vector bosons. Figure 1 shows the production cross section for various modes as a function of Higgs mass.

Figure 2 shows the branching ratios of the Standard Model Higgs as a function of Higgs mass. In the range below about 135 GeV Higgs mass, the decay to $b\bar{b}$ dominates, and for larger masses the decay to W pairs dominates.

In the gluon fusion case, for low mass Higgs, there is an overwhelming background from QCD production of $b\bar{b}$ pairs. The WH and ZH modes, however, have been extensively studied[2] and lead to several distinct signatures in which a Higgs signal can be observed with sufficient integrated luminosity.

LOW-MASS HIGGS

For low mass (< 135 GeV) Higgs, the most sensitive signatures arise from the leptonic decays of the W and Z, and are denoted $\ell \nu b \bar{b}$, $\nu \bar{\nu} b \bar{b}$, and $\ell^+ \ell^- b \bar{b}$. Hadronic decays of the W and Z lead to the $q \bar{q} b \bar{b}$ final state which suffers from large backgrounds from QCD multijet production.

In Run 1 in CDF, all four of these channels were studied, and led to limits on the Higgs cross section times branching ratio to $b \bar{b}$ as depicted in Figure 3. As the plot shows, the Run 1 limits are more than an order of magnitude above the expected Standard Model cross section, naturally provoking the question of whether and how this search can be carried out in Run 2.

Improvements to the detector, coupled with much higher instantaneous luminosity in Run 2 lead to greatly enhanced sensitivity in the Standard Model Higgs search. Unlike the Run 1 detector, the CDF Run 2 detector has a silicon vertex detector covering the entire luminous region, and has measurements of the z coordinates of tracks. Overall, the tracking coverage out to nearly $|\eta| = 2$ and the new muon chambers lead to greatly improved acceptance for Higgs. For the missing E_T channel ($\nu \bar{\nu} b \bar{b}$) channel, the trigger efficiency can be improved by using the silicon vertex trigger (SVT) to tag the jets. Coupled with the fact that the accelerator is expected to deliver a data sample over a hundred times larger than that in Run 1, the overall sensitivity of the Higgs search is dramatically improved in Run 2.

Beyond the improvements to the detector itself, maximizing the sensitivity of the search for the Higgs depends most critically on attaining the best possible $b \bar{b}$ mass resolution, and attaining the best possible b jet tagging efficiency and purity, and understanding and controlling the main irreducible backgrounds from vector boson plus heavy flavor production.

In Run 1 the top quark discovery and subsequent determination of its mass demonstrated that one could use jet information, even jets from b quarks, which have a significant semileptonic branching ratio, to determine the top mass. The case of the Higgs is simpler than that of the top, which suffers from large combinatorics. For the Higgs, the mass resolution is limited by basic physics (missing energy from neutrinos and gluon radiation) and detector resolution.

The benefit of making corrections for missing neutrinos is illustrated by CDF's search in Run 1 for $Z \to b \bar{b}$. Figure 4 shows the successive effects of correcting for overall missing energy, and muon p_T, and more general jet energy corrections. The mass resolution attained in this analysis was 13.5%; for a 120 GeV Higgs (in the background-dominated process $Z \to b \bar{b}$) the resolution predicted is 12%.

One can improve upon the jet energy corrections employed in most Run 1 analyses by making the best possible use of all detector information, including tracking, shower max, calorimeter, and muon chambers. Figure 5 shows the improvement to jet energy resolution possible by determining jet energy from an optimum linear combination of all jet information. Using all information results in a 30% improvement in jet energy resolution.

A great deal of simulation and calibration work remains and is presently underway. Optimistically, by putting together all the best kinematic corrections with optimal jet energy corrections, we hope to eventually achieve 10-12% mass resolution for the Higgs in the main low-mass search channels. (This is not as good as the $Z \to b\bar{b}$ case because there is additional missing energy in the Higgs channels due to neutrinos from W and Z decay.)

Figure 6 shows the raw mass distribution and Figure 7 shows the background-subtracted signal in the $\ell\nu b\bar{b}$ case, for a 120 GeV SM Higgs, combining data from both CDF and DØ representing 15 fb^{-1} integrated luminosity, assuming a 10% $b\bar{b}$ mass resolution, which is what was assumed (optimistically) in the Tevatron Run 2 Higgs report. The figure clearly illustrates that even with the best resolution attainable, discovering the Higgs remains a major challenge.

HIGH-MASS HIGGS

For larger Higgs masses (> 135 GeV), the Higgs decays predominantly to $WW^{(*)}$. Two modes have been shown[2] to be sensitive in this mass range: $\ell\nu\bar{\ell}\bar{\nu}$ (from gluon fusion production of single Higgs) and $\ell^{\pm}\ell^{\pm}jj$ (from tri-vector-boson final states).

The critical issues in these search modes are accurate estimation of the very large (~ 10 pb) WW background in the $\ell\nu\bar{\ell}\bar{\nu}$ case and channel and estimation of the $t\bar{t}$ and W/Z+jets backgrounds in the like-sign dilepton channel.

SM HIGGS REACH IN RUN 2

The integrated luminosity required to discover or exclude the Standard Model Higgs, combining all search channels and combining the data from CDF and DØ, is shown in figure 8. The lower edge of the bands is the nominal estimate of the Run 2 study, and the bands extend upward with a width of about 30%, indicating the systematic uncertainty in attainable mass resolution, b tagging efficiency, and other parameters.

The figure clearly shows that discovering a SM (or SM-like) Higgs at the 5-sigma level requires a very large data sample: even with 15 fb^{-1}, the mass reach is about 120 GeV at best. A 95% CL exclusion can, however, be attained over the entire mass range 115-190 GeV with the integrated luminosity foreseen in Run 2b.

The $b\bar{b}$ mass resolution assumed in making these estimates is 10% in the central part of the distribution. This represents a significant improvement over the 14-15% resolution achieved in this analysis in Run 1, which did not benefit from the more detailed corrections described above and developed after the analysis was completed. A great deal of effort, presently underway, is needed to understand the jet energy corrections to the level required to attain 10% resolution. The required integrated luminosity for Higgs discovery scales linearly with this resolution.

The estimates of required integrated luminosity assume that the b tagging efficiency and purity are essentially the same as in Run 1 in CDF, per taggable jet. The better geometric coverage of the Run 2a and 2b silicon systems, however, is taken into account and leads to a much larger taggable jet efficiency. Since the required integrated luminosity scales inversely with the *square* of the tagging efficiency (assuming constant mistagging rates), however, there is a potentially great payoff for developing high-efficiency

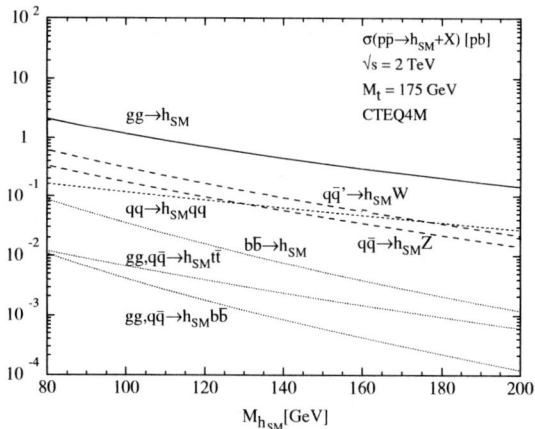

FIGURE 1. Production cross section for Standard Model Higgs at the Tevatron as a function of Higgs mass.

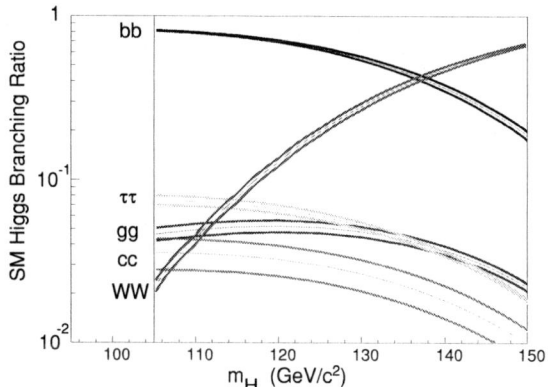

FIGURE 2. Branching ratios for Standard Model Higgs.

algorithms for b-tagging. Any such algorithms depend crucially on the quality of the information coming from the silicon vertex tracking system; the Run 2b silicon system has indeed been designed to optimize the performance in high-E_T b jet tagging.

SUSY HIGGS

In the context of the minimal supersymmetric standard model (MSSM) the Higgs sector has two doublets, one coupling to up-type quarks and the other to down-type quarks and leptons. There are five physical Higgs boson states, denoted h, A, H, and $H^{\pm}$. The masses and couplings of the Higgses are determined by two parameters, usually taken to

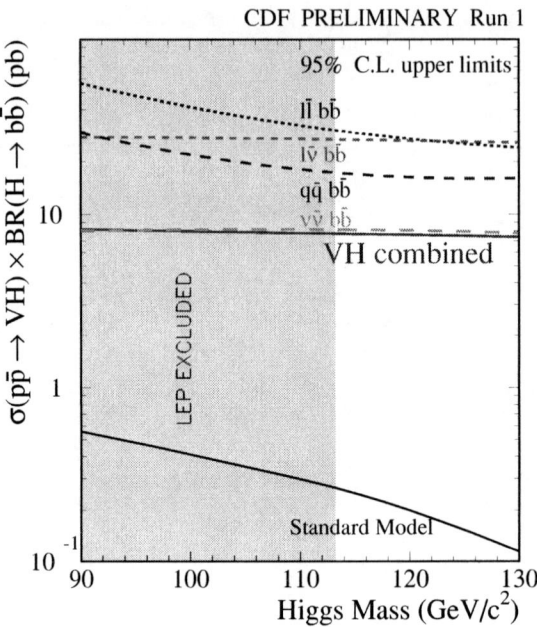

FIGURE 3. Limits on SM Higgs cross section times branching ratio to $b\bar{b}$ from CDF in Run 1.

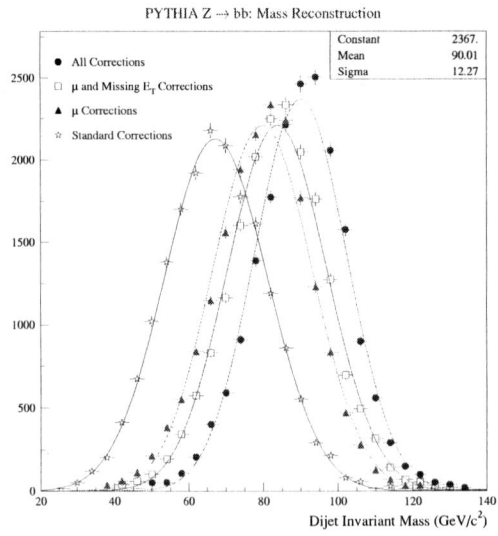

FIGURE 4. Mass resolution improvement for $Z \to b\bar{b}$ events as successive corrections are applied. After all corrections the resolution is 12%.

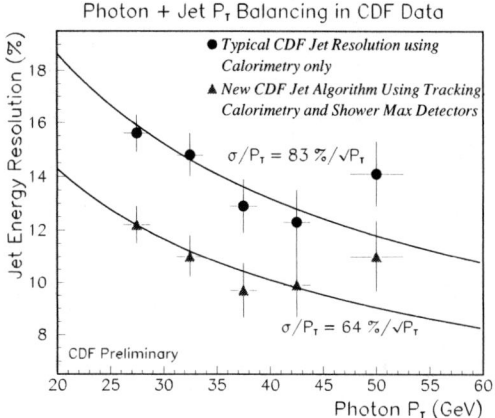

FIGURE 5. Jet energy resolution as a function of jet E_T, comparing standard corrections based on calorimeter only with energy determination combining information from tracking detetors, calorimetry, and shower max.

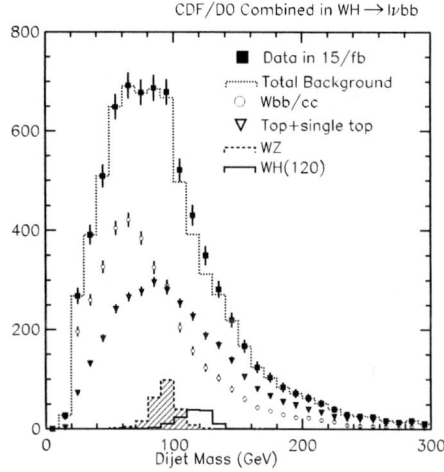

FIGURE 6. Distribution of $b\bar{b}$ mass in the $\ell\nu b\bar{b}$ Higgs search channel, showing expected background sources and expected signal from 120 GeV SM Higgs, combining 15 fb^{-1} of data from CDF and DØ.

be m_A and $\tan\beta$ (the ratio of the vacuum expectation value of the two Higgs doublets), with corrections from the scalar top mixing parameters.

The light scalar h can appear very Standard-Model-like or nearly so over a larger range of MSSM parameter space. In this scenario the results of the search for the SM Higgs produced in the WH and ZH modes are directly interpretable. Figure 9 shows the range in the space of m_A versus $\tan\beta$ in which a 5-sigma discovery can be made, as a

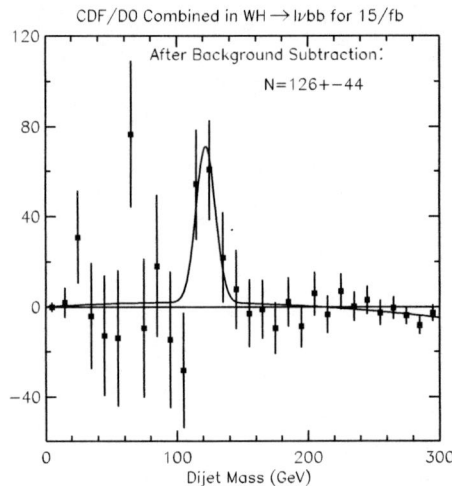

FIGURE 7. Background subtracted $b\bar{b}$ mass distribution in the $\ell\nu b\bar{b}$ channel, showing expected signal from 120 GeV SM Higgs, combining 15 fb^{-1} of data from CDF and DØ.

function of integrated luminosity, for one choice of stop mixing.

More interesting is the case of large $\tan\beta$. Since the coupling of the neutral Higgses ($h/A/H$) to down-type quarks is proportional to $\tan\beta$, there is an enhancement factor of $\tan^2\beta$ for the production of $b\bar{b}\phi$, $\phi = h, A, H$ relative to the SM rate appearing in figure 1. This leads to distinct final states with four b jets; if we demand that at least three of the jets be tagged, the background from QCD multijet processes is relatively small. In Run 1, CDF searched for this process, and from the null result excluded a large swath of MSSM parameter space inaccessible to LEP, as shown in figure 10.

Based on the Run 1 analysis, and taking into account the improved b-tagging efficiency, Figure 11 shows the regions of m_A versus $\tan\beta$ that CDF can cover for different integrated luminosities. It is interesting to note that the sensitive region in this analysis includes the region which is difficult to cover using the results of the SM Higgs search (shown in Figure 8). For this analysis the Run 2b silicon vertex system plays an absolutely crucial role: the accepted signal rate is proportional to the cube of the b tagging efficiency!

SUMMARY

With an upgraded detector and more than an order of magnitude larger instantaneous luminosity the CDF experiment, combined with DØ, has a significant chance of discovering a SM (or SM-like) Higgs boson in Run 2. If the Higgs mass is larger than about 130 GeV, the experiment is sensitive to the WW decay modes in two main channels. The

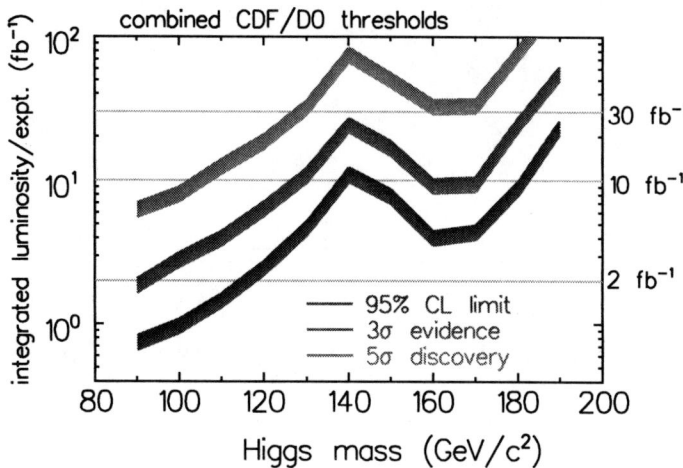

FIGURE 8. The integrated luminosity required per experiment to either exclude a SM Higgs boson at 95% CL or discover it at the 3σ or 5σ level, as a function of the Higgs mass. These results are based on the combined statistical power of both CDF and DØ and combining all search channels.

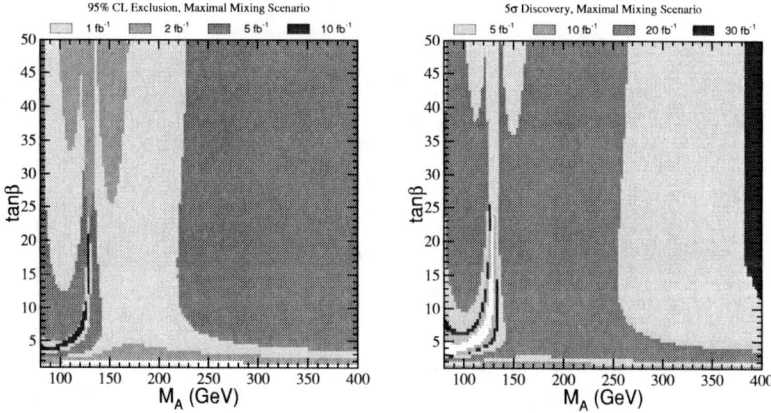

FIGURE 9. Regions of MSSM Higgs parameter space where 95% exclusion can be attained (left) and where 5σ discovery is possible (right), using SM Higgs search results.

experiment also has the chance to discover the Higgs in the MSSM, if $\tan\beta$ is large, via the striking four-b-quark final state.

The key experimental issues are maintaining the excellent secondary vertex tagging efficiency throughout the run, and working hard to understand and improve the dijet mass resolution.

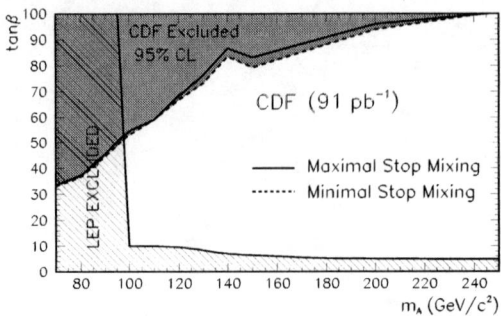

FIGURE 10. CDF limits on MSSM Higgs using $b\bar{b}b\bar{b}$ final state.

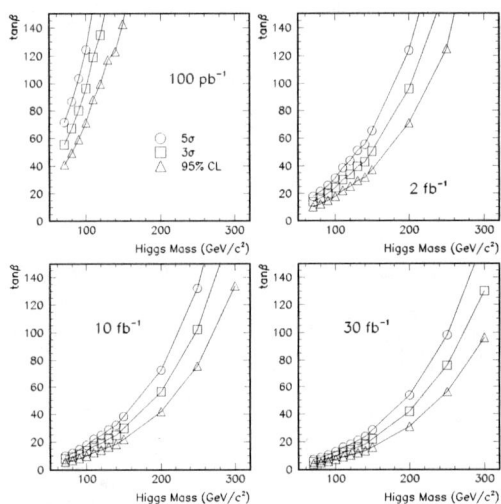

FIGURE 11. Anticipated limits in the plane of $\tan\beta$ versus $m(A)$ using $b\bar{b}b\bar{b}$ final state.

REFERENCES

1. LEP Electroweak Working Group at CERN (see http://lepewwg.web.cern.ch/LEPEWWG/).
2. Report of the Higgs Working Group of the Tevatron Run 2b SUSY/Higgs Workshop, M. Carena *et al.*, eds., unpublished. (See hep-ph/9910338 at the LANL preprint server.)

Higgs: Standard Model and Beyond

C.-P. Yuan

Department of Physics and Astronomy, East Lansing, MI 48824, USA

Abstract. In this talk, I will discuss possible new physics effects that modify the interaction of Higgs boson(s) with top and bottom quarks, and discuss how to detect such effects in current and future high energy colliders.

INTRODUCTION

Two of the great mysteries in the elementary particle physics are the cause of the electroweak symmetry breaking (that generates masses for the weak gauge bosons $W^\pm$ and Z) and the origin of the flavor symmetry breaking (that generates masses for quarks and leptons). In the Standard Model (SM), both symmetry breaking mechanisms are explained by introducing a single Higgs boson doublet field. The W and Z bosons gain their masses from Goldstone boson mechanism and the fermions gain their massed from Yukawa interactions. In the SM, the mass of the Higgs boson can receive a large radiative correction that is proportional to the square of the cutoff scale beyond which new physics effect has to take place. In order for the SM to be a valid field theory all the way up to the Planck's scale (about 10^{19} GeV), the mass of the SM Higgs boson has to be somewhere around 130 GeV to 180 GeV (to satisfy the naturalness condition). Furthermore, to explain the diverse fermion mass spectrum, every fermion has to be assigned a different Yukawa coupling. The fact that the mass of the top quark is so large as compared to the other fermions and is close to the vacuum expectation value suggests that top is a *special* quark and it may play a role in the electroweak symmetry breaking. If that is the case, then the flavor symmetry breaking and the electroweak symmetry breaking mechanisms may be related. In this talk I will discuss two different classes of models – one is the strongly interacting models and another is the weakly interacting models. (For a brief review on the SM Higgs boson, please see the talk by J. Conway in this Workshop.) The typical strongly interacting models are Technicolor, top-condensate [1], topcolor [2] and top-seesaw models [3], and the weakly interacting models are supersymmetry models, particularly, the minimal supersymmetric standard model (MSSM) [4]. In the former class of models, the electroweak symmetry breaking is generated dynamically and usually results in composite (in contrast to elementary) Higgs bosons. On the contrary, in the latter class of models, the electroweak symmetry breaking is generated spontaneously and results in elementary Higgs bosons. I shall take the topcolor model and the MSSM as two examples to discuss the phenomenology predicted by these two classes of models and to identify a few experiments at high energy colliders that can distinguish these models assuming some new physics signals are found.

MODELS

Topcolor model

In the topcolor model, the mass of the top quark is generated by topcolor dynamics that also contributes to the electroweak symmetry breaking and induces two composite Higgs boson doublets in its low energy effective theory. The physical (composite) scalars are t-Higgs (H_t^0), top-pions (π_t^0, $\pi_t^\pm$) and b-Higgs (H_b^0, A_b^0, $H_b^\pm$). Because the topcolor dynamics has to be strong enough to make top quark and antiquark to form condensate to generate the large top quark mass as well as to contribute to part of the weak boson masses, the Yukawa couplings of t-Higgs (or top-pions) with top quarks have to be large (at the order of 1). Furthermore, because the bottom quark is the isospin partner of the top quark, the strong topcolor dynamics that a left-handed top quark experiences will also affect the bottom quark, Hence, the Yukawa coupling of b-Higgs with bottom quarks have to be large as well. This should be compared with the SM in which the Yukawa coupling of top quark to the SM Higgs boson is about 1 while the coupling of bottom quark is much less than 1 (about 1/50 at the 100 GeV scale). In this model, tau lepton does not involve the topcolor dynamics so that it does not directly couple to the composite scalars and its Yukawa coupling vanishes. (Its mass has to be generated by other mechanics, such as the technicolor dynamics. Again, this is different from the SM prediction which is equal to $\sqrt{2}m_\tau/v$ where v is the vacuum expectation value ($\sim 246\,\text{GeV}$) and m_τ is the mass of tau. Hence, it is expected that the collider phenomenology of this model will be significantly different from the SM.

As noted above, there are also charged Higgs bosons and top-pions predicted in this model. Their couplings to the top, bottom, and charm quarks are shown in the equation below:

$$\mathcal{L}_{\pi_t^\pm}^{tc} = \frac{m_t \tan\beta}{v} \left[iK_{UR}^{tt} K_{UL}^{tt*} \overline{t_L} t_R \pi_t^0 + \sqrt{2} K_{UR}^{tt*} K_{DL}^{bb} \overline{t_R} b_L \pi_t^+ + iK_{UR}^{tc} K_{UL}^{tt*} \overline{t_L} c_R \pi_t^0 + \sqrt{2} K_{UR}^{tc*} K_{DL}^{bb} \overline{c_R} b_L \pi_t^+ + \text{h.c.} \right], \quad (1)$$

where $\tan\beta = \sqrt{(v/v_t)^2 - 1}$ and $v_t \simeq O(60-100)\,\text{GeV}$ is the top-pion decay constant; $K_{UL,R}$ and $K_{DL,R}$ are rotation matrices that diagonalize the up- and down-quark mass matrices M_U and M_D, i.e., $K_{UL}^\dagger M_U K_{UR} = M_U^{\text{dia}}$ and $K_{DL}^\dagger M_D K_{DR} = M_D^{\text{dia}}$, from which the CKM matrix is defined as $V = K_{UL}^\dagger K_{DL}$. As shown in Ref. [5], a typical topcolor model, that is consistent with all the precision low energy data, gives

$$K_{UR}^{tt} \simeq 0.99-0.94, \quad K_{UR}^{tc} \lesssim 0.11-0.33, \quad K_{UL}^{tt} \simeq K_{DL}^{bb} \simeq 1. \quad (2)$$

As to be discussed later, the large flavor mixing between t_R and c_R can lead to very distinct collider signatures. One example is to induce a large single-top event rate at hadron colliders [6].

MSSM

In the MSSM, the electroweak symmetry is radiatively broken due to the contribution of the heavy top quark in loops. (It would not have worked if the top quark were not heavy enough.) Therefore, in this model, top quark also plays a special role in the electroweak symmetry breaking. Two Higgs doublet fields are required in the MSSM by the requirement of supersymmetry. Among the eight real fields, three of them are the Goldstone bosons which generate the masses of the weak gauge bosons and five of them are the two CP-even Higgs bosons (h and H), one CP-odd Higgs boson (A) and two charged Higgs bosons ($H^\pm$). Their couplings to the fermions are derived by demanding one Higgs doublet couple to the up-type fermions and another to the down-type fermions, which is similar to a type-II two-Higgs-doublet model [4]. For example, the tree level Yukawa couplings of A-b-b and A-t-t are $\sqrt{2}m_b \tan\beta/v$, and $\sqrt{2}m_t \cot\beta/v$, respectively, where $\tan\beta$ is the ratio of the two vacuum expectation values. For a large $\tan\beta$, the coupling of A-b-b can become $O(1)$, while the coupling of A-t-t becomes very small. This pattern of the Yukawa couplings is not only different from that in the SM (where the top Yukawa coupling is much larger than the bottom Yukawa coupling) but also different from the topcolor model (where both the top and bottom Yukawa couplings are $O(1)$). This difference is the crucial element that allows us to distinguish different classes of electroweak symmetry braking models by carefully examining the experimental data. We shall come back to this point in the next section.

A perfect supersymmetric theory cannot describe the Nature. (Otherwise, we would have seen various supersymmetric partners of the observed particles.) Hence, supersymmetry has to be broken. To incorporate the effect from the yet-to-be found supersymmetry breaking mechanism, the MSSM contains the soft-breaking sector in its Lagrangian to parameterize all such possibilities. One interesting possibility induced by a general soft-breaking sector is that the top-squark and the charm-squark can be largely mixed to yield a sizable flavor mixing between top and charm. One such model can be generated by imposing a horizontal $U(1)_H$ symmetry, which is called the Type-B supersymmetry model in Ref. [7]. Another way to generate the up-type squark mixings is to have a non-diagonal A_u in the flavor space. Motivated by the charge-color-breaking (CCB) and vacuum stability (VS) bounds, we define at the weak scale

$$A'_u = \begin{pmatrix} 0 & 0 & 0 \\ 0 & 0 & x \\ 0 & y & 1 \end{pmatrix} A,$$

where $(x, y) = O(1)$ and A'_u is A_u in the super-CKM basis (where the quark mass matrix M_u is diagonal). Hence, we have generated large flavor-mixings in the $\tilde{t} - \tilde{c}$ sector, which are consistent with all low energy experimental flavor changing neutral current (FCNC) data and theoretical CCB/VS bounds. Without losing generality, we define the Type-A1 model as $x \neq 0$, $y = 0$ and Type-A12 model as $x = 0$, $y \neq 0$. It is obviously that in the former model $\tilde{c}_L$ decouples, and in the latter model $\tilde{c}_R$ decouples. (Here, we assume $M_{LL}^2 \simeq M_{RR}^2 \simeq \tilde{m}_0^2 I_{3\times 3}$, for simplicity.) It has been shown in Ref. [7] that both Type-A and Type-B models can radiatively generate large flavor-mixing Yukawa couplings to quarks.

$GG, Q\bar{Q} \to B\bar{B}\phi^0$ AND $B\bar{B} \to \phi^0$

In Ref. [8], we calculated the cross sections of the $gg, q\bar{q} \to b\bar{b}\phi^0$ processes at the Tevatron and the LHC, where ϕ^0 denotes a (pseudo-) scalar predicted in the strongly interacting topcolor model, the two Higgs doublet extension of top-condensate model [9], and the MSSM. To suppress the large SM QCD backgrounds in the detection mode of $\phi^0 \to b\bar{b}$, a set of kinematic cuts has to be applied together with 3 or more b-tags. For the class of strongly interacting models, the Tevatron Run II and the LHC are able to either exclude an entire model or a large part of the model parameters if the $\phi^0 b\bar{b}$ signal is not found experimentally. Likewise, for the class of weakly interacting models, such as the MSSM, a large portion of the supersymmetry parameters on the $\tan\beta$ versus m_A plane can be probed at the Tevatron and the LHC. However, because the coupling of A-b-b can receive large (about a factor of 2) radiative corrections from the supersymmetric particle threshold effect and the QCD interaction, the precise region of $\tan\beta$ as a function of m_A that can be studied via the above processes will depend on other supersymmetry parameters, such as the μ parameter and the top-squark masses [8]. Nevertheless, the constraint provided by this process on the $\tan\beta - m_A$ plane is complementary to that provided by the associated production of the weak gauge boson and the Higgs boson.

If the mass of ϕ^0 is large, it can be dominantly produced from the s-channel fusion process $b\bar{b} \to \phi^0$ at hadron colliders. The cross sections of this process for various models were also given in Ref. [10].

As noted in the previous section, to distinguish the strongly interacting from the weakly interacting models, one should also examine the production of $t\bar{t}\phi^0$ in addition to the $b\bar{b}\phi^0$ channel. Furthermore, to distinguish those two classes of models in the $b\bar{b}\phi^0$ channel, one can examine the tau lepton decay mode of ϕ^0. (In the topcolor model, b-Higgs do not couple strongly to the tau lepton, while in the MSSM with a large $\tan\beta$, the coupling of ϕ^0-τ-τ is large.)

$C\bar{B} \to H^+$

If the mass of a charged Higgs boson is lees than $\sim 170\,\text{GeV}$, it can be studied via the decay process $t \to H^+ b$ using the large data sample of the $t\bar{t}$ pairs expected at the Tevatron and the LHC [11]. For a heavy charged Higgs boson, the H^+H^- pair production rate is usually small unless enhanced by some resonant effect. One such example is to have have a heavy neutral Higgs boson with mass larger than twice of m_{H^+} in the MSSM. It can also be associated produced with a top quark via $gb \to H^{\pm}t$ [12], but again with a small rate. In [7, 5], it was pointed out that H^+ can be produced via the s-channel process $cs \to H^{\pm}$ because of the large parton luminosities of charm and strange quarks at the LHC. Furthermore, if the flavor-mixing coupling of c-b-H^+ can be large, then H^+ can also be produced via $cb \to H^{\pm}$.

In the topcolor model, the mass of top-pion $\pi_t^{\pm}$ is expected to be around the weak scale. For a given m_{π_t}, the allowed range of the Yukawa couplings has to be checked by comparing with low energy precision data. A recent study for the topcolor assisted technicolor model can be found in Ref. [13]. To test this model's prediction, we can study

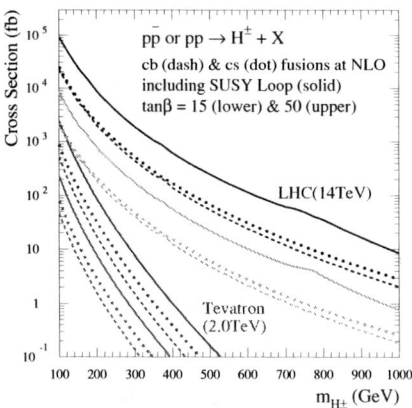

FIGURE 1. $H^{\pm}$ production via $c\bar{b}$ (and $c\bar{s}$) fusions at hadron colliders.

the single-top event signature from $c\bar{b} \to H^+ \to t\bar{b}$. The invariant mass distribution of the $t\text{-}\bar{b}$ system can reveal the existence of such a resonant [10]. To truly test this model, one should also check that the polarization of the final state top quark is right-handed because of its couplngs, cf. Eq. (1). In contrast, the top quark produced from the SM single top processes, either the s-channel process $q\bar{q}' \to W^* \to t\bar{b}$ or the t-channel process $qb \to q't$, are almost one hundred percent left-handedly polarized [6].

Similarly, in the MSSM, a sizable flavor-mixing $c\text{-}b\text{-}H^+$ coupling can be radiatively generated through radiative correction arising from the large stop and scharm mixings. At the tree level, the coupling of $c\text{-}b\text{-}H^+$ is suppressed by the CKM matrix element V_{bc} which is about 0.04. Hence, even with the large enhancement factor from a large $\tan\beta$, the tree level rate of $c\bar{b} \to H^+$ is still smaller than that of $c\bar{s} \to H^+$ at the Tevatron and the LHC because the parton luminosity of the strange quark is much larger than that of the bottom quark. As shown in Ref. [7], the Type-A supersymmetry models with the non-diagonal scalar trilinear A-term for the up-type squarks can enhance the $c\text{-}b\text{-}H^+$ coupling from the contribution of stop, scharm and gluino in loops. For $y = 0$ and $x \sim O(1)$ (i.e. Type-A1 model), the production rate of $c\bar{b} \to H^+$ can be increased by a factor of 2 to 5 as compared to its tree level rate, depending on the value of x. In Fig. 1, we show the single charged Higgs boson production rate for a typical choice of the supersymmetry parameters $(m_{\tilde{g}}, \mu, \widetilde{m}0)) = (300, 300, 600)\,\text{GeV}$, $(A, -A_b) = 1.5\,\text{TeV}$, and $\tan\beta = (15, 50)$ with $x = 0.75$.

$$T \to C\phi^0$$

The flavor-mixing dynamics predicted by models can also be tested from the FCNC decay of the top quark, such as $t \to c\phi^0, c\gamma, cg$.

It is known that the SM branching ratio of the flavor-changing top decay $t \to ch^0$ is extremely small ($\lesssim 10^{-13} - 10^{-14}$ [14]), so that this process provides an excellent window for probing new physics. In the topcolor model The low energy precision data requires the mass of the top-pions not to be too small as compared to the top quark mass [13]. In case that $m_\pi^0 < m_t - m_c$, the decay process $t \to c\phi^0$ can occur at tree level, cf. Eq. (1), which can impose further constraint on the model if such a signal is not found experimentally. A few recent studies on the other FCNC processes predicted by the topcolor model can be found in Ref. [15].

In the MSSM, the loop induced t-c-h^0 coupling can be greatly enhanced depending on the detailed parameters of the model. [7]. Assuming that the only dominant decay mode of the top quark is its SM decay mode, i.e. $t \to bW$, the decay branching ratio of $t \to ch^0$ is given by $\text{Br}[t \to ch^0] \simeq \Gamma[t \to ch^0]/\Gamma[t \to bW]$. As summarized in Table 1, $\text{Br}[t \to ch^0]$ can be as large as $10^{-3} - 10^{-5}$ over a large part of the supersymmetry parameter space where the mass of the lightest Higgs boson h^0 is around $110 - 130$ GeV. Since the LHC with an integrated luminosity of $100\,\text{fb}^{-1}$ can produce about 10^8 $t\bar{t}$ pairs, it can have a great sensitivity to discover this decay channel and test the model predictions, by demanding one top decaying into the usual $bW^\pm$ mode and another to the FCNC ch^0 mode.

TABLE 1. $\text{Br}[t \to ch^0] \times 10^3$ is shown for a sample set of Type-A1 inputs with $(\widetilde{m}_0, \mu, A) = (0.6, 0.3, 1.5)$ TeV and Higgs mass $M_{A^0} = 0.6$ TeV. The three numbers in each entry correspond to $x = (0.5, 0.75, 0.9)$, respectively.

$m_{\tilde{g}}$	tanβ = 5	20	50
100 GeV	(.011, .10, .81)	(.015, .19, 4.6)	(.016, .21, 7.0)
500 GeV	(.011, .09, .41)	(.015, .13, 1.0)	(.016, .14, 1.2)

CONCLUSION

Because the mass of the top quark is close to the weak scale, it may play an essential role in the breaking of the electroweak symmetry. Two classes of models – strongly interacting and weakly interacting models — are considered. In the topcolor model, the Yukawa couplings of the top quark are large. As an isospin partner of the top quark, the bottom quark can also experience large Yukawa interactions. With the possibility of having a large flavor-mixing between the (right-handed) top- and charm-quarks, the charged top-pions can be copiously produced via the s-channel cb-fusion process at high energy colliders, due to its large Yukawa coupling and the sizable parton luminosities. In the MSSM, the flavor symmetry is tightly connected to the supersymmetry breaking through the introduction of the soft breaking sector in the Lagrangian. To carefully study the flavor-mixing and the flavor changing neutral current processes can advance our knowledge on the supersymmetry breaking mechanism. This point was demonstrated in some production and decay processes. Although through out this talk, I only concentrated on the phenomenology at hadron colliders, some similar effects are also expected in the future Linear Colliders. For example, a polarized $\gamma\gamma$ collide can test the chirality of the Yukawa coupling of b-c-H^+ by studying the single charged Higgs boson production [16]. There are two other aspects of the physics of Higgs boson which I do not have space to write on this report. First, the associated production of A and $H^\pm$ from the quark fusion process can constrain MSSM as a function of the only one supersymmetry parameter m_A, because the coupling of $W^\mp$-A-$H^\pm$ is determined by the gauge interaction and M_A and M_{H^+} are related by M_W in the MSSM [17]. This mass relation however does not generally hold in a 2HDM. Second, in a general 2HDM, the CP-odd scalar can be very light so that it can register into detectors as a "single-photon" signature (which originates from a pair of highly boosted photons produced by the decay of A). In that case, the production process $q\bar{q}' \to W^\pm h$ with $h \to AA$ and $A \to \gamma\gamma$ appear as a $W^\pm + \gamma\gamma$ signal event at colliders [18].

ACKNOWLEDGMENTS

I would like to thank J.L Diaz-Cruz and M.A. Perez for their support, and my collaborators C. Balazs, L.J. Diaz-Cruz, H.-J. He, S. Kanemura, F. Larios, G. Tavares-Velasco, and T. Tait for their invaluable contributions. This work was supported in part by the NSF grant PHY-9802564.

REFERENCES

1. G. Cvetic, Rev. Mod. Phys. **71** (1999) 513, and the references therein.
2. C. T. Hill, Phys. Lett. **B345** (1995) 483; Phys. Lett. **B266** (1991) 419.
3. H.-J. He, C. Hill and T. Tait, hep-ph/0108041.
4. See, for instance, reviews in "Perspectives on Supersymmetry", ed. G. L. Kane, World Scientific Publishing Co., 1998.
5. H.-J. He and C.-P. Yuan, *Phys. Rev. Lett.* **83**, 28 (1999).
6. T. Tait and C.-P. Yuan, Phys. Rev. **D63**, 014018 (2001), and the references therein.
7. J. L. Diaz-Cruz, H.-J. He, C.-P. Yuan, hep-ph/0103178.
8. C. Balazs, J.L. Diaz-Cruz, H.-J. He, T. Tait, C.-P. Yuan, Phys. Rev. **D59** (199) 055016, and the references therein.
9. M.A. Luty, Phys. Rev. **D41** (1990) 2893; M. Suzuki, Phys. Rev. **D41** (1990) 3457.
10. C. Balazs, H.-J. He, C.-P. Yuan, Phys. Rev. **D60** (1999) 114001.
11. DØ Collaboration, hep-ex/0102039, and the references therein.
12. F. Maltoni, K. Paul, T. Stelzer, and S. Willenbrock, Phys. Rev. **D64** (2001) 094023; L.-J. Jin, C.-S. Li, R.J. Oakes, and S-H. Zhu, Phys. Rev. **D62** (2000) 053008.
13. C.-X. Yue, G.-R. Lu, Q.-J. Xu, G.-L. Liu, J.Phys.**G27** (2001) 1043; C.-X. Yue, Y.-P. Kuang, X.-L. Wang, and W.-B. Li, Phys. Rev. **D62** (2000) 055005.
14. B. Mele, S. Petrarca, and A. Soddu Phys. Lett. **B435** (1998) 401.
    G. Eilam, J.L. Hewett and A. Soni, Phys. Rev. **D59**(E) (1999) 039901.
15. C.-X. Yue, G.-R. Lu, G.L. Liu, and Q-J. Xu, Phys.Rev.**D64** (2001) 095004.
16. H.-J. He, S. Kanemura, and C.-P. Yuan, in preparation.
17. S. Kanemura and C.-P. Yuan, hep-ph/0112165.
18. F. Larios, G. Tavares-Velasco, and C.-P. Yuan, Phys. Rev. **D64** (2001) 055004, and the references therein.

Summary talk

G. López Castro

Departamento de Física, Cinvestav del IPN, Apdo. Postal 14-740, México, D.F., México

Abstract. Highlights at this conference are summarized in the context of two fundamental problems in today's particle physics: the electroweak symmetry breaking and the origin of fermionic flavors.

INTRODUCTION

The electroweak symmetry breaking and the origin of fermionic flavors are perhaps the more fundamental problems in today's particle physics. In the standard model (SM) of elementary particles, the Higgs mechanism provides the simplest scheme to accommodate (but not to explain) an empirical solution to both problems. This mechanism for electroweak symmetry breaking endows to all the particles of their masses and, at the same time, it allows a flavor structure for fermions through the presence of arbitrary Yukawa couplings.

Since the symmetries of the standard model do not help to fix the Yukawa couplings, a large diversity of theoretical (beyond the SM) models have been put forward to try to explain the observed pattern of fermionic masses and mixings. Despite this enormous effort, there is now a consensus that a detailed experimental scrutiny of the flavor structure of the SM would provide eventually concrete clues and guiding ideas to the flavor problem.

Two remnants of the Higgs mechanism that can be investigated experimentally are the existence of the (yet unobserved) neutral Higgs boson and the flavor structure contained in the fermion masses and mixings. While the masses of charged leptons are known with pretty good accuracy, the information about masses and mixing of quarks has not reached a good accuracy yet. Furthermore, only until very recently we have been accumulating good evidence that also neutrinos are massive and get mixed, although the detailed structure of such pattern is still poorly known. Hopefully, a precise information about masses and mixings of fermions either, would signal possible deviations from the SM or eventually will provide important clues about the underlying theory beyond the SM.

The topics presented at this workshop converge into different aspects of these two important problems. Most of the talks focused on crucial phenomenological and experimental implications of the flavor structure of the SM, on the strategies to search for the Higgs boson, and on recent results about the masses and mixings of neutrinos. In this summary, we review some of the highlights addressed by these contributions. Unfortunately, our choice of topics of this talk forces us to left out the discussion of several important topics, and we refer the readers to those contributions in these proceedings.

FLAVOR MIXING OF QUARKS

In the quark sector, the Cabibbo-Kobayashi-Maskawa (CKM) [1] mechanism of flavor mixing provides a framework that describes in a consistent and unified form a rich variety of weak decays involving hadrons [2]. Its predictions for rare decays and the phenomena of CP violation (with the possible exception of the generation of baryon asymmetry in the universe [3]) have been confirmed by experiments in a rather impressive form [4] (see also the talk by P. Cooper at this meeting). Despite its phenomenological success, we think that the CKM mechanism does not *explain* the flavor mixing and CP violation phenomena because it is linked to two of the main mysteries of the SM, namely the existence of three generations and the origin of masses. An intense research activity aiming to reproduce the information on the mixing angles by postulating new symmetries or ansatz for the Yukawa couplings, is currently under way. Two of these interesting proposals were presented at this meeting [5, 6].

Among the most promising tests of the CKM picture, recent efforts have focused on the unitarity triangle (UT) relation $\sum_{i=u,c,t} V_{id} V_{ib}^* = 0$. Undergoing experiments at the B factories can provide overconstrained information on this relation (see the lectures by A. Buras). In this regard, one of the undoubted highlights of the year 2001 in particle physics was the confirmation of CP violation in the B_d^0 system by more than 3σ [7]. When these measurements of the CP asymmetry in $B_d^0(\overline{B}_d^0) \to J/\psi K_S$ and related decays are combined with previous results of CDF and ALEPH experiments, the direct measurements of one of the CP violating angles of the unitarity triangle averages to $\sin 2\beta = 0.79 \pm 0.10$ [2]. This information is consistent with indirect constraints obtained in the SM analysis of the UT which uses the data on the CP violation in $K \to \pi\pi$ decays, the mixing of the neutral $B_{d,s}$ mesons and the value of $|V_{ub}|/|V_{cb}|$ extracted from semileptonic B decays [2]. The consistency between the information from direct and indirect constraints on the CP-violating angle β of the UT, at least within present uncertainties, indicates that the SM picture of flavor mixing, namely the CKM mechanism, provides the dominant source for CP violation [2].

However, to get better constraints on the CKM picture, it is necessary to complement refined experimental measurements with improved theoretical predictions. Present theoretical predictions for weak decays are still plagued of long-distance uncertainties associated to the evaluation of hadronic matrix elements [2]. In some cases it has been possible to avoid these uncertainties by relating the weak hadronic matrix elements to other quantities that benefit from precise measurements. This is the case of the rare semileptonic decay $K^+ \to \pi^+ \nu \bar{\nu}$ [2]. Current measurements of this decay and prospects of improving it have been discussed at this workshop in [8, 9]. The loop-induced kaon semileptonic decays $K^+ \to \pi^+ \nu \bar{\nu}$ and $K_L \to \pi^0 \nu \bar{\nu}$ decays as promising probes of flavor mixing in the SM were discussed from an experimental [8] and theoretical [2] points of view at this meeting. In particular, the decay $K^+ \to \pi^+ \nu \bar{\nu}$ is expected to provide a clean measurement of the $|V_{td}|$ CKM matrix element because theoretical uncertainties associated to the long-distance effects are very small [2]. At present, the SM prediction for this decay is $B(K^+ \to \pi^+ \nu \bar{\nu}) = (7.5 \pm 2.9) \cdot 10^{-11}$ [2], where the (large) quoted uncertainty comes essentially from current uncertainties in the relevant CKM parameters. It was reported at this meeting that a second event for this decay mode was reported by the BNL-

E787 collaboration [10], which confirms the central value previously reported by the same group. This new measurement translates into $B(K^+ \to \pi^+ \nu \bar{\nu}) = (15.7^{+17.5}_{-8.2}) \cdot 10^{-11}$ [10], namely consistent with the SM prediction but with a bigger central value. The CKM experiment [8, 9] already approved at Fermilab, will provide around one hundred $K^+ \to \pi^+ \nu \bar{\nu}$ events in around two years of data taking which will be sufficient to match the accuracy of current theoretical predictions.

Other interesting recent measurements on rare kaon decays and prospects to improve some of them, where also reported at this meeting (see ref. [8] for details). Some of them are very interesting because involve simultaneously the interplay of all fundamental interactions and of long- and short-distance dynamics in a controlable fashion to allow some tests of flavor dynamics.

MASSES AND OSCILLATIONS OF NEUTRINOS

Another highlight of the year 2001 was the experimental evidence that the deficit of neutrinos coming from the sun is due to oscillations of electron-type neutrinos into other active flavors [11, 12]. Oscillations are an indication that neutrinos have mass. Massive neutrinos are welcome for the particle physics standard model since there is no reason for $m_\nu = 0$ in that framework (this is not the case for QED, where gauge invariance forbids the photon to become massive). Once we would have established the massive nature of neutrinos, it will remain an important task of experimental physics to measure the absolute values of their masses [13] and to elucidate whether those particles are of Dirac or Majorana type. The relevance of massive neutrinos as components of the hot dark matter in the universe was discussed at this meeting in ref. [14].

The search strategies for nonzero masses and mixings of neutrinos traditionally fall into three categories [4]. First, the results of oscillation experiments provide constraints in the region of the $\Delta m \equiv m_i - m_j$ vs. $\sin^2 \theta$ plane that are allowed by those results (see for example [12, 15, 16]). The second category consists of looking at the kinematical effects of neutrino masses in the decays of other particles [13]. Finally, the observation of neutrinoless double-beta decays would provide evidence for the Majorana nature of neutrinos.

Van de Water [12] has discussed at this meeting the recent results of the Sudbury National Observatory which confirm previous ideas that the solar neutrino deficit (in the high-energy region of the neutrino spectrum) has its origin in neutrino oscillations. The SNO experiment is designed to be sensitive to solar neutrinos from ^{8}B reactions through charged current (CC) and elastic scattering (ES) reactions in their heavy water detector. The CC process $\nu_e d \to p p e^-$ can detect only electron-type neutrinos, while the ES reaction $\nu_x e^- \to \nu_x e^-$ is sensitive to all three flavors ($x = e, \mu, \tau$). Clearly, if solar ν_e can turn into other kind of neutrinos, the flux $\Phi_{CC}(\nu_e)$ measured through the CC reactions should be smaller than the one $\Phi_{ES}(\nu_x)$ detected through ES reactions. The results obtained by the SNO collaboration [11], $\Phi_{CC}(\nu_e) = (1.75 \pm 0.14_{exp} \pm 0.05_{th}) \cdot 10^6$ cm^{-2}s^{-1} and $\Phi_{ES}(\nu_x) = (2.39 \pm 0.37_{exp}) \cdot 10^6$ cm^{-2}s^{-1} indisputably confirm this hypothesis. Furthermore, their result for the ES reaction is in nice agreement with the more precise value $\Phi_{ES}^{SK}(\nu_x) = (2.32 \pm 0.08_{exp}) \cdot 10^6$ cm^{-2}s^{-1} measured previously by the SuperKamiokande collaboration [17]. Thus, by combining the results of both collaborations they obtain $\Phi_{ES}^{SK}(v_x) - \Phi_{CC}(v_e) = (0.57 \pm 0.17) \cdot 10^6 \text{cm}^{-2}\text{s}^{-1}$ [11, 12], which establishes the oscillations of solar neutrinos into other active flavors ($v_{\mu,\tau}$) at the 3.3σ level. Therefore, the *total* flux of active neutrinos from ^{8}B solar reactions which includes all the components of the neutrino flux is determined to be $\phi(v_x) = (5.44 \pm 0.99) \cdot 10^6 \text{cm}^{-2}\text{s}^{-1}$ [11], which is now in excellent agreement with the calculation based on the solar standard model $\phi(v_x) = (5.05^{+0.1}_{-0.8}) \cdot 10^6 \text{cm}^{-2}\text{s}^{-1}$ [18].

Neutrino oscillation experiments can provided information on the difference of neutrino masses, but not on the masses themselves. Furthermore, the tiny mass differences required to explain the solar and atmospheric neutrino oscillation experiments can well arise from a degenerate or hierarchical mass scenarios [13]. The only way to disentangle among these possible scenarios is with the information about the individual mass parameters. This in turn can be achieved by looking at the kinematical effects of neutrino masses in some weak decay processes.

Measurements at the end-point region of the electron spectrum in tritium β decay $^3\text{H} \rightarrow ^3\text{He}e^-v$ can provide such information [13]. Those experiments can in fact measure the "electron mass" parameter $m_{v_e}^2 = \sum |U_{ei}|^2 m_i^2$, where m_i correspond to the neutrino mass eigenstates. Those measurements are complementary to the searches in neutrinoless double-beta decay experiments, which are sensitive to the mass parameter $\langle m_{v_e} \rangle = \sum U_{ei}^2 m_i$. By using the data sets of 1998 and 1999, the Mainz group has been able to set an upper limit of $m_{v_e} < 2.2$ eV/c^2 [13] which seems to be at the sensitivity limit of this experiment and it is the best constraint available at present for this mass from direct kinematical searches. As it was remarked in [13], this limit is getting close to the interesting region (~ 1 eV/c^2) for cosmology in order to assess the contribution of neutrinos to the hot dark matter in the universe. A full proposal to improve this limit in tritium beta decay (the KATRIN experiment) is under study at present [13]. It will allow to reduce the upper limit on the mass parameter of the electron-neutrino down to 0.35eV/c^2 and, eventually, to find a positive evidence for a non-zero neutrino mass[13].

SM HIGGS BOSON SEARCHES

The neutral scalar Higgs boson, possible the most important remnant of electroweak symmetry breaking, remains the only particle in the SM to be discovered. The mass of the SM Higgs boson can not be predicted by the SM and thus, it should be addressed by experiments. However, there are several hints for the existence of a light ($m_H \leq 200$ GeV) Higgs boson. Direct searches at LEP2 have excluded a Higgs boson mass smaller than 114.1 GeV (at 95% c.l.) [19]. A possible signal of the Higgs boson at $m_H = 115.6$ GeV has been reported from an analysis of the four LEP collaborations [19] at the very edge of the sensitivity limit of LEP2. Also, one of the most recent fits to the electroweak precision data yields $m_H = (88^{+53}_{-35})$ GeV (or $m_H \leq 196$ GeV at 95% c.l.) [20]. On the other hand, by assuming the validity of the SM up to the Planck scale energies one can bound the Higgs boson mass in the range 130 GeV $\leq m_H \leq$ 190 GeV (see for example [21] at these proceedings).

The capabilities of the Run 2 at Tevatron experiments to extend the search for the

Higgs boson were discussed at this meeting by J. Conway [22]. For a low mass Higgs boson ($m_H \leq 135$ GeV), the most promising channels are those where the Higgs boson is produced in association with a W or Z boson and then detected through his $b\bar{b}$ decays accompanied by a lepton pair from gauge boson decays. In the most optimistic scenario (even with the highest expected luminosity at the Run 2b), the discovery of the SM Higgs boson up to masses of 120 GeV using the combined data of CDF and D0 experiments remains a challenging task [22]. A heavier Higgs boson ($m_H \geq 135$ GeV), which decays predominantly to a $WW^{(*)}$ pair, can be sensitive only to the $l\nu\bar{l}\bar{\nu}$ channel when the Higgs is produced by the gluon-gluon fusion and in the like-sign dilepton channel $l^\pm l^\pm$-jet-jet when produced in association with a W or Z gauge boson. However, the requirement of an accurate knowledge of the large backgrounds to these channels makes more difficult the observation of a heavier Higgs boson [22].

It is a bit discouraging that, even in the more optimistic scenario, the SM (or SM-like) Higgs boson could be discovered at the Run 2b of Tevatron experiments only if it is lighter than 120 GeV. Nevertheless, interesting 95% exclusion limits would be possible in the range of Higgs boson masses that are allowed at present (115 GeV $\leq m_H \leq$ 190 GeV). Other dynamical features of the SM Higgs boson, such as its coupling to fermions, the couplings to gauge bosons and the HHH coupling only seems to be at the reach of a future linear collider [23]. In particular, with a typical center of mass energy $\sqrt{s} \approx 350$ GeV and an integrated luminosity of $\mathcal{L} \sim 500$ fb^{-1} some branching ratios for Higgs boson decays into fermions could be measured at a few of percent level [23]. Similarly, with the same linear collider parameters, the trilinear HHH coupling could be extracted with a 20% accuracy [23]. Finally, let us comment that unexpected (non-SM) properties of the Higgs boson couplings would eventually improve the sensitivity to them. One of these interesting scenarios where the Higgs couplings to the quarks of the third generation is special, was discussed in detail by C. P. Yuan [21].

NON-PERTURBATIVE METHODS IN HADRON PHYSICS

Several talks at this meeting were devoted to the study of spectroscopic, static, and decay properties of light hadrons (namely, hadrons made of u, d and s quarks) both, in the vacuum [24, 25, 26, 27, 28] and in a medium [29, 30]. The understanding of these hadron properties requires of models that incorporate approximate symmetries of QCD in the non-perturbative regime. The variety of models that have been elaborated to explain different hadron properties since more than forty years, and their confrontation to new data, maintains the vitality of this field. The emergence of powerful and systematic calculational techniques based on fundamental symmetries of QCD have also renewed the interest in the calculation of hadron properties.

In particular, the identification of chiral and spin-flavor symmetries of QCD underlying the quark and gluon dynamics, have provided powerful constraints in the modelling of hadrons and their interactions. Based on the chiral symmetry of QCD for massless quarks, chiral perturbation theory has emerged as a systematic perturbative method to describe the low-energy interactions of hadrons [31]. On the other hand, the quark-gluon dynamics with large number of colors N_c gives rise to a spin-flavor symmetry for baryons that allows the use of the $1/N_c$ expansion technique to compute baryon properties in a systematic way [32]. Since both symmetries have their origin in different aspects of QCD, there is not *a priori* double-counting ambiguities when computing hadron properties. Furthermore, a combination of both techniques in the case of baryons seems to be very useful to compute baryon properties such as magnetic moments, mass spectra, mass spittings and corrections to baryonic axial currents [28]. The formalism of the spin-flavor symmetry and the $1/N_c$ technique for baryons and its applications to understand quantitatively the mass spectra, the axial couplings, the magnetic moments, and the SU(3) breaking in the Gell-Mann Okubo relation for baryons were introduced at this meeting in a set of lectures by E. Jenkins [24]. An extension of the chiral perturbation theory techniques to describe the pion properties in a medium composed of baryons was discussed in detail by A. Wirzba in ref. [29].

SUMMARY OF THE SUMMARY

Most of the talks at this meeting addressed important issues of two of the most important current problems in particle physics: the origin of electroweak symmetry breaking and of fermionic flavors. Recent and forthcoming results in the sector of flavor mixing and CP violation will provide us in a not very far future with overconstrained information to test in detail the CKM picture and, eventually, will signal possible deviations from the SM. Planned and undergoing neutrino experiments hopefully will furnish a more detailed picture of the mass and mixing schemes for these particles. Although it remains a challenging task, the hunting of the Higgs boson at the Tevatron collider has more chances to be successful if the Higgs boson is light enough. The rich structure that will be collected in all this information will be very useful to guide the building of a theory beyond the SM that could address the flavor problem and the breaking of the electroweak symmetry at a more fundamental level.

ACKNOWLEDGEMENTS

I am very grateful to the organizers of this workshop for inviting me to give this summary talk. I greatly appreciate the patience of several of the speakers to explain me the essence of their contributions. This work has been partially supported by Conacyt and SNI (México).

REFERENCES

1. Kobayashi, M. and Maskawa, T., *Prog. Theor. Phys.* **49**, 652 (1973); Cabibbo, N. *Phys. Rev. Lett.* **10**, 531 (1963).
2. Buras, A., *lectures at this meeting*.
3. Ayala, A., *invited talk at this meeting*.
4. Groom, D. E., Particle Data Group, *Eur. J. Phys.* **C15**, (2000).
5. Mondragón, A., *invited talk at this meeting*.
6. Gupta, V., *contributed talk at this meeting*.

7. Aubert, B. et al, *Phys. Rev. Lett.* **87**, 091801 (2001); Abe, K. et al, *Phys. Rev. Lett.* **87**, 091802 (2001).
8. Cooper, P., *invited talk at this meeting*.
9. Morelos, A., *poster session at this meeting*
10. Adler, S. et al, *Phys. Rev. Lett.* **88**, 041803 (2002).
11. Ahmad, Q.R. et al, SNO Collaboration, *Phys. Rev. Lett.* **87**, 071301 (2001).
12. Van de Water, R., *invited talk at this meeting*.
13. Bonn, J., *invited talk at this meeting*.
14. Fornengo, N., *invited talk at this meeting*.
15. Miranda, O., *invited talk at this meeting*.
16. Pastor, S., *invited talk at this meeting*.
17. Fukuda, S. et al, SuperK Collaboration, *Phys. Rev. Lett.* **86**, 5651 (2001).
18. Bachcall, J. N., Pinsonneault, M. H. and Basu, S., *Astrophys. J.* **555**, 990 (2001)
19. ALEPH, DELPHI, L3 and OPAL Collaborations + LEP Higgs Working Group, eprint hep-ex/0107029.
20. Zwirner, F., Invited talk at the 20th Lepton-Photon Symposium, Rome, Italy 23-28 July 2001 and eprint hep-ph/0112130.
21. Yuan, C.-P., *invited talk at this meeting*.
22. Conway, J., *invited talk at this meeting*.
23. Wagner, A., *invited talk at this meeting*.
24. Jenkins, E., *lectures at this meeting*.
25. Napsuciale, M., *invited talk at this meeting*.
26. Moreno, G., *invited talk at this meeting*.
27. Weber, H. J., *invited talk at this meeting*.
28. Flores Mendieta, R., *invited talk at this meeting*.
29. Wirzba, A., *invited talk at this meeting*.
30. Wambach, J.,*invited talk at this meeting*.
31. Gasser, J. and Leutwyler, H., *Annals Phys.* **158**, 142 (1984); *Nucl. Phys.* **B250**, 465 (1985); *ibid* **250**, 517 (1985); *ibid* **250**, 539 (1985).
32. Dashen, R. and Manohar, A. V., *Phys. Lett.* **B315**, 425 (1993); *ibid* **B315**, 438 (1993); Jenkins, E., *Phys. Lett.* **B315**, 431 (1993); *ibid* **B315**, 441 (1993); *ibid* **B315**, 447 (1993).

CONTRIBUTED TALKS

Vertex integrals in heavy-particle theories

Antonio O. Bouzas

Departamento de Física Aplicada, Cinvestav-IPN, Apdo. Postal 73 "Cordemex", Mérida 97310, Yucatán, México

Abstract. We give the results of complete analytical computations of two- and three-point loop integrals ocurring in heavy particle theories, with and without velocity change, for arbitrary values of external momenta and masses.

INTRODUCTION

In this talk we consider a class of one-loop integrals occurring in heavy-particle theories [1], with arbitrary real values for the external masses and residual momenta. We give the results of complete analytical computations of three-point loop integrals with and without velocity change, and two-point loop integrals. The details of the calculations are given in [2], and in a forthcoming paper.

LOOP INTEGRALS

The loop integrals we consider are of the form,

$$\begin{aligned}
\mathcal{J}_2^{\alpha_1\cdots\alpha_n} &= \frac{i\mu^{4-d}}{(2\pi)^d}\int d^d\ell \frac{\ell^{\alpha_1}\cdots\ell^{\alpha_n}}{(2v\cdot(\ell+k)-\delta M+i\varepsilon)(\ell^2-m^2+i\varepsilon)}. \\
\mathcal{J}_3^{\alpha_1\cdots\alpha_n} &= \frac{i\mu^{4-d}}{(2\pi)^d}\int d^d\ell \frac{\ell^{\alpha_1}\cdots\ell^{\alpha_n}}{(2v_1\cdot(\ell+k_1)-\delta M_1+i\varepsilon)(2v_2\cdot(\ell+k_2)-\delta M_2+i\varepsilon)} \\
&\quad \times \frac{1}{(\ell^2-m^2+i\varepsilon)} \\
\mathcal{H}_3^{\alpha_1\cdots\alpha_n} &= \frac{i\mu^{4-d}}{(2\pi)^d}\int d^d\ell \frac{\ell^{\alpha_1}\cdots\ell^{\alpha_n}}{(2v\cdot\ell-\delta M+i\varepsilon)((\ell-k_1)^2-m^2+i\varepsilon)} \\
&\quad \times \frac{1}{((\ell-k_2)^2-m'^2+i\varepsilon)}
\end{aligned} \tag{1}$$

Here v_i^μ, $i=1,2$, are the velocities of the external heavy legs, k_i^μ their residual momenta, and δM_i their mass splittings relative to the common heavy mass of the corresponding heavy quark symmetry multiplet. m and m' are the masses of the light particles within the loops, which in chiral theories are light pseudoscalar mesons. These integrals are defined in $d=4-\varepsilon$ dimensions, μ being the mass scale of dimensional regularization.

Their degrees of divergence are $n+d-3$ for $\mathcal{J}_2^{\alpha_1\cdots\alpha_n}$, $n+d-4$ for $\mathcal{J}_3^{\alpha_1\cdots\alpha_n}$ and $n+d-5$ for $\mathcal{H}_3^{\alpha_1\cdots\alpha_n}$. The factor of 2 in front of v_i^μ corresponds to our normalization of the heavy-particle propagators.

Our method of calculation [2] is to obtain the integrals (1) as large-mass limits of ordinary loop integrals. We closely follow the approach of [3] for the computation of scalar integrals, which are greatly simplified in the large-mass limit, and the method of [4] to express tensor integrals in terms of scalar ones.

TWO-POINT INTEGRALS

Two-point integrals with one heavy propagator have been given in [2, 5, 6]. The scalar two-point integral $\mathcal{J}_2$ is a function of m and $\Delta = \delta M - 2v\cdot k$. We write it in terms of $\xi = \Delta/(2m)$,

$$\mathcal{J}_2(\Delta,m) = \frac{\Delta}{32\pi^2}\left(\frac{2}{\varepsilon} + \log\left(\frac{\bar{\mu}^2}{m^2}\right) + 2\right) + \frac{m}{16\pi^2}\mathcal{F}(\xi)$$

with $\mathcal{F}(x) = \sqrt{x^2-1+i\varepsilon}\left[\log\left(x-\sqrt{x^2-1+i\varepsilon}\right) - \log\left(x+\sqrt{x^2-1+i\varepsilon}\right)\right]$. The coefficient of the dimensional regularization pole vanishes when $\Delta = 0$. This is due to the fact that the real part of the integrand in $\mathcal{J}_2$ is parity-odd when $\Delta = 0$.

The vector two-point integral $\mathcal{J}_2^\mu(v^\alpha,\Delta,m)$ is given in terms of only one form factor, $\mathcal{J}_2^\mu(v^\alpha,\Delta,m) = F(\Delta,m)v^\mu$, with $F(\Delta,m) = v_\mu\mathcal{J}_2^\mu(v^\alpha,\Delta,m) = 1/2A_0(m) + \Delta/2\mathcal{J}_2(\Delta,m)$, where A_0 is the standard one-point scalar integral (see the appendix of [2]). The second-rank tensor integral is computed analogously, explicit results being given in [2].

THREE-POINT INTEGRALS WITH VELOCITY CHANGE

The scalar three-point integral with velocity change, $\mathcal{J}_3 = \mathcal{J}_3(v_1\cdot v_2,\Delta_1,\Delta_2,m)$, where $\Delta_j = \delta M_j - 2v_j\cdot k_j$, can be expressed in terms of four dilogarithms [2],

$$\mathcal{J}_3 = \frac{1}{64\pi^2}\frac{1-\Omega^2}{\Omega}\left\{\frac{2}{\varepsilon}\log(\alpha) + \log^2(\alpha) + \sum_{k,\sigma}(-1)^{k+1}\left[\frac{1}{2}\log^2\left(\frac{-z_{k\sigma}}{\bar{\mu}}\right) + \text{Li}_2\left(\frac{-y_0}{z_{k\sigma}}\right)\right]\right\}.$$

The notation is as follows, $\omega = v_1\cdot v_2$, $\Omega = \sqrt{(\omega-1)/(\omega+1)}$, $\alpha = \omega + \sqrt{\omega^2-1} = (1+\Omega)/(1-\Omega)$ and $\bar{\mu}$ is the mass unit in the $\overline{\text{MS}}$ scheme. The sum extends over $k = 1,2$ and $\sigma = \pm$, with

$$y_0 = -\frac{1+\Omega}{2\Omega}\left(\frac{1+\Omega}{2}\Delta_1 - \frac{1-\Omega}{2}\Delta_2\right)$$

$$z_{1\pm} = \frac{1}{2}\left(\frac{1+\Omega^2}{2\Omega}\Delta_1 - \frac{1-\Omega^2}{2\Omega}\Delta_2 \pm \sqrt{\Delta_1^2 - 4m^2 + i\varepsilon}\right)$$

$$z_{2\pm} = \frac{\alpha}{2}\left(\frac{1-\Omega^2}{2\Omega}\Delta_1 - \frac{1+\Omega^2}{2\Omega}\Delta_2 \pm \sqrt{\Delta_2^2 - 4m^2 + i\varepsilon}\right).$$

This expression for $\mathcal{J}_3$ is equivalent to the result given in eq. (30) of [2], though it has been written in a more compact form by means of the identity

$$\frac{1}{2}\log^2(z) - \log(z)\log(-z) = -\frac{\pi^2}{2} - \frac{1}{2}\log^2(-z),$$

valid on the first Riemann sheet of the logarithm, and the identity (A.2) of [3] for the dilogarithm.

In order to compute the vector integral $\mathcal{J}_3^\mu$ we define two sets of form factors as $\mathcal{J}_3^\mu = I_1 v_1^\mu + I_2 v_2^\mu$ and $F_{1,2} = v_{1,2} \cdot \mathcal{J}_3$. These form factors are given by,

$$I_1 = \frac{1-\Omega^2}{4\Omega^2}\left[-(1-\Omega^2)F_1 + (1+\Omega^2)F_2\right], \quad I_2 = \frac{1-\Omega^2}{4\Omega^2}\left[(1+\Omega^2)F_1 - (1-\Omega^2)F_2\right],$$

with $F_{1,2} = 1/2\mathcal{J}_2(\Delta_{2,1},m) + \Delta_{1,2}/2\mathcal{J}_3(v_1 \cdot v_2, \Delta_1, \Delta_2, m)$. These equations give an explicit expression for $\mathcal{J}_3^\mu$. For the sake of brevity, we omit here the results for the tensor integral $\mathcal{J}_3^{\mu\nu}$, which can be found in [2].

THREE-POINT INTEGRALS WITH ONE HEAVY PROPAGATOR

The scalar three-point integral $\mathcal{H}_3 = \mathcal{H}_3(v \cdot q, q^2, \Delta, m, m')$, with $q = (k_2 - k_1)/2$ and $\Delta = \delta M - v \cdot (k_1 + k_2)$, can be expressed in terms of eight dilogarithms as,

$$\mathcal{H}_3 = \frac{1}{(4\pi)^2}\frac{1}{4|\mathbf{q}|}\sum_{j=1,2}\left(\mathcal{F}_1(y_j) + \mathcal{F}_2(x_j) - \mathcal{F}_3(z_j)\right)$$

$$\mathcal{F}_1(x) = \text{Li}_2\left(\frac{z_0 - 4|\mathbf{q}|\alpha}{z_0 - 4|\mathbf{q}|x}\right) - \text{Li}_2\left(\frac{z_0}{z_0 - 4|\mathbf{q}|x}\right)$$

$$\mathcal{F}_2(x) = -\text{Li}_2\left(\frac{z_0 - 4|\mathbf{q}|\alpha}{z_0 - 4|\mathbf{q}|x}\right) - \frac{1}{2}\log^2\left(\frac{z_0 - 4|\mathbf{q}|x}{\mu^2}\right)$$

$$\mathcal{F}_3(x) \equiv \mathcal{F}_1(x) + \mathcal{F}_2(x) = -\text{Li}_2\left(\frac{z_0}{z_0 - 4|\mathbf{q}|x}\right) - \frac{1}{2}\log^2\left(\frac{z_0 - 4|\mathbf{q}|x}{\mu^2}\right).$$

The quantities entering these equations are $|\mathbf{q}| = \sqrt{(v \cdot q)^2 - q^2}$ ($|\mathbf{q}|$ is assumed to be real), $\alpha \equiv \alpha_+$ with $\alpha_\pm = 2(v \cdot q \pm |\mathbf{q}|)$, $z_0 = -(m'^2 - m^2) - \alpha_+(\Delta - 2|\mathbf{q}|)$, and,

$$x_{1,2} = v \cdot k_2 + 2|\mathbf{q}| - \frac{\delta M}{2} \pm \sqrt{\left(v \cdot k_1 - \frac{\delta M}{2}\right)^2 - m^2 + i\varepsilon}$$

$$y_{1,2} = \frac{1}{2\alpha_-}\left(4q^2 + m'^2 - m^2 \pm \sqrt{(4q^2 - (m'+m)^2)(4q^2 - (m'-m)^2)} + i\varepsilon\sigma\right)$$

$$z_{1,2} = v \cdot k_2 - \frac{\delta M}{2} \pm \sqrt{\left(v \cdot k_2 - \frac{\delta M}{2}\right)^2 - m'^2 + i\varepsilon},$$

where in the expression for y_j we denoted $\sigma \equiv \text{sgn}(q^2)$. μ is a positive constant with dimension of mass, analogous to the mass unit in dimensional regularization. It is not difficult to show that $\mathcal{H}_3$ does not depend on μ, it appears there for purely dimensional reasons.

Tensor three-point integrals $\mathcal{H}_3^{\alpha_1\cdots\alpha_n} = \mathcal{H}_3^{\alpha_1\cdots\alpha_n}(v,k_1,k_2;\delta M,m,m')$ can be given in terms of integrals with smaller ranks and fewer points, by the well-known method of [4]. We will consider integrals of standard form $\mathsf{H}_3^{\alpha_1\cdots\alpha_n}(v,q;\Delta,m,m') = \mathcal{H}_3^{\alpha_1\cdots\alpha_n}(v,-q,q;\Delta+v\cdot(k_1+k_2),m,m')$ in terms of which we can express $\mathcal{H}_3$ as,

$$\mathcal{H}_3^{\alpha_1\cdots\alpha_n} = \mathsf{H}_3^{\alpha_1\cdots\alpha_n}(v,q;\Delta,m,m') + \sum_{j=1}^{n} r^{\{\alpha_1}\cdots r^{\alpha_j}\mathsf{H}_3^{\alpha_{j+1}\cdots\alpha_n\}}(v,q;\Delta,m,m'),$$

where $r^\mu \equiv 1/2(p'+p)^\mu$, and $A^{\{\alpha_1\alpha_2\cdots\alpha_s\}} \equiv A^{\alpha_1\alpha_2\cdots\alpha_s} + A^{\alpha_2\cdots\alpha_s\alpha_1} + \cdots + A^{\alpha_s\alpha_1\cdots\alpha_{s-1}}$. Clearly, for the scalar integral we have $\mathcal{H}_3 = \mathsf{H}_3$.

For the vector integral we write $\mathsf{H}_3^\alpha(v,q;\Delta,m_1,m_2) = \mathsf{V}v^\alpha + \mathsf{Q}q^\alpha$, with $\mathsf{V} = \mathsf{V}(v\cdot q, q^2, \Delta, m_1, m_2)$ and similarly Q. If $(v\cdot q)^2 - q^2 = |\mathbf{q}|^2 = 0$, then $q^\alpha \propto v^\alpha$ and we can set $\mathsf{Q} = 0$. If $|\mathbf{q}|^2 \neq 0$,

$$|\mathbf{q}|^2 \mathsf{Q} = v\cdot q\, v_\alpha \mathsf{H}_3^\alpha - q_\alpha \mathsf{H}_3^\alpha, \quad |\mathbf{q}|^2 \mathsf{V} = -q^2 v_\alpha \mathsf{H}_3^\alpha - v\cdot q\, q_\alpha \mathsf{H}_3^\alpha,$$

with,

$$v_\alpha \mathsf{H}_3^\alpha = \frac{1}{2}B_0(4q^2,m_1,m_2) + \frac{\Delta}{2}\mathsf{H}_3$$

$$q_\alpha \mathsf{H}_3^\alpha = \frac{1}{4}\mathcal{J}_2(\Delta - 2v\cdot q, m_2) - \frac{1}{4}\mathcal{J}_2(\Delta + 2v\cdot q, m_1) + \frac{m_1^2 - m_2^2}{4}\mathsf{H}_3.$$

We have omitted the arguments $(v,q;\Delta,m_1,m_2)$ of H_3^α on boths sides of these equations for brevity. B_0 is the standard scalar two-point integral, as given in the appendix of [2]. Higher-rank tensor integrals can be computed analogously.

The results presented in this section were obtained in collaboration with R. Flores Mendieta. A detailed derivation will be given elsewhere.

ACKNOWLEDGEMENTS

This work has been partially supported by Conacyt and SNI.

REFERENCES

1. See, e.g., A.Manohar, M.Wise, *"Heavy Quark Physics,"* Cambridge Univ. Press, 2000, and references therein.
2. A.Bouzas, Eur.Phys.J. **C12**, (2000), 643
3. G.'t Hooft, M.Veltman, Nucl.Phys. **B153**, (1979), 365.
4. G.Passarino, M.Veltman, Nucl.Phys. **B160**, (1979), 151.
5. D.Broadhurst, A.Grozin, Phys.Lett. **B267**, (1991), 105.
6. I.W.Stewart, Nucl.Phys. **B529**, (1998), 62.

Stationary Effective Field Theory of Heterotic String vs Einstein–Maxwell Theory

A. Herrera–Aguilar

Instituto de Física y Matemáticas, UMSNH. Apdo. Postal 2–82, Morelia, Michoacán, México

Abstract. We show how the Matrix Ernst Potential formalism allows one to express the stationary action of the heterotic string effective field theory in an Einstein–Maxwell form. The relation between both theories allows one to extrapolate the results obtained for the Einstein–Maxwell theory to the heterotic string realm.

INTRODUCTION

An interesting feature of the toroidally compactified down to three dimensions effective field theory of the heterotic string is that it can be written as a sigma model possessing the $O(d+1,d+n+1)$ group symmetry (U–duality). A similar situation takes place for the stationary Einstein–Maxwell (EM) theory, where the symmetry group is the eight-parameter $SU(1,2)$. Due to this fact, one could think of certain relationship between the mathematical structure of both theories that could help us to generalize the results obtained in EM theory to the heterotic string one. Such a relationship is estalished and its applications are discussed.

HETEROTIC STRING THEORY AND ERNST POTENTIALS

We consider the effective action of heterotic string theory in D dimensions:

$$S^{(D)} = \int d^{(D)}x |G^{(D)}|^{\frac{1}{2}} e^{-\phi^{(D)}} (R^{(D)} + \phi^{(D)}_{;M}\phi^{(D);M} - \frac{1}{12}H^{(D)}_{MNP}H^{(D)MNP} - \frac{1}{4}F^{(D)I}_{MN}F^{(D)IMN}) \quad (1)$$

where

$$F^{(D)I}_{MN} = \partial_M A^{(D)I}_N - \partial_N A^{(D)I}_M, \quad H^{(D)}_{MNP} = \partial_M B^{(D)}_{NP} - \frac{1}{2}A^{(D)I}_M F^{(D)I}_{NP} + \text{cycl. perms. of M, N, P.}$$

Here $G^{(D)}_{MN}$ is the metric, $B^{(D)}_{MN}$ is the anti–symmetric Kalb-Ramond field, $\phi^{(D)}$ is the dilaton and $A^{(D)I}_M$ is a set of $U(1)$ Abelian vector fields ($I = 1, 2, ..., n$). In the consistent critical case $D = 10$ and $n = 16$. After the compactification down to three dimensions on a d–torus we get the following fields [1]–[2]
a) scalar fields

$$G = \left(G_{pq} = G^{(D)}_{p+3,q+3} \right), \quad B = \left(B_{pq} = B^{(D)}_{p+3,q+3} \right),$$

$$A = \left(A_p^I = A_{p+3}^{(D)I}\right), \qquad \phi = \phi^{(D)} - \frac{1}{2}\ln|\det G|, \tag{2}$$

where the subscripts $p, q = 1, 2, ..., d$.
b) tensor fields

$$g_{\mu\nu} = e^{-2\phi}\left(G_{\mu\nu}^{(D)} - G_{p+3,\mu}^{(D)}G_{q+3,\nu}^{(D)}G^{pq}\right),$$

$$B_{\mu\nu} = B_{\mu\nu}^{(D)} - 4B_{pq}A_\mu^p A_\nu^q - 2\left(A_\mu^p A_\nu^{p+d} - A_\nu^p A_\mu^{p+d}\right), \tag{3}$$

(we set $B_{\mu\nu} = 0$ to remove the effective three-dimensional cosmological constant from our consideration).

c) vector fields $A_\mu^{(a)} = \left((A_1)_\mu^p, (A_2)_\mu^{p+d}, (A_3)_\mu^{2d+l}\right)$ $(a = 1, ..., 2d+n)$

$$(A_1)_\mu^p = \frac{1}{2}G^{pq}G_{q+3,\mu}^{(D)}, \qquad (A_3)_\mu^{l+2d} = -\frac{1}{2}A_\mu^{(D)l} + A_q^l A_\mu^q,$$

$$(A_2)_\mu^{p+d} = \frac{1}{2}B_{p+3,\mu}^{(D)} - B_{pq}A_\mu^q + \frac{1}{2}A_p^l A_\mu^{l+2d}. \tag{4}$$

In three dimensions all vector fields can be dualized on-shell:

$$\nabla \times \vec{A_1} = \frac{1}{2}e^{2\phi}G^{-1}\left(\nabla u + (B + \frac{1}{2}AA^T)\nabla v + A\nabla s\right),$$

$$\nabla \times \vec{A_3} = \frac{1}{2}e^{2\phi}(\nabla s + A^T\nabla v) + A^T\nabla \times \vec{A_1},$$

$$\nabla \times \vec{A_2} = \frac{1}{2}e^{2\phi}G\nabla v - (B + \frac{1}{2}AA^T)\nabla \times \vec{A_1} + A\nabla \times \vec{A_3}. \tag{5}$$

Thus, the resulting three-dimensional theory describes the scalars G, B, A and ϕ and pseudoscalars u, v and s coupled to the metric $g_{\mu\nu}$.

We define two matrix Ernst potentials as follows [3]

$$X = \begin{pmatrix} -e^{-2\phi} + v^T X v + v^T A s + \frac{1}{2}s^T s & v^T X - u^T \\ Xv + u + As & X \end{pmatrix},$$

$$\mathcal{A} = \begin{pmatrix} s^T + v^T A \\ A \end{pmatrix}, \tag{6}$$

where $X = G + B + \frac{1}{2}AA^T$. They have dimensions $(d+1) \times (d+1)$ and $(d+1) \times n$, respectively, and contain all the information about the matter content of the effective field theory. The physical meaning of these variables is as follows: the relevant information concerning the gravitatoinal field is encoded in X, whereas its rotational character is hidden in u; v is related to multidimensional components of the Kalb–Ramond field, and, finally, A and s stand for electric and magnetic potentials, respectively. In terms of MEP the effective three–dimensional Lagrangian of the theory takes the form

$$\mathcal{L}^{(3)} = -R + \text{Tr}[\frac{1}{4}(\nabla X - \nabla \mathcal{A}\mathcal{A}^T) \mathcal{G}^{-1} (\nabla X^T - \mathcal{A}\nabla \mathcal{A}^T) \mathcal{G}^{-1} + \frac{1}{2} \nabla \mathcal{A}^T \mathcal{G}^{-1} \nabla \mathcal{A}] \qquad (7)$$

where $\mathcal{G} = \frac{1}{2}(X + X^T - \mathcal{A}\mathcal{A}^T)$. The field equations for the matter part of this theory read

$$\nabla^2 X - 2(\nabla X - \nabla \mathcal{A}\mathcal{A}^T)(X + X^T - \mathcal{A}\mathcal{A}^T)^{-1} \nabla X = 0,$$
$$\nabla^2 \mathcal{A} - 2(\nabla X - \nabla \mathcal{A}\mathcal{A}^T)(X + X^T - \mathcal{A}\mathcal{A}^T)^{-1} \nabla \mathcal{A} = 0. \qquad (8)$$

STATIONARY EINSTEIN–MAXWELL THEORY

On the other hand, it is well–known that the stationary formulation of the four–dimensional Einstein–Maxwell theory can be expressed in terms of two complex Ernst potentials, namely, the gravitational $\mathcal{E}$ and the electromagnetic Φ ones [4]–[5]:

$$\mathcal{L}_{EM} = \frac{1}{2f^2} |\nabla \mathcal{E} + \bar{\Phi} \nabla \Phi|^2 - f^{-1} |\nabla \Phi|^2. \qquad (9)$$

where $f = \frac{1}{2}(\mathcal{E} + \bar{\mathcal{E}} + \Phi \bar{\Phi})$. The corresponding pair of field equations for the matter part of the theory is

$$\nabla^2 \mathcal{E} - f^{-1}(\nabla \mathcal{E} + \bar{\Phi} \nabla \Phi) \cdot \nabla \mathcal{E} = 0,$$
$$\nabla^2 \Phi - f^{-1}(\nabla \mathcal{E} + \bar{\Phi} \nabla \Phi) \cdot \nabla \Phi = 0. \qquad (10)$$

Thus, we see that the Lagrangian (7) is very similar to that of the stationary Einstein–Maxwell theory (9). In fact, we can establish a relationship between both theories through the "map"

$$X \longleftrightarrow -\mathcal{E}, \qquad \mathcal{A} \longleftrightarrow \Phi,$$

$$\text{matrix transposition} \longleftrightarrow \text{complex conjugation}. \qquad (11)$$

CONCLUSION AND DISCUSSION

The "map" (11) establish some sort of matrix generalization of the stationary EM theory and allows us to extrapolate results of the EM theory to the heterotic string realm with the aid of the MEP formalism. Some of these results have already been obtained [6]–[7]. With respect to further work we could mention the construction of a full non–extremal charged rotating black hole–type solution by applying either the symmtery generation technique or by directly solving the equations of motion; the construction of interacting black hole solutions as well as soliton configurations in this framework are interesting issues under current investigation.

ACKNOWLEDGMENTS

This work was supported by grants CONACYT–J34245–E and CIC–UMSNH–4.18.

REFERENCES

1. Maharana, J., and Schwarz, J.H., *Nucl. Phys.* **B390** 3 (1993).
2. Sen, A., *Nucl. Phys.* **B434** 179 (1995).
3. Herrera-Aguilar, A., and Kechkin, O., *Int. J. Mod. Phys.* **A13** 393 (1998).
4. Ernst, F.J., *Phys. Rev.* **167** 1175 (1968).
5. Mazur, P.O., *Acta Phys. Pol.* **14** 219 (1983).
6. Herrera-Aguilar, A., and Kechkin, O., *Phys. Rev.* **D59** 124006 (1999).
7. Herrera-Aguilar, A., and Kechkin, O., *Class. Quant. Grav.* **16** 1745 (1999).

Resonances in $\Lambda_c^+ \to pK^-\pi^+$

Juan Medellin Z., Jürgen Engelfried, Antonio Morelos
For the SELEX Collaboration

*Instituto de Física, Universidad Autónoma de San Luis Potosí,
Álvaro Obregon 64, Zona Centro, San Luis Potosí, S.L.P. 78000, México*

Abstract. We report very preliminary results of a Dalitz-plot analysis [1] of Λ_c^+ in the decay to $pK^-\pi^+$ with the helicity formalism. We used the data from the fixed target experiment SELEX [2] (E781) in Fermilab. We report about branching-ratios of the resonant states involved, and a possible initial state polarization.

INTRODUCTION AND ANALYSIS METHOD

In our search for resonances in the 3-body decay of $\Lambda_c^+ \to pK^-\pi^+$ we are considering in addition to the non-resonant mode the following resonant decay modes: $\Lambda_c^+ \to \bar{K}^{*0}p$, $\bar{K}^{*0} \to K^-\pi^+$; $\Lambda_c^+ \to \Delta^{++}K^-$, $\Delta^{++} \to p\pi^+$; and $\Lambda_c^+ \to \Lambda(1520)\pi^+$, $\Lambda(1520) \to pK^-$.

As a search tool for the resonances we calculate the invariant masses of pairs of daughter particles M_{ij}^2, M_{ik}^2 and fill with them into a two-dimensional histogram (Dalitz plot). A Monte-Carlo simulation of what could be expected is shown in fig. 1 (right).

The data sample used for this analysis is the same as in a previous SELEX publication [3], where more details can be found. We applied the following cuts to get a clean Λ_c^+ signal: good fits for tracks and vertices; momentum $> 8.0\,\text{GeV}/c$ for all tracks; proton and kaon identified in the RICH [4]; secondary vertex outside material; separation L of primary and secondary vertices $L > 8\sigma$, where σ denotes the combined error of the vertices; $\sigma < 1.7\,\text{mm}$; at least two tracks with a miss distance to the primary vertex of more than $20\mu m$; the momentum vector of the reconstructed Λ_c^+ has to point back to the primary vertex. The obtained $pK^-\pi^+$ invariant mass distribution can be found in fig. 1 (left).

To eliminate the contribution of background under the Λ_c^+-peak, we produce a Dalitz-plot of the sidebands shown in fig. 1, with a proper mapping of the allowed phase space. After normalizing to the correct number of events, we subtract this contribution bin-by-bin from a Dalitz plot of the signal region. We verified this procedure by comparing the plots obtained by different sideband regions on the left and right side of the signal region. In fig. 2 we show the Dalitz plots for the sideband and the signal regions. In the background subtracted signal region one can still see some enhancement on the lower left of the phase space, also present in the sideband; we are still investigating the nature of this.

To extract the information from the Dalitz plot, we used an helicity formalism [5, 6]. We will describe the fit functions used in the following sections.

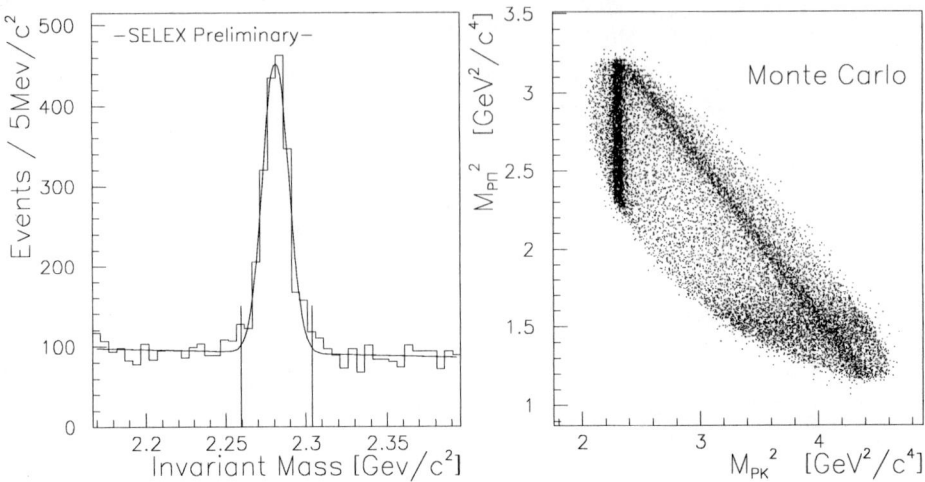

FIGURE 1. Left: Invariant mass distribution for $pK^-\pi^+$. A signal of approximately 1500 Λ_c^+ can be clearly seen. The signal and the sideband regions used in this analysis are indicated. Right: Dalitz plot from Monte Carlo with a mixture of non-resonant and three resonant decay modes.

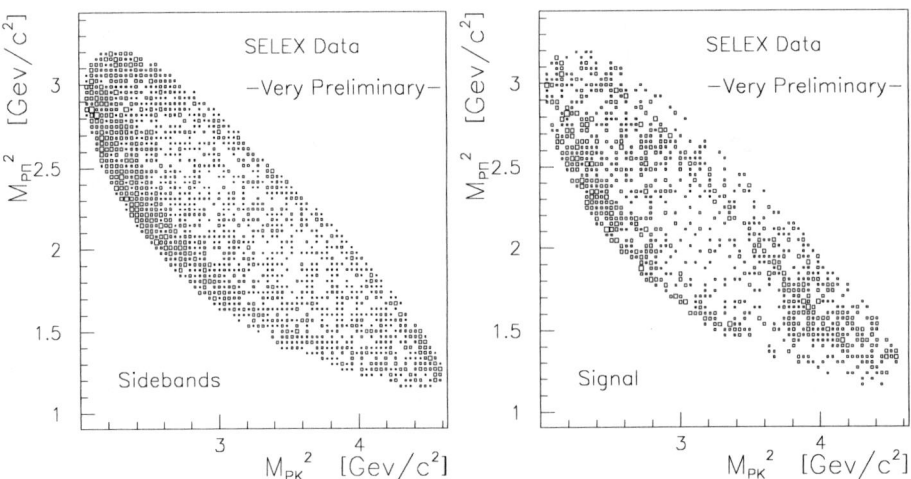

FIGURE 2. Left: Dalitz plot for the sideband region indicated in fig. 1 with phase space mapping. Right: Dalitz plot of signal region after sideband subtraction.

HELICITY FORMALISM FOR TWO BODY DECAYS

From Fermi's Golden Rule the partial decay width is given by $d\Gamma = \frac{(2\pi)^4}{2M}|\Omega|^2 d\Phi_n$, with $d\Gamma \sim |\Omega|^2 = |<BC|T|A>|^2$, where A and BC denote the initial and final states. We are working in the eigenstate base $|M, S, p^\mu, \lambda>$ of the operators $M^2 = P_\mu P^\mu$, $S = -\omega_\mu \omega^\mu$

with $\omega^\sigma = \frac{1}{2}\varepsilon^{\sigma\mu\nu\lambda} M_{\mu\nu} P_\lambda$, and $P^\mu, \omega^0 = J_x P_x + J_y P_y + J_z P_z$; ω^0 represents the helicity operator with eigenvalue λ. For simplicity we assume only $|p^\mu, \lambda>$.

In the rest frame of the mother particle A we can express the spin states as $|j_A m_A>$ and the final state as $|\theta_B \phi_B \lambda_B \lambda_C>$, with

$$|\theta_B \phi_B \lambda_B \lambda_C> = \sum_{JM} \sqrt{\frac{2J+1}{4\pi}} D^J_{M\lambda_1}(\phi_B, \theta_B, -\phi_B) |JM\lambda_B \lambda_C>.$$

Applying angular momentum conservation, we can write the transition amplitude as

$$<BC|T|A> = \sqrt{\frac{2j_A+1}{4\pi}} D^{\star j_A}_{m_A \lambda_1}(\phi_B, \theta_B, -\phi_B) <\lambda_B \lambda_C|T|j_A m_A>$$

With $D^j_{m\lambda}(\phi, \theta, -\phi) = e^{-i\phi(m-\lambda)} d^j_{m\lambda}(\theta)$, $\lambda_1 = \lambda_B - \lambda_C$ ($|\lambda_1| \leq m_A$), and summing over all spin and helicity projections and we obtain

$$d\Gamma \sim \sum_{m_A} \sum_{\lambda_B \lambda_C} |\alpha_{\lambda_B \lambda_C} e^{i\phi_B(m_A - \lambda_1)} d^{j_A}_{m_A \lambda_1}(\theta_B)|^2$$

HELICITY FORMALISM FOR THREE BODY RESONANT DECAYS

The decay width in this case is given by $d\Gamma \sim |<DE|T_2|B><BC|T_1|A>|^2$. In the rest frame of the resonance B, with the z-axes pointing into the direction of motion of the resonance B in the rest system of the mother particle A, we can write

$$<DE|T_2|B> = \sqrt{\frac{2j_B+1}{4\pi}} e^{i\phi'_D(\lambda_B - \lambda_2)} d^{j_B}_{\lambda_B \lambda_2}(\theta'_D) <\lambda_D \lambda_E|T_2|B>.$$

Summing over all resonances we obtain

$$d\Gamma \sim \sum_{m_A \lambda_C \lambda_D \lambda_E} |\sum_{\lambda_B} \sum_B BW(m_r) \alpha_{\lambda_B \lambda_C} \alpha_{\lambda_D \lambda_E} e^{i\phi_B(m_A - \lambda_1)} e^{i\phi'_D(\lambda_B - \lambda_2)} d^{j_A}_{m_A \lambda_1}(\theta_B) d^{j_B}_{\lambda_B \lambda_2}(\theta'_D)|^2 \tag{1}$$

where we also consider the finite width of the resonance with a Breit-Wigner distribution $BW(m_r) \sim \frac{m_0 \Gamma_r}{m_r^2 - m_0^2 + i m_r \Gamma_r}$. In the parity conserving decay of the resonance we can write $\alpha_{\lambda_D \lambda_E} = (-1)^{S_D + S_E - S_B} \eta_B \eta_D \eta_E \alpha_{-\lambda_D -\lambda_E}$.

In this formalism we can naturally introduce an initial polarization of the mother particle P_A, given by $d\Gamma \sim \frac{1}{2}(1+P_A)\sum_{\lambda_C \lambda_D \lambda_E}|\sum_{\lambda_B}\sum_B BW(m_B)\xi_{B,\frac{1}{2},\lambda_B,\lambda_C,\lambda_D,\lambda_E}|^2$
$+\frac{1}{2}(1-P_A)\sum_{\lambda_C \lambda_D \lambda_E}|\sum_{\lambda_B}\sum_B BW(m_B)\xi_{B,-\frac{1}{2},\lambda_B,\lambda_C,\lambda_D,\lambda_E}|^2$.

Integrating the contribution of a resonance over the phase space gives

$$F_r = \frac{\int \sum_{m_A,\lambda_B} |BW(m_r)\xi_{r,m_A,\lambda_B}|^2 d\vec{x}}{\int \sum_{m_A,\lambda_B} |\sum_B BW(m_B)\xi_{B,m_A,\lambda_B}|^2 d\vec{x}} \tag{2}$$

from where, after applying weight factors for isospin conservation, we can extract the branching ratio for the resonance.

PRELIMINARY RESULTS

We performed an unbinned maximum likelihood fit with the functions described in the previous section, with 23 free parameters in the fit. The preliminary results are shown in the following tables. Only statistical errors are shown.

	Parameter	Amplitude	Phase
	N_1	1. fixed	0. fixed
nonres.	N_2	320 ± 82	$.2 \pm 1.9$
	N_3	26 ± 56	4.7 ± 1.9
	N_4	200 ± 135	3.2 ± 0.2
$\bar{K}^{*0}$	A_1	495 ± 143	5.6 ± 0.2
	A_2	70 ± 91	3.9 ± 10^3
$\bar{K}^{*0}$	A_3	362 ± 91	3.3 ± 0.1
	A_4	95 ± 69	2.9 ± 481
	B_1	11 ± 23	3.3 ± 10^4
Δ^{++}	B_2	196 ± 87	$3.5 \pm .9$
	C_1	115 ± 142	2.6 ± 18
$\Lambda(1520)$	C_2	644 ± 144	$0.8 \pm .2$

Polarization	$P_A = 0.1 \pm 0.4$

Branching Ratio	
nonres.	0.73 ± 0.29
$\bar{K}^{*0}$	0.14 ± 0.17
Δ^{++}	0.09 ± 0.14
$\Lambda(1520)$	0.04 ± 0.09

The two most significant features in the Dalitz plot are not properly taken into account in this analysis method. The fit function does not include a resonance describing the feature seen in the lower left of the phase space, leading to an over-estimate of the non-resonant contribution, and the central region with a small number of entries is not taken into account in an unbinned fit. At this moment, we are finalizing the analysis procedure, by optimizing the cuts used for extracting the Λ_c^+ signal, a study if additional resonances have to be included, and on a binned fit to included the information about the center region. Also some systematic studies are under way to quantify the significance of the results. This work was supported by CONACyT Mexico under Grant 28435-E.

REFERENCES

1. Juan Medellin Zapata, *Investigación del decaimiento de* $\Lambda_c^+ \to pK^-\pi^+$. Master Thesis, Instituto de Fisica, UASLP, (2002), in preparation.
2. SELEX is a collaboration of: Ball State University, Bogazici University, Carnegie-Mellon University, Centro Brasileiro de Pesquisas Físicas, Fermilab, Institute for High Energy Physics (Protvino), Institute of High Energy Physics (Beijing), Institute of Theoretical and Experimental Physics (Moscow), Max-Planck-Institut für Kernphysik, Moscow State University, Petersburg Nuclear Physics Institute, Tel Aviv University, Universidad Autónoma de San Luis Potosí, Universidade Federal da Paraíba, University of Bristol, University of Iowa, University of Michigan-Flint, University of Rome "La Sapienza" and INFN, University of São Paulo, University of Trieste and INFN.
3. SELEX Collaboration, A. Kushnirenko et al., *Precision measurements of the* Λ_c^+ *and* D^0 *lifetimes*. Phys. Rev. Letter **86**, 5243-5246 (2001).
4. J. Engelfried et al., *The SELEX Phototube RICH Detector*. Nucl. Instr. and Meth. **A431**, 53-69 (1999).
5. George F. Fox, *Multidimensional Resonance Analysis of* $\Lambda_c^+ \to pK^-\pi^+$. Ph.D. Thesis, University of South Carolina (1999).
6. G. Otter, A.M. Freire Endler, *Formalismo da Helicidade e suas Aplicacoes*. Report Number CBPF-MO-002/81 (1981).

Large B–Fields and Noncommutative Solitons[1]

Jorge Moreno[2]

Departamento de Física, Centro de Investigación y de Estudios Avanzados del I.P.N., Apdo. Postal 14-740, 07000, México D.F. México

Abstract. The purpose of this talk is to review a few issues concerning noncommutativity arising from String Theory. In particular, it is shown how in Type IIB Theory, the annihilation of a $D3 - \overline{D3}$ brane pair yields a $D1$–string. This object, in the presence of a large B–field and fermions, happens to be a complex noncommutative soliton endowed with superconductivity.

INTRODUCTION

Type IIB Superstring Theory allows two interesting ingredients: stable-BPS RR-charged Dp-branes (p odd) and a massless antisymmetric $B_{\mu\nu}$ field. Although a $D3 - D3$ brane system is stable; a $D3 - \overline{D3}$ case is not, due to the presence of a tachyon in its spectrum [1]. The job of the B-field in the low energy limit is to introduce noncommutativity [2]. In the $D3 - \overline{D3}$ brane configuration, we may turn on a large B-field along two worldvolume spatial coordinates. The effect of this is that the complex tachyon allows a GMS soliton [3]. This object appears to be superconducting in the presence of fermions arising from the open string sector [4].

B-FIELDS AND NONCOMMUTATIVITY

Consider an open string attached to a D-brane. The OPE has the form

$$e^{ik_1 \cdot X} e^{ik_2 \cdot X} \sim \left(\tau - \tau'\right)^{2\alpha' g^{\mu\nu} k_{1\mu} \cdot k_{2\nu}} \times e^{i(k_1+k_2) \cdot X} + \cdots. \tag{1}$$

However, the introduction of a large B-field alters this to

$$e^{ik_1 \cdot X} e^{ik_2 \cdot X} \sim \left(\tau - \tau'\right)^{2\alpha' G^{\mu\nu} k_{1\mu} \cdot k_{2\nu}} \times \left[e^{-\frac{i}{2}\Theta^{\mu\nu} k_{1\mu} \cdot k_{2\nu}}\right] \times e^{i(k_1+k_2) \cdot X} + \cdots, \tag{2}$$

where $G^{\mu\nu} = \left(\frac{1}{g+2\pi\alpha' B} g \frac{1}{g-2\pi\alpha' B}\right)^{\mu\nu}$ is the effective metric seen by the open string modes, and $\Theta^{\mu\nu} = -(2\pi\alpha')^2 \left(\frac{1}{g+2\pi\alpha' B} B \frac{1}{g-2\pi\alpha' B}\right)^{\mu\nu}$ is the noncommutativity parameter matrix [3]. Likewise, the new term in brackets is known -in configuration space- as the Moyal $*$ product. Thus, in the low energy effective theory, fields get $*$-multiplied.

---

[1] This work is dedicated to my parents Juan Moreno and Martha Soto.
[2] E-mail address: jmoreno@fis.cinvestav.mx

NONCOMMUTATIVE SOLITONS

The idea of GMS solitons was cleverly used in [5] to construct real solutions to the tachyon in bosonic D-branes. In this work, complex tachyons in RR D-brane pairs are considered instead. For a $D3 - \overline{D3}$ brane pair in the presence of a large B-field along the $x-y$ plane, the action of the tachyon is

$$S^{(\Sigma_{3+1})} = \int_{\Sigma_{3+1}} dxdydzdt \left[\overline{D^\mu T} * D_\mu T - V_* \left(T, \overline{T} \right) \right], \tag{3}$$

where $[x,y] = i\theta$, and z and t are commutative coordinates. Also, $D_\mu T = \partial_\mu T - iA_\mu * T + i\widetilde{A_\mu} * T$, where A_μ and $\widetilde{A_\mu}$ are the respective gauge fields in each of the D-brane's Chan-Paton $U(1)$ symmetry groups.

We make three assumptions:

- The potential is a polynomial:

$$V_* \left(T, \overline{T} \right) = \sum_{k=1}^{n} a_k \left(\overline{T} * T \right)^k. \tag{4}$$

- The gauge field $R_\mu \equiv A_\mu - \widetilde{A_\mu}$ has the form:

$$R_\mu = R_\mu(z,t). \tag{5}$$

- We'll focus on solutions of the form:

$$T = T(x,y), \overline{T} = \overline{T}(x,y). \tag{6}$$

Eventually, it is shown that the solution is

$$T = t_* T_0, \overline{T} = \overline{t_*} \overline{T_0}, \tag{7}$$

where t_* and $\overline{t_*}$ solve the equations of motion in the commutative case, while $T_0, \overline{T_0}$ satisfy

$$\left(\overline{T_0} * T_0 \right)^k = \overline{T_0} * T_0. \tag{8}$$

Since $[x,y] = i\theta$ is analogous to $[\hat{q}, \hat{p}] = i$ in quantum mechanics, we identify $T_0 \leftrightarrow i|0\rangle\langle 0|$ and $\overline{T_0} \leftrightarrow -i|0\rangle\langle 0|$ in the Simple Harmonic Oscillator basis.

Applying the Weyl-Wigner-Moyal correspondence yields the following lowest energy solitonic solution:

$$T(x,y) = 2it_* e^{-r^2}, \overline{T}(x,y) = -2i\overline{t_*} e^{-r^2}, \tag{9}$$

where $r^2 = x^2 + y^2$ [4].

THE NONCOMMUTATIVE SUPERCONDUCTING STRING

Given a $(3+1)$ Dirac spinor Ψ, with two-component entries ψ_R and ψ_L obeying $\vec{\sigma} \cdot \hat{p}\psi_R = \psi_R$ and $\vec{\sigma} \cdot \hat{p}\psi_L = -\psi_L$, the action for the spinor coupled to the complex soliton has the form:

$$S^{(\Sigma_{3+1})} = \int_{\Sigma_{3+1}} dt dz dx dy \left[f\left(\overline{T}\right) * \overline{\Psi} * g(T) \slashed{D} \Psi \right], \tag{10}$$

where f and g are polynomials, and $\slashed{D}\Psi = \gamma^\mu \left(\partial_\mu \Psi - i R_\mu * \Psi \right)$.

In terms of operators, we may express our spinors as [6]:

$$\widehat{\psi}_{L,R}(x^\mu) = \sum_{m,n \geq 0} \psi^{L,R}_{mn}(z,t) |m\rangle \langle n|. \tag{11}$$

In order to find the effective theory along the *string* (the z,t coordinates,) we make two assumptions:

- $\theta \longrightarrow \infty$, which means that the noncommutative kinetic part is negligible.
- $\psi \longrightarrow \psi^L$ (as in Witten's superconducting string [7]).

Therefore, the action for the *Noncommutative D-string* is

$$S^{(\Sigma_{1+1})} = -2\pi i \theta f\left(\overline{t_*}\right) g\left(t_*\right) \int_{\Sigma_{1+1}} \left[i \overline{\psi}^L \sigma^a D_a \psi^L - \overline{\psi}^L m \psi^L \right], \tag{12}$$

where ψ^L denotes ψ^L_{00} in (11), m is a "mass" matrix ($m(z,t) = \sigma^\alpha R_\alpha$).[3]

SUPERCONDUCTIVITY

In the massless case (after rescaling the action and getting rid of the unnecessary L subscript)

$$S^{(\Sigma_{3+1})} = \int_{\Sigma_{3+1}} dz dt \left(i \overline{\psi} \sigma^a D_a \psi \right). \tag{13}$$

According to the bosonization technique [7], in two dimensions we can equivalently describe the theory by either bosons or fermions.

This is done by introducing a scalar field $\zeta(z,t)$ such that

$$\overline{\psi} \sigma^a \psi = \frac{1}{\sqrt{\pi}} \varepsilon^{ab} \partial_b \zeta. \tag{14}$$

It can be shown that the kinetic term corresponds to

$$i \overline{\psi} \sigma^a D_a \psi = \frac{1}{2} (\partial_a \zeta)(\partial^a \zeta) - \frac{1}{\sqrt{\pi}} R_a \varepsilon^{ab} \partial_b \zeta, \tag{15}$$

---

[3] $a = z, t$ and $\alpha = x, y$.

which is associated to a conserved current

$$J^a = \partial^a \zeta + \frac{1}{\sqrt{\pi}} \varepsilon^{ab} R_b. \tag{16}$$

However, this current may be expressed in terms of another scalar:

$$J^a = \varepsilon^{ab} \partial_b \varphi. \tag{17}$$

Thus, $\partial_b \varphi = -\varepsilon_{ba} J^a$ and

$$\partial^b \partial_b \varphi = -\partial^b \varepsilon_{ba} \left(\partial^a \zeta + \frac{1}{\sqrt{\pi}} \varepsilon^{ab} R_b \right) = -\frac{1}{\sqrt{\pi}} \partial^b R_b. \tag{18}$$

This means that from $\partial_a \varphi = -\varepsilon_{ac} J^c$, we get that $-\varepsilon_{ac} \partial^a J^c = -\frac{1}{\sqrt{\pi}} \partial^a R_a$. In other words, for J^3 (the current along the string):

$$\frac{dJ^3(z,t)}{dt} = \frac{1}{\sqrt{\pi}} \frac{dR_0(z,t)}{dz}, \tag{19}$$

which has the following solution:

$$J^3(z,t) = \frac{1}{\sqrt{\pi}} [R_0(z,\tau) - R_0(z,\tau_i)]. \tag{20}$$

This means that the current is nondecaying as long as the gauge fields has different values when it's turn on ($t = \tau_i$) and turn off ($t = \tau$).

ACKNOWLEDGMENTS

It is a pleasure to thank my advisor Dr. H. García-Compeán for useful discussions and enjoyable collaboration. I'm very grateful to J.M.L. Bauche A. for helping me write this manuscript. I also thank the organizers of the *VIII Mexican Workshop on Particles and Fields* for their hospitality. This work was supported in part by the CONACyT Fellowship No. 33951E.

REFERENCES

1. J. H. Schwarz, "TASI Lectures on Non-BPS D-Brane Systems", hep-th/9908144.
2. R. Gopakumar, S. Minwalla, and A. Strominger, "Noncommutative Solitons," JHEP **0005** (2000) 020, hep-th/0003160.
3. N. Seiberg and E. Witten, "String Theory and Noncommutative Geometry," JHEP **09** (1999) 032, hep-th/9908142.
4. H. García-Compeán and J. Moreno, "Remarks on Noncommutative Solitons," hep-th/0110119.
5. J. Harvey, P. Kraus, F. Larsen, and E.J. Martinec, "D-branes and Strings as Noncommutative Solitons," JHEP **0007** (2000) 042, hep-th/0005031.
6. L. Pilo and A. Riotto, "The Non-commutative Brane World," JHEP **03** (2001) 015, hep-ph/0012174
7. E. Witten, "Superconducting Strings," Nucl. Phys. B **249** (1985) 557.

Finite size effects in colour superconductivity

P. Amore, N. R. Walet* and M. C. Birse, J. A. Mc Govern[†]

*Dept. of Physics, UMIST, P.O. Box 88, Manchester M60 1QD, UK
[†]Theoretical Physics Group, Department of Physics and Astronomy, University of Manchester, Manchester M13 9PL, U.K.

Abstract. Recently it has been speculated by several authors that quark matter at large density and low temperature can become a color superconductor. This prediction relies on the asymptotically free character of the strong interaction, QCD, and on the general property of fermionic systems, for which the Fermi surface is unstable with respect to an attractive interaction among the fermions, however weak. In our work we have studied the effects of finite size on the two flavour colour superconducting state, by taking into account three different aspects: the boundary conditions felt by the quark fields, which are now constrained in a finite volume, the exact projection of the BCS state over a definite baryon number and the projection of the BCS state over a color singlet. These effects have been evaluated numerically.

INTRODUCTION

In recent years it has been proposed that quark matter at large density can become a colour superconductor: as a matter of fact, the one-gluon exchange in the $\bar{3}$ quark-quark channel or the instanton mediated interaction between quarks provide an attractive interaction, which can make the Fermi surface unstable at large density and lead to the formation of Cooper pairs of quarks.

Due to the richness of the colour $SU(3)$ gauge group compared to the electromagnetic $U(1)$ gauge group, different patterns for the emergence of colour superconductivity are available: these include the two flavor colour superconductor, for brevity **2SC** [1], the colour flavor locked phase, **CFL** [2], and the crystalline colour superconducting phase [3]. A sizable colour superconducting gap, of the order of 100 MeV, is predicted in these calculations. The density at which the pairing is maximum exceeds several times the nuclear density.

We have investigated the effects of finite size on the colour superconducting state. This question is particularly relevant for assessing the properties of small chunks of superconducting matter, which could possibly form — for example — in relativistic heavy ion collisions. We have kept the temperature fixed to zero in this calculation and limited our study to the simplest state, i.e. the **2SC**, using the model of ref. [1]. We have carried out the projection of the BCS wave function onto states with definite baryon number and onto the colour singlet state. The first is needed because the BCS only fixes the average baryon number, without constraining its fluctuations, which indeed become large when the condensate contains few particles; the second is needed to ensure that the superconducting condensate is in a colour singlet state, in compliance with our knowledge of QCD, that only "white", i.e. colour singlet states, are physically observable states.

THE MODEL

Our analysis is based on the model used by Alford, Rajagopal and Wilczek in ref. [1]; we have considered quarks of two flavors, which experience an attractive interaction of the form:

$$\hat{H}_I = -K\,\Xi_{kl;\alpha\beta\gamma\delta}\int d^3x\,\overline{\psi}_{R1\alpha}(x)\,\psi_{Lk\gamma}(x)\,\overline{\psi}_{R2\beta}(x)\,\psi_{Ll\delta}(x) + h.c. \tag{1}$$

where $\Xi_{kl;\alpha\beta\gamma\delta} = \varepsilon_{kl}\,(3\,\delta_{\alpha\gamma}\delta_{\beta\delta} - \delta_{\alpha\delta}\delta_{\beta\gamma})$ is a tensor in the flavor and colour indices (latin and greek indices respectively).

At large density the interaction (1) is responsible for the condensation of Cooper pairs of quarks; in these conditions the system can be described by the BCS wave function:

$$|\Psi\rangle = \hat{G}_L\,\hat{G}_R\,|p_F\rangle, \tag{2}$$

where $\hat{G}_{L,R}$ are the operators introduced in ref. [1]. $|p_F\rangle$ is a Fermi sea of quarks, which is filled up to a Fermi momentum p_F. We assume that $p_F^{red} = p_F^{green} = p_F^{blue}$.

The wave function of eq. (2) describes the condensation of pairs of quarks with the same helicity, but antisymmetric in flavor and in colour, as can be seen by inspection of $\hat{G}_{L,R}$. Given the form of eq. (2), only two colours, which here are chosen, by convention, to be the 1 and 2, enter in the BCS functional, leaving the third colour inactive.

The colour-superconducting gap is then found by minimizing the thermodynamical potential $\hat{H} - \mu\hat{N}$ with respect to the BCS parameters $\theta_{A,B,C}^{L,R}(p)$ and $\xi_{A,B,C}^{L,R}(p)$.

Unfortunately, this approach is inadequate in a finite system. In fact, the BCS wave function of eq. (2) *does not have a definite baryon number* and it *is not in a colour singlet state*. For example, while eq. (2) describes a state with a fixed average baryon number, the fluctuations around this value become larger and larger as the number of particles in the state decreases. In these conditions a projection of the state onto the correct quantum numbers is necessary.

The projections onto definite baryon number and onto a singlet state have been combined in the following form [4]:

$$|\Psi_n\rangle = C_n \int d\Omega_g\,\hat{U}_g \oint \frac{d\zeta}{\zeta^{n+1}}\,\hat{G}_L(\zeta)\,\hat{G}_R(\zeta)\,|p_F\rangle. \tag{3}$$

n here is the number of particle pairs minus the number of antiparticle and hole pairs (which lower the total baryon number). C_n is the normalization constant, which is determined through the requirement $\langle\Psi_n|\Psi_n\rangle = 1$. Notice that in the colour neutral state, n is fixed to 0, as it is explained in ref. [4].

The first integral in eq. (3) takes care of the projection onto the colour singlet. The operator $\hat{U}_g$ induces a rotation in the colour space (which is described by 8 angles), $d\Omega_g$ being the volume element of the $SU(3)$ manifold [6]. The second integral projects the state onto a definite baryon number, $B = 6\,n + 3\,N_F$, where N_F is the number of quarks in the Fermi state $|p_F\rangle$.

The operators $\hat{G}_{L(R)}(\zeta)$ are a generalization of the Bogolubov operators for left (right) handed quarks which have been previously encountered in the BCS treatment. For example, the left-handed operators are given by the expression:

$$\hat{G}_L(\zeta) = \prod_{\alpha,\beta,p} \left[\cos\theta_A^L(p) + \zeta\, \varepsilon_{\alpha\beta 3}\, \sin\theta_A^L(p) e^{i\xi_A^L(p)}\, a_{L1\alpha}^\dagger(p)\, a_{L2\beta}^\dagger(-p)\right]$$

$$\times \prod_{\alpha,\beta,p} \left[\cos\theta_B^R(p) + \frac{1}{\zeta}\, \varepsilon_{\alpha\beta 3}\, \sin\theta_B^R(p) e^{i\xi_B^R(p)}\, b_{R1\alpha}^\dagger(p)\, b_{R2\beta}^\dagger(-p)\right]$$

$$\times \prod_{\alpha,\beta,p} \left[\cos\theta_C^R(p) + \frac{1}{\zeta}\, \varepsilon_{\alpha\beta 3}\, \sin\theta_C^R(p) e^{i\xi_C^R(p)}\, c_{R1\alpha}^\dagger(p)\, c_{R2\beta}^\dagger(-p)\right]$$

$$\equiv \prod_{\alpha,\beta,p} \hat{G}_{LA\alpha\beta}(\zeta,p) \prod_{\alpha,\beta,p} \hat{G}_{LB\alpha\beta}(\zeta,p) \prod_{\alpha,\beta,p} \hat{G}_{LC\alpha\beta}(\zeta,p)\,, \qquad (4)$$

and act on the evenly filled Fermi gas of quarks, $|p_F\rangle$. Full details of this calculation can be found in ref. [4].

RESULTS

Let us now explore the effects of the projections in eq. (3). We consider a system of quarks, enclosed in a finite (cubic) box, of size L, and we fix the coupling constant K to give a constituent mass for the quarks $M = 400$ MeV.

The numerical solutions have been found variationally, assuming the same functional form as in infinite volume and minimizing the total energy with respect to the gap Δ.

In Fig. 1 we have studied the finite size effects upon the superconducting gap (left) and upon the energy per particle (right), while keeping the density fixed and changing the size of the box. The density has been chosen to correspond roughly to the values for which the pairing is strongest in the infinite volume.

At each volume we have considered three different cases: the effect of the boundary conditions alone (circles), the effect of baryon number projection alone (triangles up) and the effect of the combined baryon number and colour projection (triangles down). This last case is described by the wave function (3) with $n = 0$.

By looking at the left plot, we observe that the projected results (triangles) differ noticeably from the corresponding unprojected ones, at small volumes ($L = 3$ fm): as a matter of fact, the fluctuations in the particle number are large in this regime and the BCS approximation breaks down. The effect of the projection onto colour, which can be appreciated by comparing the two cases (triangles up and down), is sensible only for small volumes.

For large volumes, $L \approx 10$ fm, the gap is seen to converge rapidly towards a constant value. The situation is somewhat different for the energy, where residual oscillations persist: these are due to the fact that in a finite box, even a large one, the Fermi distribution of the unpaired quarks has discontinuities.

If Fig. 2 we have kept the volume fixed and varied the number of particles in the condensate. We have compared these results with the corresponding cases in the infinite volume (continuous line) and with the finite volume, unprojected case (dashed line). The chemical potential μ is chosen to yield for the finite system a BCS condensate with the same number of particles (on average) as the Fermi gas, upon which it is built on. This ensures that the unprojected BCS state has a substantial overlap with the projected state which we want to build.

We notice that the effect of the projection is large at small volumes ($L = 3$ fm) and yields a sizable gap (≈ 50 MeV). At larger volumes ($L = 6$ fm) some of these effects persist, to a smaller extent, in the region of low density/small number of particle.

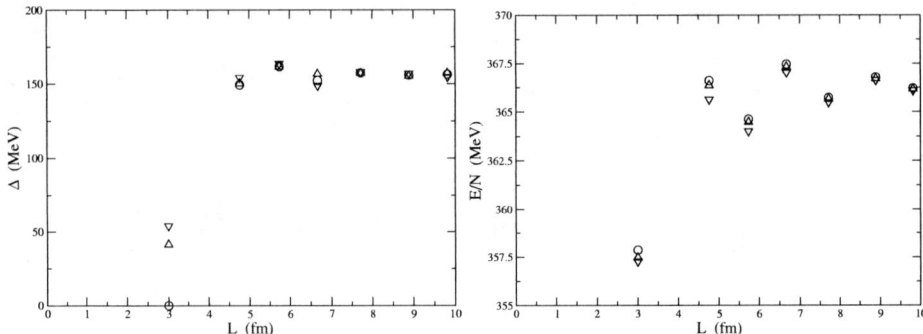

FIGURE 1. Left: Superconducting gap at a fixed density of $\rho = 3.5$ fm^{-3} for the finite box (circles), number projected (triangle up) and number-colour projected (triangle down). Right: Energy per particle at a fixed density of $\rho = 3.5$ fm^{-3} for the finite box (circles), number projected (triangle up) and number-colour projected (triangle down).

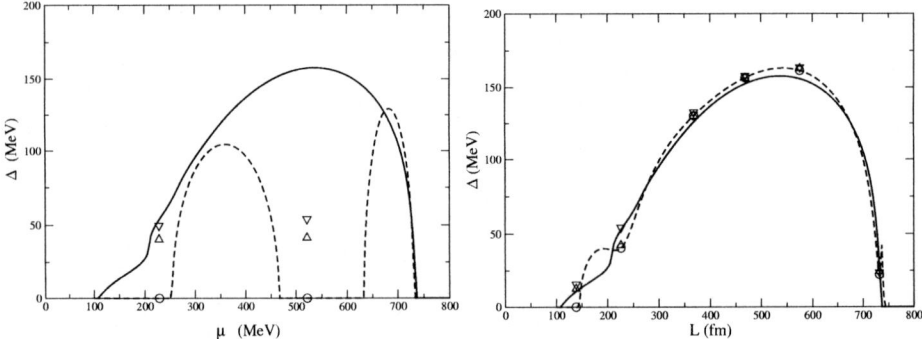

FIGURE 2. Left: Gap at a fixed size of $L = 3$ fm for the finite box (circles), number projected (triangle up) and number-colour projected (triangle down). The solid line is the infinite volume result. Right: same for $L = 6$ fm.

REFERENCES

1. M. Alford, K. Rajagopal and F. Wilczek, Phys. Lett. B **422**, 247 (1998)
2. M. Alford, K. Rajagopal and F. Wilczek, Nucl. Phys. B **537**, 443 (1999)
3. M. Alford, J. Bowers, K. Rajagopal, Phys. Rev. D **63**, 074016 (2001)
4. P. Amore, M. C. Birse, J. A. McGovern and N. Walet, submitted to Phys. Rev. D, hep-ph/0110267
5. K. Dietrich, H. Mang and J. Pradal, Phys. Rev. **135B**, 22 (1964)
6. M.S. Byrd and E.C.G. Sudarshan, J.Phys. **A31**,(1998) 9255

Deconfined matter and Λ^0 polarization in ultra-relativistic heavy-ion collision

Alejandro Ayala[1], Eleazar Cuautle[1], Gerardo Herrera[2], Luis Manuel Montaño[2]

[1]*Instituto de Ciencias Nucleares, Universidad Nacional Autónoma de México, Apartado Postal 70-543, México Distrito Federal 04510, México.*
[2]*Centro de Investigación y de Estudios Avanzados, Apartado Postal 14-740, México Distrito Federal 07300, México.*

Abstract. Polarization of Λ^0's produced in ultra-relativistic heavy-ion collisions has been calculated and the results compared to the polarization observed in proton-proton collisions and studied as a signal for the production of deconfined matter. The Λ^0's produced in this environment come from two different regions within the interaction zone, in the first one, where the density of participants in the collision is larger than a certain critical density, quarks are free and the Λ^0's are produced unpolarized. The second zone, where the density is below a certain critical density, Λ^0's are produced polarized. In the model we also incorporate the effects from multiple scattering within the nuclear environment.

INTRODUCTION

The main goal to study collisions at relativistic and ultra-relativistic energies is the expectation of a transition from ordinary nuclear matter to deconfined matter, *i.e.*, a Quark Gluon Plasma (QGP), which should be observed when conditions of sufficiently high baryonic densities and/or temperature are achieved during the collisions. A combination of experimental observables has been proposed [1] to identify this transition. One of the first proposed signatures to unveil the production of QGP in relativistic collisions was to study the change in the polarization properties of Λ^0 hyperons as compared to that observed in proton-proton collision [2, 3]. As is well known, in high-energy proton-induced reaction, the produced hyperons exhibit a strong polarization. A quantitative description of hyperon polarization properties has been attained by a semi-classical model put forward by DeGrand and Miettinen [4]. In this model, the hyperon polarization is due to a Thomas precession effect during the quark recombination process of slow (sea) s-quark and fast (valence) *ud*-diquarks.

We study a model with a collision region of high enough density for the production of deconfined matter where the quarks are free and consequently the chaotic quark spin orientation produces Λ^0's unpolarized, while in the rest of the interaction zone Λ^0's are produced polarized in the same way as they are in free $N - N$ interactions. Polarization in this environment is treated with De Gran-Miettinen Model [4]. Depolarization effects arising from multiple scattering are taken into account with a sequential model of collisions [5, 6].

NUCLEUS + NUCLEUS COLLISIONS

In the nucleus + nucleus collisions, the effects that can possibly produce a diminishing of the Λ^0 polarization are: secondary Λ^0's produced by pion-nucleon scattering, secondary scattering of leading Λ^0's with nucleons within the interaction zone and Λ^0 production from a QGP formation. Secondary scattering with nucleons can influence the final polarization measurements by producing a shift in the Λ^0 longitudinal and transverse momentum and a flip in the original Λ^0 spin direction. Using a sequential model for collisions [5, 6] and in the high-energy limit ($p_{\Lambda^0} \gg p_T$) the average values for the Λ^0 fraction of longitudinal and transverse momentum are (see [7]):

$$\langle x(z,b) \rangle = x e^{-I\bar{N}(z,b)}, \tag{1}$$

$$\langle p_T(z,b) \rangle \simeq p_T e^{-I\bar{N}(z,b)} \cos\left(\Gamma \sqrt{\bar{N}(z,b)}\right), \tag{2}$$

where z is the distance traveled by the Λ^0 within the nucleus, $\bar{N}$ the average number of collision suffered by Λ^0, b the impact parameter, I the inelasticity coefficient, Γ the dispersion angle per collision and x and p_T are the original longitudinal and transverse momentum.

The degree of polarization in a single Λ^0-nucleon collision is quantified in terms of the polarization transfer coefficient D. In the case of forward scattering, the final average polarization can be expressed as

$$\langle \mathcal{P}'(z,b) \rangle = \mathcal{P} e^{-\bar{N}(z,b)(1-D)} \tag{3}$$

where D is a function of $\bar{V}$, S_Λ, S_N and Δ, which represent the amplitudes for the spin-independent, Λ^0 spin-orbit, nucleon spin-orbit and spin-spin interactions, respectively, appearing in the expression for the two-body Λ^0-nucleon potential.

Λ^0 PRODUCTION AND POLARIZATION

Λ^0's produced in nuclear collisions come from two different sources: from the zone with high enough density of participants (n_c) for QGP formation, where the differential cross sections is [7]

$$\frac{d^2\sigma_{\Lambda^0}^{QGP}}{d^2b} = c \left[\int n_p(\mathbf{s},\mathbf{b}) \theta[n_p(\mathbf{s},\mathbf{b}) - n_c] d^2s \right]^2 \tag{4}$$

and from a zone with density of participants (n_p) below n_c, where Λ^0's are produced by recombination with a differential cross section

$$\frac{d^2\sigma_{\Lambda^0}^{REC}}{d^2b} = \sigma_{\Lambda^0}^{NN} \int T_B(\mathbf{b}-\mathbf{s}) T_A(\mathbf{s}) \theta[n_c - n_p(\mathbf{s},\mathbf{b})] d^2s, \tag{5}$$

where c is the normalization constant obtained from experimental results and θ is the step function.
The polarization defined as

$$\mathcal{P}(b) = \left[\frac{d^2\sigma_{\Lambda^0\uparrow}}{d^2b} - \frac{d^2\sigma_{\Lambda^0\downarrow}}{d^2b}\right] \bigg/ \left[\frac{d^2\sigma_{\Lambda^0\uparrow}}{d^2b} + \frac{d^2\sigma_{\Lambda^0\downarrow}}{d^2b}\right], \qquad (6)$$

within our model receives contributions only from Λ^0 produced in the latter region. Defining the function $f(b)$ and polarization from recombination $\mathcal{P}^{\text{REC}}(b)$ as

$$f(b) = \left[\frac{d^2\sigma_{\Lambda^0}^{\text{QGP}}}{d^2b}\right] \bigg/ \left[\frac{d^2\sigma_{\Lambda^0}^{\text{REC}}}{d^2b}\right], \qquad (7)$$

$$\mathcal{P}^{\text{REC}}(b) = \left[\frac{d^2\sigma_{\Lambda^0\uparrow}^{\text{REC}}}{d^2b} - \frac{d^2\sigma_{\Lambda^0\downarrow}^{\text{REC}}}{d^2b}\right] \bigg/ \left[\frac{d^2\sigma_{\Lambda^0\uparrow}^{\text{REC}}}{d^2b} + \frac{d^2\sigma_{\Lambda^0\downarrow}^{\text{REC}}}{d^2b}\right], \qquad (8)$$

the final polarization can be written as

$$\mathcal{P}(b) = \frac{\mathcal{P}^{\text{REC}}(b)}{[1 + f(b)]}, \qquad (9)$$

When we incorporate the Λ^0 shift in the momentum for densities below the critical density for QGP formation, the Λ^0's produced within a given phase space cell (x, p_T), are scattered into a different phase space cell, which, on average, is $(\langle x \rangle, \langle p_T \rangle)$. This shift in the momentum is reflected in the polarization, as is shown in Fig. 1.

CONCLUSIONS

We have studied the change of Λ^0 polarization in high-energy heavy-ion collisions with respect to the polarization observed in proton-proton collision, as a probe for production of deconfined matter. Multiple scattering effects have been taken into account because for those Λ^0's traveling inside the nuclear matter. These effects are a shift in the momentum and flip in the original spin of the Λ^0 hyperon.

Depolarization effects are expressed in terms of four parameters: inelasticity coefficient I, average scattering angle per collision Γ, depolarization coefficient D and impact parameter b. The first two were calculated from experimental results and extrapolated to relativistic energies.

ACKNOWLEDGMENTS

This work has been supported in part by the PAPIIT-UNAM under project number IN10800 and by CONACyT under project ICM, 35792-E.

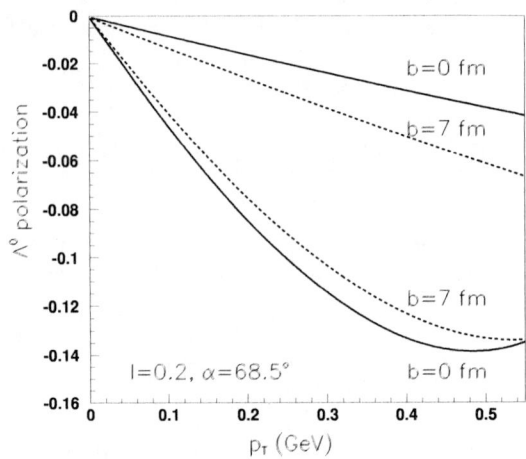

FIGURE 1. Λ^0 polarization as function of p_T for two values of impact parameter b without (upper curves) and with (lower curves) multiple scattering effects, for angle per collision of 68.5^0 and inelasticity parameter, $I = 0.2$, obtained from experimental results.

REFERENCES

1. For a recent review on the subject, see U.W. Heinz, *Hunting Down the Quark Gluon Plasma in Relativistic Heavy-Ion Collisions*, Proceedings of the Conference on Strong and Electroweak Matter (SEWM 98), Copenhagen Denmark, Eds. J. Ambjorn, P. Damgaard, K. Kainulainen and K. Rummukainen (World Scientific Publ. Co., Singapore, 1999), 81.
2. N. Angert *et al.*, Proceedings of the Conference on Quark Matter Formation and Heavy Ion Collisions, Eds. M. Jacob and H. Satz (World Scientific Publ. Co., Singapore, 1982), 557.
3. A.D. Panagiotou, *Phys. Rev.* **C 33**, 1999 (1986), Int. J. Mod. Phys. **A 5**, 1197 (1990).
4. T.A. DeGrand and H.I. Miettinen, *Phys. Rev.* **D 23**, 1227 (1981), *ibid* **24**, 2419 (1981); T. Fujita and T. Matsuyama, *ibid* **38**, 401 (1988); T.A. DeGrand, *ibid* **38**, 403 (1988).
5. R.C. Hwa, *Phys. Rev. Lett.* **52**, 492 (1984).
6. L.P. Csernai and J.I. Kapusta, *Phys. Rev.* **D 29**, 2664, *ibid* **31** (1985).
7. A. Ayala, E. Cuautle, G. Herrera, L. M. Montaño preprint: nucl-th/0110027, to appear in *Phys. Rev. C*

Boundary and expansion effects on two-pion correlation function in relativistic heavy-ion collisions

Angel Sánchez and Alejandro Ayala

Instituto de Ciencias Nucleares, Universidad Nacional Autónoma de México, Apartado Postal 70-543, México Distrito Federal 04510, México.

Abstract. We examine the effects that a confining boundary together with hydrodynamical expansion play on two-pion distribution in relativistic heavy-ion collisions. We show that the effects arise from the introduction of further correlations due both to collective motion and the system's finite size. As is well know, the former leads to a reduction in the apparent source radius with increasing and large average pair momentum K. However, for small K, the presence of boundary leads to a decrease of the apparent source radius with decreasing K. These two competing effects produce a maximum at a finite value of the effective source radius as a function of K.

INTRODUCTION

The most abundantly produced hadrons in relativistic heavy-ion collisions are pions. Typically, the number of pions produced one unit around central rapidity in central Au+Au collisions at energies of order $10A$ GeV is $dN_\pi/dy \sim 300$ [1]. Under the assumption that the transverse dimensions of the system formed at freeze out are of order of the transverse size of an Au nucleus and that the typical pion formation time is of order 1 fm, this multiplicity implies that the average pion separation d at freeze out in the central rapidity region is of order $d \sim 0.6$ fm which is less that the average range of the pion strong interaction, $d_s \sim 1.4$ fm. The pions in such system are strongly interacting with each other and they are moving in an attractive mean field potential which extends over the pion system. The attractive mean-field leads to a quasibound pion system which can be considered to be in a liquid phase with a surface boundary, rather than in a gas phase [2]. An important missing ingredient in the description of the transverse pion spectra within a boundary model was the proper inclusion of hydrodynamical expansion. The phenomenological description in terms of a bound, expanding pion system has been named the *expanding pion liquid* model and was development in Refs. [3], [4] and successfully applied to the description of the experimental mid-rapidity, transverse pion spectra in central Au+Au collisions at 11.6A GeV/c [1]. A further natural test ground for the model is the study of multiparticle correlations in particular two-pion correlations.

THE EXPANDING PION LIQUID.

To incorporate the effects of hydrodynamical flow, we notice that this ordered motion can be represented by a four-velocity field $u^\mu = \gamma(r)[1, \mathbf{v}(r)]$, where γ is the Lorentz-gamma factor and $\mathbf{v}(r)$ is the velocity vector. This field represents a redistribution of the momentum in each of the fluid cells, as viewed from a given reference frame (the center of mass in this case), becoming centered around the momentum associated with the velocity of the fluid element. This behavior can be described by the substitution $p^\mu \to p^\mu - mu^\mu$ where m is the pion mass. The term mu^μ represents the collective momentum of the given fluid element.

The stationary states for an spherical symetric source are given by (see Ref [4] for more details).

$$\psi_{nlm'}(\mathbf{r},t) = \frac{A_{nl}}{\sqrt{2E_{nl}}} e^{-iE_{nl}t} e^{im\beta r^2/(2R)} Y_l^{m'}(\hat{\mathbf{r}}) e^{-\alpha_{nl}^2 r^2/2} r^l$$
$$\times \, _1F_1\left(\frac{(l+3/2)}{2} - \frac{\varepsilon_{nl}^2}{4\alpha_{nl}^2}, l+3/2, \alpha_{nl}^2 r^2\right) \quad (1)$$

where β represents the surface fireball hydrodynamical velocity, $_1F_1$ is a confluent hypergeometric function and $Y_l^{m'}$ is a spherical harmonic. The system's finite size and the strength of the collective expansion are given in terms of the parameters R and β, respectively.

TWO-PION CORRELATION FUNCTION

The two-pion correlation function C_2 can be written as (see Ref [5])

$$C_2(\mathbf{p_1},\mathbf{p_2}) = 1 + \frac{\left|\sum_\lambda 2E_\lambda N_\lambda \psi_\lambda^*(\mathbf{p_1})\psi_\lambda(\mathbf{p_2})\right|^2}{\sum_\lambda 2E_\lambda N_\lambda |\psi_\lambda(\mathbf{p_1})|^2 \sum_\lambda 2E_\lambda N_\lambda |\psi_\lambda(\mathbf{p_2})|^2} \quad (2)$$

where P_1 and P_2 are the one and two-pion momentum distributions, respectively (see Refs. [5] and [6] for more details). Notice that, as a consequence of the factorization assumption, the correlation function in such that $C_2(\mathbf{p},\mathbf{p}) = 2$. This is usually referred to as the completely chaotic pion production scenario [7], which is the situation expected to occur in a heavy ion collision, given the considerable rescattering experienced by pions in the production region. In contrast, if the particles were produced completely coherently, they would occupy a pure quantum state and the two-pion momentum distribution would be simply the product of two single-pion momentum distributions, leading to the absence of the **HBT** effect. For the spherically symmetric situation described here, C_2 depends on the angle between the momenta p_1 and p_2. For simplicity, we choose p_2 parallel to p_1. The behavior of C_2 as we vary the paramenters in the model is shown in Fig. 2 of Ref. [6].

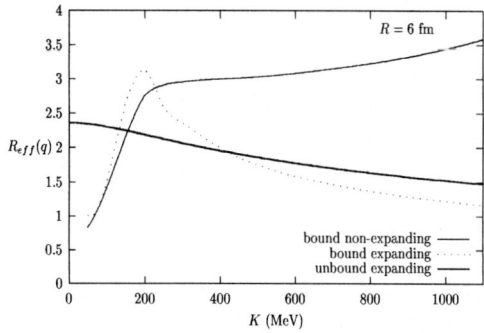

FIGURE 1. R_{eff} for a fixed value $R = 6$ fm as a function of K for an expanding and bound system (solid line), and exxpanding and unbound system (dotted line) and a bound and non-expanding system (thick solid line).

THE EFFECTIVE RADIUS.

Another property of the bound and expanding system of pions that can be extracted from the correlation function $C_2(q)$ is the behavior of the system's effective radius R_{eff} as a function of K, the average momentum of the pion pair. The relevant quantity to pay attention to is the ratio $\eta = T/\gamma(R)m\beta$ of the energy scale asssocieted with random motion, *i.e.* T, to the energy scale associated with ordered motion, *i.e.* $\gamma(R)m\beta$. For $K > \eta T$, the collective motion dominates over the thermal component in K and the relevant physical effects that dictates the behavior of R_{eff} is the correlation between the spatial region of emission of pions and the pair momentum introduced by the collective expansion, in such a way that faster pions are more likely to be emitted from points close in space [8] leading to a reduction in size of the apparent region of particle emission. The behavior of R_{eff} as a function of K is shown in Fig. 1 (solid line). Notice that the curve shows a maximum for a value of $K \sim \eta T$. The curve is obtained by fitting the correlation functions $C_2(q)$ to Gaussians $g(q) = 1 + \lambda e^{-q^2 R_{eff}^2}$.

For comparison, also shown in Fig. 1 is the behavior of R_{eff} for an expanding system without a boundary (dotted line) and for a bound system without the effects hydrodynamical expansion (thick solid line). For the bound and non-expanding system, the eigenfunctions are given in terms of Bessel functions of the first kind [3] (see also Ref. [5]). The corresponding expression for R_{eff} is obtained from that of $C_2(q)$ by also fitting Gaussians.

CONCLUSIONS

We have studied the effects that a confining boundary together with hydrodynamical expansion at freeze out, play on the two-pion correlation function, in the context of relativistic heavy-ion collisions. We have argued that the confining boundary could be produced as a consequence of the high pion density that can be achieved at freeze out in central collisions.

Since the importance of the correlation analyses rests basically on the information that in can provide about the physical size of the system produced during the collision, a main result of the present work is the functional dependence of the effective system radius R_{eff} with the magnitude of the average pair momentum K. We have shown that the interplay of the energy scales associated with collective and random motion, $\gamma(R)m\beta$ and T, respectively, produce a maximum for R_{eff} at a value $K \sim \eta T$. The physical origins of this behavior are the combined effects of the confining boundary and hydrodynamical expantion.

This behavior signals that the pion system feels the presence of the boundary at freeze out and can be used to experimentaly determine such effects in relativistic heav-ion collisions.

ACKNOWLEDGMENTS

Support for this work has been received by DGAPA-UNAM under grant number IN108001.

REFERENCES

1. E-802/866 Collacboration, L. Ahle *et al.*, Phys. Rev. C **57**, R466 (1998).
2. E.V. Shuryak, Phys. Rev. D **42**, 1764 (1990).
3. A. Ayala and A. Smerzi, Phys. Lett. **B405**, 20 (1997).
4. A. Ayala, J. Barreiro and L.M. Montaño, Phys. Rev. C **60**, 014904 (1999); A. Ayala. Rev. Mex. Fis. **45** S2, 116 (1999).
5. Q.H. Zhang and S.S. Padula, Phys. Rev. C **62**, 024902 (2000).
6. A. Sánchez and A. Ayala, Phys. Rev. C. **63**, 064901 (2001).
7. M. Gyulassy, S. Kauffmann and L. Wilson, Phys. Rev. C **20** 2267 (1979).
8. S. Pratt, Phys. Rev. Lett. **53**, 1219 (1984)

Ferromagnets and antiferromagnets in the effective Lagrangian perspective

Christoph P. Hofmann

Instituto de Física, Universidad Autónoma de San Luis Potosí, Alvaro Obregón 64, San Luis Potosí, S.L.P. 78000, Mexico

Abstract. Nonrelativistic systems exhibiting collective magnetic behavior are analyzed within the framework of effective Lagrangians. The method, which formulates the dynamics of the system in terms of Goldstone bosons, allows to investigate the consequences of spontaneous symmetry breaking from a unified point of view. Analogies and differences with respect to the Lorentz-invariant situation (chiral perturbation theory) are pointed out. We then consider the low-temperature expansion of the partition function both for ferro- and antiferromagnets, where the spin waves or magnons represent the Goldstone bosons of the spontaneously broken symmetry $O(3) \to O(2)$. In particular, the low-temperature series of the staggered magnetization for antiferromagnets and the spontaneous magnetization for ferromagnets are compared with the condensed matter literature.

In condensed matter physics, spontaneous symmetry breaking is a common phenomenon and effective field theory methods are widely used in this domain. Only recently, however, has chiral perturbation theory – the effective theory of the strong interactions – been extended to such nonrelativistic systems.[1] The method applies to any system where the Goldstone bosons are the only excitations without energy gap. The essential point is that the properties of these degrees of freedom and their mutual interactions are strongly constrained by the symmetry inherent in the underlying model – the specific nature of the underlying model itself, however, is not important.

In the following presentation, our interest is devoted to the low-energy analysis of nonrelativistic systems, which exhibit collective magnetic behavior. The Heisenberg Hamiltonian is invariant under a simultaneous rotation of the spin variables, described by the symmetry group G = O(3), whereas the ground states of ferro- and antiferromagnets break this symmetry spontaneously down to H = O(2). The corresponding Goldstone modes are referred to as spin waves or magnons. Note that, in contrast to the relativistic version of Goldstone's theorem, the theorem does now neither specify the exact form of the dispersion relation at large wavelengths, nor does it determine the number of different Goldstone *particles*: these features of the Goldstone degrees of freedom are not fixed by symmetry considerations alone – rather, in the case of a Lorentz-noninvariant ground state, they depend on the specific properties of the corresponding nonrelativistic systems. Only the number of real Goldstone *fields* turns out to be universal, given by the dimension of G/H.

Indeed, it is well known that the structure of the ferromagnetic dispersion relation is quite different from the antiferromagnetic one: at large wavelengths, the former takes a *quadratic* form, whereas the latter follows a *linear* law. The mechanism which leads to this pattern and, at the same time, explains the different number of independent magnon states – *one* for a ferromagnet, *two* for an antiferromagnet – is understood. Remarkably, in the framework of the effective description, the difference is related to the value of a single observable, the spontaneous magnetization Σ.[1, 2]

In the leading order effective Lagrangian of a *ferromagnet*, the spontaneous magnetization shows up as a coupling constant associated with a topological term involving a single time derivative,

$$\mathcal{L}_{eff}^{F} = \Sigma \frac{\partial_0 U^1 U^2 - \partial_0 U^2 U^1}{1+U^3} + \Sigma f_0^i U^i - \tfrac{1}{2} F^2 D_r U^i D_r U^i. \tag{1}$$

In the above notation, the two real components of the magnon field, $U^a (a=1,2)$ have been collected in a three-dimensional unit vector $U^i = (U^a, U^3)$. The quantity f_0^i involves the magnetic field H: $f_0^i = \mu H \delta_3^i$. At leading order of the low-energy expansion, the ferromagnet is thus characterized by two low-energy coupling constants, Σ and F. Note that the corresponding equation of motion (Landau-Lifshitz equation) is of the Schrödinger type: first order in time, but second order in space. As only positive frequencies occur in its Fourier decomposition, a complex field is required to describe one particle – in a ferromagnet there exists only *one* type of spin-wave excitation exhibiting a quadratic dispersion law.

The ground state of an antiferromagnet, on the other hand, does not exhibit spontaneous magnetization, such that the leading order effective Lagrangian takes the form

$$\mathcal{L}_{eff}^{AF} = \tfrac{1}{2} F_1^2 D_0 U^i D_0 U^i - \tfrac{1}{2} F_2^2 D_r U^i D_r U^i + \Sigma_s \mu h^i U^i, \quad D_\mu U^i = \partial_\mu U^i + \varepsilon_{ijk} f_\mu^j U^k. \tag{2}$$

Note that the anisotropy field $\vec{h}$ couples to the staggered magnetization Σ_s. The corresponding equation of motion now is of second order both in space and in time, its relativistic structure determining the number of independent magnon states: the Fourier decomposition contains both positive and negative frequencies, such that a single real field suffices to describe one particle. Accordingly, there exist *two* different types of spin-wave excitations in an antiferromagnet – as is the case in Lorentz-invariant theories, Goldstone fields and Goldstone particles are in one-to-one correspondence. These low-energy excitations follow a linear dispersion relation, with the velocity of light replaced by the spin-wave velocity $v = F_2/F_1$. As is commonly done with the velocity of light in relativistic theories, we may put the spin-wave velocity to one. In this "$\hbar = v = 1$"-system the two coupling constants F_1 and F_2 then coincide: $F_1 = F_2 \equiv \mathcal{F}$.

Accordingly, the low-energy properties of ferromagnets and antiferromagnets are quite different. As an illustration, let us consider the low-temperature expansion for the corresponding order parameters, the spontaneous and staggered magnetization, respectively.

The order parameter for an O(N) antiferromagnet, the staggered magnetization, is given by the derivative of the free energy density with respect to the anisotropy field,[3] $\Sigma_s(T) = -\partial z/\partial h$:

$$\Sigma_s(T) = \Sigma_s \left\{ 1 - \frac{N-1}{24} \frac{T^2}{\mathcal{F}^2} - \frac{(N-1)(N-3)}{1152} \frac{T^4}{\mathcal{F}^4} \right. \tag{3}$$
$$\left. - \frac{(N-1)(N-2)}{1728} \frac{T^6}{\mathcal{F}^6} \ln\frac{T_\Sigma}{T} + O(T^8) \right\}.$$

The terms of order T^0, T^2, T^4 and T^6 arise from tree-, one-loop, two-loop and three-loop graphs, respectively. Up to and including T^6, the coefficients are determined by the constant $\mathcal{F}$ which thus sets the scale of the expansion. The logarithm only shows up at order T^6: the scale T_Σ involves next-to-leading order coupling constants.

Let us first consider the particular case N=4. The two groups O(4) and O(3) are locally isomorphic to SU(2) × SU(2) and SU(2), respectively. Hence, the above three-loop formula referring to the order parameter of an O(4) antiferromagnet in zero external field in fact describes the low-temperature expansion of the quark condensate of massless QCD with two flavors. This nicely illustrates the concept of *universality*: in the construction of effective Lagrangians only the mathematical structure of the groups G and H, associated with the spontaneously broken symmetry, is relevant, whereas the specific properties of the underlying model merely manifest themselves in the numerical values of the coupling constants.

Remarkably, for N=3, the T^4-term in the above formula drops out, such that we end up with the following low-temperature series for the staggered magnetization of the O(3) antiferromagnet:

$$\Sigma_s(T) = \Sigma_s \left\{ 1 - \frac{1}{12} \frac{(k_B T)^2}{\hbar v \mathcal{F}^2} - \frac{1}{864 \hbar^3 v^3 \mathcal{F}^6} (k_B T)^6 \ln \frac{T_\Sigma}{T} + O(T^8) \right\}. \quad (4)$$

Note that we have restored the dimensions: k_B is Boltzmann's constant and v is the spin-wave velocity.

The microscopic calculation agrees with the above effective expansion up to order T^2, provided that the two coupling constants $\mathcal{F}$ and Σ_s are identified as

$$\mathcal{F}^2 = \frac{S - \sigma}{\sqrt{2z}} \frac{\hbar v}{a^2} = 2S(S - \sigma) \frac{|J|}{a}, \quad \Sigma_s = \frac{g\mu_B (S - \sigma)}{a^3}. \quad (5)$$

The expression involves the following quantities: the exchange integral (J), the highest eigenvalue of the spin operator S_n^3 (S), the number of nearest neighbors of a given lattice site (z), the length of the unit cell (a), the Landé factor (g), the "Anderson factor" (σ) and the Bohr magneton (μ_B). Note that the spin-wave velocity v is given by the following combination of microscopic quantities,

$$v = 2|J|S\sqrt{2z}a/\hbar. \quad (6)$$

The scale of the low-temperature expansion is set by $\mathcal{F}\sqrt{\hbar v}$ – let us briefly estimate its value. Written in terms of the exchange integral J, we obtain

$$\mathcal{F}\sqrt{\hbar v} = 2|J|S\sqrt{(S-\sigma)\sqrt{2z}}. \quad (7)$$

Now, for a simple cubic lattice ($z=6, \sigma=0.078$) and for $S=1/2$, the double square root on the right hand side is approximately equal to one, such that we end up with $\mathcal{F}\sqrt{\hbar v} \approx |J|$. Typically, the exchange integral for antiferromagnets is around $|J| \approx 10^{-3} eV$, and the scale $\mathcal{F}\sqrt{\hbar v}$ thus of the same order of magnitude. This is to be contrasted with the situation in QCD, where the relevant quantity, $F_\chi \sqrt{\hbar c}$, takes the value $92 MeV$ – the respective scales in the two theories thus differ in about eleven orders of magnitude.

As far as subleading terms in the expansion of the staggered magnetization are concerned, it is well known that a T^4-contribution is absent: the spin-wave interaction only manifests itself at higher orders. However, the logarithmic dependence on the temperature is not found in a microscopic calculation. We conclude that it is extremely difficult to calculate the corrections of order T^6 in the framework of a microscopic theory.

Let us now turn to the ferromagnet. The low-temperature expansion for the spontaneous magnetization is given by the derivative of the free energy density with respect to the magnetic field and takes the form[4]

$$\Sigma(T)/\Sigma = 1 - \alpha_0 T^{\frac{3}{2}} - \alpha_1 T^{\frac{5}{2}} - \alpha_2 T^{\frac{7}{2}} - \alpha_3 T^4 + O(T^{\frac{9}{2}}). \tag{8}$$

The coefficients α_n are independent of the temperature and involve the various coupling constants occurring in the effective Lagrangian, which phenomenologically parametrize the microscopic detail of the system.

In the above series, half-integer powers of the temperature correspond to *noninteracting* magnons; these contributions can be absorbed into a redefinition of the dispersion relation. Remarkably, the leading term describing the magnon-magnon *interaction* (two-loop graph) is of order T^4, and clearly confirms Dyson's microscopic calculation.[5] The effective Lagrangian technique, however, proves to be more efficient than conventional condensed matter methods as the analysis can be carried to higher orders: the calculation shows that the next interaction term, which arises at the three-loop level, is of order $T^{9/2}$.

The effective Lagrangian method is also more transparent, since it addresses the problem from a model-independent point of view based on symmetry – at large wavelengths, the microscopic structure of the system only manifests itself in the numerical values of a few coupling constants.

REFERENCES

1. H. Leutwyler, Phys. Rev. D **49**, 3033-3043 (1994).
2. C.P. Hofmann, Phys. Rev. B **60**, 388-405 (1999).
3. C.P. Hofmann, Phys. Rev. B **60**, 406-413 (1999).
4. C.P. Hofmann, cond-mat/0106492, to appear in Phys. Rev. B.
5. F.J. Dyson, Phys. Rev. **102**, 1217-1230, 1230-1244 (1956).

Weak Interactions Effect on the P-N Mass Splitting and the Principle Of Equivalence

N. Chamoun* and H. Vucetich[1][†]

*Department of Physics, Higher Institute for Applied Sciences and Technology, Box 31983, Damascus, Syria.
†IFUNAM, UNAM, México

Abstract. The neutron-proton mass difference is computed from a model-indepentent sum rule. When this contribution is included in the analysis of the Eötvös experiment, the bound for possible weak interactions violations to the equivalence principle is improved by one order of magnitude from 10^{-2} to 10^{-3}.

The Principle of Equivalence (EP) is the physical basis of General Relativity. It loosely states that any freely falling reference frame is locally equivalent to an inertial reference frame [1]. This is a very strong statement: its unrestricted validity leads to General Relativity as the unique theory for the gravitational field [3] and experimental tests of its consequences probe deeply the structure of gravitation.

One of the consequences of EP is the Universality of Free Fall (UFF) which states that the world line of a body submerged in a gravitational field is independent of its composition and structure [2]. UFF, among the consequences of EP, is one of the strongest tests of its validity. For instance, it has been shown that sufficiently sensitive related experiments can provide strict tests on superstring theories (see, eg. [4]) or Kaluza-Klein theories (eg. [5]), thus exhibiting the presence of "new physics". Indeed, the STEP satellite experiment [6] will improve these tests sensitivity by many orders of magnitude.

One of the profound consequences of UFF is that all forms of non-gravitational energy should couple in the same way to the gravitational field. Any violation of UFF should break equation $m_I = m_P$ and the difference between inertial m_I and passive gravitational mass m_P of a body could be expressed via phenomenological parameters Γ_i specific to each form of interaction and reflecting its degree of violation to the equivalence principle:

$$m_I - m_P = \Delta m = \sum_i \Gamma_i E_i \qquad (1)$$

where the binding energies E_i are usually estimated with the semiempirical mass formula [9] or, in the case of weak interactions, a suitable generalization [10, 11].

---

[1] On leave of absence, Observatorio Astronómico, Universidad Nacional de La Plata, Paseo del Bosque S/N, CP 1900 La Plata, Argentina.

Eötvös experiments [2, 7, 8] set an upper limit on the difference of acceleration in a gravitational field for different materials and so impose an upper bound on the violation parameters Γ_i. The parameter Γ_W, measuring the degree of violations of weak interactions, has a quite large bound (10^{-2}), not only due to the tiny contribution of weak interactions to the total mass but also largely because the binding energy per nucleon due to weak interactions is a very slowly varying function across the periodic table which then leads to a large cancellation in the analysis of Eötvös experiments [10, 11]. In order to improve the significance of Eötvös experiments with respect to weak interactions, one should include the individual nucleons contribution to the nucleus mass since it changes much faster along the periodic table.

There is a model-independent approach to the weak contribution to the proton-neutron mass difference: the development of a sum rule that gives the nucleon self mass in terms of observable quantities. This was first done for the electromagnetic interactions in [12] and will be called the generalized Cottingham formula. This sum rule (which, by the way, shows that the electromagnetic and weak contributions to the nucleon self-mass are finite in the Born approximation) is a rigorous model-independent way for computing the proton-neutron mass difference. Besides, it has been generalized to strong interactions [13]. Detailed proofs can be found in the above references.

We shall develop a sum rule corresponding to the weak interactions similar to Cottingham's formula. Because of weak isospin symmetry neither charged currents nor the axial part of the neutral current will contribute to the neutron-proton mass difference and only the vector neutral current will give a nonzero contribution for the difference. This current, however, has the same structure as the electromagnetic current and so the assumptions involved in the derivation of Cottingham's formula are still valid and following the steps in its derivation one gets the similar result:

To first order in the Fermi constant, the neutral current contribution to the self-energy of the nucleon may be written as:

$$\Delta M_N^{em} = \frac{iG_F}{2(2\pi)^4} \int d^4 q G_Z^{\mu\nu}(q^2) T_{\mu\nu}^{em,N}(\mathbf{q},q_0) \qquad (2)$$

where $G_Z^{\mu\nu} \simeq \eta^{\mu\nu}$ is the Z propagator (the denominator, $1/M_Z^2$, has been absorbed in G_F) and $T_{\mu\nu}^{em,N}(\mathbf{q},q_0)$ is the Compton scattering amplitude of a virtual Z with momentum q by a nucleon N at rest. In the Born approximation this amplitude reduces to:

$$T_{\mu\nu}^{Z,N}(\mathbf{q},q_0) = \frac{(2\pi)^4}{2} \frac{4Mq^2}{q^4 - 4M^2 q_0^2} \left(1 + \frac{q^2}{2M^2}\right)$$
$$\sum_{\text{spin}} [\langle N | J_\mu^{em}(0) | N' \rangle \langle N' | J_\nu^{em}(0) | N \rangle + \mu \leftrightarrow \nu] \qquad (3)$$

where M is the mass of the nucleon N at rest, N' indicates a nucleon with four-momentum $(\mathbf{q}, q_0 + M)$ and the sum is over both its spin states.

In the same approximation, the neutral weak current matrix elements between two nucleons of momenta p, $p' = p + q$ and spin α and α' can be expressed in the form:

$$\langle N(p,\alpha) | J_\mu^{em}(0) | N'(p',\alpha')\rangle = \bar{u}^{(\alpha)}(p) \left[F_1^N(q^2)\gamma_\mu + iF_2^N(q^2)\sigma_{\mu\nu}q^\nu\right] u^{(\alpha')}(p') \qquad (4)$$

where $u(p)$ are Dirac spinors and F_1, F_2 are the Dirac and Pauli form factors of the nucleon.

Plugging (4) into (3) and doing a Wick rotation, one can get, after some algebra, the expression for the weak, neutral current induced, nucleon self energy:

$$\Delta M_N^{W-NV} = -\frac{1}{\pi M^2} \int_0^\infty q\,dq \int_0^q dv\sqrt{q^2-v^2}\frac{4Mq^2}{q^4+4M^2v^2} \\ [3q^2 f_1^Z(q^2) - (q^2+2v^2)f_2^Z(q^2)] \quad (5)$$

where the quantities $f_1^Z(q^2), f_2^Z(q^2)$ are related to the neutral weak form factors:

$$f_1^Z(q^2) = \frac{\alpha_W}{\pi}\frac{[G_M^Z(q^2)]^2 - [G_E^Z(q^2)]^2}{q^2+4M^2} \quad (6)$$

$$f_2^Z(q^2) = \frac{\alpha_W}{\pi}\frac{q^2[G_M^Z(q^2)]^2 + 4M^2[G_E^Z(q^2)]^2}{q^2(q^2+4M^2)} \quad (7)$$

and where

$$\alpha_W = \frac{\sqrt{2}G_F M^2}{\pi} = 0.463 \times 10^{-5} \quad (8)$$

The sum rule (5) is the contribution to the self mass of the nucleon coming from the isospin-breaking part of the weak interaction which is, as we said above, related to the vector part of the weak neutral current. The weak contribution to the proton-neutron mass difference is obtained, then, by straightforward subtraction of the proton and neutron weak neutral vector self masses $\Delta M_{p-n}^W = \Delta M_p^{W-NV} - \Delta M_n^{W-NV}$.

The weak form factors, except for isolated points, have not been measured [16]. However, using CVC, they can be related to the electromagnetic form factors [17]:

$$G^{pZ} = \frac{1}{2}(G^p - G^n) - 2\sin^2\theta_W G^p - \frac{1}{2}G^{sZ} \quad (9)$$

$$G^{nZ} = -\frac{1}{2}(G^p - G^n) - \frac{1}{2}G^{sZ} \quad (10)$$

where we have normalized them to the weak isospin values $G_E^{p,nZ}(0) = t_{3L}$ and where G^s is the contribution of the s-quark sea to the weak form factor. Measurements show that this latter quantity is very small and we shall neglect it [16].

To compute the $P-N$ mass differences the weak Cottingham formula, we use the "Galster parameterization" [18, 19] for the electromagnetic form factors. With these values we obtain the results:

$$\frac{(m^N - m^P)^W}{m} = (-5.0 \pm 1.0) \times 10^{-9} \quad (11)$$

The error was estimated from the known discrepancies of the Galster parameterization with experiment, plus a generous allowance for the largely unknown strange contribution.

TABLE 1. Upper bounds for the UFF violation parameters.

Γ	$\Delta M^{n-p} = 0$	ΔM_{Ct}^{n-p}
Γ^S	1.0×10^{-9}	1.1×10^{-9}
Γ^E	1.2×10^{-9}	1.2×10^{-9}
Γ^W	2.8×10^{-2}	1.0×10^{-3}

Comparison with the results of Eötvös experiments, summarized in references [2, 7, 8], are shown in table 1. The first two columns show the upper bounds obtained assuming that a single interaction breaks the equivalence principle. The first column ($\Delta M = 0$) excludes the nucleon structure contribution while the second column (ΔM_{Ct}) includes it, as computed with the generalized Cottingham formulae. We find that while the inclusion of individual nucleons effect does not change much the upper limit on the strong and electromagnetic violation parameters (10^{-8}), it lowers the bound on Γ^W from (3×10^{-2}) to $(\times 10^{-3})$: an order of magnitude increase in sharpness.

As a final remark, let us observe that while proton-neutron weak mass splitting originates in the neutral currents, the "nuclear" contribution of weak interactions is dominated by the charged ones [10, 11]. Hopefully, the STEP experiment, with its larger accuracy and better cover of the periodic table may help to put bounds on the separate currents. Even though we should interpret our results with caution, they confirm that Eötvös experiments do test weak interactions effect with an accuracy, at least one order of magnitude, better than previous studies.

REFERENCES

1. A. Einstein, *Ann. Phys.* **35**, 898 (1911)
2. C. M. Will, *Theory and experiment in gravitational physics*, Cambridge University Press, Cambridge (1981)
3. S. Weinberg, *Gravitation and Cosmology*, Wiley, New York (1972)
4. T. Damour and A. M. Polyakov, *Nucl. Phys. B* 423:532 (1994)
5. J. M. Overduin and P. S. Wesson, *Phys. Rep.* 283:303 (1997)
6. R. Reinhard, Y. Jafry, and R. Laurence. *ESA J.*, 17:251, 1993.
7. H. Vucetich, *Bol. Acad. Nac. Cs. (Córdoba)* **61**, 1 (1996)
8. T. Damour in *Proceedings of the Workshop on the Scientific Applications of Clocks in Space* Kluwer (1997).
9. R. Eisberg and R. Resnick, *Fisica Cuantica,* Limusa, Mexico (1978)
10. M.P. Haughan and C.M. Will, *Phys. Rev. Lett.* 37:1 (1976)
11. E. Fishbach, M.P. Haughan, D. Tadic and H. Cheng, *Phys. Rev. D,* 32:154 (1985)
12. W. N. Cottingham, *Ann. Phys. N. Y.*, 25:424 (1963)
13. L. N. Epele, H. Fanchiotti, C. A. García Canal and R. Méndez Galain, in *Frontier Phisics: Essays in Honour of Jayme Tiomno* S. MacDowell, H. M. Nussenzveig and R. A. Salmeron eds. (1991)
14. H. R. Christiansen, L. N. Epele, H. Fanchiotti and C. A. García Canal, *Phys. Lett.* B267:164 (1991)
15. R. Méndez Galain *La diferencia de masa neutrón-protón* Ph. D. Thesis. U. Nac. de La Plata. (1989).
16. U.-G. Meißner *Baryon form factors: Model-independent results* hep-ph/9907323.
17. N. C. Mukhopadhyay *Weak form factors of the nucleon* hep-ph/9810039.
18. M. J. Musolf *et. al. Phys. Rep.* **239** 1 (1994).
19. S. Galster et al. *Nucl. Phys. B* **32** 221 (1971)

A nonperturbative fermion-boson vertex

A. Bashir and A. Raya

Instituto de Física y Matemáticas, Universidad Michoacana de San Nicolás de Hidalgo, Apartado Postal 2-82, Morelia, Michoacán 58040, México.

Abstract. We calculate the massive fermion propagator at one-loop order in QED3. The Ward-Takahashi identity (WTI) relates the propagator to the vertex. This allows us to split the vertex into its longitudinal and transverse parts. The former is fixed by the WTI. Following the scheme of Ball and Chiu later modified by Kızılersü et. al., we calculate the full vertex at one-loop order. A mere subtraction of the longitudinal part of the vertex gives us the transverse part. The α dependence in the transverse vertex can be eliminated by making use of the perturbative expressions for the wavefunction renormalization function and the mass function of complicated arguments of the incoming and outgoing fermion momenta. This leads us to a vertex which is nonperturbative in nature. We also calculate an effective vertex for which the arguments of the unknown functions have no angular dependence, making it particularly suitable for numerical studies of dynamical symmetry breaking.

INTRODUCTION

QED3 is an attractive model to study the intricacies of Schwinger-Dyson equations due to its simplicity as compared to QED4 and QCD. The knowledge of the three-point vertex is crucial in this study. In this respect, perturbation theory is a powerful point of reference, since we expect that physically meaningful solutions of the Schwinger-Dyson equations must agree with perturbative results in the weak-coupling regime. In this paper, we study the massive QED3 based on [1, 2]. The Ward-Takahashi identity (WTI) relates the vertex to the propagator. This identity allows us to split the full vertex into its longitudinal and transverse parts [3]. The former is fixed by the WTI, while the later needs to be claculated, subtracting the longitudinal part from the full vertex. We do so at one-loop level by calculating the fermion propagator and the full vertex at this order following [4]. Using perturbative constraints as a guide, we carry out a construction of the non-perturbative vertex, which has no dependence on the coupling α. For practical purposes of numerical study of dynamical chiral symmetry breaking, we also construct an effective vertex that shifts the angular dependence from the unknown fermion propagator functions to the known basic functions, without changing its perturbative properties at the one-loop level.

THE LONGITUDINAL VERTEX TO ONE-LOOP

The Fermion Propagator

The full fermion propagator can be written in its most general form as:

$$S_F(p) = \frac{F(p^2)}{\not{p} - \mathcal{M}(p^2)}. \tag{1}$$

At one-loop level, it is [1]:

$$\frac{1}{F(p^2)} = 1 - \frac{\alpha\xi}{2p^2}\left[m - (m^2 + p^2)I(p^2)\right],$$
$$\frac{\mathcal{M}(p^2)}{F(p^2)} = m\left[1 + \alpha(\xi + 2)I(p^2)\right]. \tag{2}$$

where m is the bare mass of the fermion, ξ is the covariant gauge parameter and α is the electromagnetic coupling. We have used the simplifying notation $I(p^2) = (1/\sqrt{-p^2})\arctan\sqrt{-p^2/m^2}$.

WTI and the Longitudinal Vertex

The full vertex satisfies the Ward-Takahashi identity

$$q_\mu \Gamma^\mu(k,p) = S_F^{-1}(k) - S_F^{-1}(p). \tag{3}$$

This relation allows us to decompose the full vertex into longitudinal $\Gamma_L^\mu(k,p)$ and transverse $\Gamma_T^\mu(k,p)$ parts. Following the work of Ball and Chiu [3], we can calculate the longitudinal component of the vertex in terms of the fermion propagator as:

$$\begin{aligned}\Gamma_L^\mu &= \left[1 + \frac{\alpha\xi}{4}\sigma_1\right]\gamma^\mu + \frac{\alpha\xi}{4}\sigma_2[k^\mu\not{k} + p^\mu\not{p} + k^\mu\not{p} + p^\mu\not{k}] \\ &+ \alpha(\xi+2)\sigma_3[k^\mu + p^\mu],\end{aligned} \tag{4}$$

where

$$\begin{aligned}\sigma_1 &= \frac{m^2 + k^2}{k^2}I(k^2) + \frac{m^2 + p^2}{p^2}I(p^2) - m\frac{k^2 + p^2}{k^2 p^2} \\ \sigma_2 &= \frac{1}{(k^2 - p^2)}\left[\frac{m^2 + k^2}{k^2}I(k^2) - \frac{m^2 + p^2}{p^2}I(p^2) + m\frac{k^2 - p^2}{k^2 p^2}\right] \\ \sigma_3 &= m\left[I(k^2) - I(p^2)\right].\end{aligned} \tag{5}$$

THE TRANSVERSE VERTEX TO ONE-LOOP

The Complete Vertex

Using the Feynman rules, the one-loop correction to the full vertex can be expressed as:

$$-ie\Lambda^\mu = \int_M \frac{d^3w}{(2\pi)^3}(-ie\gamma^\alpha)iS_F^0(p-w)(-ie\gamma^\mu)iS_F^0(k-w)(-ie\gamma^\beta)i\Delta^0_{\alpha\beta}(w), \qquad (6)$$

where w is the loop momentum, and k and p are the momenta of the incoming and outgoing fermions. Subtracting Eq. (4) from Eq. (6), we obtain the transverse vertex which can conveniently be written as:

$$\Gamma_T^\mu(k,p) = \sum_{i=1}^{8} \tau_i(k^2, p^2, q^2) T_i^\mu(k,p), \qquad (7)$$

where we use the basis poposed by Kızılersü, Reenders and Pennington [4]. After a tedious but straightforward algebra, the coefficients τ_i can be identified. All of them have the form:

$$\tau_i = \alpha g_i \left[\sum_{j=1}^{5} a_{ij}(k,p) I(l_j^2) + \frac{a_{i6}(k,p)}{k^2 p^2} \right]. \qquad (8)$$

The coefficients a_{ij}, a_{i6}, the factors g_i and the functions l_j^2 have simple dependence on k and p [1].

Non-perturbative vertex

Fortunately, in the case of QED3, tha α dependence in Eq. (8) can be completely eliminated by making use of Eq. (2). The non-perturbative τ_i can then be written as:

$$\tau_i = g_i \Bigg\{ \sum_{j=1}^{5} \left(\frac{2a_{ij}(k,p)l_j^2}{\xi(m^2+l_j^2)} \left[\frac{\xi}{2(\xi+2)l_j^2 I(l_j^2)} \left(\frac{\mathcal{M}(l_j^2)}{F(l_j^2)} - m \right) - \left(1 - \frac{1}{F(l_j^2)}\right) \right] \right)$$
$$+ \frac{2a_{i6}(k,p)}{\xi[k^2(m-(m^2+p^2)I(p^2)) - p^2(m-(m^2+k^2)I(k^2))]} \left[\frac{1}{F(k^2)} - \frac{1}{F(p^2)} \right] \Bigg\}. \qquad (9)$$

The arguments of $\mathcal{M}$ and F have explicit angular dependence, which makes them harder for numerical studies. We can circumvent this problem by noting that we can write the perturbative τ_i as

$$\tau_i(k,p) = \alpha g_i \left[b_{i1}(k,p) I(k^2) + b_{i2} I(p^2) + \frac{a_{i6}}{k^2 p^2} \right], \qquad (10)$$

where b_{i1} and b_{i2} are functions of a_{ij}. The angular dependence can now be absorbed into the known functions to obtain the effective τ_i:

$$\tau_i = g_i \left\{ \sum_{j=1}^{2} \left(\frac{2b_{ij}(k,p)\kappa_j^2}{\xi(m^2+\kappa_j^2)} \left[\frac{\xi}{2(\xi+2)\kappa_j^2 I(\kappa_j^2)} \left(\frac{\mathcal{M}(\kappa_j^2)}{F(\kappa_j^2)} - m \right) - \left(1 - \frac{1}{F(\kappa_j^2)} \right) \right] \right) \right. $$
$$\left. + \frac{2a_{16}(k,p)}{\xi[k^2(m-(m^2+p^2)I(p^2)) - p^2(m-(m^2+k^2)I(k^2))]} \left[\frac{1}{F(k^2)} - \frac{1}{F(p^2)} \right] \right\}, \quad (11)$$

where $\kappa_1^2 = k^2$ and $\kappa_2^2 = p^2$. Eq. (7) and (11) form the nonperturbative vertex in conjunction with the longitudinal Ball-Chiu component [3].

CONCLUSIONS

In this paper, we calculate the massive fermion propagator at one-loop level. We also calculate the longitudinal and the full vertex. Making use of the WTI we calculate the transverse vertex following [3, 4]. We find a way to write the vertex in a non-perturbative fashion using the propagator of complicated functions of the incoming and outgoing fermion momenta. For practical purposes, we also construct an effective vertex which has no angular dependence on the fermion momenta. This vertex could be helpful in studying dynamical chiral symmetry breaking.

ACKNOWLEDGEMENTS

A.B. and A.R. acknowledge the CIC and Conacyt grants under the Projects No. 4.12 and 32395-E respectively.

REFERENCES

1. A. Bashir and A. Raya, Phys. Rev **D64** 105001 (2001)
2. A. Raya, XV annual meeting of the division of particles and fields, México, (2001).
3. J.S. Ball and T.-W. Chiu, Phys. Rev. **D22**, 2542 (1980).
4. A. Kızılersü, M. Reenders and M.R. Pennington, Phys. Rev. **D52**, 1242 (1995).

Higher order correction to the neutrino self-energy

Sarira Sahu

Instituto de Ciencias Nucleares, Universidad Nacional Autonoma de Mexico, Circuito Exterior C.U., A. Postal 70-543, 04510 Mexico D.F.

NEUTRINO DISPERSION RELATION IN A MEDIUM

The particle propagation in the presence of a heat bath is discussed elaborately in the literature. It is well known that particle properties changes in the heat bath[1, 2, 3]. Here we will discuss about the neutrino propagation in a heat bath, containing both particles and anti-particles.

The Dirac equation for a fermion is given by $[\not{k} - \Sigma(k)]\psi = 0$. where k is the neutrino four momentum and $\Sigma(k)$ is its self energy. The lowest order contribution to the neutrino self energy can be calculated from the one-loop contribution to the self energy from the W and Z bosons exchange. Because of the self energy term the pole of the neutrino propagator will shift. The neutrino self energy for W exchange is given by

$$-i\Sigma_W(k) = \int \frac{d^4p}{(2\pi)^4} \left(-\frac{ig_W}{\sqrt{2}}\gamma_\mu L\right) iS_e(p) \left(-\frac{ig_W}{\sqrt{2}}\gamma_\nu L\right) D^{\mu\nu}(k-p). \quad (1)$$

where $L, R = (1 \pm \gamma_5)/2$. We are interested to calculate the real part of it, which corresponds to the propagation. Now let us define $\Sigma_W = R\tilde{\Sigma}_W L$. The real part of the self-energy can be written as $-Re\tilde{\Sigma}_W = a_W \not{k} + b_W \not{u}$, where the four vector $u = (1,0)$, is the four velocity of the heat bath, which is at rest. Now we can define $A_W = T_k = -\frac{Tr[\not{k}Re\tilde{\Sigma}_W]}{4} = a_W k^2 + b_W k_0$ and $B_W = T_u = -\frac{Tr[\not{u}Re\tilde{\Sigma}_W]}{4} = a_W k_0 + b_W$. Then from this we obtain $a_W = \frac{k_0}{\mathbf{k}^2}B_W - \frac{A_W}{\mathbf{k}^2}$ and $b_W = \frac{k_0}{\mathbf{k}^2}A_W - \frac{k_0^2 - \mathbf{k}^2}{\mathbf{k}^2}B_W$. Thus we obtain

$$a_W = -\sqrt{2}G_F \left[\frac{k_0^2}{M_W^2}(N_e - \bar{N}_e) - \frac{14}{3M_W^2}\Phi_1 - \frac{m^2}{3M_W^2}\Phi_2\right], \quad (2)$$

and

$$b_W = -\sqrt{2}G_F \left[\left(1 + \frac{3m^2}{2M_W^2}\right)(N_e - \bar{N}_e) - \frac{8k_0}{3M_W^2}\Phi_1 + \frac{2k_0^2}{3M_W^2}\Phi_2\right], \quad (3)$$

where we have defined $\Phi_1 = <E_e>N_e + <E_{\bar{e}}>\bar{N}_e$ and $\Phi_2 = \frac{N_e}{<E_e>} + \frac{\bar{N}_e}{<E_{\bar{e}}>}$. For relativistic non-degenerate fermions, the average energy per particle is $<E> = \frac{\rho}{n} = \frac{7\pi^4 T}{180\xi(3)} = \frac{7}{2}\frac{\xi(4)}{\xi(3)}T$ where $\xi(4) = \frac{\pi^4}{90}$, and in the relativistic limit (for which $T \gg m$)

the number density N_i (i=fermion and boson) is given by $N_f = \frac{3}{4}\frac{\xi(3)}{\pi^2}gT^3$ and $N_b = \frac{\xi(3)}{\pi^2}gT^3$, where g is the degeneracy factor, which is 2 for electrons, positrons, protons and neutrinos and 1 for neutrinos and anti neutrinos. In terms of the number density of photon N_γ, the lepton asymmetry is defined as, $L_e = \frac{(N_e - \bar{N}_e)}{N_\gamma}$ where $N_\gamma = \frac{2}{\pi^2}\xi(3)T^3$. In Eqs.(2) and (3), k_0 is the neutrino energy. Replacing it by its average, in the relativistic limit we obtain

$$a_W = \sqrt{2}G_F N_\gamma \left[\left(\frac{28}{3} - L_e\right)\frac{7\xi(4)}{2\xi(3)}T + \frac{2m^2}{3T^2}\frac{2\xi(3)}{7\xi(4)}\right], \quad (4)$$

and

$$b_W = -\sqrt{2}G_F N_\gamma \left[\left(1 + \frac{3m^2}{2M_W^2}\right)L_e - 4\left(\frac{7\xi(4)}{2\xi(3)}T\right)^2 \frac{T^2}{M_W^2} + \frac{2m^2}{M_W^2}\right]. \quad (5)$$

Now let us consider the **Z** boson exchange (bubble) diagram. In the **W** diagram we have to make the following changes to obtain the **Z** exchange contribution. $g_W \to \frac{g_Z}{\sqrt{2}\cos\theta_W}$ and $M_W \to M_Z$. Then we obtain $a_Z = -\frac{G_F N_\gamma}{\sqrt{2}M_Z^2}\left(\frac{7\xi(4)}{2\xi(3)} + 2L_\nu\right)$ and $b_Z = -\frac{G_F N_\gamma}{\sqrt{2}}\left(2L_\nu - 4\frac{T^2}{M_Z^2}\left(\frac{7\xi(4)}{2\xi(3)}\right)^2\right)$, where the neutrino asymmetry is defined as $L_\nu = \frac{(N_\nu - N_{\bar\nu})}{N_\gamma}$. Now let us consider the tadpole diagram with **Z** propagator. In this there is no momentum transfer, which corresponds to the forward scattering of the propagating neutrino from the background particles. We obtain, $a_{Z(T)} = 0$ and $b_{Z(T)} = \sum_f \sqrt{2}G_F X_f(N_f - N_{\bar f})$, where X_f has different values for electrons, protons, neutrons and neutrinos. Here $Z(T)$ corresponds to the tadpole contribution due to Z exchange. Using the charge neutrality condition we obtain, $b_{Z(T)} = -\frac{G_F}{\sqrt{2}}(N_n - N_{\bar n}) + \sqrt{2}G_F(N_\nu^L - N_{\bar\nu}^L)$. The effective potential of neutrino in the background is given by $b_\nu = V_\nu = b_W + b_Z + b_{Z(T)}$. For different species of neutrinos we obtain $V_{\nu_e} = b_W + b_Z + b_{Z(T)}$ and $b_{\nu_\mu,\nu_\tau} = b_Z + b_{Z(T)}$. As the μ and τ neutrinos do not have charge current interaction, b_W does not contribute to their respective potentials. Putting the values of b_W and b_Z in the above equations we obtain

$$V_{\nu_e} = -\sqrt{2}G_F N_\gamma \left(L_e + \frac{L_n}{2} - 4\left(\frac{7\xi(4)}{2\xi(3)}T\right)^2 \frac{T^2}{M_W^2}(1 + \cos^2\theta_W)\right) \quad (6)$$

and similarly

$$V_{\nu_\mu,(\nu_\tau)} = -\frac{G_F N_\gamma}{\sqrt{2}}\left(L_n - 4\left(\frac{7\xi(4)}{2\xi(3)}T\right)^2 \frac{T^2}{M_Z^2} - 2(L_{\nu_e} + L_{\nu_\tau,(\nu_\mu)})\right) \quad (7)$$

In the above one we consider the background neutrinos and electrons as relativistic and non degenerate. Now let us consider a general situation, where the background particles have masses and chemical potentials. Then we have

$$N_f = g\frac{m^3}{2\pi^2}\sum_{l=1}^{\infty}(-1)^{l+1}e^{\beta\mu l}\frac{K_2(\beta m l)}{\beta m l}. \quad (8)$$

Similarly we can write

$$J_2 = g \int \frac{d^3p}{(2\pi)^2} \frac{n(p.u)}{E} = g\frac{m^2}{2\pi^2} \sum_{l=1}^{\infty} (-1)^{l+1} e^{\beta \mu l} \frac{K_1(\beta ml)}{\beta ml}, \qquad (9)$$

$$\begin{aligned} J_3 &= g \int \frac{d^3p}{(2\pi)^2} E n(p.u) \\ &= g\frac{m^4}{2\pi^2} \sum_{l=1}^{\infty} (-1)^{l+1} e^{\beta \mu l} \left[\frac{3}{(\beta ml)^2} K_0(\beta ml) + \frac{K_1(\beta ml)}{\beta ml}\left(1 + \frac{6}{(\beta ml)^2}\right)\right]. \end{aligned} \qquad (10)$$

Then for anti-particles we can obtain by taking $\mu \to -\mu$ in the above equations. Then we obtain

$$N_f - N_{\bar{f}} = g\frac{m^3}{\pi^2} \sum_{l=1}^{\infty} (-1)^{l+1} sinh(\beta\mu l) \frac{K_2(\beta ml)}{\beta ml}, \qquad (11)$$

$$\Phi_2 = J_2 + \bar{J}_2 = g\frac{m^2}{\pi^2} \sum_{l=1}^{\infty} (-1)^{l+1} cosh(\beta\mu l) \frac{K_1(\beta ml)}{\beta ml}, \qquad (12)$$

$$\Phi_1 = J_3 + \bar{J}_3 = g\frac{m^4}{\pi^2} \sum_{l=1}^{\infty} (-1)^{l+1} cosh(\beta\mu l) \left[\frac{3}{(\beta ml)^2} K_0(\beta ml) + \frac{K_1(\beta ml)}{\beta ml}\left(1 + \frac{6}{(\beta ml)^2}\right)\right] \qquad (13)$$

From Eq.(2) and (3) we can see that the first order in G_F contribution is proportional to the difference between the number density of particle and anti-particle and higher order contributions are the sum of the number densities along with their energies. For relativistic limit the simplifications in Eq.(2) and (3) are good approximations. But for the temperature near to the electron mass, it is not a good approximation. On the other hand Eqs.(11), (12) and (13) are exact. So for any temperature one can use these.

APPLICATION

The cosmological gamma ray bursts (GRB) release about 10^{52} erg energy in a few seconds. The fireball model of the GRBs is that, a large concentration of energy, mostly radiation, in a small region of space in which there are very few baryons[4]. The sudden release of a large amount of gamma-ray photons into a compact region can lead to an opaque photon-lepton fireball through the production of electron-positron pairs. If the initial temperature of the fireball is high enough (assume to be 2 to 6 MeV) then pair creation will take place and the radiation can not escape, forming a fireball consists of photons, electrons and positions, which has a typical radius $10^7 - 10^8$ cm. Neutrinos produced by collision of protons with photons can have energy of order 10^{14} eV[5]. So these neutrinos when propagate through the fireball (which we assume to be only, photons, electrons and positions) may change their flavor. If the fireball is free from baryons contamination, then it behaves like a CP symmetric system, having only electrons and positrons. So only the electron neutrino will experience a potential and all other neutrinos will not. So depending on the density and size of the fireball, electron neutrino may convert to other neutrinos.

REFERENCES

1. J. C. Dt'Olivo, Manuel Torres and J. F. Nieves, Phys. Rev. **D 46**, 1172 (1992).
2. D. Notzold and G. Raffelt, Nucl. Phys. B307, 924, (1988).
3. K. Enqvist, K. Kainulaien and J. Maalampi, Nucl. Phys. **B349**, 754 (1991).
4. Tsavi Piran, Phys. Reports **314**, 575, (1999).
5. E. Waxman, J. N. Bahcall, Phys. Rev. Lett. **78**, 2292 (1997).

Ultra High Energy Neutrinos and their Detection in the Pierre Auger Observatory

A. Carreño, J.C. D'Olivo, and L. Nellen

Instituto de Ciencias Nucleares, UNAM, Ap. Postal 70-543, 04510 México DF, México

Abstract. The cosmic ray spectrum has been shown to extend well beyond $10^{20} eV$. In all production models neutrinos and photons are part of the cosmic ray flux. The Pierre Auger Observatory will be the largest air shower array ever built. Neutrinos can be detected and identified in Auger.

INTRODUCTION

The origin of Ultra High Energy Cosmic Rays (UHECR) observed on the Earth is a long standing mystery. UHECR are those cosmic rays with energies above 10^{18} eV. At those extreme energies the Cosmic Microwave Background Radiation (CMB Radiation) makes the Universe essentially opaque to protons, nuclei, and photons, which suffer energy losses from pion photo-production, photo-disintegration, or pair production. These processes led Greisen, Zatsepin, and Kuz'min [1] to predict a spectral cutoff around $5 \times 10^{19} eV$, the GZK cutoff. However, the observed energy spectrum extends beyond 10^{20} eV. Mechanisms producing or accelerating particles near or above 10^{21} eV are still uncertain. At these high energies the Galactic magnetic field can not confine protons in the Galaxy, so they likely originate in extragalactic source.

There are severe limitations to the properties of astrophysical candidates for accelerating cosmic rays beyond 10^{20} eV. Only very powerful astrophysical objects can, in principle, produce these energies through conventional acceleration. The Fermi acceleration mechanism of stochastic process uses ordinary interaction of particles with electromagnetic fields. In this mechanism the particle crosses the shock region in a plasma repeatedly, gaining energy at each crossing. Many interactions like this can accelerate a particle to high energy. The maximum energy is limited by the life time of the source, the size of the source respect the Larmor radius of the particle, and energy losses mainly by the following process:

$$pp \longrightarrow N's + \pi's,$$
$$p\gamma \longrightarrow \Delta^+ \longrightarrow N + \pi's.$$

At the end, the π decays will produce neutrinos.

The size of the acceleration region R is assumed to be comparable to the Larmor radius of a particle in a magnetic field, which furthermore must be sufficiently weak so that synchrotron losses are not greater than the energy gain. This means that the maximum energy attainable is given by $E = kZeBR\beta c$, where B is the magnetic field in the region of the shock, βc is the shock speed, and $k < 1$. Only a few astrophysical sources fulfill the requirements to accelerate particles to the highest energies observed. Some possible candidates are Active Galactic Nuclei (AGN), Gamma Ray Bursts, and Neutron Stars [2]. However, the environment of the source itself generally prevents the accelerated particle to escape from the site. Only neutrinos have a chance to escape and they can carry information from inside the object.

Because of the difficulty to explain UHECR with conventional astrophysics, alternative scenarios based on new particle physics have been proposed. Some examples are the collapse of Topological Defects (TD) or the decay of Super Massive Relic Particles (SMRP). They are suited to produce particles above 10^{20} eV, but we have no evidence for their existence. In this case, protons are produced in the hadronization of quarks accompanied by the production of mesons, principally pions, which decay into photons, electrons, and neutrinos.

So both astrophysical and exotic models predict the existence of high-energy neutrinos. In this framework, neutrinos are an invaluable probe of the nature and the distribution of the potential sources. Essentially unaffected on their journey to Earth they may allow us to disentangle the source characteristics from the propagation distortions.

DETECTION OF THE ULTRA HIGH ENERGY NEUTRINOS

In the Standard Model of Elementary Particles Physics (SMPP), the properties of the neutrinos are zero rest mass, spin = 1/2, weak and gravitational interactions, and three flavors $(\nu_e, \nu_\mu, \nu_\tau)$. In extensions of the SMPP, neutrinos can have non-zero masses, in which case, they can exhibit the phenomenon of flavor oscillations. Neutrinos interact weakly but their cross sections increase with energy. Thus it is easier to detect neutrinos at high energies.

UHECR are detected using the Extensive Air Showers (EAS) they produce, that is the particle cascade following the interaction of a CR particle with a nucleus of the upper atmosphere. An EAS is essentially a thin disk of particles moving at the speed of light. The position of the shower maximum will strongly depend on the primary type (proton or neutrinos). The longitudinal and the lateral development as well as the time structure of the shower are also characteristics of its nature. The UHE neutrinos may be detected and distinguished from ordinary hadrons by the shape of the horizontal EAS they produce. We know that the ordinary hadrons interact at the top of the atmosphere and at large zenith angle (above 80 deg) the distance from the shower maximum to the ground becomes large than the distance between the first interaction and the shower maximum (100 km). At ground level the electromagnetic part of the shower is totally extinguished and only high energy muons survive. In addition, the shower front is very flat. Unlike hadrons, neutrinos may interact deeply in the atmosphere and can initiate

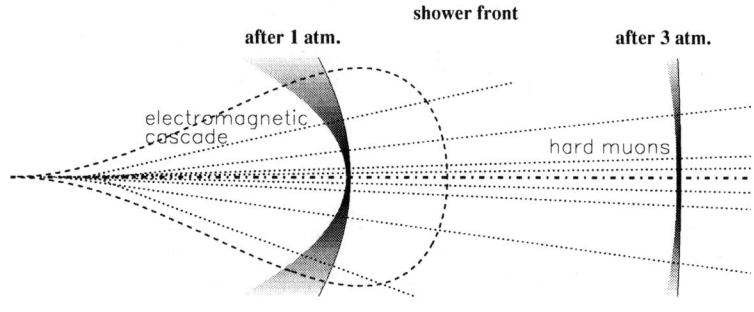

FIGURE 1. Horizontal shower development.

a shower in the volume of air immediately above the ground detectors. This shower will appear as a "normal" one, although horizontal, with a curved front and a large electromagnetic component (Fig. 1). With such important differences neutrinos can be detected and identified [3].

There are two majors techniques used to detect EAS [4]. The first and the most frequent, is to build an array of sensors (scintillators, water Čerenkov tanks) spread over a large area. The detectors count the particles densities sampling the EAS particles hitting the ground. The surface of the array is chosen to match the incident flux and the energy range one wants to explore. The second technique consist in studying the longitudinal development of the EAS by detecting the fluorescence light produced by the interactions of the charged secondaries. The first successful detectors based on these ideas were built by a team of the University of Utah, under the name of "Fly's Eye"[7].

The Auger Observatory combines both techniques [5, 6]. The detector is designed to be fully efficient for showers above $10^{18}eV$, with a duty-cycle of 100% for the ground array, and 10 to 15 % for the fluorescence telescope. The ground array consist of ≈ 1600 cylindrical Čerenkov tanks of 10 m^2 surface and 1.2 m height, filled with filtered water; they provide a sizable cross-section even for inclined showers: 10 m^2 at 0° and 4.3 m^2 at 90° [4]. With a very large area and non zero acceptance to horizontal showers, the Auger could have the possibility to observe τ-leptons induced showers from charged current interactions of the ν_τ with the ground surrounding the array. Additional Monte Carlo simulations of the detection capabilities of neutrinos in The Pierre Auger Observatory using the standard packages AIRES [8] and CORSIKA [9] are currently being done.

ACKNOWLEDGMENTS

This work has been partially supported by CONACyT grant 32279 E

REFERENCES

1. K.Greisen, *Phys. Rev. Lett.* **16** (1966) 748; G.T. Zatsepin, V.A. Kuz'min, *JETP Lett.* **4** (1966) 78.
2. M.Nagano, A.A.Watson, (2000), *Rev. Mod. Phy.***72**, 689.
3. X.Bertou *et al.*, **astro-ph/0104452 v2**.
4. A. Letessier-Selvon, *et al.*, *ICRC 2001*.
5. Pierre Auger Proyect Design Report, 2nd version, March (1997) **(http://www.auger.org/admin/Design Report/)**.
6. Pierre Auger Observatory Technical Design Report; in preparation.
7. D. J. Bird *et al.*, Ap. J. **424** (1994) 491.
8. S. J. Scuitto, *"Air Shower Simulations with AIRES System"* a system for air shower simulations, *version 2.2,1 (2000)*
9. J. N. Capdevielle *et al.*, Kernforschungszentrum Karlsruhe prepint KfK 4998 (1992). J. Knapp, D. Heck, preprint KfK 5196 (1993).

Mixing and Instability

Marek Nowakowski

Instituto de Física, Universidad de Guanajuato Loma del Bosque # 103, Lomas del Campestre, 37150 León, Guanajuato; México

Abstract. Given a Lagrangian which captures all essential features of the $K^0 - \bar{K}^0$ system, we argue that the analysis of the time evolution of the unstable states which are linear superposition of other, observable, states can, in principle, be carried out in two different ways. Staying close to the $K^0 - \bar{K}^0$ system, we compare both methods pointing out some of their shortcomings and advantages.

INTRODUCTION

Consider a single unstable state $|\lambda\rangle$ characterized by some quantum numbers denoted by λ. Although not an eigenstate to the full Hamiltonian, the state has a definite mass. Viewing this state as an open quantum system, its time evolution can be generally given as $|\lambda(\tau)\rangle = p_\lambda(\tau)|\lambda\rangle$ where the effective Hamiltonian $\mathcal{H}_{\text{eff}}$ governing this system is non-hermitian and in the Weisskopf-Wigner approximation [1] one recovers the exponential decay law $p_\lambda(\tau) = e^{-iE\tau}e^{-\Gamma/2\tau}$. Suppose now that we combine this mixing and the instability analysis into one single equation. With a two level quantum system, the non-hermitian Hamiltonian of the single particle case becomes a non-hermitian effective 2×2 mass matrix $\mathcal{M}_{\text{eff}}$. Obviously, the diagonalization of $\mathcal{M}_{\text{eff}}$ is generally not given by a unitary transformation and as a result the norm is not automatically preserved. The two emerging mass eigenstates, say $|\lambda_{1,2}\rangle$, can turn out to be non-orthogonal i.e. $\langle\lambda_1|\lambda_2\rangle \neq 0$. Let us now assume that the mixing is well defined within a Lagrangian. Performing only unitary transformations we can define two orthogonal mass eigenstates and proceed to analyze their time behaviour for each one of them separately. Obviously, the outcomes will be different. Also obviously, we follow the first method in the neutral kaon system [2], [3] (and related mesonic systems) whereas for massive unstable neutrinos the mixing is done within the Langrangian framework [4] and hence we opt here for the second possibility. However, there exist no argument which would restrict the applicability of the first method only to kaons or more generally to composite objects. Indeed, regardless whether the particles are fundamental or composite, this method can be applied to any unstable two level system, in particular to neutrinos.

REVIEW OF THE LEE-OEHME-YANG THEORY

The theory of instability and mixing of neutral kaon is usually formulated within the Lee-Oehme-Yang (LOY) scheme [2], [3]. Although we will formulate it for the neutral kaon system, it is important to stress again that this theory is applicable to many unstable systems with mixing, be it elementary or composite. In particular, it would also apply to massive neutrinos and the theory whose Lagrangian we presented in section 3. The LOY theory examines the mixing and instability simultaneously and as a result of this analysis we end up with an effective non-hermitian mass matrix $\mathcal{M}_{\text{eff}}$. given by

$$(\mathcal{M}_{\text{eff}})_{ij} = m_K \delta ij + \langle K_i^0|H_{\text{weak}}|K_j^0\rangle + \sum_n \frac{\langle K_i^0|H_{\text{weak}}|n\rangle\langle n|H_{\text{weak}}|K_j^0\rangle}{m_K - E_n + i\varepsilon} \quad (1)$$

where K_i^0 can be either K^0 or $\bar{K}^0$. Parametrizing the off-diagonal elements of the mass matrix by p^2 and q^2 the mass-eigenstates can be calculated to be $|K_{S/L}\rangle = p|K^0\rangle \pm q|\bar{K}^0\rangle$, $|p|^2 + |q|^2 = 1$. Because $\mathcal{M}_{\text{eff}}$ is in general non-hermitian, it is not necessary that we have $\langle K_S|K_L\rangle = \langle K_S|K_L\rangle = 0$ or, in other words, that $|p|^2 = |q|^2$. Indeed, if CP is a good symmetry, it follows that $p^2 = e^{-2i\xi}q^2$ where the phase ξ comes from the CP-transformation $CP|K^0\rangle = e^{i\xi}|\bar{K}^0\rangle$. We conclude that if $|p|^2 - |q|^2 \neq 0$, we have CP-violation in the LOY theory and, indeed, this is taken as a genuine signal of broken CP. However, if $p^2 = e^{i\beta}q^2$, with some phase β, we cannot decide whether CP is broken or conserved. As apparent from the Lagrangian example in section 2, a system can still exhibit CP-violation even if $|p|^2 = |q|^2$. Quite similarly, if $K_{S/L}$ are eigenstates to CP, then $|p|^2 = |q|^2$ (or if $|p|^2 = |q|^2$, then $K_{S/L}$ are eigenstates to CP). As said above with $|p|^2 = |q|^2$ CP can still be violated in the Lagrangian. In conclusion, in the LOY theory analyzing mixing and instability in a single step, one has to handle a non-hermitian mass-matrix. Its effect, in the presence of CP-violation, is the non-orthogonality of the mass-eigenstates i.e.

$$\langle K_L|K_S\rangle = \langle K_L|K_S\rangle = |p|^2 - |q|^2 \neq 0 \quad (2)$$

It should not come now as a surprise that in LOY theory CP-violation is correlated with the width of $K_{S/L}$. The best place to see it, is by means of the Bell-Steinberger relation which reads [6]

$$(\lambda_L - \lambda_S^*)\langle K_S|K_L\rangle = \sum_f \langle f|T|K_S\rangle^* \langle f|T|K_L\rangle, \quad \lambda_{S/L} \equiv m_{S/L} - \frac{i}{2}\Gamma_{S/L} \quad (3)$$

where T is the transition operator. Evidently, if the transition matrix elements containing K_S or K_L vanish (which is equivalent to having Γ_S or Γ_L zero) and maintaining $m_S \neq m_L$, we are forced to assume that $\langle K_S|K_L\rangle = 0$ i.e. no CP-violation.

A TOY LAGRANGIAN FOR TWO LEVEL MIXING

To compare the results of the LOY theory with a Lagrangian scheme, we start with a simple Lagrangian involving two complex scalar fields ϕ and χ, one scalar neutral field φ and possibly other fields η_i which we do not need to specify further. We have

$$\mathcal{L} = \mathcal{L}_\varphi^{\text{kin}} + \mathcal{L}_\phi^{\text{kin}} + \mathcal{L}_\chi^{\text{kin}} + \mathcal{L}_{\phi\varphi\chi}^{\text{int}} + \mathcal{L}_{\text{CP-violat.}}^{\text{mix}} + \mathcal{L}_{\phi\varphi\chi\eta_i}^{\text{int}} \quad (4)$$

where $\mathcal{L}^{\text{kin}}$ are the usual kinetic terms. We choose for the interaction between ϕ, φ and χ the simple expression

$$\mathcal{L}^{\text{int}}_{\phi\varphi\chi} = \frac{\lambda_{00}}{\sqrt{2}}\phi\varphi^2 + \frac{\lambda_{+-}}{\sqrt{2}}\phi\chi\chi^* - i\frac{\lambda_{000}}{\sqrt{2}}\phi\varphi^3 - i\frac{\lambda_{+-0}}{\sqrt{2}}\chi\chi^*\varphi + \text{h.c.} \quad (5)$$

where λ_{00}, λ_{+-}, λ_{000} and λ_{+-0} are real coupling constants and the factor i in front of λ_{000} and λ_{+-0} is for convenience. We demand now that $\mathcal{L}^{\text{int}}_{\phi\varphi\chi\eta_i}$ be for itself CP-invariant such that φ has the CP quantum numbers of a pseudoscalar. Furthermore, we impose on $\mathcal{L}^{\text{int}}_{\phi\varphi\chi\eta_i}$ a global $U(1)$ symmetry with respect to a global $U(1)$ transformation of only the ϕ field. Then $\mathcal{L}^{\text{mix}}_{\phi\varphi\chi}$ breaks this symmetry and even with $\mathcal{L}^{\text{mix}}_{\text{CP-violat.}}$ being zero we cannot identify the ϕ and ϕ^* as mass-eigenstates. Writing therefore $\phi = (\phi_1 + i\phi_2)/\sqrt{2}$ where $\phi_{1,2}$ are now the proper mass eigenstates (no mixing yet) we obtain easily

$$\mathcal{L}^{\text{int}}_{\phi\varphi\chi} = \lambda_{00}\phi_1\varphi^2 + \lambda_{+-}\phi_1\chi\chi^* + \lambda_{000}\phi_2\varphi^3 + \lambda_{+-0}\phi_2\chi\chi^*\varphi \quad (6)$$

The Lagrangian $\mathcal{L}^{(1)} = \mathcal{L} - \mathcal{L}^{\text{mix}}_{\text{CP-violat.}}$ in which the mass-eigenstates ϕ_1 and ϕ_2 are defined, describes the interaction of four mass-eigenstates fields: the neutral pseudoscalar fields φ, the charged field χ and the scalar ϕ_1 as well as the pseudoscalar ϕ_2. Obviously, since we were able to assign CP quantum numbers to all fields $\mathcal{L}^{(1)}$ is still invariant under CP-transformations. How can we break the CP-invariance through mixing of ϕ_1 and ϕ_2 [5]? To this end we introduce

$$\mathcal{L}^{\text{mix}}_{\text{CP-violat.}} = -\mu^2\phi\phi - \mu^{*2}\phi^*\phi^* = -\sqrt{2}\Re e\mu^2\phi_1^2 + \sqrt{2}\Re e\mu^2\phi_2^2 - \sqrt{2}\Im m\mu^2\phi_1\phi_2 \quad (7)$$

with a complex parameter μ^2. With $\Im m\mu^2 \neq 0$, the fields ϕ_1 and ϕ_2, previously carrying the quantum numbers of scalar and pseudoscalar, respectively, will mix which evidently leads to CP-violation. With $\Im m\mu^2 = 0$, there is no CP-violation, but the mixing remains. Hence, ϕ_1 and ϕ_2 are no longer mass-eigenstates. The mixing due to (7) leads after diagonalization to two new mass-eigenstates, Λ_1 and Λ_2 viz. $\Lambda_{1/2} = \frac{1}{\sqrt{2}}e^{i\theta}[\phi \pm e^{-2i\theta}\phi^*]$. The masses and the mixing angle given by $m^2_{\Lambda_{1/2}} = m^2 \pm 2|\mu^2|$, $\tan 2\theta = \frac{\Im m\mu^2}{\Re e\mu^2}$. In terms of the new mass-eigenstates $\Lambda_{1,2}$ the interaction Lagrangian $\mathcal{L}^{\text{int}}_{\phi\varphi\chi}$ reads

$$\begin{aligned}\mathcal{L}^{\text{int}}_{\phi\varphi\chi} &= \lambda_{00}(\cos\theta\Lambda_1 + \sin\theta\Lambda_2)\varphi^2 + \lambda_{+-}(\cos\theta\Lambda_1 + \sin\theta\Lambda_2)\chi\chi^* \\ &+ \lambda_{000}(-\sin\theta\Lambda_1 + \cos\theta\Lambda_2)\varphi^3 + \lambda_{+-0}(-\sin\theta\Lambda_1 + \cos\theta\Lambda_2)\chi\chi^*\varphi \end{aligned} \quad (8)$$

Performing once again a CP-transformation, but now on mass-eigenstates, we see that the CP-symmetry is broken as long as $\sin\theta \neq 0$ i.e. $\Im m\mu^2 \neq 0$. With CP-violation and provided that $m_{\Lambda_{1,2}}$ are bigger than $2m_\varphi$ and $m_\varphi + 2m_\chi$, both states Λ_1 and Λ_2 will simultaneously decay into two, i.e. $\varphi\varphi$ and $\chi\chi^*$, and three, i.e. $\varphi\varphi\varphi$ and $\chi\chi^*\varphi$, spinless states. The Lagrangian (4) captures, at least theoretically, all important features of the $K^0 - \bar{K}^0$ system and is applicable to fundamental as well as composite fields (in the latter case we should replace the coupling constants by form factors, but the essential conclusions would remain unchanged). Indeed, we can identify ϕ and ϕ^* with

K^0 and $\bar{K}^0$, respectively and the mass m in $\mathcal{L}_\phi^{\text{kin}}$ with the kaon mass parameter usually denoted by m_K. The field φ represents then the neutral pion and χ the charged pion field. The other fields η_i in $\mathcal{L}_{\phi\varphi\chi\eta_1}^{\text{int}}$ stand, of course for more hadronic states. Hence $\mathcal{L}_{\phi\varphi\chi\eta_1}^{\text{int}}$ is the strong interaction part which conserves CP and strangeness, our imposed $U(1)$ symmetry. The latter is broken by weak interaction i.e. by $\mathcal{L}_{\phi\varphi\chi}^{\text{int}}$. If $\Im m\mu^2 = 0$ i.e. there is no CP-violation in the system, Λ_1 and Λ_2 have the quantum numbers of scalar and pseudoscalar and as such can be identified with K_1 and K_2 which do not decay simultaneously into two and three pions. If $\Im m\mu^2 \neq 0$, the CP-violation makes, however, both decays possible. A direct CP-violation can be also switched on by making the coupling constants in $\mathcal{L}_{\phi\varphi\chi}^{\text{int}}$ complex. Anticipating our later discussion, we point out two essential features of the Lagrangian (4). Unlike the $K^0 - \bar{K}^0$ system described in the LOY-theory, the mass-eigenstates in the Lagrangian remain orthogonal and this in spite of CP-violation. Secondly, Λ_1 (Λ_2) has a well-defined anti-particle, namely it is anti-particle to itself. The above is only a comparison. It would be too early to take (4) as a viable phenomenological description of the $K^0 - \bar{K}^0$ system.

CONSEQUENCES OF NON-ORTHOGONALITY

The first curious problem has to do with definition of anti-particles to $K_{S/L}$. If $\Theta \equiv CPT$ transforms the strangeness eigenstates as $\Theta|\bar{K}^0\rangle = e^{-i\delta}|K^0\rangle$ [7], then the CPT-transformed mass-eigenstates $|K_{S/L}^\Theta\rangle$ do not have a difinite mass and lifetime. If, up to a phase, we would demand that $|K_S^\Theta\rangle = |K_S\rangle$ (which is reasonable as the mass-eigenstates kaon do not carry any other quantum number except mass and spin 0), we would end up with $e^{-2i\delta} = 1$ and $|p|^2 = |q|^2$ which obviously is not what one would like to have in the LOY theory. Working with non-hermitian operators, there is still a different possibility to define the CPT-transformation, viz. [8] $\Theta \mathcal{M}_{\text{eff}} \Theta^{-1} = \mathcal{M}_{\text{eff}}^\dagger$ Then one can prove the following theorem [8]: provided that the non-hermitian Hamiltonian $\mathcal{H}$ is normal i.e $[\mathcal{H}, \mathcal{H}^\dagger] = 0$, there exist for every state $|\Psi\rangle$ with definite mass and lifetime a CPT-transformed state defined by $|\Psi^\theta\rangle \equiv \Theta^{-1}|\Psi\rangle$ with the same mass and lifetime as $|\Psi\rangle$. Certainly, for normal Hamiltonians this is a good definition of CPT and anti-particle states. Unfortunately in our case we get $[\mathcal{M}_{\text{eff}}, \mathcal{M}_{\text{eff}}^\dagger] = |p|^2 - |q|^2$ which brings us back to the very source of the problem. It seems therefore that we cannot unambiguously define anti-particle states to the kaon mass-eigenstates as long as the latter are non-orthogonal. The second curious consequence of the non-orthogonality has to do with time evolution. There are at least three different proofs of the following statement [9], [10], [11]: as long as $|p|^2 - |q|^2 \neq 0$, the time evolution beyond the Weisskopf-Wigner approximation is strictly given by $|K_{S/L}(\tau)\rangle = p_{SS/LL}(\tau)|K_{S/L}\rangle + p_{SL/LS}(\tau)|K_{L/S}\rangle$ where the coefficients $p_{S/L} = -p_{L/S}$ are non-zero. Hence, as long as we insist on (2) (say,as a signal as CP-violation), we will get a time evolution which looks like a vacuum regeneration of the mass eigenstates. It has been estimated that the coefficient p_{SL} is tiny [10],[12], indeed too small to be detected experimentally, but as matter of principle we should be worried about the interpretation of this effect. The two last examples display already the unconventional properties of the non-orthogonality of the mass-eigenstates emerging form the LOY theory. We point out once again that they would apply also to the dynamics of the Lagrangian in section 2, had we applied the LOY theory to the mixing and instability.

ACKNOWLEDGMENTS

This work was supported by Conacyt-Mexico.

REFERENCES

1. V. F. Weisskopf and E. P. Wigner, Z. Phys. **63** (1930) 54
2. T. D. Lee, R. Oehme and C. N. Yang, Phys. Rev. **106** (1957) 340
3. P. K. Kabir, *The CP Puzzle*, Academic Press, London 1968
4. S. M. Bilenky and S. T. Petcov, Rev. Mod. Phys. **59** (1987) 671
5. For a different scenario of CP-violation with bosons in the Higgs sector see G. Cvetic, M. Nowakowski and A. Pilaftsis, Phys. Lett. **B301** (1993) 77
6. J. S. Bell and J. Steinberger in Proc. Intern. Conf. on Elementary Particles, Oxford 1965
7. T. D. Lee, *Particle Phyisics and Introduction to Field Theory*, Harwood Academic Press 1981
8. V. S. Mathur and S. G. Rajeev, Mod. Phys. Lett. **A6** (1991) 2741
9. L. A. Khalfin, University of Texas at Austin, CPT-Report no. 211 (1990); *ibid* CPT-Report no. 246 (1991)
10. C. B. Chiu and E. C. G. Sudarshan, Phys. Rev. **D42** (1990) 3712
11. P. K. Kabir and A. Pilaftsis, Phys. Rev. **A53** (1996) 66
12. M. Nowakowski, Int. J. Mod. Phys. **A14** (1999) 589

New approach to the parametrization of the quark and neutrino mixing matrices

Virendra Gupta

Departamento de Fisica Aplicada, CINVESTAV-IPN, Unidad Merida
A.P.73 Cordemex 97310, Merida, Yucatan, Mexico

Abstract. The quark and neutrino mixing matrices, V and V_ν, are written as a linear combination of the unit matrix I and a hermitian unitary matrix (U and U_ν). Thus, $V = \cos\theta I + i\sin\theta U$ and $V_\nu = \cos\theta_\nu I + i\sin\theta_\nu U_\nu$. In general, the matrix V (V_ν) depends on only 3 real parameters including θ (θ_ν). Our *ansatz* gives a good fit of the avaliable data on the CKM-matrix (V) for $\theta = \pi/4$. The neutrino oscillation data requires $\theta_\nu = \pi/4$ for maximal ν_μ and ν_τ mixing with U_ν depending on only one small parameter. Even though V and V_ν are very different, in our approach the remarkable equality $\theta = \theta_\nu = \pi/4$ emerges which suggests an underlying quark-lepton symmetry in the mixing matrices.

INTRODUCTION

Flavor mixing in both the quark and lepton sectors has been firmly established experimentally for a long time. However, still there is no deep theoretical understanding of the observed mixings. In the standard model, these phenomena are described by the CKM matrix V for the quark sector [1]. In the lepton sector, the mixing manifests itself in neutrino oscillations and are described by the MMS matrix V_ν [2]. Recently, a new approach to the parametrization of V and V_ν for three generations of the quarks and leptons was considered [3,4,5] in wich the mixing matrix is divided into two parts. That is, a trivial or "diagonal part" proportional to the unit matrix (I) and a "non-trivial" part represented by a non-diagonal matrix wich causes flavor mixing. Thus, we write the mixing matrices as a linear combination of two matrices.

QUARK SECTOR

We write the CKM matrix

$$V(\theta) = \cos\theta I + i\sin\theta U \tag{1}$$

the value of θ determines the relative importance of the two parts and also the magnitude of CP- violation. For $0 < \theta < \pi/2$, for V to be unitary, U (which is independent of θ) has to be hermitian and unitary. For three generations, mathematically, the 3×3 matrix U depends on at most 4 real parameters, namely two moduli and two phases (see ref. 3).

Explicity

$$U = I - 2 \begin{pmatrix} |a|^2 + |b|^2 & b^*c & a^*c^* \\ bc^* & |a|^2 + |c|^2 & ab^* \\ ac & a^*b & |b|^2 + |c|^2 \end{pmatrix} \quad (2)$$

with the constraints: $|a|^2 + |b|^2 + |c|^2 = 1$ and $\phi_a - \phi_b + \phi_c = \pi/2$, where ϕ_a, ϕ_b and ϕ_c are the phases of the complex numbers a, b and c. Using the freedom to make re-phasing transformation on V (without affecting its physical predictions) the phases in U can be eliminated to make it a real matrix depending on only two positive parameters [4]. Thus, V depends on 3 parameters (including θ), one less than the usual parametrization. The above *ansantz* for V has some interesting features:

(i) the 3 parameters are enough to fit the avaliable data including CP-violation. Also, they can be extracted without ambiguity;
(ii) the matrix V is moduli symmetric, that is $|V_{ij}| = |V_{ji}|$, as a consequence of the hermiticity of U;
(iii) the physical relevant phase for CP-violation $\Phi \equiv \phi_{12} + \phi_{23} - \phi_{13}$ (here ϕ_{ij} is the phase of V_{ij}), is automatically fixed to be equal to $\pi/2$ [3]. The value $|\Phi| = (\pi/2) \bmod 2\pi$ has been suggested [6] as a possible criterion for maximal CP-violation.

As shown before [4], the fit to the avaliable data on the CKM matrix requires a value of $\theta = \pi/4$, implying equal importance of the two parts. This determines the angles of the triangle corresponding to the unitarity constraint

$$V_{11}V_{13}^* + V_{21}V_{23}^* + V_{31}V_{33}^* = 0 \quad (3)$$

The fit with $\theta = \pi/4$, requires [4] that, in standard notation, the values of the three angles of the triangle are $\alpha = 88.46°$, $\beta = 45.06°$ and $\gamma = 46.5°$. The unitarity triangle is predicted to be approximately a right-angled isosceles triangle. The near isosceles nature of the triangle follows from the fact that $|V_{13}| = |V_{31}|$ in our parametrization and experimentally $|V_{11}|$ and $|V_{33}|$ are nearly equal. A consequence of our fit, $\sin 2\beta = 1$ is in excellent agreement with the value $0.99 \pm 0.14 \pm 0.06$ reported by the BELLE Collaboration [7]. For the numerical fits and other details see references 3 and 4.

LEPTON SECTOR

The neutrino mixing matrix

$$V_\nu(\theta_\nu) = \cos\theta_\nu I + i\sin\theta_\nu U_\nu \quad (4)$$

where the angle θ_ν and the non-diagonal matrix U_ν corrrespond to θ and U in the quark case. The general remarks above for V and U apply *mutatis mutandis* to V_ν and U_ν. As shown earlier [5], the atmospheric neutrino data requires that $\theta_\nu = \pi/4$ for the maximal mixing of ν_τ and ν_μ. The solar neutrino problem is solved via the MSW effect [8] with a small mixing angle, with U_ν depending on one small parameter $\varepsilon \sim (1-3.5) \times 10^{-2}$.

Here, V_ν, with only 2 parameters, can fit the avaliable data! The explicit form of V_ν, for $\theta_\nu = \pi/4$, is

$$V = \frac{1}{\sqrt{2}} \begin{pmatrix} 1 - i(1 - 2\varepsilon^2) & \sqrt{2}\varepsilon & -\sqrt{2}\varepsilon\sqrt{1 - 2\varepsilon^2} \\ -\sqrt{2}\varepsilon & 1 & -i\sqrt{1 - 2\varepsilon^2} \\ \sqrt{2}\varepsilon\sqrt{1 - 2\varepsilon^2} & -i\sqrt{1 - 2\varepsilon^2} & 1 - i2\varepsilon^2 \end{pmatrix} \quad (5)$$

For other consequences see ref. 5. It is interesting to look at the corresponding unitarity triangle for the neutrino mixing matrix. Here, again we obtain a right-angled isosceles triangle but with $\alpha_\nu = \gamma_\nu = 45°$ and $\beta_\nu = 90°$. The reason is that moduli of the last two terms in the unitarity constraint for V_ν are equal. It is noteworthy that $\gamma = \gamma_\nu \simeq 45°$.

SUMMARY

In summary, it is remarkable that our *ansantz* works for the two cases and moreover reveals an underlying quark-lepton symmetry in the fact that $\theta = \theta_\nu = \pi/4$, even though V and V_ν are quite different. One may speculate that the mixing matrices considered here and the suggested quark-lepton symmetry may emerge naturally in a grand unified model.

ACKNOWLEDGEMENT

This work was supported by CONACYT Project No. 32598PE.

REFERENCES

1. M. Kobayashi and T. Maskawa, *Prog. Theor. Phys.* **49**, 652, (1973)
2. Z. Maki, M. Nakagawa and S. Sakata, *Prog. Theor. Phys.* **28**, 247, (1962); V.N. Gribov and B.M. Pontecorvo, *Phys. Lett.* **28B**, 493, (1969)
3. V. Gupta, *Int. J. Mod. Phys.* **A16**, 1645, (2001)
4. S. Chaturvedi and V. Gupta, **hep-ph/0110239**, (2001)
5. V. Gupta and X-G He, *Phys. Rev.* **D64**, 117301, (2001)
6. M. Gronau and J. Schechter, *Phys. Rev. Lett.* **54**, 385, (1985); M. Gronau, R. Johnson and J. Schechter, *Phys. Rev. Lett.* **54**, 2176, (1985). Others references can be found in these.
7. BELLE Collaboration, *Phys. Rev. Lett.* **87**, 091802, (2001). Others references to measured value of $\sin 2\beta$ are: CDF Coll. *Phys. Rev.* **D61**, 072205 (2000); ALEPH Coll. *Phys. Lett.* **86**, 2509, (2001); BaBar Coll. *Phys. Rev. Lett.* **87**, 091801, (2001).
8. L. Wolfenstein, *Phys. Rev.* **D17**, 2369, (1978); S. Mikheyev and A. Smirnov, *Sov. J. Nucle. Phys.* **42**, 913, (1985)

Models of Flavor with Discrete Symmetries

Alfredo Aranda

Department of Physics, Boston University, 590 Commonwealth Ave, Boston, MA 02215

Abstract. In an attempt to understand the observed patterns of lepton and quark masses, models invoking a flavor symmetry G_f, under which the Standard Model generations are charged, have been proposed. One particularly successful symmetry, U(2), has been extensively discussed in the literature. The Yukawa matrices in models based on this symmetry reproduce the observed mass ratios in the lepton and quark sectors. The features of the symmetry that determine the texture of the Yukawa matrices can be found in other symmetries as well. We present a model based on a minimal, non-Abelian discrete symmetry that reproduces the Yukawa matrices associated with U(2) theories of flavor. In addition to reproducing the mass and mixing angle relations obtained in such theories, the different representation structure of our new horizontal symmetry allows for solutions to the solar and atmospheric neutrino problems.

INTRODUCTION

In this talk we discuss the possibility of using discrete symmetries to construct models of flavor. We start with the observation that U(2) symmetry has been used [1] to construct a successful model of quarks and charged leptons, where all the mass ratios and mixing angles are generated via two small parameters associated with the breaking of the flavor symmetry. It is interesting to ask whether there is a smaller symmetry that can reproduce the results of the U(2) model and can also be extended to incorporate the recent results on neutrino mixing. This symmetry does indeed exist and it was discussed in [2]. There it was shown that using $T' \times Z_3$ symmetry one can construct a viable and minimal model of flavor. In the next section we review the basic features of the U(2) model and point out the key ingredients that a symmetry must have to generate the desired Yukawa textures. We then present the necessary steps to construct a model using the new local discrete symmetry. This is followed by a review of the minimal model presented in [2], and finally some comments on possible implementations of T' beyond the minimal model are presented before concluding.

U(2) MODEL

The flavor symmetry group is $G_f = U(2)$. Quarks and leptons of the first two generations are assigned to the two-dimensional representation, while the third generation fields are singlets. The assignment of the third generation as a singlet is motivated by the heaviness of the top quark, while putting the first two generations in a doublet yields degenerate scalar masses and thus the model is safe from FCNC contributions. The model contains three flavon fields: ϕ transforming as a doublet, A a singlet, and S a triplet. When these flavons acquire vevs they break the flavor symmetry thus generating the Yukawa textures. The breaking occurs sequentially in two steps, the first one is generated by the vevs of ϕ and S. This happens in such a way that a U(1) symmetry that rotates first generation fields by a phase is left unbroken. This remaining U(1) is then broken down to nothing at a somewhat lower scale by the vev of A. The result is a set of Yukawa textures described by two parameters, ε which is related to the vevs of ϕ and S, and ε' related to the vev of A (and therefore $\varepsilon' < \varepsilon$). For a detailed description of the model see [1]. The ingredients that are key in obtaining the U(2) Yukawa textures are: the **1**, **2**, and **3** representations of U(2) are used in the model; the multiplication rule $\mathbf{2} \otimes \mathbf{2} = \mathbf{3} \oplus \mathbf{1}$ puts the vevs of A in the right place and with the right sign; the existence of a U(1) subgroup that rotates first generation fields by a phase. These are the key ingredients that a smaller symmetry must contain in order to reproduce the successful textures of the U(2) model.

T'

The group T' is the smallest with **1**, **2**, and **3** dimensional representations with the desired multiplication rule [2], therefore we use T' to construct the minimal model. The T' representations are denoted by $\mathbf{1}^0$, $\mathbf{1}^{\pm}$, $\mathbf{2}^0$, $\mathbf{2}^{\pm}$, and **3**, where the superscripts add modulo 3. The remaining step is to determine whether or not it is possible to find a subgroup that allows the breaking of the symmetry sequentially and generate the desired textures. T' has a Z_3 subgroup which can be used as the remaining symmetry during the first breaking, namely, a symmetry that rotates first generation fields by a phase. The two-dimensional representation matrix of the element that generates this subgroup and that corresponds to the desired rotation turns out to correspond to $\mathbf{2}^-$. Unlike the U(2) model however, there is an additional condition that must be satisfied which did not exist before. We argued that it would be interesting to find a smaller symmetry, hence a discrete symmetry is desirable. Furthermore, we are interested in the possibility of having a "local" discrete symmetry. This is motivated by several arguments that global symmetries are violated by quantum gravitational effects [3]. If this is the case we need to make sure the model is anomaly free. This can be accomplished by noting that T' is a subgroup of SU(2) and thus can be embedded in it. If we do this, then the only constraint on the model is that the matter fields fill out complete SU(2) representations, which correspond to the $\mathbf{2}^0$ and $\mathbf{1}^0$ reps of T' (see [2] for details). This, together with the fact that the desired subgroup must rotate first generations fields by a phase, leads us to extend the flavor symmetry to $G_f = T' \times Z_3$, this is the smallest group that has the desired features. The two step breaking now can take place, where the middle step symmetry is the diagonal Z_3^D subgroup of G_f. In passing we note that we have used the linear Ibáñez-Ross condition [4] for the cancellation of anomalies. We also assume that the Z_3 factor may be embedded in a U(1) gauge symmetry whose anomalies are canceled by the Green-Schwarz mechanism [5].

A MODEL

The three generations of matter fields are assigned to the representations $\mathbf{2^{0-}} \oplus \mathbf{1^{00}}$ (the second triality corresponds to the Z_3 and also adds modulo 3), and the Higgs fields $H_{U,D}$ transform as singlets. The Yukawa mass matrices can now be obtained and we introduce three flavons A, ϕ, and S with the representations $\mathbf{1^{0-}}$, $\mathbf{2^{0+}}$, and $\mathbf{3^{-}}$ respectively. Again, the vevs of S and ϕ are assumed to break $T' \times Z_3$ down to Z_3^D putting entries of $O(\varepsilon)$ in the Yukawa matrices, and then finally the vev of A breaks the remaining Z_3^D down to nothing yielding entries of $O(\varepsilon')$. These considerations yield the textures

$$Y_{U,D,L} \sim \begin{pmatrix} 0 & \varepsilon' & 0 \\ -\varepsilon' & \varepsilon & \varepsilon \\ 0 & \varepsilon & 1 \end{pmatrix}, \tag{1}$$

where $O(1)$ coefficients have been omitted. These are the same textures of the $U(2)$ model, as desired. As a note we mention that in order to differentiate between the up-type and down-type quarks it is possible to embed both the $U(2)$ model and the $T' \times Z_3$ model into a GUT, for example into an SU(5) GUT [2]. Now the flavons may have non-trivial transformation properties under the GUT symmetry and the textures are accordingly modified. From now on the discussion will concentrate on this "GUT-model" version. Now that we have reproduced the $U(2)$ model, neutrinos are introduced into the model. Three generations of right-handed neutrinos are introduced with the assignment $\mathbf{2^{0-}} \oplus \mathbf{1^{-+}}$. This assignment leads to Dirac and Majorana mass matrices that allow the introduction of flavons that do not contribute at all to the charged fermion mass matrices. Two such flavons are introduced transforming as $\mathbf{2^{+0}}$ and yielding the following Dirac and Majorana mass matrices:

$$M_{LR} \approx \begin{pmatrix} 0 & l_1\varepsilon' & l_5 r_2 \varepsilon' \\ -l_1\varepsilon' & l_2\varepsilon^2 & l_3 r_1 \varepsilon \\ 0 & l_4\varepsilon & 0 \end{pmatrix} \langle H_U \rangle, \quad M_{RR} \approx \begin{pmatrix} r_3\varepsilon'^2 & r_4\varepsilon\varepsilon' & r_2\varepsilon' \\ r_4\varepsilon\varepsilon' & r_5\varepsilon^2 & r_1\varepsilon \\ r_2\varepsilon' & r_1\varepsilon & 0 \end{pmatrix} \Lambda_R, \tag{2}$$

where $O(1)$ coefficients have been introduced and Λ_R is the right-handed neutrino scale. Using the seesaw mechanism one obtains the texture

$$M_{LL} \sim \begin{pmatrix} (\varepsilon'/\varepsilon)^2 & \varepsilon'/\varepsilon & \varepsilon'/\varepsilon \\ \varepsilon'/\varepsilon & 1 & 1 \\ \varepsilon'/\varepsilon & 1 & 1 \end{pmatrix} \frac{\langle H_U \rangle^2}{\Lambda_R}. \tag{3}$$

This texture leads naturally to large mixing between second- and third-generation neutrinos. The $1-2$ mixing is of $O(\varepsilon'/\varepsilon)$, which can be accommodated to give the bimaximal solution with the use of the $O(1)$ coefficients (in order to determine the mixings accurately one computes the CKM matrix for the lepton sector). In [2] we presented a detailed numerical analysis of this model consisting of a fit to the experimental data, namely the quark and charged lepton masses, the entries of the CKM matrix, and the neutrino oscillation parameters. This fit contained a renormalization group analysis and a χ^2 minimization in order to prove that a set of $O(1)$ coefficients could be found that reproduced the experimental data.

ALTERNATIVE USES

Here we comment on the possibility of using the group T' in different models of flavor. In particular, it can be used as a global symmetry. In this case there is no need for an extra Z_3 and it is possible to have a model that reproduces the U(2) model and accommodates the solutions to the atmospheric and solar neutrino deficits [2]. This is an interesting result when one notes that T' is a subgroup of SU(2), which is known not to lead to a good theory of flavor unless flavor universality is assumed. Another example in which T' can be used is to consider the model based on the local $T' \times Z_6$ symmetry presented in [2]. In this model it is not necessary to have a GUT in order to explain the differences between the up- and down-type quark sectors of the theory. Furthermore, this model also predicts the ratio of m_t/m_b, which in the models described above it is put in by hand. This model also accommodates the neutrino results.

CONCLUSION

Models based on T' flavor symmetry were discussed. In particular a minimal model with $G_f = T' \times Z_3$ that reproduces the U(2) textures for fermion masses was reviewed. This model can also accommodate the results on neutrino oscillations. The main ingredients in the construction of the model were discussed.

ACKNOWLEDGMENTS

This work was done in collaboration with Christopher D. Carone and Richard F. Lebed. The author is supported by the Department of Energy under grant DE-FG02-91ER40676.

REFERENCES

1. R. Barbieri, G. Dvali, and L.J. Hall, Phys. Lett. B **377**, 76 (1996); R. Barbieri, L.J. Hall, and A. Romanino, Phys. Lett. B **401**, 47 (1997); R. Barbieri, L.J. Hall, S. Raby and A. Romanino, Nucl. Phys. **B493**, 3 (1997).
2. A. Aranda, C. D. Carone and R. F. Lebed, Phys. Lett. B **474**, 170 (2000); A. Aranda, C. D. Carone and R. F. Lebed, Phys. Rev. D **62**, 016009 (2000); A. Aranda, C. D. Carone and R. F. Lebed, arXiv:hep-ph/0010144.
3. S. R. Coleman, Nucl. Phys. B **310**, 643 (1988); S. B. Giddings and A. Strominger, Nucl. Phys. B **307**, 854 (1988); G. Gilbert, Nucl. Phys. B **328**, 159 (1989).
4. L.E. Ibáñez and G.G. Ross, Phys. Lett. B **260**, 291 (1991)
5. M. Green and J. Schwarz, Phys. Lett. B **149**, 117 (1984).

On the Implications of Recent SNO Results

Alexis A. Aguilar and J.C. D'Olivo

Instituto de Ciencias Nucleares, Universidad Nacional Autonoma de Mexico, Circuito Exterior C.U., Apartado Postal 70-543, 04510, Mexico D.F.

Abstract. Model-dependent and model-independent implications of the Super Kamiokande and SNO results are outlined. Related deductions about the deformation of the solar neutrino spectrum, and the existence of oscillations between two active types of neutrinos, are revisited.

INTRODUCTION

The recent measurement of the reaction $v_e + d \to p + p + e^-$ by the SNO collaboration [1] can be interpreted as the presence of a non-v_e active neutrino component in the solar neutrino flux observed by Super-Kamiokande (SK). As noted in a previous analysis [2], and also discussed in the SNO paper [1], the comparison between the two experiments can be made in a model-independent way by choosing adequately the energy threshold of SK. Here, we show that this comparison can also be made by taking into account that the observed distortion of the ^{8}B neutrino spectrum is small.

ENERGY SPECTRUM OF SK AND SNO

The neutrino energy spectrum at SK can be seen as the count-rate per energy interval. This will be related to the true spectrum of solar neutrinos arriving at the Earth $\phi(E_v)$, as follows:

$$\frac{dR_{SK}}{dE_v} = \phi(E_v)\left(\sigma^e_{SK}(E_v)P_e(E_v) + \sigma^a_{SK}(E_v)P_a(E_v)\right), \qquad (1)$$

where $\sigma^{e,a}_{SK}(E_v)$ ($a = \mu, \tau$) is the cross section for $v_{e,a}\,e$ elastic scattering, and $P_{e,a}(E_v)$ is the ratio $\phi^{v_{e,a}}(E_v)/\phi(E_v)$. The quantities $P_x(E_v)$ ($x = e, a, s$) satisfy $\sum_x P_x(E_v) = 1$, with s denoting an sterile neutrino. According to the Standard Solar Model (SSM) [3], the only neutrinos that are produced in the Sun are v_e, therefore the energy spectrum measured at SK should be equal to

$$\frac{dR^{SSM}_{SK}}{dE_v} = \phi^{SSM}(E_v)\,\sigma^e_{SK}(E_v), \qquad (2)$$

where $\phi^{SSM}(E_v)$ is the neutrino energy spectrum predicted by the SSM.

We will say that there is *no deformation* of the solar v spectrum *at the Sun* with respect to the SSM prediction if $\phi(E_v) = f\,\phi^{SSM}(E_v)$, with f a constant factor. In making this definition we assume that only v_e are produced in the Sun. This would imply that the ratio of the true total neutrino flux to the predicted total flux of the SSM is constant. In a more general situation, we will have

$$\frac{\phi(E_\nu)}{f\,\phi^{SSM}(E_\nu)} = \zeta(E_\nu) \neq 1 \,. \tag{3}$$

In general, the ratio $r_{SK}(E_\nu)$ of the observed to predicted spectra at SK will be energy-dependent:

$$r_{SK}(E_\nu) = \frac{dR_{SK}/dE_\nu}{dR_{SK}^{SSM}/dE_\nu} = f\left(\mathcal{P}_e(E_\nu) + \frac{\sigma_{SK}^a(E_\nu)}{\sigma_{SK}^e(E_\nu)}\,\mathcal{P}_a(E_\nu)\right), \tag{4}$$

where $\mathcal{P}_{e,a}(E_\nu) = \zeta(E_\nu)\,P_{e,a}(E_\nu)$. For the energy range of SK and SNO, to a good approximation, $\frac{\sigma_{SK}^a(E_\nu)}{\sigma_{SK}^e(E_\nu)} \approx 0.154$ for all E_ν [4]. If the solar neutrino spectrum has no deformation at the Sun then the function $\zeta(E_\nu)$ in (3) is equal to one for all energies. In this case, $\mathcal{P}_{e,a} = P_{e,a}(E_\nu)$, and therefore $\sum_x \mathcal{P}_x(E_\nu) = 1$.

For the SNO experiment the ratio of the observed to the predicted spectra can also be written as

$$r_{SNO}(E_\nu) = \frac{dR_{SNO}/(E_\nu)}{dR^{SSM}/(E_\nu)} = \frac{\phi(E_\nu)\,\sigma_{SNO}^{cc}(E_\nu)\,P_e(E_\nu)}{\phi^{SSM}\sigma_{SNO}^{cc}(E_\nu)} = f\mathcal{P}_e(E_\nu), \tag{5}$$

where $\sigma_{SNO}^{cc}(E_\nu)$ is the cross-section for the charged current reaction $\nu_e + d \to p + p + e^-$. Relations (4) and (5) are model independent. They make no assumption on f or the number of active neutrinos, nor require the quantities $\mathcal{P}_{e,a}(E_\nu)$ to be considered as probabilities.

According to SK [5], the ratio $r_{SK}(E_\nu)$ remains practically constant for the energies above the threshold of 5 MeV, that is, $r_{SK}(E_\nu) = r = constant$. From Eq. (4) this means that

$$\mathcal{P}_e(E_\nu) + 0.154\,\mathcal{P}_a(E_\nu) = \frac{r}{f}\,. \tag{6}$$

From the previous expressions we notice that $\mathcal{P}_{e,a}(E_\nu) = const \Rightarrow r_{SK} = const$, but the converse is not necessarily true. If we consider oscillations between two active neutrinos with no deformation of the spectrum at the Sun ($\zeta(E_\nu) = 1$), then we have $\mathcal{P}_e(E_\nu) = 1 - \mathcal{P}_a(E_\nu)$. In this case, it happens that $r_{SK} = const \Rightarrow \mathcal{P}_{e,a}(E_\nu) = const$. The experimental results also seem to indicate that $r_{SNO}(E_\nu)$ is practically constant. From relation (5) we see that this implies that $\mathcal{P}_e$ is a constant. Using this fact altogether with Eq. (6), we conclude that $\mathcal{P}_a$ is also constant.

Using Eq. (3) and the definition of $\mathcal{P}_{e,a}(E_\nu)$, the ν_e component of the solar neutrino flux $\phi^{\nu_e}(E_\nu) = P_e(E_\nu)\,\phi(E_\nu)$ can be written as

$$\phi^{\nu_e}(E_\nu) = f\,\mathcal{P}_e(E_\nu)\,\phi^{SSM}(E_\nu)\,. \tag{7}$$

On the other hand, we will say that the solar neutrino spectrum has *no deformation at the Earth* when $\phi^{\nu_e}(E_\nu) = k\,\phi^{SSM}(E_\nu)$, where k is a constant. Notice that a constant $\mathcal{P}_e$ implies that there is no distortion of the solar neutrino spectrum at the Earth.

SK AND SNO FLUXES

As indicated in Eq.(1) the SK measured flux of elastic-scattered neutrinos ϕ_{SK}^{ES} is composed of a fraction comming from the scattering of ν_e and a fraction comming from the scattering of the other active flavors ν_a. Therefore, the total count rate measured by this experiment (through which the flux of solar ν 's is determined), can be written as follows:

$$R_{SK} = \overline{\sigma}_{SK}^e \, \phi_{SK}^{ES} \quad , \quad \phi_{SK}^{ES} = \phi_{SK}^{\nu_e} + \frac{\overline{\sigma}_{SK}^a}{\overline{\sigma}_{SK}^e} \, \phi_{SK}^{\nu_a} \,, \tag{8}$$

where we have defined

$$\begin{aligned}
\phi_{SK}^{\nu_{e,a}} &= \phi \, \langle \mathcal{P}_{e,a} \rangle_{SK} \,, \\
\overline{\sigma}_{SK}^{e,a} &= \int dE_\nu \, \varphi^{SSM}(E_\nu) \, \sigma_{SK}^{e,a}(E_\nu) \,, \\
\langle \mathcal{P}_{e,a} \rangle_{SK} &= \frac{1}{\overline{\sigma}_{SK}^{e,a}} \int dE_\nu \, \varphi^{SSM}(E_\nu) \, \sigma_{SK}^{e,a}(E_\nu) \, \mathcal{P}_{e,a}(E_\nu) \,.
\end{aligned} \tag{9}$$

Here ϕ is the true total solar neutrino flux. The quantity $\varphi^{SSM}(E_\nu) = \phi^{SSM}(E_\nu)/\phi^{SSM}$ is the normalized solar neutrino spectrum as predicted by the SSM, which satisfies the relation $\varphi^{SSM} \mathcal{P}_{e,a}(E_\nu) = \varphi(E_\nu) P_{e,a}(E_\nu)$, where $\varphi(E_\nu) = \phi(E_\nu)/\phi$ is the true (and unknown) normalized solar neutrino spectrum. We must keep in mind that the cross section $\sigma_{SK}^e(E_\nu)$ depends on the SK detector resolution [5]

With similar definitions, the count rate for SNO is given by

$$R_{SNO}^{cc} = \overline{\sigma}_{SNO}^{cc} \, \phi_{SNO}^{cc} \,, \tag{10}$$

with

$$\begin{aligned}
\phi_{SNO}^{cc} &= \phi \, \langle \mathcal{P}_e \rangle_{SNO} \,, \\
\overline{\sigma}_{SNO}^{cc} &= \int dE_\nu \, \varphi^{SSM}(E_\nu) \, \sigma_{SNO}^{cc}(E_\nu) \,, \\
\langle \mathcal{P}_e \rangle_{SNO} &= \frac{1}{\overline{\sigma}_{SNO}^{cc}} \int dE_\nu \, \varphi^{SSM}(E_\nu) \, \sigma_{SNO}^{cc}(E_\nu) \, \mathcal{P}_e(E_\nu) \,.
\end{aligned} \tag{11}$$

Here, ϕ_{SNO}^{cc} is the ν_e flux as measured by SNO through the charged current reaction. From (9) and (11) we get

$$\frac{\phi_{SNO}^{cc}}{\phi_{SK}^{ES}} = \frac{\phi_{SK}^{\nu_e}}{\phi_{SK}^{ES}} \times \frac{\langle \mathcal{P}_e \rangle_{SNO}}{\langle \mathcal{P}_e \rangle_{SK}} \,. \tag{12}$$

If the ratio $\phi_{SK}^{\nu_e}/\phi_{SK}^{ES}$ is less than one, then from Eq. (8) this necessarily implies the presence of a non-ν_e active neutrino in the solar neutrino flux. What is actually done in recent analyses is to calculate the ratio $\phi_{SNO}^{cc}/\phi_{SK}^{ES}$. As Eq. (12) shows, in principle, it could be possible to have the ratio $\phi_{SK}^{\nu_e}/\phi_{SK}^{ES}$ equal to one, and still be in agreement with the experimental results from SNO and SK, by having $\langle \mathcal{P}_e \rangle_{SNO}/\langle \mathcal{P}_e \rangle_{SK} < 1$. However, as we explained at the end of the previous section, the combined measurements of SNO and SK suggest that $\mathcal{P}_e(E_\nu)$ has an approximately constant value $\mathcal{P}_e$. If this is true, then form Eqs. (9) and (11) we see that $\langle \mathcal{P}_e \rangle_{SK} \approx \langle \mathcal{P}_e \rangle_{SNO} \approx \mathcal{P}_e$. Therefore, in (12), $\langle \mathcal{P}_e \rangle_{SNO}/\langle \mathcal{P}_e \rangle_{SK} \approx 1$, and $\phi_{SNO}^{cc}/\phi_{SK}^{ES} < 1$ implies that $\phi_{SNO}^{\nu_e}/\phi_{SK}^{ES} < 1$.

As detailed in ref. [2], the response functions of SK and SNO, $\sigma^e_{SK}(E_v)/\overline{\sigma}^e_{SK}\,\varphi^{SSM}(E_v)$ and $\sigma^{cc}_{SNO}(E_v)/\overline{\sigma}^{cc}_{SNO}\,\varphi^{SSM}(E_v)$, that appear in equations (9) and (11) respectively, can be made approximately equal by an appropiate selection of the SK energy threshold. By doing this one also obtains $\langle \mathcal{P}_e \rangle_{SK} \approx \langle \mathcal{P}_e \rangle_{SNO}$, although they do not need to be constants.

CONCLUSIONS

In this work we have explicitly shown that the condition $\langle \mathcal{P}_{e,a} \rangle_{SNO} \approx \langle \mathcal{P}_{e,a} \rangle_{SK}$, which leads to the conclusion that the recent experimental results from SNO can be interpreted as the presence of ν_μ and/or ν_τ in the solar neutrino flux, can be made valid using the observed non-deformation of the ^{8}B neutrino spectrum. This result is consistent with the two neutrino oscillations picture, and it is worthnoting that the fits in the two neutrino picture for the parameters f and $\langle \mathcal{P}_e \rangle$ obtained in [2] give reasonable results. In particular, they obtain the value $f = 1.03^{+0.50}_{-0.58}$, which indicates that such assumptions as those presented in this paper regarding the spectra at the Sun and at the Earth hold true.

ACKNOWLEDGMENTS

This work was partially supported by the CONACyT grants 32279E and 35792E.

REFERENCES

1. SNO Collaboration, Q. R. Ahmad et al, *Phys. Rev. Lett.*, **87**, 071301 (2001).
2. Fogli, G. L., Lisi, E., Montanino, D., and Palazzo, A., *Phys. Rev. D*, **64**, 093007 (2001).
3. J. Bahcall, S. B., M. H. Pissoneault, *Astrophys. J.*, **55**, 990 (2001).
4. Sirlin, A., Bachall, J., and Kamionowski, M., *Phys. Rev. D*, **51**, 6146 (1995).
5. Super-Kamiokande Collaboration, S. Fukuda et al, *Phys. Rev. Lett.*, **86**, 5651 (2001).

Systematic Study of 331 Models

William A. Ponce*, Yithsbey Giraldo* and Luis A. Sánchez[†]

*Instituto de Física, Universidad de Antioquia, A.A. 1226, Medellín, Colombia.
[†]Escuela de Física, Universidad Nacional de Colombia, A.A. 3840, Medellín, Colombia.

Abstract. We carry a systematic study of possible models based on the local gauge group $SU(3)_c \otimes SU(3)_L \otimes U(1)_X$. Old and new models emerge from the analysis.

The Standard Model (SM) based on the local gauge group $SU(3)_c \otimes SU(2)_L \otimes U(1)_Y$ [1] can be extended in different ways: first, by adding new fermion fields (adding a right-handed neutrino constitute its simplest extension, and has deep consequences as the implementation of the see-saw mechanism, and the enlarging of the number of local abelian symmetries that can be gauged simultaneously); second, by augmenting the scalar sector to more than one higgs representation, and third by enlarging the local gauge group. In this last direction, $SU(3)_L \otimes U(1)_X$ as a flavor group has been studied previously by several authors [2]-[6], who have explored possible fermion and higgs-boson representation assignments.

In what follows we carry a systematic analysis of the local gauge model based on the group $SU(3)_c \otimes SU(3)_L \otimes U(1)_X$ which we call the 331 theory.

We assume that the electroweak group is $SU(3)_L \otimes U(1)_X \supset SU(2)_L \otimes U(1)_Y$. We also assume that the left handed quarks (color triplets), left-handed leptons (color singlets) and scalars, transform under the two fundamental representations of $SU(3)_L$ (the 3 and 3*). Two classes of models will be discussed: one family models where the anomalies cancel in each family as in the SM, and family models where the anomalies cancel by an interplay between the families. $SU(3)_c$ is vectorlike as in the SM.

The most general expression for the electric charge generator in $SU(3)_L \otimes U(1)_X$ is a linear combination of the three diagonal generators of the gauge group

$$Q = aT_{3L} + \frac{2}{\sqrt{3}}bT_{8L} + xI_3, \tag{1}$$

where $T_{iL} = \lambda_{iL}/2$, being λ_{iL} the Gell-Mann matrices for $SU(3)_L$ normalized as $\mathrm{Tr}(\lambda_i \lambda_j) = 2\delta_{ij}$, $I_3 = Dg.(1,1,1)$ is the diagonal 3×3 unit matrix, and a and b are arbitrary parameters to be determined anon. Notice that we can absorb an eventual coefficient for x in its definition.

If we assume that the usual isospin $SU(2)_L$ of the SM is such that $SU(2)_L \subset SU(3)_L$, then $a = 1/2$ and we have just one parameter set of models, all of them characterized by the value of b. So, Eq. (1) allows for an infinite number of models in the context of the 331 theory, each one associated to a particular value of the parameter b, with characteristic signatures that make one different from other, as we will see.

There are a total of 17 gauge bosons in the group under consideration, they are: one field B^μ associated with $U(1)_X$, the 8 gluon fields associated with $SU(3)_c$ which remain massless after breaking the symmetry, and another 8 associated with $SU(3)_L$ that we may write in the way

$$\frac{1}{2}\lambda_{\alpha L} A_\mu^\alpha = \frac{1}{\sqrt{2}} \begin{pmatrix} D^0_{1\mu} & W^+_\mu & K^{(1/2+b)}_\mu \\ W^-_\mu & D^0_{2\mu} & K^{-(1/2-b)}_\mu \\ K^{-(1/2+b)}_\mu & K^{(1/2-b)}_\mu & D^0_{3\mu} \end{pmatrix},$$

where $D^\mu_1 = A^\mu_3/\sqrt{2} + A^\mu_8/\sqrt{6}$, $D^\mu_2 = -A^\mu_3/\sqrt{2} + A^\mu_8/\sqrt{6}$, and $D^\mu_3 = -2A^\mu_8/\sqrt{6}$. The upper indices in the gauge bosons in the former expression stand for the electric charge of the corresponding particle, some of them functions of the b parameter as they should be. Notice that the gauge bosons have integer electric charges only for $b = \pm 1/2, \pm 3/2, ; \pm 5/2, ..., \pm(2n+1)/2, n = 1, 2, 3....$ A deeper analysis shows that the negative values for b can be related to the positive ones just by taking the complex conjugate in the covariant derivative of each model, which in turn is equivalent to replace $3 \leftrightarrow 3^*$ in the fermion content of each particular model. So, our first conclusion is that, if we do not want exotic electric charges in the gauge sector of our theory, then b must be equal to 1/2. We will see next that this is also the condition for excluding exotic electric charges in the fermion sector.

Contrary to the SM where only the abelian $U(1)_Y$ factor is anomalous, in the 331 theory both, $SU(3)_L$ and $U(1)_X$ are anomalous ($SU(3)_c$ is vectorlike). So, special combination of multiplets must be used for each particular model in order to cancel the possible anomalies and end with renormalizable models. The triangle anomalies we must take care of are: $[SU(3)_L]^3$, $[SU(3)_c]^2 U(1)_X$, $[SU(3)_L]^2 U(1)_X$, $[grav]^2 U(1)_X$ and $[U(1)_X]^3$.

In order to present specific examples, let us see how the charge operator in Eq.(1) acts on the representations 3 and 3^* of $SU(3)_L$:

$$Q[3] = Dg.(\frac{1}{2} + \frac{b}{3} + x, -\frac{1}{2} + \frac{b}{3} + x, -\frac{2b}{3} + x)$$

$$Q[3^*] = Dg.(-\frac{1}{2} - \frac{b}{3} + x, \frac{1}{2} - \frac{b}{3} + x, \frac{2b}{3} + x).$$

Notice from these expressions that, if we accommodate the known left-handed quark and lepton isodoublets in the two upper components of 3 and 3^* (or 3^* and 3), and forbid the presence of exotic electric charges, then the electric charge for the third component in those representations must be equal either to the charge of the first or second component, which in turn implies $b = \pm 1/2$. Since the negative value is equivalent to the positive one, $b = 1/2$ is a necessary and sufficient condition in order to exclude exotic electric charges in the fermion sector. Let us see some examples.

THE PLEITEZ-FRAMPTON MODEL

As a first example let us take $b = 3/2$, consecuently $Q[3] = Dg.(1 + x, x, x - 1)$ and $Q[3^*] = Dg.(x - 1, x, 1 + x)$. Then the following multiplets are associated with the respective $(SU(3)_c, SU(3)_L, U(1)_X)$ quantun numbers: $(e^-, \nu_e, e^+)_L^T \sim (1, 3^*, 0)$; $(u, d, j)_L^T \sim (3, 3, -1/3)$ and $(d, u, s)_L^T \sim (3, 3^*, 2/3)$, where j and s are isosinglets exotic quarks of electric charges $-4/3$ and $5/3$ respectively. This multiplet structure is the basis of the Pleitez-Frampton model [2] for which the anomaly-free arrangement for three families is given by:

$$q_L^i = (u^i, d^i, j^i)_L^T \sim (3, 3, -1/3); \quad q_L^1 = (d^1, u^1, s)_L^T \sim (3, 3^*, 2/3);$$
$$u_L^{ca} \sim (3, 1, -2/3); \quad d_L^{ca} \sim (3, 1, 1/3); \quad s_L^c \sim (3, 1, -5/3);$$
$$j_L^{ci} \sim (3, 1, -4/3); \quad \psi_L^a = (e^a, \nu^a, e^{ca})_L^T \sim (1, 3^*, 0),$$

where the upper c symbol stands for charge conjugation, $a = 1, 2, 3$ is a family index and $i = 2, 3$ is related to two of the 3 families (in the 331 basis). As can be seen, there are six triplets of $SU(3)_L$ and six anti-triplets, which ensures cancellation of the $[SU(3)_L]^3$ anomaly. A power counting shows that the other four anomalies also vanish.

OTHER 331 FAMILY MODELS IN THE LITERATURE

Let us analyze other two 331 models for three families present in the literature, for which $b = 1/2$. Then: $Q[3] = Dg.(2/3 + x, -1/3 + x, -1/3 + x)$ and $Q[3^*] = Dg.(-2/3 + x, 1/3 + x, 1/3 + x)$. We thus get the following multiplets associated with the given quantun numbers: $(u, d, D)_L^T \sim (3, 3, 0)$, $(e^-, \nu_e, N^0)_L^T \sim (1, 3^*, -1/3)$ and $(d, u, U)_L^T \sim (3, 3^*, 1/3)$, where D and U are exotic quarks with electric charges $-1/3$ and $2/3$ respectively. With this gauge structure we may construct the following anomaly free model for three families:

$$q_L^{'i} = (u^i, d^i, D^i)_L^T \sim (3, 3, 0); \quad q_L^{'1} = (d^1, u^1, U)_L^T \sim (3, 3^*, 1/3);$$
$$u_L^{ca} \sim (3, 1, -2/3); \quad d_L^{ca} \sim (3, 1, 1/3); \quad U_L^c \sim (3, 1, -2/3);$$
$$D_L^{ci} \sim (3, 1, 1/3); \quad \psi_L^{'a} = (e^a, \nu^a, N^{0a})_L^T \sim (1, 3^*, -1/3),$$

where, as before, $a = 1, 2, 3$ and $i = 2, 3$. This model has been analyzed in the literature in Ref. [3]. If needed, the lepton structure can be augmented with an undetermined number of neutral Weyl states $N_L^{0b} \sim (1, 1, 0)$, $b = 1, 2, ...$, without violating the anomaly constraint relations. We call this **Model A**.

The other model has the same quark multiplets used in the previous one arranged in a different way, and uses the new lepton multiplet $\psi"_L = (\nu_e, e^-, E^-)_L^T \sim (1, 3, -2/3)$. The structure of this new anomaly-free model for three families is:

$$q"_L^1 = (u^1, d^1, D)_L^T \sim (3, 3, 0); \quad q"_L^{'i} = (d^i, u^i, U^i)_L^T \sim (3, 3^*, 1/3);$$
$$u_L^{ca} \sim (3, 1, -2/3); \quad d_L^{ca} \sim (3, 1, 1/3); \quad D_L^c \sim (3, 1, 1/3); \quad U_L^{ci} \sim (3, 1, 2/3);$$
$$\psi"_L^a = (\nu^a, e^a, E^a)_L^T \sim (1, 3, -2/3); \quad e^{ca} \sim (1, 1, 1), \quad E^{ca} \sim (1, 1, 1).$$

This model has been analyzed in the literature in Ref. [4]. We call this **Model B**.

TABLE 1. Anomalies for S_i.

Anomalies	S_1	S_2	S_3	S_4	S_5	S_6
$[SU(3)_c]^2 U(1)_X$	0	0	0	0	0	0
$[SU(3)_L]^2 U(1)_X$	$-2/3$	$-1/3$	1	0	0	-1
$[grav]^2 U(1)_X$	0	0	0	0	0	0
$[U(1)_X]^3$	10/9	8/9	$-12/9$	$-6/9$	6/9	12/9
$[SU(3)_L]^3$	1	-1	-3	3	-3	3

OTHER MODELS

Now we want to consider other possible 331 models without exotic electric charges ($b = 1/2$). Let us start first defining the following closed set of fermions (closed in the sense that they include the antiparticles of the charged particles):

$S_1 = [(\nu_\alpha, \alpha^-, E_\alpha^-); \alpha^+; E_\alpha^+]$ with quantum numbers $[(1,3,-2/3);(1,1,1);(1,1,1)]$.

$S_2 = [(\alpha^-, \nu_\alpha, N_\alpha^0); \alpha^+]$ with quantum numbers $[(1,3^*,-1/3);(1,1,1)]$.

$S_3 = [(d,u,U); u^c; d^c; U^c]$ with quantum numbers $(3,3^*,1/3);(3^*,1,-2/3);(3^*,1,1/3)$ and $(3^*,1,-2/3)$, respectively.

$S_4 = [(u,d,D); d^c; u^c; D^c]$ with quantum numbers $(3,3,0);(3^*,1,1/3);(3^*,1,-2/3)$ and $(3^*,1,1/3)$, respectively.

$S_5 = [(e^-, \nu_e, N_1^0); (E^-, N_2^0, N_3^0); (N_4^0, E^+, e^+)]$ with quantum numbers $(1,3^*,-1/3); (1,3^*,-1/3)$ and $(1,3^*,2/3)$, respectively.

$S_6 = [(\nu_e, e^-, E^-); (E_2^+, N_1^0, N_2^0); (N_3^0, E_2^-, E_3^-); e^+; E_1^+; E_3^+]$ with quantum numbers $[(1,3,-2/3); (1,3,1/3); (1,3,-2/3); (1,1,1); (1,1,1); (1,1,1)]$.

The anomalies for the former sets are presented in **Table 1**.
Notice that **Model A** is given by $(3S_2 + S_3 + 2S_4)$ and **Model B** by $(3S_1 + 2S_3 + S_4)$. But they are not the only anomaly-free structures we may build. Let us see:

One family models.
There are two anomaly-free one family structures that can be extracted from the Table. They are:

Model C: $(S_4 + S_5)$. This model is associated with an E_6 subgroup and has been analyzed in Ref. [5].

Model D: $(S_3 + S_6)$. This model is associated with an $SU(6)_L \otimes U(1)_X$ subgroup and has been analyzed in Ref. [6].

The former two models can become realistic models (for 3 families) just by carbon copy each family as in the SM, that is, taking $3(S_4 + S_5)$ and $3(S_3 + S_6)$.

Two family models.
There are three two family models. They are given by: $(S_1 + S_2 + S_3 + S_4)$, $2(S_4 + S_5)$ and $2(S_3 + S_6)$. These three models are not realistic.

Three family models.
Besides models **A** and **B** we have two more. They are: **Model E**: $(S_1 + S_2 + S_3 + 2S_4 + S_5)$ and **Model F**: $(S_1 + S_2 + 2S_3 + S_4 + S_6)$.

The main feature of these last two models is that, contrary to all the other models, each one of the three families is treated in a different way. As far as we know, these two models have not been studied in the literature so far.

We may construct now four, five, etc. family models (a four family model is given for example by: $2(S_1 + S_2 + S_3 + S_4)$), but as for the two family case, they are not realistic.

THE SCALAR SECTOR

Even though the fermion content for each model is different, all $SU(3)_L \otimes U(1)_X$ models presented have a gauge boson sector which depends only on the b parameter. In what follows we are going to reffer only to models for which $b = 1/2$ (models **A-F**). For that particular value of b there are only two Higgs scalars which may develop a nonzero Vacuum Expectation Value (VEV), they are $\phi_1(1,3^*,-1/3) = (\phi_1^-, \phi_1^0, \phi_1^{'0})$, with (VEV) $\langle \phi_1 \rangle = (0,v,V)^T$ and $\phi_2(1,3^*,2/3) = (\phi_2^0, \phi_2^+ \phi_2^{'+})$ with VEV $\langle \phi_2 \rangle = (v',0,0)^T$, with the hierarchy $V > v \sim v' \sim 250$ GeV, the electroweak mass scale. We aim to break the symmetry in the way

$$SU(3)_c \otimes SU(3)_L \otimes U(1)_X \longrightarrow SU(3)_c \otimes SU(2)_L \otimes U(1)_Y \longrightarrow SU(3)_c \otimes U(1)_Q,$$

and produce mass terms for the fermion fields at the same time.

In some models it is more convenient to work with a different arrangement of Higgs fields. For example, in the model in Ref. [5] the following three scalars were used: $\phi_1(1,3^*,-1/3)$ with $\langle \phi_1 \rangle = (0,0,V)^T$, $\phi_2(1,3^*,-1/3)$ with $\langle \phi_2 \rangle = (0,v/\sqrt{2},0)^T$ and $\phi_3(1,3^*,2/3)$ with $\langle \phi_3 \rangle = (v'/\sqrt{2},0,0)^T$. For that particular case we get the following mass terms for the charged gauge bosons in the electroweak sector: $M_{W^\pm}^2 = (g^2/4)(v^2 + v'^2)$, $M_{K^\pm}^2 = (g^2/4)(2V^2 + v'^2)$ and $M_{K^0(\bar{K}^0)}^2 = (g^2/4)(2V^2 + v^2)$. For the neutral gauge bosons a mass term of the form

$$M = V^2 \left(\frac{g'B^\mu}{3} - \frac{gA_8^\mu}{\sqrt{3}}\right)^2 + \frac{v^2}{8}\left(\frac{2g'B^\mu}{3} - gA_3^\mu + \frac{gA_8^\mu}{\sqrt{3}}\right)^2 + \frac{v'^2}{8}\left(gA_3^\mu - \frac{4g'B^\mu}{3} + \frac{gA_8^\mu}{\sqrt{3}}\right)^2,$$

is obtained. Diagonalizing M and defining $Z_1^\mu = Z_\mu \cos\theta + Z'_\mu \sin\theta$, $Z_2^\mu = -Z_\mu \sin\theta + Z'_\mu \cos\theta$, and

$$-\tan(2\theta) = \frac{\sqrt{12}C_W(1-T_W^2/3)^{1/2}[v'^2(1+T_W^2) - v^2(1-T_W^2)]}{3(1-T_W^2/3)(v^2+v'^2) - C_W^2[8V^2 + v^2(1-T_W^2)^2 + v'^2(1+T_W^2)^2]},$$

we get that the photon field A^μ and the neutral fields Z_μ and Z'_μ are given by

$$\begin{aligned}
A^\mu &= S_W A_3^\mu + C_W \left[\frac{T_W}{\sqrt{3}} A_8^\mu + (1 - T_W^2/3)^{1/2} B^\mu\right], \\
Z^\mu &= C_W A_3^\mu - S_W \left[\frac{T_W}{\sqrt{3}} A_8^\mu + (1 - T_W^2/3)^{1/2} B^\mu\right], \\
Z'^\mu &= -(1 - T_W^2/3)^{1/2} A_8^\mu + \frac{T_W}{\sqrt{3}} B^\mu.
\end{aligned}$$

S_W and C_W are, respectively, the sine and cosine of the electroweak mixing angle ($T_W = S_W/C_W$) defined by $S_W = \sqrt{3}g'/\sqrt{3g^2+4g'^2}$. The Y hypercharge associated with the SM gauge boson is:

$$Y^\mu = \left[\frac{T_W}{\sqrt{3}}A_8^\mu + (1-T_W^2/3)^{1/2}B^\mu\right].$$

In the limit $\theta \longrightarrow 0$, $M_Z = M_{W^\pm}/C_W$, and $Z_1^\mu = Z^\mu$ is the gauge boson of the SM. This limit is obtained either by demanding $V \longrightarrow \infty$ or $v'^2 = v^2(C_W^2 - S_W^2)$. In general θ may be different from zero although it takes a very small value.

CONCLUSIONS

In this paper we have studied the theory of $SU(3)_c \otimes SU(3)_L \otimes U(1)_X$ in detail. By restricting the fermion field representations to particles without exotic electric charges we end up with six different realistic models, two one family models and four models for three families which are relatively new in the literature, with two of them (models **E** and **F**) introduced here for the first time, as far as we know.

If we allow for particles with exotic electric charges an infinite number of models can be constructed, where the model in Ref. [2] is just one of them.

The low energy predictions of the six models are not the same. All of them have in common a new neutral current which mixes with the SM neutral current which is also included as part of each model (see Refs. [3]-[6]).

The most remarkable result of our analysis is that, contrary to what is stated in Ref. [2], the 331 theory can be used to construct either one family models or multi-family models, with the number of families being a free number. Conspicuously enough are the existence of models **E** and **F**, where the three families are treated different.

Acknowledgments: Work partially supported by BID and Colciencias in Colombia.

REFERENCES

1. $SU(2)_L \otimes U(1)_Y$ was introduced in: S.L. Glashow, Nucl. Phys. **22**, 579 (1961); S. Weinberg, Phys. Rev. Lett. **19**, 1264 (1967); A. Salam, in *Elementary Particles Theory: Relativistic Groups and Analyticity (Nobel Symposium No.8)*, edited by N. Svartholm (Almqvist and Wiksell, Stockholm, 1968), p.367. For a review of $SU(3)_c$ see for example: W. Marciano and H. Pagels, Phys. Rep. **36C**, 137 (1978).
2. F. Pisano and V. Pleitez, Phys. Rev. **D46**, 410 (1992); P.H. Frampton, Phys. Rev. Lett. **69**, 2887 (1992).
3. M. Singer, J.W.F. Valle and J. Schechter, Phys. Rev. **D22**, 738 (1980); R. Foot, H.N. Long and T.A. Tran, Phys. Rev. **D50**, R34 (1994); H.N. Long, Phys. Rev. **D53**, 437 (1996).
4. M. Ozer, Phys. Rev. **D54**, 1143 (1996); W.A. Ponce, J.B. Flórez and L.A. Sánchez, *"Analysis of the $SU(3)_c \otimes SU(3)_L \otimes U(1)_X$ local gauge theory"*; hep-ph/0103100, IJMPA (to appear).
5. L.A. Sánchez, W.A. Ponce and R. Martínez, Phys. Rev. **D64**, 075013 (2001), hep-ph/0103244.
6. R. Martínez, W.A. Ponce and L.A. Sánchez, *"$SU(3)_c \otimes SU(3)_L \otimes U(1)_X$ as an $SU(6) \otimes U(1)_X$ subgroup"*, Phys. Rev. **D65**, 05xxxx (2002), hep-ph/0110246.

Dirac Neutrino Anapole Moment

L. G. Cabral-Rosetti[a], M. Moreno[b] and A. Rosado[b,c]

[a]*Instituto de Ciencias Nucleares, Universidad Nacional Autónoma de México (ICN-UNAM),*
Circuito Exterior, C.U., Apartado Postal 70-543, 94510 México, D.F., México.
[b]*Instituto de Física, Universidad Nacional Autónoma de México, (IF-UNAM).*
Circuito de la Investigación Científica, C.U., Apartado Postal 20-364, 01000 México, D.F., México.
[c]*Instituto de Física, Benemérita Universidad Autónoma de Puebla, (IF-BUAP).*
Apartado Postal J-48, Colonia San Manuel, Puebla, Puebla. 72570, México.

Abstract. We calculate the Dirac neutrino anapole moment (a_{ν_l}) in the context of the Standard Model (SM) making use of the Dirac form factor $F_D(q^2)$ introduced recently by J. Bernabéu, L. G. Cabral-Rosetti, J. Papavassiliou y J. Vidal by using the Pinch Technique (PT) formalism, working in two different gauge-fixing schemes (R_ξ guage and the electroweak BFM), at the one loop level. We show that the neutrino anapole form factor $F_A(q^2)$ and Dirac form factor $F_D(q^2)$ are related as follows: $F_A(q^2) = \frac{1}{q^2} F_D(q^2)$. Hence, the Dirac neutrino charge radius $\langle r^2_{\nu_l} \rangle$ and the anapole moment satisfy the simple relation $a_{\nu_l} = \frac{1}{6} \langle r^2_{\nu_l} \rangle$. Therefore, we show that the anapole moment (as the charge radius) of the neutrino is a physical quantity, which only gets contribution from the proper neutrino electromagnetic vertex (in electroweak BFM), and that a_{ν_l} is of the order $10^{-34}\, cm^2$.

INTRODUCTION

In 1987, M. Abak and C. Aydin [1] calculated a_{ν_l} in the context of the Standard Model of the electroweak interaction (SM) [2], using the 't Hooft-Feynman gauge [3] and conclude that a_{ν_l} is too small to be measured. In 1992, A. Góngora-T and R. G. Stuart, defined the charge radius and anapole moment of a free fermion as being its vector and axial-vector contact interactions with an external electromagnetic current and they got, at one loop in the SM, a finite and gauge invariant expression for these quantities [4]. In 1987, H. Czyz et al. [5] discussed the anapole moment of charged leptons in the context of the SM and showed that this quantity is gauge dependent and therefore is not a physical quantity in this model. In 1991, M. J. Musolf and B. R. Holstein analized the neutrino matrix element of the electromagnetic current for low q^2 and showed that the anapole moment is just $\frac{1}{6}$ times the charge radius [6]. In 2000, A. Rosado presented a calculation of the neutrino anapole moment in the linear R_ξ gauge at the one loop level in the context of the SM and showed explicitly that it is an infinite and gauge dependent quantity in this model. He also introduced, the electroweak anapole moment $a^{EW}_{\nu_l}$ through the elastic scattering $\nu_l l'$ which is finite, gauge independent, and independent of the lepton l' used to define it [7, 8].

In this work, we calculate the neutrino anapole moment a_{ν_l} in the context of the SM of the electroweak interactions $SU(2)_L \otimes U(1)_Y$ making use of the Dirac form factor introduced by J. Bernabéu, L. G. Cabral-Rosetti, J. Papavassiliou y J. Vidal [9]. This

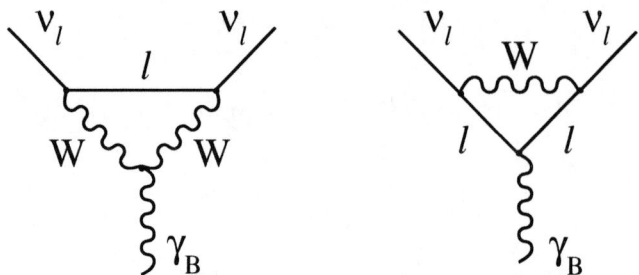

FIGURE 1. Only two proper vertex diagrams contribute to the neutrino charge radius $\langle r_{\nu_l}^2 \rangle$ and therefore neutrino anapole form factor $F_A(q^2)$ at the one-loop level after that the massive gauge cancellation, which takes place when the Pinch Technique is used, in the electroweak Background Field Method.

electromagnetic form factor was obtained through the process $e^+e^- \to \nu_l \bar{\nu}_l$ at the one loop level, by using the Pinch Technique (PT) formalism [10, 11, 12, 13, 14], working in two different gauge-fixing schemes (R_ξ guage and the electroweak BFM), at the one loop level, in the context of SM and becomes for arbitrary momentum transfer q^2: finite, independent on the gauge-fixing parameter, on the gauge-fixing scheme employed, on the Higgs and quark sector of the theory, and on the properties of the charged lepton used to define it. This paper is organized as follows. In the next section, we show that the neutrino anapole form factor $F_A(q^2)$ and $F_D(q^2)$ are related as follows: $F_A(q^2) = \frac{1}{q^2} F_D(q^2)$. Hence, the neutrino charge radius $\langle r_{\nu_l}^2 \rangle$ and the anapole moment satisfy the simple relation $a_{\nu_l} = \frac{1}{6} \langle r_{\nu_l}^2 \rangle$. Therefore, we show that the anapole moment (as the charge radius) of the neutrino is a physical quantity. Also in this section, we give the expression of the anapole form factor and the numerical values of the anapole moment for the three different neutrino species. Finally, we present our conclusions.

THE NEUTRINO ANAPOLE MOMENT

The matrix element of the electromagnetic current in the frame of the SM can be expressed at the lowest order in α, *for all* q^2, where $q = p - p'$, in terms of only one form factor $F_D(q^2)$ as [15, 16, 17, 18]

$$\mathcal{M}_\mu = F_D(q^2) \bar{u}_{\nu_l}(p') \gamma_\mu (1 - \gamma_5) u_{\nu_l}(p). \quad (1)$$

In Fig. 1, we show only the two one-loop diagrams, which contribute to the Dirac form factor, after that the massive gauge cancellation, which takes place when the PT is used [9], in the electroweak Background Field Method [19]. For a massless Dirac neutrino we can rewrite Eq. (1), for all q^2, as follows

$$\mathcal{M}_\mu = \bar{u}_{\nu_l}(p') \{ \gamma_\mu f_1(q^2) - \gamma_\lambda \gamma_5 [g^\lambda{}_\mu q^2 - q^\lambda q_\mu] f_3(q^2) \} u_{\nu_l}(p) \quad (2)$$

where[20] $f_1(q^2) = F_D(q^2)$ and $f_3(q^2) = F_A(q^2) = \frac{1}{q^2} F_D(q^2)$ are the Dirac and anapole form factor of the neutrino, respectively, with

$$f_1(0) = F_D(0) = 0, \tag{3}$$

$$\langle r_{\nu_l}^2 \rangle = -6 \left. \frac{\partial f_1(q^2)}{\partial q^2} \right|_{q^2=0} = -6 \left. \frac{\partial F_D(q^2)}{\partial q^2} \right|_{q^2=0}, \tag{4}$$

and

$$a_{\nu_l} = f_3(0) = \left. \frac{F_D(q^2)}{q^2} \right|_{q^2=0} = \left. \frac{\partial F_D(q^2)}{\partial q^2} \right|_{q^2=0}. \tag{5}$$

That is,

$$a_{\nu_l} = \frac{1}{6} \langle r_{\nu_l}^2 \rangle. \tag{6}$$

Finally, according to the results given for $F_D(q^2)$ in Eq. (7.10) of Ref. [9],

$$F_D(q^2) = -\frac{\alpha e}{8\pi s_W^2} \left\{ 1 + \left(\frac{1}{2} + \frac{M_W^2}{q^2} \right) \left[B_0(q^2; m_l^2, m_l^2) - B_0(q^2; M_W^2, M_W^2) \right] \right.$$
$$\left. + M_W^2 \left(2 + \frac{M_W^2}{q^2} \right) C_0(0, q^2, 0; m_l^2, M_W^2, M_W^2) + \frac{(q^2 + M_W^2)^2}{q^2} C_0(0, q^2, 0; M_W^2, m_l^2, m_l^2) \right\}. \tag{7}$$

Hence, we conclude that the anapole moment, as the charge radius, of the neutrino is a physical quantity, which only gets contribution from the proper neutrino electromagnetic vertex Fig. (1). Taking into account the relations among scalar, two-points B_0 and three-points C_0 Passarino-Veltman functions reported in the Refs. [17, 18, 21] we get

$$a_{\nu_l} = \frac{G_F}{24\sqrt{2}\pi^2} \left\{ 3 - 2\log\left(\frac{m_l^2}{M_W^2}\right) \right\}, \tag{8}$$

The numerical evaluation of the above expression for the three different neutrino species yields: $a_{\nu_e} = 6.8 \times 10^{-34}\, cm^2$, $a_{\nu_\mu} = 4.0 \times 10^{-34}\, cm^2$ and $a_{\nu_\tau} = 2.5 \times 10^{-34}\, cm^2$.

CONCLUSIONS

The main goal in this paper has been to calculate the anapole moment of the neutrino, making use of the Dirac form factor introduced by J. Bernabéu, L. G. Cabral-Rosetti, J. Papavassiliou y J. Vidal [9]. This form factor was obtained through the physical process $e^+e^- \to \nu_l \bar{\nu}_l$ working in the linear R_ξ gauge and using the electroweak BFM in the context of the standard model of the electroweak interactions, at the one loop level. We showed in frame of the SM, that the neutrino anapole form factor $F_A(q^2)$ and the Dirac form factor $F_D(q^2)$ are related as follows: $F_A(q^2) = \frac{1}{q^2} F_D(q^2)$. Therefore the neutrino anapole moment a_{ν_l} and the neutrino charge radius $\langle r_{\nu_l}^2 \rangle$ satisfy the simple relation

$a_{\nu_l} = \frac{1}{6}\langle r_{\nu_l}^2 \rangle$. Hence, we showed that the anapole moment (as the charge radius) of the neutrino becomes a physical quantity, which has the following properties: (i) it only gets contribution from the proper neutrino electromagnetic vertex, (ii) it is finite, (iii) it is independent on the gauge-fixing parameter, (iv) it is independent on the gauge-fixing scheme employed, (v) it does not depend on the Higgs or quark sector of the theory, (vi) it does not depend on the properties of the charged lepton used to define it. The numerical values of a_{ν_l} for the three different neutrinos are of the order $10^{-34}\,cm^2$.

ACKNOWLEDGMENTS

This work was supported in part by the *Programa de Apoyo a Proyectos de Investigación e Inovación Tecnológica* (**PAPIIT**) de la **DGAPA-UNAM** *No. de Proyecto*: IN109001.

REFERENCES

1. M. Abak and C. Aydin, *Europhys. Lett.* **4**, 881 (1987).
2. S.L. Glashow, *Nucl. Phys.* **22**, 579 (1961);
   S. Weinberg, *Phys. Rev. Lett.* **19**, 1264 (1967);
   A. Salam: *Proc. 8th Nobel Symposium*, p. 367, edited by N. Svartholm, (Almqvist and Wiksell, Stockholm, 1968).
3. G. 't Hooft, *Nucl. Phys.* **B35**, 167 (1971).
4. A. Góngora-T and Robin G. Stuart, *Z. Phys.* **C55**, 101 (1992).
5. H. Czyz, K. Kolodziej, M. Zralek and P. Christova, *Can. J. Phys.* **66**, 132 (1988);
   H. Czyz and M. Zralek, *Can. J. Phys.* **66**, 384 (1988).
6. M. J. Musolf and B. R. Holstein, *Phys. Rev.* **D43**, 2956 (1991).
7. A. Rosado, *Phys. Rev.* **D61**, 013001 (2000).
8. A. Rosado, *Rev. Mex. Fís.* **47** (2), 132 (2001).
9. J. Bernabéu, L. G. Cabral-Rosetti, J. Papavassiliou and J. Vidal, *Phys. Rev.* **D62**, 113012 (2000).
10. J. M. Cornwall, in *Proceeding of French-American Seminar on Theoretical Aspects of Quamtun Chromodynamics*, Marseille, France, 1981, edited by J. W. Dash (Centre de Physique Théorique, Marseille, 1982). J. M. Cornwall, *Phys. Rev. D* **26**, 1452 (1982).
11. J. M. Cornwall and J. Papavassiliou, *Phys. Rev.* **D4**, 3474 (1989).
12. J. Papavassiliou, *Phys. Rev.* **D41**, 3179 (1990).
13. G. Degrassi and A. Sirlin, *Phys. Rev.* **D46**, 3104 (1992).
14. N. J. Watson, To appear in the proceedings of the *International Workshop on Physical Variables in Gauge Theories*, Dubna, Russia, 21-25 Sep. 1999., `hep-ph/9912303`.
15. Chung Wook Kimm and Aihud Pevsher, *Neutrinos in Physics and Astrophysics*, Contemporary Concepts in Physics Volume 8, Harwood Academic Publishers 1993, in special see pp. 247-252.
16. Rabindra N. Mohapatra and Palas B. Pal, *Massive Neutrinos in Physics and Astrophysics*, Second Edition, World Scientific Lectures Notes in Physics Vol. 60, 1998, in especial see pp. 193-196.
17. L. G. Cabral-Rosetti, J. Bernabéu, J. Vidal and A. Zepeda, *Eur. Phys. J. C.* **12**, 633 (2000).
18. L. G. Cabral-Rosetti, Ph. D. Thesis *"Factores de Forma del Neutrino e Invariancia Gauge Electrodébil: El Radio de Carga"*, Departament de Física Teòrica, Facultad de Fisiques, Universitat de Valencia, Estudi General, 11 de Diciembre de 2000, Valencia, España.
19. A. Denner, G. Weiglein and S. Dittmaier *Nucl. Phys.* **B440**, 95 (1995).
20. Theory of weak interactions in particle physics: R. E. Marshak, Riazuddin and C. P. Ryan, Willey-Interscience 1998, p. 231.
21. Luis G. Cabral-Rosetti and Miguel A. Sanchis-Lozano, *J. Comp. Appl. Math.* **115**, 93 (2000).

Weinberg's angle in standard-model particles from spin 9+1 dimensional space

J. Besprosvany

Instituto de Física, Universidad Nacional Autónoma de México, Apartado Postal 20-364, México 01000, D. F., México

Abstract. We consider field equations set on an extended spin space that contain fermion and boson solutions, and determined scalar symmetries and representations at given dimension. At 9+1, the $SU(3) \times SU(2)_L \times U(1)$ gauge groups emerge, as well as solution representations with quantum numbers of related gauge bosons, leptons, quarks, Higgs-like particles, and others. The fields' configuration determine a Weinberg's angle of $sin^2(\theta_W) = 3/13 \approx .23078$, under electroweak breaking conditions.

The current theory of elementary particles, the standard model (SM) is successful in describing their behavior, but it is phenomenological. The origin of the interaction groups, the particles' spectrum and representations, and parameters has remained largely unexplained. Still, the generalization of features of the model into larger structures with a unifying principle has suggested connections among the observables. Thus, additional dimensions in Kaluza-Klein theories are associated with gauge interactions, and larger groups in grand-unified theories[1] put some restrictions on them. Spin is a physical manifestation of the fundamental representation of the Lorentz group and it is more so in relation to space, which uses the vector representation. Poincaré-invariant equations formulated on an extended spin space [2] determine boson and spin-1/2 particle solutions with specific symmetry groups and representations. The $SU(3) \times SU(2)_L \times U(1)$ gauge groups emerge as scalar symmetries within the 9+1 dimensional spin space, the minimal symmetry generator space that contains them. We use information from their vector solutions to calculate Weinberg's angle, at electroweak breaking scales.

The Dirac equation [3]

$$\gamma_0(i\partial_\mu\gamma^\mu - M)\Psi = 0, \qquad (1)$$

uses an extended spin space when Ψ represents a matrix instead of, as traditionally, a four-entries (column) spinor. Eq. 1 contains four conditions over four spinors in a 4×4 matrix. There are, then, additional possible transformations and symmetry operations that further classify Ψ. The Dirac-operator transformation $(i\partial_\mu\gamma^\mu - M) \to U(i\partial_\mu\gamma^\mu - M)U^{-1}$ induces the left-hand side of the transformation

$$\Psi \to U\Psi U^\dagger, \qquad (2)$$

and Ψ is postulated to transform as indicated on the right-hand side.

The solutions of Eq. 1 are further characterized with an additional equation [2], consistent with the transformation in Eq. 2, (the Dirac operator transforming accordingly), and both consistent with the Klein-Gordon equation. Indeed, the solutions can be generally characterized as bosonic since Ψ can be understood to be formed of spinors as $\sum_{i,j} a_{ij} |w_i\rangle \langle w_j|$.

U and Ψ can be classified in terms of Clifford algebras. In four dimensions (4-d) U is conventionally a 4×4 matrix containing symmetry operators as the Poincaré generators, but it can contain others, although, e. g., in the chiral massless case it can only carry an additional $U(2)$ scalar symmetry [2]. More symmetry operators appear if Eq. 1, $\mu = 0, ..., 3$, is assumed within the larger Clifford algebra $\{\gamma_\mu, \gamma_\nu\} = 2g_{\mu\nu}, \mu, \nu = 0, ..., N-1$, where N is the (assumed even) dimension, whose structure is helpful in classifying the available symmetries, and which is represented by $2^{N/2} \times 2^{N/2}$ matrices. The usual 4-d Lorentz symmetry, generated in terms of $\sigma_{\mu\nu} = \frac{i}{2}[\gamma_\mu, \gamma_\nu], \mu, \nu = 0, ..., 3$, is maintained and U contains also $\gamma_a, a = 4, ..., N-1$, and their products as possible symmetry generators, whose space we represent by S_N.

Generalized operators acting on this tensor-product space (spinor × spinor × configuration or momentum space) further characterize the solutions. Positive-energy solutions, according to Eq. 1 are interpreted as negative-energy solutions from the right-hand side. This problem is overcome if we assume the hole interpretation for the $\langle w_j|$ components, which amounts to the requirement that operators generally acting from the right-hand side acquire a minus, and that the commutator be used for operator evaluation. Thus, the 4-d plane-wave solution combination $\frac{1}{4}[(1-\gamma_5)\gamma_0(\gamma_1 - i\gamma_2)]e^{-ikx}$, with $k^\mu = (k,0,0,k)$, is a massless vector−axial ($V-A$) state propagating along $\hat{\mathbf{z}}$ with left-handed circular polarization, normalized covariantly according to $\langle \Psi_A | \Psi_B \rangle = tr\Psi_A^\dagger \Psi_B$, the generalized inner product for the solution space. In fact, combinations of solutions of Eqs. 1 and its counterpart can be formed with a well-defined Lorentz index: vector $\gamma_0\gamma_\mu$, pseudo-vector $\gamma_5\gamma_0\gamma_\mu$, scalar γ_0, pseudoscalar $\gamma_0\gamma_5$, and antisymmetric tensor $\gamma_0[\gamma_\mu, \gamma_\nu]$. For example, $A^C_\mu(x) = \frac{i}{2}\gamma_0\gamma_\mu e^{-ikx}$ is a combination that transforms under parity into $A^{C\mu}(\tilde{x})$, $\tilde{x}_\mu = x^\mu$, that is, as a vector.

Solutions contain also products of γ_a matrices that define their scalar-group representation. For given N, there are variations of the symmetry algebra depending on the chosen Poincaré generators and Dirac equation, respectively, through the projection operators $\mathcal{P}_P, \mathcal{P}_D \in S_{N-4}, [\mathcal{P}_P, \mathcal{P}_D] = 0$. $\mathcal{P}_P$ acts as in, e.g., $\mathcal{P}_P\sigma_{\mu\nu}$, and $\mathcal{P}_D$ modifies Eqs. 1 through $\mathcal{P}_D\gamma_0(i\partial_\mu\gamma^\mu - M)$. Together, they characterize the Lorentz and the scalar-group symmetries and solution representations. We require rank$\mathcal{P}_D \leq$ rank$\mathcal{P}_P$, for otherwise pieces of the solution space exist that do not transform properly. For $\mathcal{P}_D \neq 1$ Lorentz operators act trivially on one side of the solutions containing $1 - \mathcal{P}_P$, since $(1 - \mathcal{P}_P)\mathcal{P}_P = 0$, so we also get fermions. Fig. 1 depicts the distribution of Lorentz-representation solutions according to the matrix space they occupy in S_6, when $\mathcal{P}_P = \mathcal{P}_D \neq 1$.

It is apparent that the minimal algebra that includes the SM groups requires $N = 9+1$, with S_6. A detailed analysis is given in [4]. The vector-field normalized polarizations provide the coupling constants [2]. The physics guides the obtention of the field configurations, pointing at overcounted degrees of freedom in K direct-product reducible representations, and gives a clue on the energy scale. The weak $W^{(3,1)}_{B,L\mu}$ interacting with corresponding baryons (B) and leptons (L) fit in 1**d**, with $\bar{W}^{(3,1)}_{B\mu}$, in direct product with $SU(3)$ space, leading to its coupling to be rescaled by $\sqrt{3}$. Assuming for

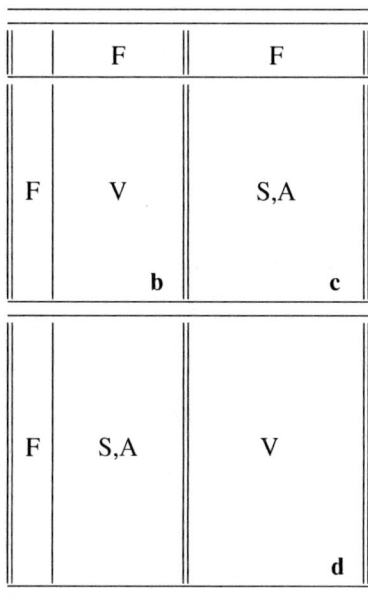

FIGURE 1. Arrangement of S_6 scalar components of $N = 9 + 1$ solutions. S_6 is divided into four 6-d 8×8 matrix blocks, **b, d** of the form $(1 \pm \gamma_5)\gamma\gamma$, and **c** as $(1 + \gamma_5)\gamma$, with fermion (F), vector (and axial-) (V), and scalar (and pseudo-) and antisymmetric (S,A) terms.

the physical weak field $W_\mu^{(3,1)}$ a single $SU(2)_L$ irreducible representation, and the hypercharge Y living in 1b and 1d (with its quantum numbers deducible[4]), we obtain $g'/g = \sqrt{N_{SU(2)}/N_Y} = \{2(2^2)/[2(2+2^2+6(\frac{1}{3})^2+3(\frac{2}{3})^2+3(\frac{4}{3})^2)]\}^{1/2} = \sqrt{\frac{3}{10}}$. From the demand that a vector particle be obtained with the correct parity, or others[2], Q is derived as the only non-trivial scalar commuting the Hamiltonian, leading to an expression for the photon $A_\mu = \frac{1}{\sqrt{g^2+g'^2}}(gB_\mu + g'W_\mu^0)$. One gets Weinberg's angle θ_W from here. It can also be consistently obtained directly from g and g'. From the SM [5] $tan(\theta_W) = g'/g$, so $sin^2(\theta_W) = 3/13 \approx .23078$. The assumed fermion massive terms and electroweak symmetry-breaking conditions suggest a comparison of these numbers with experimental values at energies of order M_Z. This is [6], one standard-deviation last-ciphers uncertainty in parenthesis, $sin^2(\theta_{Wex}) = .23117(16)$. Unified-generator fields are obtained when assuming that Y and I_3 belong to the same group, thus having the same normalization convention, so g' needs to be rescaled to $g_1^{uni} = \sqrt{\frac{5}{3}}g'^{uni}$. From the assumption that $g_1 = g$ and that I_3 acts equally on massless chiral leptons or quarks, we recover the $SU(5)$ unification result[7] $sin^2(\theta_W^{uni}) = 3/8$.

ACKNOWLEDGEMENT

The author acknowledges support from DGAPA, project IN118600, at the Universidad Nacional Autónoma de México.

REFERENCES

1. H. Georgi and S. L. Glashow, Phys. Rev. Lett. **32**, 438 (1974).
2. J. Besprosvany, Int. J. Theor. Phys. **39**, 2797 (2000).
3. Dirac, P. A. M., *The Principles of Quantum Mechanics* (Claredon Press, Oxford, 1947).
4. J. Besprosvany, Submitted, (2001).
5. S. Glashow, Nucl. Phys. **22**, 579 (1961); Weinberg, S., Phys. Rev. Lett. **19**, 1264 (1967); A. Salam, in W. Svartholm (Ed.), *Elementary Particle Theory*, (Almquist and Wiskell, Stockholm, 1968).
6. D.E. Groom, et al., Euro. Phys. Jour. C **15**, 73 (2000).
7. H. Georgi, H. R. Quinn, and S. L. Glashow, Phys. Rev. Lett. **33**, 451 (1974).

The condensate in QED3

A. Bashir, A. Huet and A. Raya

Instituto de Física y Matemáticas, Universidad Michoacana de San Nicolás de Hidalgo, Apartado Postal 2-82, Morelia, Michoacán 58040, México.

Abstract. We study three dimensional quenched QED in the bare vertex approximation. We investigate the gauge dependence of the chiral condensate for a wide range of values of the gauge parameter ξ. The effect of the wavefunction renormalization and the Ward-Green-Takahashi identity is quantitaively studied in restoring the gauge invariance of the said quantity.

INTRODUCTION

At the level of physical observables, gauge symmetry reflects as the fact that all physical observables be independent of the gauge parameter. Perturbation theory respects these requirements at every level of approximation. However, such is not the case in general in the non-perturbative study of dynamical symmetry breaking in gauge field theories through Shwinger-Dyson (SD) equations. Quantum Electrodynamics in 2+1 dimensions (QED3) provides us with a neat laboratory to study dynamical chiral symmetry breaking as it is ultraviolet well-behaved. Burden and Roberts, [1], studied the gauge dependence of the chiral condensate in quenched QED3 and proposed a vertex ansatz which appreciably reduces this gauge dependence in the range $0 - 1$ of the covariant gauge parameter ξ. Unfortunately, the choice of their vertex does not transform correctly under the operation of charge conjugation. Moreover, the selected range of values for ξ is very narrow close to the vicinity of the Landau gauge. We undertake the calculation of the condensate for a wide range of values of ξ within the bare vertex approximation. In this connection, we study the quantitative effects of incorporating wavefunction renormalization and employing the Ward-Green-Takahashi identity on the gauge dependence of the above mentioned physical observable.

WAVEFUNCTION RENORMALIZATION

The SD equation for the fermion propagator in the quenched approximation is :

$$S_F^{-1}(p) = S_F^{0-1}(p) - ie^2 \int \frac{d^3k}{(2\pi)^3} \Gamma^\nu(k,p) S_F(k) \gamma^\mu \Delta^0_{\mu\nu}(q), \qquad (1)$$

where $S_F^0(p) = 1/(\not{p})$ and $\Delta^0_{\mu\nu}(q) = -\left[q^2 g_{\mu\nu} + (\xi-1)q_\mu q_\nu\right]/q^4$ are the bare fermion and photon propagators. $\Gamma^\nu(k,p)$ is the full fermion-boson vertex and $S_F(p)$ is the full fermion propagator which we conveniently parametrize as $S_F(p) = F(p^2)/(\not{p}-M(p^2))$.

In this section, we shall work in the bare vertex approximation so that $\Gamma^\nu(k,p) = \gamma^\nu$. On Wick rotating Eq. (1) to the Euclidean space and carrying out angular integration, we then arrive at the following equations to be solved simultaneously :

$$\frac{1}{F(p^2)} = 1 - \frac{\alpha\xi}{\pi p^2} \int_0^\infty dk \frac{k^2 F(k^2)}{k^2 + M^2(k^2)} \left[1 - \frac{k^2 + p^2}{2kp} \ln\left|\frac{k+p}{k-p}\right| \right], \quad (2)$$

$$\frac{M(p^2)}{F(p^2)} = \frac{\alpha(2+\xi)}{\pi} \int_0^\infty dk \frac{k}{p} \frac{F(k^2)M(k^2)}{k^2 + M^2(k^2)} \ln\left|\frac{k+p}{k-p}\right|, \quad (3)$$

where $\alpha = e^2/4\pi$. A trivial solution to Eq. (3) is $M(p^2) = 0$, which we expect in perturbation theory, where, if the bare mass of the fermion is zero, then at no order of approximation, mass can get generated. However, we shall look for a non-trivial solution which would indicate dynamical symmetry breaking. We select $e = 1$. The form of Eq. (2) suggests that if we are close to the Landau gauge $\xi = 0$, we may be able to put $F(p^2) \approx 1$, an approximation used frequently in the literature. Working in a broad range of ξ, we check the validity of this approximation in calculating the gauge dependence of the condensate, defined $4p^2M(p^2)/(2+\xi)$, [2]. We find that the said approximation introduces a large amount of unwanted gauge dependence as depicted in Fig. (1).

WARD GREEN TAKAHASHI IDENTITY

Another source of gauge dependence of the physical observables can be attributed to the fact that the bare vertex does not obey WGT identity beyond the lowest order. In order to at least partly include the effect of the WGT identity, we can re-write Eq. (1) as

$$\begin{aligned} S_F^{-1}(p) &= S_F^{0-1}(p) - ie^2 \int \frac{d^3k}{(2\pi)^3} \Gamma^\nu(k,p) S_F(k) \gamma^\mu \left(\Delta_{\mu\nu}^{0T} - \xi \frac{q_\mu q_\nu}{q^4} \right) \\ &= S_F^{0-1}(p) - ie^2 \int \frac{d^3k}{(2\pi)^3} \Gamma^\nu(k,p) S_F(k) \gamma^\mu \Delta_{\mu\nu}^{0T} \\ &\quad + ie^2 \xi \int \frac{d^3k}{(2\pi)^3} q_\nu \Gamma^\nu(k,p) S_F(k) \frac{\not{q}}{q^4}, \end{aligned} \quad (4)$$

where $\Delta_{\mu\nu}^{0T}(q) = -[q^2 g_{\mu\nu} - q_\mu q_\nu]/q^4$. We now make use of the WGT identity, i.e., $q_\nu \Gamma^\nu(k,p) = S_F^{-1}(k) - S_F^{-1}(p)$, in the last term of the above equation[1]. Carrying out the angular integration as before, we arrive at the following two equations :

$$F(p^2) = 1 - \frac{\alpha\xi}{\pi p^2} \int_0^\infty dk \frac{k^2 F(k^2)}{k^2 + M^2(k^2)} \left\{ \frac{p^2}{k^2 - p^2} + \frac{p}{2k} \ln\left|\frac{k+p}{k-p}\right| \right.$$

---

[1] An analogous study in QED4 reveals that such a use of the WGT identity eliminates a spurious term, an artifact of the gauge dependent cut-off regulator.

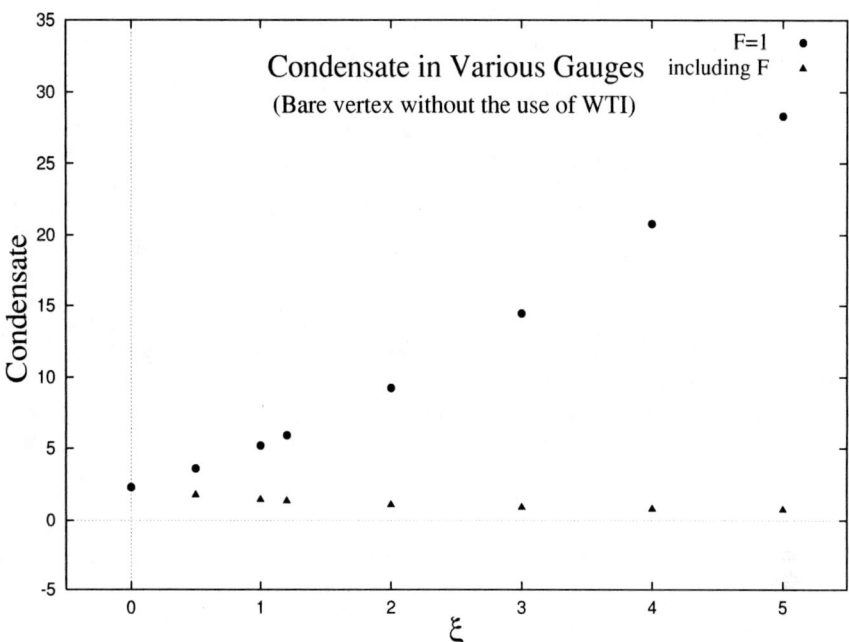

FIGURE 1. Wavefunction renormalization and the gauge dependence of the condensate.

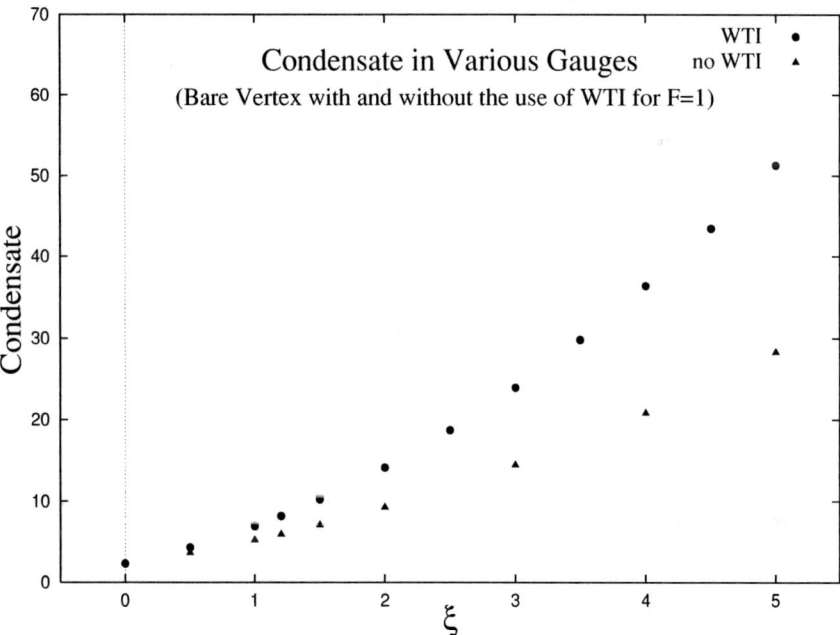

FIGURE 2. WGT identity and the gauge dependence of the condensate.

$$+M(k^2)M(p^2)\left(\frac{1}{k^2-p^2}-\frac{1}{2kp}\ln\left|\frac{k+p}{k-p}\right|\right)\Bigg\},(5)$$

$$\frac{M(p^2)}{F(p^2)} = \frac{\alpha}{\pi}\int_0^\infty dk \frac{k^2 F(k^2)}{k^2+M^2(k^2)}\Bigg\{M(k^2)\left[\frac{2}{kp}\ln\left|\frac{k+p}{k-p}\right|\right]$$

$$-\frac{\xi}{F(p^2)}\left[\frac{M(k^2)-M(p^2)}{k^2-p^2}-\frac{M(k^2)+M(p^2)}{2kp}\ln\left|\frac{k+p}{k-p}\right|\right]\Bigg\}.(6)$$

These equations are much more involved than Eq. (3) and Eq. (6), and are harder to solve. In order to see the effect of using the WGT identity alone, we hold $F(p^2)=1$ and solve Eq. (6). Contrary to our expectations, we see that the use of the WGT identity worsens the gauge dependence of the condensate, as depicted in Fig. (2). A similar effect in QED4 was also observed, [3]. However, there it was thought to be the artifact of the gauge dependent regulator.

CONCLUSIONS

We conclude that in the calculation of the chiral condensate, an observable associated with dynamical symmetry breaking, wavefunction renormalization plays an important role in restoring its gauge invariance. On the other hand, a selective use of the WGT identity seems to increase its gauge dependence.

ACKNOWLEDGEMENTS

We acknowledge the CIC and the CONACyT grants under the projects 4.12 and 32395-E, respectively.

REFERENCES

1. C.J. Burden and C.D. Roberts, Phys. Rev. **D44** 540 (1991).
2. H.D. Politzer, Nucl. Phys. **B117**, 397 (1976).
3. V.P. Gusynin, A.W. Schreiber, T. Sizer y A.G. Williams, Phys. Rev. **D60** 065007 (1999).

Pulsar motions from VEP neutrino oscillations

M. Barkovich[a], H. Casini[b], J.C. D'Olivo[a], and R. Montemayor[c]

[a]*Instituto de Ciencias Nucleares, UNAM, Ap. Postal 70-543, 04510 México DF, Mexico*
[b]*Centre de Physique Teorique - CNRS, Campus de Luminy, Case 907, F-13288 Marseille, France*
[c]*Instituto Balseiro and CAB, Universidad Nacional de Cuyo and CNEA, 8400 Bariloche, Argentina*

Abstract. We show that a violation of the equivalence principle (VEP) can explain pulsar motions. We find that both the translational and rotational velocities can be accounted by VEP induced anisotropies in the linear and angular momentum of the neutrinos emitted by the protoneutron star. The violation needed to obtain the observed motions is compatible with existing boundaries.

INTRODUCTION

Neutron stars are observed as pulsars, and have very characteristic proper motions. Translational velocities of pulsars have a significant component from kicks at the protostar stage [1], that are not satisfactorily explained by proposed mechanisms [2, 3, 4]. Angular velocities are also much larger than expected for the cores of the proto stars [5]. Besides, observational evidence seems to indicate a polarization of the motion along a direction near the plane of the galaxy [6]. This correlation could mean that kicks involve a characteristic length of the order of the galaxy radius, which is very difficult to explain.

A simple explanation for the translational kick could be a 1% anisotropy in the neutrino momenta emitted during pulsar formation. According to the KS mechanism [3], such an anisotropy could be generated by a deformation of the resonance surface when neutrinos undergo matter oscillations in a magnetic field. However, the field must be relatively high, $B \geq 10^{15}$ G [3, 7, 8], and $m_\nu \sim 100$ eV. Such heavy neutrinos are cosmologically ruled out unless they are unstable. The same effect can also be produced by a violation of the equivalence principle(VEP). In this case, neutrino oscillations are due to a flavor dependent coupling of neutrinos to gravity [9], and can happen even for massles neutrinos [10, 4].

In this work, we discuss VEP effects in protoneutron stars, using a generalized parametrized post-Newtonian (PPN) formalism [11, 12]. We show that VEP effects consistent with the present boundaries could generate the necessary kicks to produce the observed pulsar motions [13].

NEUTRINOS IN THE GRAVITATIONAL BACKGROUND

We assume that deviations from a metric theory are small. In presence of VEP all the PPN parameters depend on the flavor numbers. Due to this dependence, distinct neutrinos will undergo different phase shifts when passing through the same space, and with neutrino mixing this leads to oscillations. We will consider only two neutrino flavors, ν_e and ν_μ. They are supposed to be linear superpositions of the gravitational eigenstates ν_1^g and ν_2^g, with a mixing angle θ_g. If electrons are the only leptons present in a medium, the flavor evolution for massless neutrinos in the presence of a static gravitational field is governed by

$$i\frac{d}{dr}\begin{pmatrix}\nu_e\\ \nu_\mu\end{pmatrix} = \left[\frac{\Delta_0}{2}\begin{pmatrix}-\cos 2\theta_g & \sin 2\theta_g\\ \sin 2\theta_g & \cos 2\theta_g\end{pmatrix} + \frac{G_F}{\sqrt{2}}N_e(r)\begin{pmatrix}1 & 0\\ 0 & -1\end{pmatrix}\right]\begin{pmatrix}\nu_e\\ \nu_\mu\end{pmatrix}, \quad (1)$$

where $N_e(r)$ is the electron number density, and $\Delta_0 = E^2 - E^1$. For an isotropic protoneutron star in rigid rotation, in terms of the PPN adimensional parameters γ, γ', Δ_1, Δ_2, Γ, ν, α_1, and α_2, we have ($G = \hbar = c = 1$):

$$\begin{aligned}\Delta_0 = & \left\{-(\delta\gamma + \delta\gamma)U - \delta\Gamma J - \delta\Gamma I(\hat{\mathbf{r}}\cdot\hat{\mathbf{p}})^2\right.\\ & + \left[(\delta\alpha_2 - \frac{1}{2}\delta\alpha_1)U - \delta\alpha_2 J\right]\mathbf{v}\cdot\hat{\mathbf{p}} - \delta\alpha_2 I(\hat{\mathbf{r}}\cdot\mathbf{v})(\hat{\mathbf{r}}\cdot\hat{\mathbf{p}})\\ & \left. - \frac{1}{2}(7\delta\Delta_1 + \delta\Delta_2)J\boldsymbol{\omega}\times\mathbf{r}\cdot\hat{\mathbf{p}}\right\}E, \end{aligned} \quad (2)$$

where $E = p$ is the neutrino energy, $\delta\gamma = \gamma^2 - \gamma^1$, and the same for the difference between the other PPN parameters. Here, ω is the angular velocity and

$$U = 4\pi\int_0^R dr' r'^2 \left[r^{-1}\theta(r-r') + r'^{-1}\theta(r'-r)\right]\rho(r'), \quad (3)$$

$$I = 4\pi r^{-3}\int_0^R dr' r'^2 (r^2 - r'^2)\theta(r-r')\rho(r'), \quad (4)$$

$$3J = 4\pi\int_0^R dr' r' \left[r'^3 r^{-3}\theta(r-r') + \theta(r'-r)\right]\rho(r'), \quad (5)$$

where $\rho(r)$ is the mass distribution of the star. The $\mathbf{v}$ parameter can be interpreted as the existence of a translational effect associated to a preferred direction. Its action is analogous to that of a magnetic field in the KS mechanism.

The Hamiltonian matrix in Eq. (1) is diagonalized by a local rotation, with the mixing angle in matter $\theta_m(r)$ given by $\sin 2\theta_m(r) = \frac{\Delta_0(r)}{\Delta(r)}\sin 2\theta_g$, with $\Delta^2(r) = \left(\Delta_0(r)\cos 2\theta_g - \sqrt{2}G_F N_e(r)\right)^2 + (\Delta_0(r)\sin 2\theta_g)^2$. There is a resonance whenever the diagonal elements of the Hamiltonian vanish, i.e., when $\sqrt{2}G_F N_e(r_R) = \Delta_0(r_R)\cos 2\theta_g$.

THE KICKS

The translational kick is produced by the integrated effect of the anisotropy in the radial momentum of the neutrinos emerging from the resonance surface, which acts as an effective ν_τ-neutrinosphere. Only the radial component of $\hat{\mathbf{p}}$ contributes to this integration. Defining χ as the angle between $\mathbf{r}$ and $\mathbf{v}$

$$\Delta_0 = [A(r) + B(r)\mathrm{v}\cos\chi]E, \tag{6}$$

where $A = -(\delta\gamma' + \delta\gamma)U - \delta\Gamma(I+J)$, and $B = (\delta\alpha_2 - \frac{1}{2}\delta\alpha_1)U - \delta\alpha_2(I+J)$.

The radius of a point on the distorted resonance surface can be written as $r_R = r_o + \delta r \cos\chi$ ($\delta r \ll r_o$). The radius of the unperturbed resonance sphere r_o is determined by the resonance condition, $EA(r_o)\cos 2\theta_g = \sqrt{2}G_F N_e(r_o)$, and

$$\delta r \simeq BA^{-1}\mathrm{v}\left(h_{N_e}^{-1} - h_A^{-1}\right)^{-1}\Big|_{r_o}, \tag{7}$$

where for any function $Z(r)$, $h_Z^{-1} = \frac{d}{dr}\ln Z(r)$.

There are different models for protoneutron stars, such as the hard neutrinosphere models in thermal equilibrium [3, 7], the Eddington neutrinosphere model [8], or models defined by a given density profile [4]. In all of them the momentum asymmetry $\Delta p/p$ is proportional to δr; for example, for a hard neutrinosphere we get $\Delta p/p \approx \frac{2}{9}h_T^{-1}\delta r$. To produce the required translational kick $\Delta p/p \simeq 0.01$. Using the resonance condition and assuming $\delta\alpha_1 \sim \delta\alpha_2 \sim \delta\alpha$, we obtain $10^{-10} \leq \mathrm{v}\delta\alpha \leq 10^{-7}$.

There is also an effect associated to the non radial component of the neutrino momentum. When $\omega = 0$, at a given point of the resonance surface the emitted neutrinos have an azimuthal symmetry respect to the radius vector, but for $\omega \neq 0$ the last term in Eq.(2) brakes this symmetry and produces an angular acceleration of the star. Using a simple hard model for the resonance surface at $r = r_0 + \delta r$, from the resonance condition we get

$$\delta r = CA^{-1}\Lambda h_{N_e}\,\omega\cdot\mathbf{r}\times\hat{\mathbf{p}}\big|_{r_o}, \tag{8}$$

where $C(r) = -\frac{1}{2}(7\delta\Delta_1 + \delta\Delta_2)J$ and $\Lambda = h_A/(h_A - h_{N_e})$.

Neutrinos emitted in different directions come from regions at different r and therefore have different energies and angular momenta. Taking the Stefan-Boltzmann law for the neutrino flux at the resonance surface, a neutrino emitted in a direction $\hat{\mathbf{p}}$ has a momentum $p = E_o(1 + 4h_T^{-1}\delta r)$, where $E_o = E(r_o)$. Therefore, it carries an angular momentum $\mathbf{l} = r_o E_o(\hat{\mathbf{r}} \times \hat{\mathbf{p}})\left(1 + 4h_T^{-1}\delta r\right)$. By integrating at each point of the resonance surface over all directions and points, we compute the angular momentum gained by the star. Because of the symmetry of the system the resulting angular acceleration points along the rotational axis. The angular velocity is given by

$$\omega(t) = \omega_o \exp\left(\frac{4r_o^2}{27}\int_{t_0}^t \frac{C\Lambda}{\Lambda I}\frac{h_{N_e}}{h_T}\dot{E}dt\right), \tag{9}$$

where I and ω_o are the momentum of inertia and the initial angular velocity of the protostar, and $\dot{E}$ is the energy carried by the neutrinos per time unit. It should be noted that the rotational kick does not require a velocity $\mathbf{v}$ associated to a preferred frame. If typical initial angular velocities are of order $\omega_o \sim 0.01\omega_f$, then the VEP parameters must be in the range $10^{-6} \leq \delta\Delta_1 + \frac{1}{7}\delta\Delta_2 \leq 10^{-8}$ to reproduce the observed values for ω_f.

FINAL REMARKS

In sum, resonant VEP neutrino oscillations may be responsible for both the translational and rotational motion of the pulsars, even for massless neutrinos. The values obtained for the VEP parameters are consistent with the present limits [14]. It is interesting to note that if VEP were the only cause for the translational velocity, then all pulsar velocities should be strongly correlated. Such a correlation would be blurred by the presence of other kick mechanisms besides the one considered here.

ACKNOWLEDGMENTS

This work has been partially supported by CONICET (Argentina) and by CONACYT and UNAM (México). M. B. also acknowledges support from SRE (México).

REFERENCES

1. A.G. Lyne, D. R. Lorimer, Nature **369**, 127 (1994); J. M. Cordes, D.F.Chernoff, Ap.J. **505**, 315 (1998); C. Fryer, A. Burrows, W. Benz, Ap.J. **496**, 333 (1998).
2. E. Kh. Akhmedov, A. Lanza, D.W. Sciama, Phys. Rev. **D56**, 6117 (1997); D. Grasso, H. Nunokawa, J.W. Valle, Phys. Rev. Lett. **81**, 2412 (1998); C. J. Horowitz, G. Lee, Phys. Rev. Lett. **80**, 3694 (1998); W. Keil, H.-Th. Janka, E. Muller, Ap.J. **473**, L111 (1996).
3. A. Kusenko, G. Segrè, Phys. Rev. Lett. **77**, 4872 (1996); ibid **79**, 2751 (1997); Phys. Lett. **B396**, 197 (1997); Phys. Rev. **D59**, 061302 (1999); J.C. D'Olivo, J.F.Nieves, P.B.Pal, Phys. Rev **D40**, 3679 (1989); J. C. D'Olivo, J. F. Nieves, Phys. Lett. **B383**, 87 (1996).
4. R. Horvat, Mod. Phys. Lett. **A13**, 2379 (1998).
5. H. Spruit, E. S. Phinney, Nature **393**, 139 (1998).
6. D. R. Lorimer, R. Ramachandran, Proceedings of the IAU 177 meeting, *Pulsar Astronomy 2000 and Beyond*.
7. Y. Z.Qian, Phys. Rev. Lett. **79**, 2750 (1997)
8. H. T. Janka, G. G. Raffelt, Phys. Rev. **D59**, 023005 (1999).
9. M.Gasperini, Phys.Rev. **D38**, 2635 (1988); A. Halprin, C. N. Leung, Phys. Rev. Lett.**67**, 1833 (1991).
10. A. Halprin, C. N. Leung, J. Pantaleone, Phys. Rev. **D53** , 5365 (1995).; J.N.Bahcall, P.I.Krastev, C. N. Leung, Phys. Rev. **D52**, 1770 (1995); J.R. Mureika, R. B. Mann, Phys. Rev. **D54**, 2761 (1996).
11. C.M. Will, *Theory and experiment in gravitational physics* (Cambridge Univ. Press, Cambridge, England, 1981).
12. H.Casini, J. C. D'Olivo, R. Montemayor, L. Urrutia, Phys. Rev. **D59**, 062001 (1999); H.Casini, J. C. D'Olivo, R. Montemayor, Phys. Rev. **D61**, 105004 (2000).
13. M.Barkovich, H.Casini, J. C. D'Olivo, R. Montemayor, Phys. Lett. **B506**, 20 (2001)
14. R. B. Mann, U. Sarkar, Phys. Rev. Lett. **76**, 865 (1996); J. Pantaleone, T. K. Kuo, S.W. Mansour, Phys. Rev. **D61**, 033011 (2000).

# POSTERS

Polarization Studies of Hyperons

C.J. Solano[1]

EFEI, Itajubá and CBPF, Rio de Janeiro, Brazil

Abstract. Using data from Fermilab fixed-target experiment E791, we have measured polarization for Ξ^- and Ξ^+ hyperons in π^- – nucleon interactions at 500 GeV/c. The polarization is measured as function of the angular distribution, transversal to the production plane, and over the ranges of Feynman-x (x_F) and p_T^2 $-0.12 \leq x_F \leq 0.12$ and $0 \leq p_T^2 \leq 4(GeV/c)^2$. We didn't find a convincing evidence for polarization non-zero. Our results only have statistical errors and are very preliminary.

INTRODUCTION

Since the unexpected measurement of substantial polarization for inclusively produced Lambdas by 300 GeV/c protons on Berylium [1], it has been determined that hyperons produced at high energy have non-zero polarization. The measurements all [2] seem to be consistent with a polarization given by:

$$\vec{P}_{a+A \to b+X} = f_{a,b}(x_F, p_T^2, A)\hat{n} \qquad (1)$$

where A is the atomic weight of the target nucleus and $\hat{n}$ the normal to the production plane.

Several theoretical ideas have been proposed to explain this phenomena. Anderson [3] proposed that polarization from the soft, semiclassical process of quark-antiquark pair production by tunneling in a confined color field. De Gran and Miettinen [4] use the parton recombination model to relate the polarization to a Thomas precession like term in the recombination process.

There are a few results of hyperon polarization in π^--nucleus interactions. For Λ^0 in $\pi^- - Cu$ interactions at 230 GeV/c, Barlag [5] found, for $x_F > 0$ and $p_T^2 > 1(GeV/c)^2$, a polarization of -0.28±0.09±0.02. For Ξ^- and Ξ^+ there are no studies with a π^- beam, but just a few results with a proton and Σ^- beam found a polarization of 10%.

---

[1] javier@cbpf.br

THE EXPERIMENT AND THE SAMPLE

As a byproduct of our charm program in Fermilab experiment E791, we collected a large sample of $\Lambda^0/\bar{\Lambda}^0$, Ξ^-/Ξ^+, and Ω^-/Ω^+ hyperons. Initially, we used the Ξ^- and Ξ^+ samples to measure the polarization production reported here, but we expect to continue the studies with the other hyperons.

Experiment E791 recorded data from 500 GeV/c π^- interactions in five thin foils (one platinum and four diamond) separated by gaps of 1.38 to 1.39 cm. Each foil had a thickness of approximately 0.4% of a pion interaction length (0.5 mm for the upstream platinum target, and 1.6 mm for each of the carbon targets). The E791 spectrometer [6] in the Fermilab Tagged Photon Laboratory was a large-acceptance, two-magnet spectrometer augmented by eight planes of multiwire proportional chambers (MWPC) and six planes of silicon microstrip detectors (SMD) for beam tracking. The magnets provided a total transverse momentum kick of 512 MeV/c. Downstream of the target there were 17 planes of SMD's for track and vertex reconstruction, 35 drift chamber planes, two MWPC's, two multicell threshold Čerenkov counters, electromagnetic and hadronic calorimeters (with apertures about 70 by 140 mr), and a muon detector. An important element of the experiment was its extremely fast data acquisition system [7] which was combined with a very open transverse-energy trigger to record a data sample of 2×10^{10} events. The trigger required that the total "transverse energy" (i.e., sum of the products of energy observed times the tangent of the angle from the target to each calorimeter segment) be at least 3 GeV.

The $\Lambda^0/\bar{\Lambda}^0$ sample was from approximately 7% of the overall sample recorded for the experiment. The total signal, taken as the sum of background subtracted signal in each bin, was $2\,587\,870 \pm 1\,780$ Λ^0's and $1\,690\,030 \pm 1\,500$ $\bar{\Lambda}^0$'s. Ξ^- were selected via the decay mode $\Xi^- \to \Lambda^0 \pi^-$ and Ω^- via the decay mode $\Omega^- \to \Lambda^0 K^-$. Beginning with a Λ^0 candidate, we added a third, distinct track as a possible pion or kaon daughter. All three tracks were required to have hits in the drift chamber region only. For these samples, we removed the requirement on the Λ^0 impact parameter. The invariant mass of the candidate hyperon (calculated from the known mass and measured momentum of the Λ^0, as determined by its two decay tracks, together with the third track) was required to be between 1.290 and 1.350 GeV/c^2 for the Ξ^- and between 1.642 and 1.702 GeV/c^2 for the Ω^-. As with the Λ^0 sample, the Ξ^-/Ξ^+ and Ω^-/Ω^+ invariant mass plots were fit to a Gaussian signal plus linear background for each interval of x_F and p_T^2. With the final selection criteria and the full E791 data set, we found $996\,180 \pm 1\,200$ Ξ^-, $706\,620 \pm 1\,020$ Ξ^+, $8\,750 \pm 110$ Ω^-, and $7\,460 \pm 100$ Ω^+ after background subtraction. Again, these numbers and their errors come from the sum of signals in all bins. We checked that the Ξ^- contamination for the Ω^-'s, after all cuts, was negligible.

Selection criteria for the particle and antiparticle samples were identical. However, geometrical acceptances and reconstruction efficiencies were not necessarily the same, mostly due to an inefficient region in the drift chambers produced by the negative pion beam. To evaluate this effect, a large sample of Monte Carlo (MC) events was created using the PYTHIA/JETSET event generators [8]. These were projected through a detailed simulation of the E791 spectrometer to simulate "data" in digitized format which was then processed through the same computer reconstruction code as that used for data from the detector. Candidate events were then subjected to the same selection criteria as that used for data. To account for correlations between x_F and p_T^2, efficiencies were determined in bins of the two parameters.

PROCEDURE

To study the polarization of the Ξ one looks at the angular distribution of the lambda in the decay $\Xi^- \to \Lambda^0 \pi^-$ with $\Lambda^0 \to p^+\pi^-$ (and the corresponding conjugate decay), with respect to the normal to the production plane. This plane is defined as the one formed by the beam direction and the momentum of the Ξ

The angular distribution of Λ's produced is not isotropic but it depends on the cosine of angle between the momentum of the Λ (in the Ξ rest frame) and the quantization direction of the spin ($\alpha_{\Xi^-} = -\alpha_{\Xi^+} = -0.456$)

$$\frac{1}{N_o}\frac{dN_\Xi}{dcos\theta_{\Lambda\hat{n}}} = \frac{1}{2}(1 + \alpha_\Xi P_\Xi cos\theta_{\Lambda\hat{n}}) \tag{2}$$

with

$$cos\theta_{\Lambda\hat{n}} = \frac{\vec{p}_{\Lambda CM_\Xi} \cdot \hat{n}}{|p_{\Lambda CM_\Xi}|} \tag{3}$$

For the actual state of our analysis, in hyperon polarization, we corrected the data by the efficiency of MC. After that, we fitted the angular distributions of polarization for Ξ^- and Ξ^+ and the results you can see in the tables.

Polarization of Ξ^-

P_{Ξ^-}	$0 \le p_T^2 \le 1$	$1 \le p_T^2 \le 2$	$2 \le p_T^2 \le 3$	$3 \le p_T^2 \le 4$
$-.12 \le x_F \le -.06$	0.058 ± 0.048	0.008 ± 0.136	0.472 ± 0.206	0.382 ± 0.284
$-.06 \le x_F \le 0.00$	0.120 ± 0.032	0.026 ± 0.090	0.350 ± 0.150	0.176 ± 0.198
$0.00 \le x_F \le 0.06$	0.008 ± 0.048	0.180 ± 0.126	0.248 ± 0.204	0.144 ± 0.282
$0.06 \le x_F \le 0.12$	0.068 ± 0.156	$-.174 \pm 0.312$	$-.436 \pm 0.518$	0.598 ± 0.606

Polarization of Ξ^+

P_{Ξ^+}	$0 \le p_T^2 \le 1$	$1 \le p_T^2 \le 2$	$2 \le p_T^2 \le 3$	$3 \le p_T^2 \le 4$
$-.12 \le x_F \le -.06$	$-.106 \pm 0.056$	$-.236 \pm 0.138$	$-.148 \pm 0.222$	$-.170 \pm 0.310$
$-.06 \le x_F \le 0.00$	$-.096 \pm 0.034$	$-.128 \pm 0.092$	$-.294 \pm 0.158$	$-.224 \pm 0.238$
$0.00 \le x_F \le 0.06$	$-.082 \pm 0.048$	$-.148 \pm 0.122$	$-.198 \pm 0.218$	0.168 ± 0.290
$0.06 \le x_F \le 0.12$	0.058 ± 0.160	$-.524 \pm 0.326$	0.062 ± 0.360	$-.786 \pm 0.784$

RESULTS AND CONCLUSIONS

Most of the intervals we checked (see tables) are compatible with no polarization. In few intervals we have non zero polarization but just with statistical errors. We need to continue the study to look for systematic effects. Another future step of our analysis is to extend the study to the other two hyperons, $\Lambda^0/\bar{\Lambda}^0$ and Ξ^-/Ξ^+.

ACKNOWLEDGMENTS

We gratefully acknowledge the assistance of the staffs of Fermilab and of all the participating institutions. This research was supported by the FAPEMIG (Brazil).

REFERENCES

1. G. Bunce et al., Phys. Rev. Lett. **36** (1976) 1113.
2. B. Lundberg et al., Phys. Rev. **D 40** (1989) 3557; A. Morelos et al., Phys. Rev. Lett. **71** (1993) 2172; P.M. Ho et al., Phys. Rev. **D 44** (1991) 3402.
3. B. Anderson et al.,Phys. Lett. **B 85** (1979) 417.
4. A. DeGran and H.I. Miettinen, Phys. Rev. **D 24** (1981) 2419.
5. S. Barlag et al.,Phys. Lett. **B 325** (1994) 531.
6. J.A. Appel, Ann. Rev. Nucl. Part. Sci. **42** (1992) 367, and references therein; E.M. Aitala, et al. (E791 Collaboration), EPJdirect **C 4** (1999) 1.
7. S. Amato, et al., Nucl. Instr. and Methods A 324 (1993) 535.
8. *PYTHIA* 5.7 and *JETSET* 7.4. Physics Manual, CERN-TH-7112/93 (1993); H.U. Bengtsson and T. Sjöstrand, Computer Physics Commun **46** (1987) 43; T. Sjostrand, CERN-TH.7112/93 (1993).

Fundamental Measurements and Instrumentation " CKM "

A. Morelos, J. Engelfried, J. Mata, I. Torres, E. Vázquez-Jáuregui
for the CKM Collaboration

Instituto de Física, Universidad Autónoma de San Luis Potosí

Abstract. The physics being pursued by CKM (E921) [1], an experiment recently approved at Fermilab, has as goal testing the description of $C\mathcal{P}$ Violation within the Standard model. Measuring the branching ratio of $K^+ \to \pi^+ \nu \bar{\nu}$ with 10% accuracy, we can extract the magnitude of V_{td} with an overall precision (including theoretical uncertainties) of 10%. Within the collaboration, the experimental high energy physics group at IF-UASLP has the responsibility for designing, testing, and building two Ring Imaging Cherenkov detectors. The present status of the experiment is shown in this poster.

INTRODUCTION

CKM, Charged Kaons at the Main Injector, is an experiment which goal is to measure the branching ratio of the ultra-rare charged kaon decay $K^+ \to \pi^+ \nu \bar{\nu}$ by observing a large sample (~ 100) of these decays with small background [2].

This measurement will play a critical role in testing the Standard Model hypothesis that the sole source of $C\mathcal{P}$ violation in nature resides in the imaginary parts of the V_{td} and V_{ub} Cabibbo, Kobayashi, Maskawa matrix elements. Attacking this question in the kaon sector is both experimentally and theoretically independent of the ongoing programs to measure these same parameters in the B meson sector, see figure 1. Each sector provides an independent test of the Standard Model description of $C\mathcal{P}$ violation. The $K \to \pi \nu \bar{\nu}$ decay mode is the theoretically cleanest laboratory in which to measure the magnitude of V_{td}. The calculation of this decay rate is relative to the $K^+ \to \pi^0 e^+ \nu$ decay mode, this cancels uncertainties due to the hadronic structure of the $K^+ \to \pi^+ X$ transition.

Long distances contributions to this decay have been calculated to be negligible. The only significant uncertainty in the relationship between the branching ratio and $|V_{td}|$ is a small contribution from the charmed quark which depends upon the poorly known charmed quark mass.

Evidence for this decay mode has recently been published by experiment E787 at Brookhaven National Laboratory [3]. They reported the observation of two events quoting a branching ratio of $[1.57^{+1.75}_{-0.82}] \cdot 10^{-10}$. This branching ratio is consistent with the current prediction of $[0.77 \pm 0.21] \cdot 10^{-10}$ [4]. The next important step is a measurement of this rate with sufficient precision to quantitatively challenge the Standard Model interpretation of the source of $C\mathcal{P}$ violation. That is the goal of this experiment.

CKM started from an Expression of Interest submitted to Fermilab in 1996, and recently, after a Proposal Technical Review, it has been approved in June 2001. The experiment expects to take data in the second half of the present decade.

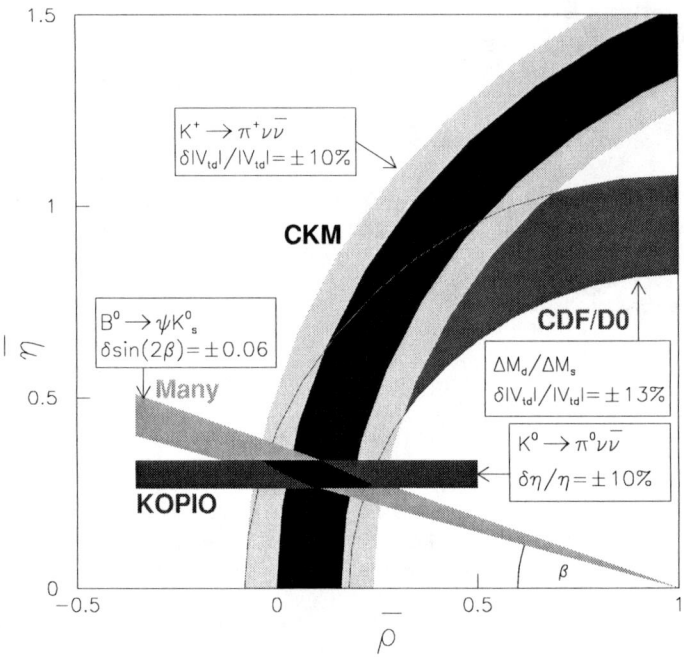

FIGURE 1. A possible outcome for $\bar{\eta}$, $\bar{\rho}$ and β obtained from future K meson and B meson experiments: $K^+ \to \pi^+ \nu \bar{\nu}$ (CKM); $K_L^0 \to \pi^0 \nu \bar{\nu}$ (KOPIO) and $\Delta m_{Bd}/\Delta m_{Bs}$; $B^0(\tilde{B}^0) \to J\psi K_s^0$. The band includes possible theoretical uncertainties in $BR(K^+ \to \pi^+ \nu \bar{\nu})$ connected with the influence of c-quarks [2].

EXPERIMENTAL APPROACH

The challenge of this measurement is clearly experimental. A signal of 100 events if the branching ratio is at the predicted level requires that the apparatus controls all backgrounds to less than the 10^{-11} level.

Redundancy in measuring the kaon and pion vector momentum is important to be able to compute with good precision the missing mass. We will use a magnet and a velocity spectrometers for completely uncorrelated momentum measurements. The velocity spectrometer is based on a pair of Ring Imaging Cherenkov counters (RICHes), one for the kaon and another for the pion. Due to the fact of the fast phototube signal response of the RICHes they are being used for the trigger logic, too. The missing mass reconstructed from the RICH detectors versus the one from the standard magnetic spectrometer is shown in figure 2. The plot is shown in logarithmic scale and also it can be appreciated that the measurements are uncorrelated.

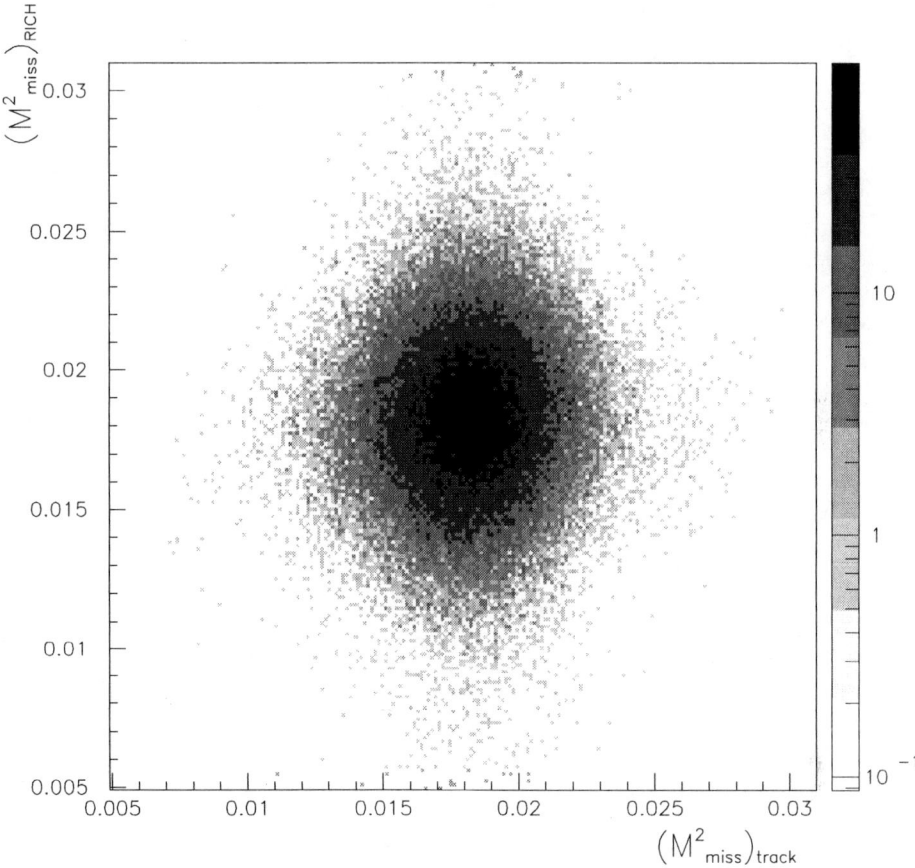

FIGURE 2. Missing mass squared from the velocity (RICH) spectrometer versus missing mass squared from the magnetic (track) spectrometer in GeV^2/c^4 for one million $K^+ \to \pi^+\pi^0$ decays in CKM_GEANT [2].

These RICHes are being designed and prototyped at IF-UASLP, also IF-UASLP forms part of the organizational CKM structure for building and managing the full scale RICHes at Fermilab. Simulation tools are being built and maintained at IF-UASLP and used by the whole collaboration [2, 5]. A laboratory at IF-UASLP is being setup for characterizing phototubes for the RICHes [6]. Recently we are buying from a mexican vendor a spherical and flat mirrors, the flat mirror is part of the kaon RICH to be installed at Fermilab, and the spherical mirror will be used for characterizing the quality of mirrors at IF-UASLP using the Ronchi technique [7].

In addition to the paramount goal of measuring the $K^+ \to \pi^+ \nu \bar{\nu}$ branching ratio we also plan a series of other measurements of rare charged kaon decay properties using the CKM apparatus. The high rate capabilities and redundant measurement capabilities of the CKM spectrometer will make it well suited to such a program of measurements.

CONCLUSION

The hypothesis that the CKM matrix is the sole source of $C\!P$ violation cannot be tested until three independent precision measurement are made. CKM has the potential for a theoretically robust measurement on the $\rho - \eta$ plane in the K system comparable to those expected in the B systems from the running B factories and Hadron Collider experiments. UASLP is significantly involved in this experiment on: detector (Research and Development), software development, and organizational aspects.

ACKNOWLEDGEMENT

This work was supported by CONACyT-México under Grants 28435-E and 34351-E and by FAI-UASLP.

REFERENCES

1. CKM is a Collaboration of: Brookhaven National Laboratory, USA; Fermi National Accelerator Laboratory, USA; Institute of High Energy Physics, Serpukov, Russia; Instituto de Física, Universidad Autónoma de San Luis Potosí, México; University of Michigan, Ann Arbor, USA; University of Texas at Austin, USA; University of Virginia, USA. http://www.fnal.gov/projects/ckm/Welcome.html
2. J.Frank et al., Charged Kaons at the Main Injector, A Proposal for a Precision Measurement of the Decay $K^+ \to \pi^+ \nu \bar{\nu}$ and Other Rare K^+ Processes at Fermilab Using the Main Injector, 2nd Edition submitted to Fermi National Accelerator Laboratory, Batavia, IL, USA, April, 2001.
3. S. Adler et al.: *Further Evidence for the Decay* $K^+ \to \pi^+ \nu \bar{\nu}$, Phys. Rev. Lett., **88**, 041803 (2002); S. Adler et al.: Phys. Rev. Lett. **84**, 3768 (2000); S. Adler et al.: Phys. Rev. Lett. **79**, 2204 (1997).
4. A.J.Buras, $C\!P$ Violation, Rare Decays and the CKM Matrix, these proceedings.
5. I. Torres Aguilar: *Simulación de decaimientos y detectores en experimentos de Kaones*, Tesis de Licenciatura, Facultad de Ciencias Físico Matemáticas, Benemérita Universidad Autónoma de Puebla (2001). Artícle in preparation, to be submitted to Revista de la SMF.
6. Eric Vázquez-Jáuregui et al.: *Avances en la caracterización de fototubos para el RICH de CKM*, Talk at XV Reunión anual de Partículas y Campos, Unidad de Seminarios Ignacio Chávez, C.U., Mexico, D.F., Junio 27-28-29, 2001.
7. L.Stutte, J.Engelfried, J.Kilmer, A Method to evaluate Mirrors for Cherenkov Counters, Nucl. Instr. and Meth. A369, 69-78, (1996).

Single Spin Asymmetries in $p_{\uparrow}\, p \rightarrow \rho + X$

G. Domínguez, G. Herrera, I. León-Monzón

Physics Department. Centro de Investigación y de Estudios Avanzados del IPN. Apdo. 14-740, México D.F, México 07300.

Abstract. We study the single spin asymmetries of ρ mesons in inclusive production in $p\uparrow p$ collisions. We use a two components model and a model for the polarized structure function .

THE TWO COMPONENTS MODEL. Protons are composed, by *uud* valence quarks and a sea of gluons and quarks. ρ mesons, ρ^+, ρ^0 and ρ^-, has a flavor composition of $u\bar{d}$, $\bar{u}d$ and $u\bar{u}+ d\bar{d}$, respectively. When a collision between $p_{\uparrow}$ p take place, ρ^+, ρ^0 and ρ^- may be produced. For example, when a u quark combines with a $\bar{d}$, a ρ^+ meson may be produced. We will separate the differential cross section in two parts, recombination and fragmentation [1]. In ρ^+, a u quark can be a valence quark or a sea quark, but $\bar{u}$ necessarily comes from the sea, so ρ^+ may come from fragmentation as well as recombination.

Recombination Model. When the collision $p\uparrow\ p \rightarrow \rho\ X$, takes place, see figure 1a), one of the valence quarks $u(d)$ combines with a $\bar{d}(u)$ to form a $\rho^+(\rho^-)$, in this process, the sea quark is accelerated [2] because of the different momentum distribution between valence and sea quarks. Is in this moment that Thomas precession effect appears. We define the differential cross section for recombination of ρ^+ when the spin of the proton is up as:

$$\frac{d\sigma^{\uparrow}}{dx_F\, dp_T} = d\sigma^{\uparrow}_{rec} \sim g_u^{\uparrow}\frac{1}{2}\left|M_{\bar{d}}^{\downarrow}\right| + g_u^{\uparrow}\frac{1}{2}\left|M_{\bar{d}}^{\uparrow}\right| + g_u^{\downarrow}\frac{1}{2}\left|M_{\bar{d}}^{\uparrow}\right| + g_u^{\downarrow}\frac{1}{2}\left|M_{\bar{d}}^{\downarrow}\right| \qquad (1)$$

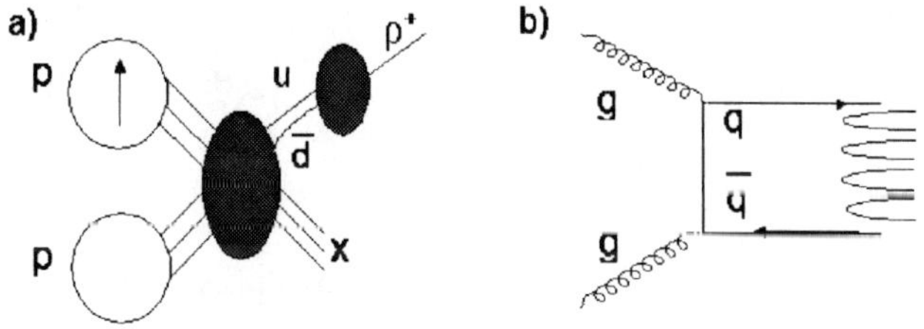

Figure 1. a) Shows the recombination scenario, where a u and $\bar{d}$ recombines to form a ρ^+, and b) shows the fragmentation scenario.

CP623, *Particles and Fields: Eighth Mexican Workshop,* edited by J. L. Diaz-Cruz et al.

Where $g_u^\uparrow$ and $g_u^\downarrow$ denote the probability of finding a u quark aligned or antialigned with the spin of the proton, respectively. Similarly, the cross section for ρ^+ when the proton comes with the spin down is given by :

$$\frac{d\sigma_{rec}^\downarrow}{dx_F dp_T} = d\sigma_{rec}^\downarrow \sim h_u^\uparrow \frac{1}{2}\left|M_d^\downarrow\right| + h_u^\uparrow \frac{1}{2}\left|M_d^\uparrow\right| + h_u^\downarrow \frac{1}{2}\left|M_d^\uparrow\right| + h_u^\downarrow \frac{1}{2}\left|M_d^\downarrow\right|, \quad (2)$$

here $h_u^\uparrow$ and $h_u^\downarrow$ denotes the probability of finding a quark u aligned or antialigned, respectively, when the proton have its spin down.

Clearly $g_u^\uparrow = h_u^\downarrow$ and $g_u^\downarrow = h_u^\uparrow$. $\left|M_d^\downarrow\right|^2$ ($\left|M_d^\uparrow\right|^2$) represent the probability of spin flip up (down) at the moment of recombination. In the recombination process the scattering amplitude, $\left|M_d^\uparrow\right|^2$ ($\left|M_d^\downarrow\right|^2$) for $p\uparrow p \rightarrow \rho X$ is inversely proportional to the energy difference between initial and final states, denoted by i and f in figure 1,

$$M_S \propto \frac{1}{\Delta E + \vec{S}\cdot\vec{\omega}} \quad (3)$$

here ΔE represents the energetic change between final and initial states without spin effects. $\vec{\omega}_T$ is the Thomas frequency [3]. For ρ^+, for example, ΔE is given by,

$$\Delta E = (p_u^2 + m_u^2)^{\frac{1}{2}} + (p_d^2 + m_d^2)^{\frac{1}{2}} - (p_{\rho^+}^2 + m_{\rho^+}^2)^{\frac{1}{2}}. \quad (4)$$

In the Breit frame ΔE can be written as:

$$\Delta E = \frac{1}{2x_F p}\left(\frac{p_{uT}^2 + m_u^2}{1-\xi}\right)^{\frac{1}{2}} + \left(\frac{p_{dT}^2 + m_d^2}{\xi}\right)^{\frac{1}{2}} - \left(p_{T\rho^+}^2 + m_{\rho^+}^2\right)^{\frac{1}{2}} \quad (5)$$

where $x_F = x_u + x_{\bar{d}}$, and $\xi = x_{\bar{d}}/x_F$, p is the momenta of the proton. $\vec{\omega}_T$ is defined as in [2],

$$\vec{\omega}_T = \frac{\langle\sin\theta\rangle}{\Delta t}\frac{\Delta p}{m}\hat{n} \quad (6)$$

where Δp represents the change in momentum of the sea quark $\bar{d}$ for ρ^+.

$$\Delta p = \left(\frac{x_F}{2} + x_{\bar{d}}\right)p \quad (7)$$

We assume that at the final state u and $\bar{d}$ share the momentum. In order to evaluate the asymmetry, ΔE and ω_T (or Δp) must be averaged over the appropriate parton distribution function. As in [2]

$$\omega_T = \frac{1}{x_F} \frac{4}{\Delta x_0} \frac{1-2\xi}{1+2\xi} p_{T\rho^0}, \qquad (8)$$

we take m_u and m_d as given in PDG [4], where $m_u/m_d = 2$. De Grand and Miettinen [2] assume a linear dependence between ξ and x_F, in [5] $\xi(x_F)$ is explicitly calculated, and the results does not change drastically if a linear dependence is used. Therefore we take

$$\xi(x_F) = \frac{1}{2}(1-x_F) + 0.1 x_F, \qquad (9)$$

So far, we have all the elements to calculate the differential cross section for recombination scenario.

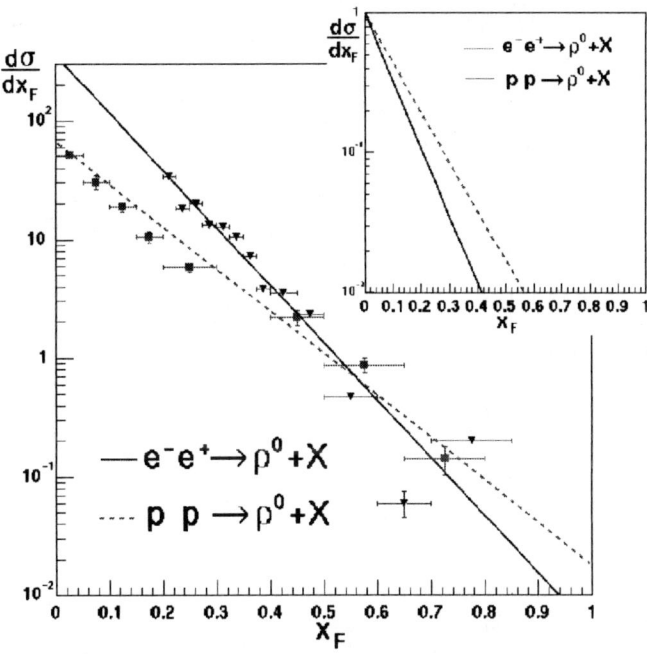

FIGURE 2. Differential cross section in e- e+ [6] and p p interaction [7].
Insert: Cross section normalized to each other. The different behavior indicates the contribution of soft processes, i.e, recombination.

The Fragmentation Model. In a deep inelastic process, ρ mesons are created, figure 1. The differential cross section for ρ mesons has been studied in $e^- e^+$ annihilations [6]. In this reaction, information about the spin polarization of the proton does not contribute. With this assumption, the differential cross section is independent of recombination process. In [6] the differential cross section is given by an exponential of the form, see figure 2,

$$\frac{d\sigma}{dx_F} = Ae^{-Bx_F}.\tag{10}$$

SINGLE SPIN ASYMMETRIES. In the framework where ρ particles are produced by the two mechanisms, the asymmetry is given by

$$A_N = \frac{d\sigma_{rec}^{\uparrow} - d\sigma_{rec}^{\downarrow}}{d\sigma_{rec}^{\uparrow} + d\sigma_{rec}^{\downarrow} + 2d\sigma_{frag}^{\downarrow}},\tag{11}$$

the asymmetry for ρ^+ is given by:

$$A_N = \frac{d\sigma_{rec}^{\uparrow} - d\sigma_{rec}^{\downarrow}}{d\sigma_{rec}^{\uparrow} + d\sigma_{rec}^{\downarrow} + 2d\sigma_{frag}^{\downarrow}},\tag{12}$$

where $\frac{\Delta u}{u} \equiv \frac{g_u^{\uparrow} - g_u^{\downarrow}}{g_u^{\uparrow} + g_u^{\downarrow}}$ and we have used Eqs. 1 and 2. Single spin asymmetry for ρ^+ can be written as

$$A_{\rho^+} \cong \frac{1}{3}\left(\frac{\Delta u}{u}\left(\frac{|M_{\bar{d}\downarrow}|^2 - |M_{\bar{d}\uparrow}|^2}{|M_{\bar{d}\downarrow}|^2 + |M_{\bar{d}\uparrow}|^2 + 2d\sigma_{frag}^{\downarrow}}\right)\right).\tag{13}$$

As we can see, A_{ρ^+} is just $-\frac{1}{3}A_{\pi^+}$, see [8], this is a prediction of our model. This prediction has been reported in [9]. Also For ρ^- and ρ^0 similar procedure is applied, but in ρ^0 we must take into account its wave function $\frac{1}{\sqrt{2}}(u\bar{u} + d\bar{d})$. Finally, asymmetries for ρ^- and ρ^0 are

$$A_{\rho^-} \cong \frac{1}{3}\frac{\Delta d}{d}\left[\frac{\omega_T}{\Delta E} - 2\frac{\omega_T}{\Delta E}\left(\frac{d\sigma_{frag}^{\downarrow}}{1/(\Delta E)^2 + 2d\sigma_{frag}^{\downarrow}}\right)\right],\tag{14}$$

and

$$A_{\rho^0} \cong -\frac{1}{3}\left(2\frac{\Delta u}{u} + \frac{\Delta d}{d}\right)\left[\frac{|M_{\bar{d}\downarrow}|^2 - |M_{\bar{d}\uparrow}|^2}{|M_{\bar{d}\downarrow}|^2 + |M_{\bar{d}\uparrow}|^2 + 2d\sigma_{frag}^{\downarrow}}\right],\tag{15}$$

as expected, the factor 2 in $\frac{\Delta u}{u}$ is due to the 2:1 ratio between the *u* and *d* valence quarks in the proton. A_{ρ^-} and A_{ρ^0} are related to A_{π^-} and A_{ρ^0} as follows, and $A_{\rho^-} \approx -1/3 A_{\pi^-}$ [8] . This is an important result of the two components model. This result has been reported too in [9].

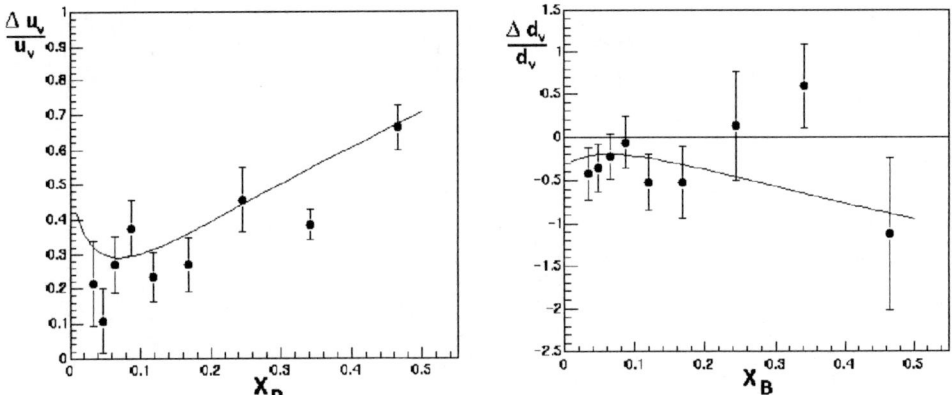

FIGURE 3. Gehrmann model for the polarized structure function. Experimental data corresponds to [9].

$\dfrac{\Delta d}{d}$ and $\dfrac{\Delta u}{u}$ were measured by the HERMES Collaboration [10]. Figure 3 shows the experimental values for $\dfrac{\Delta u}{u}$ and $\dfrac{\Delta d}{d}$.

RESULTS

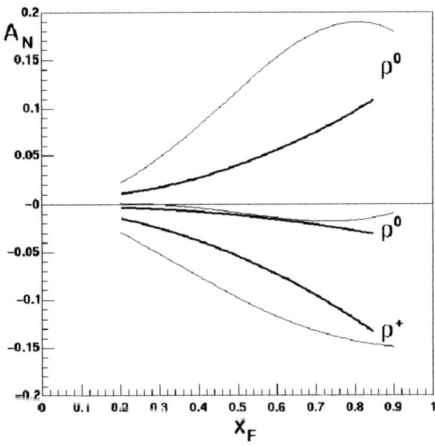

FIGURE 4. Single spin asymmetries using the Gehrmann polarized structure as in figure 3. ρ^-, ρ^0 and ρ^+ are shown, in thin line our model. The thick lines corresponds to the asymmetries obtained in [9].

CONCLUSIONS. We studied the single spin asymmetries in the inclusive production of vector mesons in proton-proton collisions. We used a two components model for the cross section. The model for the polarized structure function, proposed by Gehrmann [11] which describes the experimental data we used. As a result, we give a prediction for the single spin asymmetries for ρ mesons. The general trend is in agreement with [9].

REFERENCES

1. G. Domínguez Zacarías. G. Herrera, Phys. Lett. B **484**, 30 (2000).
2. Thomas A. DeGrand et al. Phys. Lett. B. **381**, 337 (1996).
3. John David Jackson, *Classical Electrodynamics* 2nd. Edition, (1975).
4. EPJC **5,** 1, (2000)
5. Luis M. Montaño, G. Herrera. Phys. Lett. B **381**, 337 (1996).
6. H. Albrecht et al. Z. Phys. C **61**, 1, (1994).
7. M. Aguilar-Benitez *et al.* Z. Phys. C **50**, 405 (1991)
8. G. Dominguez Zacarías. G. Herrera, Ildefonso León-Monzón Eur. Phys. J. C. **20**, 677, (2001).
9. Acta Phys. Polon. **27** 1750, (1996).
10. Ackerstaff et al. Phys.Lett. B **464** 123 (1999).
11. T. Gehrmann and W. J Stirling Phys. Rev. D **5** 6100 (1996).

Numerical analysis of the quark mass matrix

Maritza de Coss and Rodrigo Huerta

Departamento de Física Aplicada, Cinvestav-IPN, Apdo. Postal 73 "Cordemex", Mérida 97310, Yucatán, México

Abstract. Starting from a weak basis in which the up (down) quark matrix is diagonal, we make a numerical analysis of the down (up) quark mass matrix. Using the data available for the quark masses and mixing angles, we find a numerical expression for these matrices at different scales. This numerical analysis could provide an indication for one specific texture. The complex phases are introduced in the mass matrix, and we also find the numerical value for these phases as a function of δ, the CP-violating parameter.

There has been several works dealing with the quark mass problem and the literature is vast. In particular, there are studies considering the mass matrix as an hermitian, symmetric [1, 2, 3] or antihermitian [4]. Due to the great number of entries in the mass matrix (18 for each one of $\mathbf{m}_u$ and $\mathbf{m}_d$), attempts have been made to reduce it, introducing zeros in the matrix elements leading to the terminology of texture [3]. Depending on the texture it is possible to find relations between masses and mixing angles.

We start from the piece of the lagrangian which is important to us,

$$\mathcal{L}_{mass} = \bar{u}'_L \mathbf{m}_u u'_R + \bar{d}'_L \mathbf{m}_d d'_R + g \bar{u}'_L \mathbf{W}^\dagger d'_L + c.h., \tag{1}$$

where the quark mass matrices are general matrices, and its size depends on the number of generations. We can rotate the weak states to get a diagonal mass matrix making the usual transformations,

$$\begin{aligned}\mathcal{L}_{mass} &= \bar{u}'_L \mathbf{R}^u_L (\mathbf{R}^u_L)^\dagger \mathbf{m}_u (\mathbf{S}^u_R)^\dagger \mathbf{S}^u_R u'_R + \bar{d}'_L \mathbf{R}^d_L (\mathbf{R}^d_L)^\dagger \mathbf{m}_d (\mathbf{S}^d_R)^\dagger \mathbf{S}^d_R d'_R \\ &\quad + g \bar{u}_L (\mathbf{R}^u_L)^\dagger \mathbf{W}^\dagger \mathbf{R}^d_L d_L + c.h. \\ &= \bar{u}_L \mathbf{M}_u u_R + \bar{d}_L \mathbf{M}_d d_R + g \bar{u}_L \mathbf{W}^\dagger \mathbf{V}_{CKM} d_L + c.h.\end{aligned} \tag{2}$$

where the unitary matrices given by $\mathbf{R}^{u,d}_L$ and $\mathbf{S}^{u,d}_R$ rotate the weak basis to the mass eigenstates,

$$\begin{aligned} u'_L &= \mathbf{R}^u_L u_L, & d'_L &= \mathbf{R}^d_L d_L \\ u'_R &= \mathbf{S}^u_R u_R, & d'_R &= \mathbf{S}^d_R d_R. \end{aligned} \tag{3}$$

In this way we obtain the biunitary transformations which finally transforms the mass matrices $\mathbf{m}_u$ and $\mathbf{m}_d$ to its diagonal form,

$$\mathbf{M}_u = (\mathbf{R}^u_L)^\dagger \mathbf{m}_u \mathbf{S}^u_R, \quad \mathbf{M}_d = (\mathbf{R}^d_L)^\dagger \mathbf{m}_d \mathbf{S}^d_R. \tag{4}$$

The CKM mixing matrix [5] is then given by

$$\mathbf{V}_{CKM}^L = (\mathbf{R}_L^u)^\dagger \mathbf{R}_L^d. \tag{5}$$

In the case of having a charged right-handed current of the type $g\bar{u}_R' \mathbf{W}^\dagger d_R'$ in the lagrangian [6], we would have

$$\mathbf{V}_{CKM}^R = (\mathbf{S}_R^u)^\dagger \mathbf{S}_R^d. \tag{6}$$

Without any loss of generality we can have one of the mass matrices diagonal [7, 8]. Assume first that $\mathbf{m}_u = Diag(m_u, m_c, m_t)$, then the matrices $\mathbf{R}_L^u$ and $\mathbf{S}_R^u$ are equal to the unit matrix. From equations (5) and (6) we obtain

$$\begin{aligned} \mathbf{V}_{CKM}^L &= \mathbf{R}_L^d, \\ \mathbf{V}_{CKM}^R &= \mathbf{S}_R^d. \end{aligned} \tag{7}$$

In this case the diagonal mass matrix d is given by,

$$\mathbf{M}_d = (\mathbf{V}_{CKM}^L)^\dagger \mathbf{m}_d \mathbf{V}_{CKM}^R. \tag{8}$$

One can make $\mathbf{V}_{CKM}^R$ equal to $\mathbf{V}_{CKM}^L$ [7] assuming left-right symmetry in the flavor sector, and then one can see that we need only to consider a single mixing matrix which is responsible of the diagonalization. This is equivalent to assume that $\mathbf{m}_d$ is hermitian, which we do. Now we can express the mixing angles in terms of the mass eigenvalues m_i and the mass matrix elements m_{ij}. To this end we take an hermitian mass matrix and choose the standard form for $\mathbf{V}_{CKM}$, so we have

$$\mathbf{V}_{CKM} \begin{pmatrix} m_1 & 0 & 0 \\ 0 & m_2 & 0 \\ 0 & 0 & m_3 \end{pmatrix} = \begin{pmatrix} m_{11} & m_{12}e^{i\delta_{12}} & m_{13}e^{i\delta_{13}} \\ m_{12}e^{-i\delta_{12}} & m_{22} & m_{23}e^{i\delta_{23}} \\ m_{13}e^{-i\delta_{13}} & m_{23}e^{-i\delta_{23}} & m_{33} \end{pmatrix} \mathbf{V}_{CKM}. \tag{9}$$

Considering the magnitudes of the elements (2,3), (1,3) and (1,2) in both sides of the equation we get the mixing angles in exact form [9]. To obtain the quark masses as a function of the matrix elements we use the fact that $\mathbf{m}_d$ satisfies the following characteristic equation,

$$\begin{aligned} det(\mathbf{m}_d - m\mathbf{1}) = & -m^3 + (m_{11} + m_{22} + m_{33})m^2 - (m_{11}m_{22} + m_{11}m_{33} \\ & + m_{22}m_{33} - m_{23}^2 - m_{13}^2 - m_{12}^2)m + m_{11}(m_{22}m_{33} - m_{23}^2) \\ & - m_{12}(m_{12}m_{33} - m_{13}m_{23}) + m_{13}(m_{12}m_{23} - m_{13}m_{22}) \\ = & \ 0. \end{aligned} \tag{10}$$

The eigenvalues m_i also satisfy the equation

$$\begin{aligned} (m_1 - m)(m_2 - m)(m_3 - m) = & -m^3 + (m_1 + m_2 + m_3)m^2 \\ & -(m_1m_2 + m_1m_3 + m_2m_3)m + m_1m_2m_3 = 0. \end{aligned} \tag{11}$$

After equating the coefficients with the same power of m in (15) and (16) we get,

$$m_1 + m_2 + m_3 = m_{11} + m_{22} + m_{33} \qquad (12)$$

$$m_1 m_2 + m_1 m_3 + m_2 m_3 = m_{11} m_{22} + m_{11} m_{33} + m_{22} m_{33} - m_{23}^2 - m_{13}^2 - m_{12}^2 \qquad (13)$$

$$m_1 m_2 m_3 = m_{11}(m_{22} m_{33} - m_{23}^2) - m_{12}(m_{12} m_{33} - m_{13} m_{23}) + m_{13}(m_{12} m_{23} - m_{13} m_{22}). \qquad (14)$$

Now, we go back to eq. (8), we look into the phase factors and compare the matrix elements (1,2), (1,3) and (2,3),

$$\tan(\delta_{12}) = \frac{-(m_3 - s_{12}^2 m_2 - m_1) s_{13} s_{23} \sin\delta}{(m_2 - m_1) s_{12} + (m_3 - s_{12}^2 m_2 - m_1) s_{13} s_{23} \cos\delta} \qquad (15)$$

$$\tan(\delta_{13}) = \frac{(m_3 - s_{12}^2 m_2 - m_1) s_{13} \sin\delta}{(m_2 - m_1) s_{12} s_{23} - (m_3 - s_{12}^2 m_2 - m_1) s_{13} \cos\delta} \qquad (16)$$

$$\tan(\delta_{23}) = \frac{(m_2 - m_1)(1 + s_{23}^2) s_{12} s_{13} \sin\delta}{[m_3 - (s_{12}^2 - s_{13}^2) m_1 - (1 - s_{12}^2 s_{13}^2) m_2] s_{23} - (m_2 - m_1)(1 - s_{23}^2) s_{12} s_{13} \cos\delta} \qquad (17)$$

Using equations (9-11), (14-16) and (17-19), we can solve them in order to know the mass matrix structure, at different scales. We use the data consisting in six quark masses and CKM parameters.

We take the values of the quark masses at different scales [10]: 1 GeV, M_Z, m_t, 10^9 GeV and M_X.

The mixing angles we consider are [11] $\sin\theta_{12} = 0.2225 \pm 0.0021$, $\sin\theta_{23} = 0.04 \pm 0.0018$ and $\sin\theta_{13} = 0.0035 \pm 0.0009$. And for the CP-phase we considerer $36° \leq \delta \leq 97°$ [12].

To calculate the mass matrix at 10^9 and M_X, we consider the evolution of the mixing angles. We use the Kielanowski formalism [13], where only magnitudes of V_{ij} are considered, to find the values of the mixing angles at different scales analytically. We assume there is no running for the δ phase [14].

We can collect all the numerical results and write the quark mixing matrix as

m_{ij}^d	1 GeV	m_Z
m_{11}	$(.0189 \pm .0029)$	$(.0092 \pm .0019)$
m_{12}	$(.0403 \pm .0023)e^{-i(1.26 \pm .50)°}$	$(.0192 \pm .0018)e^{-i(1.11 \pm .43)°}$
m_{13}	$(.0235 \pm .0066)e^{-i(70.05 \pm 31.22)°}$	$(.0097 \pm .0027)e^{-i(70.58 \pm 31.31)°}$
m_{22}	$(.1992 \pm .0308)$	$(.0938 \pm .0160)$
m_{23}	$(.2795 \pm .0192)e^{i(0.03 \pm 0.01)°}$	$(.1162 \pm .0061)e^{(0.03 \pm 0.01)°}$
m_{33}	$(7.167 \pm .6089)$	$(2.995 \pm .1155)$

m_{ij}^d	m_t	$10^9\ GeV$
m_{11}	$(.0088 \pm .0019)$	$(.0050 \pm .0011)$
m_{12}	$(.0184 \pm .0018)e^{-i(1.10\pm.42)°}$	$(.0107 \pm .0010)e^{-i(1.31\pm.49)°}$
m_{13}	$(.0092 \pm .0026)e^{-i(70.61\pm31.32)°}$	$(.0055 \pm .0002)e^{-i(71.04\pm31.38)°}$
m_{22}	$(.0898 \pm .0155)$	$(.0527 \pm .0090)$
m_{23}	$(.1104 \pm .0059)e^{i(0.03\pm0.01)°}$	$(.0670 \pm .0036)e^{-i(0.03\pm0.01)°}$
m_{33}	$(2.845 \pm .1155)$	$(1.507 \pm .0633)$

m_{ij}^d	$2 \times 10^{16}\ GeV$
m_{11}	$(.0039 \pm .0006)$
m_{12}	$(.0011 \pm .0009)e^{-i(15.89\pm6.70)°}$
m_{13}	$(.0050 \pm .0012)e^{-i(66.95\pm30.60)°}$
m_{22}	$(.0052 \pm .0006)$
m_{23}	$(.0590 \pm .0030)e^{i(0.04\pm0.01)°}$
m_{33}	$(1.067 \pm .0393)$

A similar analysis can be done in the case of having m_d diagonal.

Conclusions: Starting with the weak basis, for which one of the quark mass matrices is diagonal, we find exact relations that are analysed numerically. We found numerical expressions at different scales. From this analysis we can say that NO texture is found from $1\ GeV$ up to $2 \times 10^{16}\ GeV$. We also found the explicit dependence of δ_{ij} in terms of the quark masses and the CKM mixing angles. The numerical evaluation shows that we have $\delta_{23} \ll \delta_{12} \ll \delta_{13}$ at lower scales and $\delta_{23} \ll \delta_{12} \sim \delta_{13}$ at GUT scales.

Acknowledgments. This work was partially supported by CONACYT (México).

REFERENCES

1. H. Fritzsch, Phys. Lett. B **70**, 436 (1977); **73**, 317 (1978).
2. H. Georgi and C. Jarlskog, Phys. Lett. B **86**, 297 (1979).
3. P. Ramond, R. G. Roberts an G.G. Ross, Nucl. Phys. **B406**, 19 (1993).
4. G. C. Branco and J. I. Silva-Marcos, Phys. Lett. B **331**, 390 (1994).
5. N. Cabibbo, Phys. Rev. Lett. **10**, 531 (1963); M. Kobayashi and T. Maskawa, Prog. Theor. Phys. **49**, 652 (1973).
6. P. Langacker and S. Uma Sankar, Phys. Rev. D **40**, 1569 (1989).
7. E. Ma, Phys. Rev. D **43**, 2761 (1991).
8. Y. Koide, H. Fusaoka and C. Habe, Phys. Rev. D **46**, R4813 (1992).
9. A. Rašin, Phys. Rev. D **58**, 96012 (1998).
10. H. Fusaoka and Y. Koide, Phys. Rev. D **57**, 3986 (1998).
11. D. E. Groom et al., Particle Data Group, Eur. Phys. J. **C15**:, 1 (2000), p. 103.
12. A. Ali and D. London, Eur. Phys. J. **C9**, 687-703 (1999).
13. P. Kielanowski, S.R. Juárez, J. C. Mora, Phys. Lett. **B**, 179 (2000).
14. C. Balzereit *et al.*, Eur. Phys. J. **C9**, 197-211 (1999).
15. H. Fritzsch and Z. Xing, Phys. Lett. B **413**, 396-404 (1997).

The ALICE Pixel Detector

Jorge Mercado-Pérez[1]

Organisation Européenne pour la Recherche Nucléaire, CERN
CH-1211 Genève 23, Suisse
and
Centro de Investigación y de Estudios Avanzados del IPN, CINVESTAV
Apartado Postal 14-740, 07000 México, D.F.

Abstract. The present document is a brief summary of the performed activities during the 2001 Summer Student Programme at CERN under the *Scientific Summer at Foreign Laboratories Program* organized by the Particles and Fields Division of the Mexican Physical Society (Sociedad Mexicana de Física). In this case, the activities were related with the ALICE Pixel Group of the EP-AIT Division, under the supervision of Jeroen van Hunen, research fellow in this group. First, I give an introduction and overview to the ALICE experiment; followed by a description of wafer probing. A brief summary of the test beam that we had from July 13th to July 25th is given as well.

THE ALICE EXPERIMENT

ALICE (A Large Ion Collider Experiment) [1] is an experiment at the Large Hadron Collider (LHC) optimized for the study of heavy-ion collisions, at a centre-of-mass energy ~5.5 TeV per nucleon. The prime aim of the experiment is to study in detail the behaviour of nuclear matter at high densities and temperatures, in view of probing deconfinement and chiral symmetry restoration.

The detector consists essentially of two main components: the central part, composed of detectors mainly devoted to the study of hadronic signals and dielectrons, and the forward muon spectrometer, devoted to the study of quarkonia behaviour in dense matter.

There is a considerable spread in the currently available predictions for the multiplicity of charged particles produced in a central Pb-Pb collision. The design of the experiment has been based on the highest value, 8000 charged particles per unit of rapidity at midrapidity. This multiplicity dictates the granularity of the detectors and their optimal distance from the colliding beams.

The central part, which covers ±45° over the full azimuth, is embedded in a large magnet with a weak solenoidal field. Outside the Inner Tracking System (ITS), there are a cylindrical Time Projection Chamber (TPC) and a large area Particle Identification System (PID) array of time-of-flight (TOF) counters. In addition, there are two small-area single-arm detectors: an electromagnetic calorimeter (Photon

---

[1] Electronic mail: jmercado@fis.cinvestav.mx

Spectrometer, PHOS) and an array of Ring Imaging Cherenkov (RICH) counters optimized for high-momentum inclusive particle identification (HMPID).

The Inner Tracking System (ITS)

Tracking in ALICE

Track finding in heavy-ion collisions at the LHC presents a big challenge, because of the extremely high track density. In order to achieve a high granularity and a good two-track separation, ALICE uses three-dimensional information, wherever feasible, with many points on each track and a weak magnetic field. The ionization density of each track is measured for particle identification. The need for a large number of points on each track has led to the choice of a TPC as the main tracking system. At smaller radii, and hence larger track densities, tracking is taken over by the ITS.

Layout of the ITS

The basic functions of the inner tracker [2] are achieved with six barrels of high-resolution detectors. The number of layers and their position has been optimized for efficient pattern recognition and impact parameter resolution. Because of the high particle density, the four innermost layers ($r \leq 24$ cm) must be two-dimensional devices, such as silicon pixel and silicon drift detectors. The outer superlayer at $r \approx 45$ cm will be equipped with double-sided silicon micro-strip detectors. With the exception of the two innermost planes, all layers will have analog readout for independent particle identification via dE/dx in the non-relativistic region, which will give the inner tracking system a stand-alone capability as a low-p_t particle spectrometer.

WAFER PROBING

The Silicon Pixel Detectors are being developed by the ALICE Pixel Group. The ALICE pixel detector contains around 1200 readout chips (pixel chips) with a size of 14 × 16 mm each. One readout chip contains 8192 readout cells (i.e. pixels) and is bump bonded [3] to a silicon detector. The pixel chips are produced in a company outside CERN and are arranged in wafers, each one containing 86 chips. However, some tests have to be made before cutting and bump-bonding.

Setup Of Wafer Probing

Wafer probing is taken place in building 14-6-030, where there is a MIC probe-station (Probe-station Karl Suess), 3 probe-cards equipped with needles, some spare cards and a VME-crate with pilot and JTAG controller. The interface between the probe-card and the VME-crate is done by MB-card (card developed by Mike Burns) which supplies the operating voltages of the chips. Before measurements with real wafers, some preparations had to be made, such as installation of the probe-card, preparation and test of cables to the MB-card and mechanical alignment of the probe-card with dummy wafers.

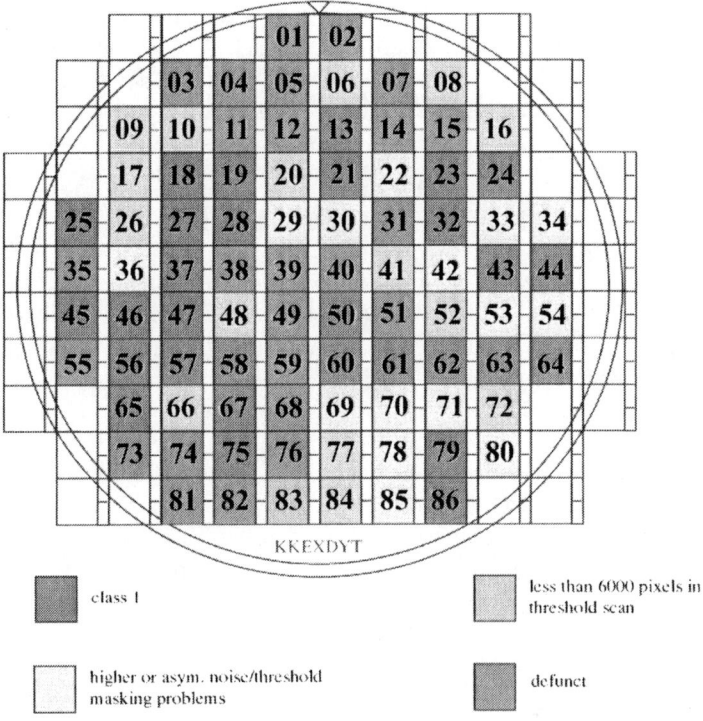

FIGURE 1. Preliminary criteria applied to each chip based on the wafer probing performed.

Development Of Wafer Probing

The software system used to carry out wafer probing, was developed by Peter Chochula and is written in LabView. At the beginning of July, we had 12 wafers. Of these, 6 belong to Lot 1. (A2A38001LF) with a thickness of 750 micron; the other 6 belonging to Lot 2. (A2B02000LF) with a thickness of 300 micron.

In order to operate the semi-automatic wafer-prober (K. Suess) in a safe and easy way, I wrote some systematic steps that allow "any" user move from one chip to another one, such that each pixel chip on the wafer can be tested electronically.

The set of results obtained from wafer probing are, obviously, different from chip to chip. In order to have an objective reference of each chip, Petra Riedler proposed the following (preliminary) criteria (Figure 1):
- Dead chips: Excessive or no current consumption, no response from the chip.
- Class III: Chips with masking problems, high/asymmetric noise or threshold.

- Class II: Fully functional chips, but less than 6000 pixels responding to the threshold scan.
- Class I: Good ones.

Based on this criteria, we could classify each chip within a wafer. For example, wafer KKEXDYT (see Figure 1) presented 25 dead chips, 15 class III chips, 16 class II chips and 30 class I chips. This means that more than 50% of chips in wafer KKEXDYT are in good enough conditions. In fact, it is very usefull to catalogue a given chip by means of its noise and threshold distributions, which can be obtained from the full threshold scan log-file by means of the ROOT software.

TESTBEAM DESCRIPTION

From July 13th to July 25th the ALICE Pixel Group (among other CERN groups) had a testbeam. Our pixel setup was located in NA57 area in the H4 beam-line of the north hall. The beam was 150 GeV/c π^-, with about 10^5 particles per spill.

The first thing to do was the installation of the setup in the area, which consisted of scintillators, power supplies, assembly, MB card and cabling from the setup place to the control room.

On July 15th, the first beam was on assembly 12; from this day until the 18th, several measurements were performed; efficiency, cluster size, timing precision, etc. At the beginning, the efficiency was ~85% due to the fact that the trigger step was not optimized. On July 19th, the time delay unit and the scintillator trigger was changed, and an efficiency of ~92% was reached. After placing a small scintillator in a new position on the setup table, an efficiency of 98% was reached.

On July 21st, the x–y table was installed. It allowed to move the assembly from the control room. The x–y table had three degrees of freedom: vertical movement (z), horizontal movement (x), and angular movement (θ).

The following measurements were performed:
- Efficiency and cluster size as function of angle.
- Efficiency and cluster size as function of different settings of the DAC.
- Efficiency and cluster size as function of detector bias voltage.

REFERENCES

1. CERN/LHCC/95-71 *ALICE: Technical Proposal for A Large Ion Collider Experiment*. CERN Publication, 15 December, 1995. ISBN 92-9083-077-8.
2. CERN/LHCC/99-12 *ALICE: Technical Design Report of the Inner Tracking System (ITS)*. CERN Publication, 18 June, 1999. ISBN 92-9083-144-8.
3. K. Wyllie. *ALICE1LHCB Documentation: Draft 3*. 5 February, 2001.

Bounds on neutrino-photon interactions from Z decays

F. Larios[a], M.A. Pérez[b] and G. Tavares-Velasco[b]

[a]*Departmento de Física Aplicada, CINVESTAV-Mérida, A.P. 73, 97310 Mérida, Yucatán, México.*
[b]*Departmento de Física, CINVESTAV, Apdo. Postal 14-740, 07000 México.*

Abstract. It is shown that the LEP bounds on the rare Z decays $Z \to \nu\bar{\nu}\gamma$ and $Z \to \nu\bar{\nu}\gamma\gamma$ are useful to put constraints on neutrino-one-photon and neutrino-two-photon interactions, *i.e.* $\nu\bar{\nu}\gamma$, $\nu\bar{\nu}\gamma\gamma$. The respective constraints are then used to bound the τ neutrino magnetic moment μ_{ν_τ} and the τ neutrino decay $\nu_\tau \to \nu_i\gamma\gamma$. Some interesting consequences are discussed.

INTRODUCTION

Neutrino-photon interactions have long received considerable attention as they may have important consequences on some physics issues. In particular, neutrino-one-photon interactions may play an important role in elucidating the mass neutrino puzzle which in turn has gained renewed interest after the Super-Kamiokande collaboration measurements [1]. As for neutrino-two-photon interactions, they may play an important role in astrophysics and cosmology [2]. For instance, it is well known that neutrino pair production in photon-photon collisions, $\gamma\gamma \to \nu\bar{\nu}$, may be an important way for stars to lose energy. There are other interesting processes involving neutrino-two-photon interactions, such as $\nu\gamma \to \nu\gamma$, $\nu\bar{\nu} \to \gamma\gamma$, and the neutrino double-radiative decay $\nu_j \to \nu_i\gamma\gamma$.

Some time ago, the L3 collaboration searched for energetic single-photons near the Z pole at the CERN e^+e^- LEP collider and set a bound for the rare decay $Z \to \nu\bar{\nu}\gamma$ [3]. It has been shown that this measurement can impose a stringent constraint on the τ neutrino magnetic moment [3, 4]. On the other hand, it has been shown that the standard model (SM) contribution to the rare decay $Z \to \nu\bar{\nu}\gamma$ is negligible small (the respective branching ratio is of the order of 10^{-10}), placing this decay as an interesting mode to search for evidences of new physics [5]. As for the rare Z decay $Z \to \nu\bar{\nu}\gamma\gamma$, the L3 and the OPAL collaborations searched for events with a lepton pair accompanied by a photon pair of large invariant mass [6]. After combining both of these searches, the OPAL collaboration put a lower bound on the cross section of the process $e^+e^- \to \nu\bar{\nu}\gamma\gamma$ and the branching ratio of the rare Z decay $Z \to \nu\bar{\nu}\gamma\gamma$. It was found that $BR(Z \to \nu\bar{\nu}\gamma\gamma) \leq 3.1 \times 10^{-6}$.

In this work we will show that the bound on the rare decay $Z \to \nu\bar{\nu}\gamma\gamma$ turns out to be very useful to put constraints on the effective vertices $\nu\bar{\nu}\gamma$ and $\nu\bar{\nu}\gamma\gamma$. We will also show that the latter bound can be used to constrain the neutrino decay $\nu_\tau \to \nu_\mu\gamma\gamma$.

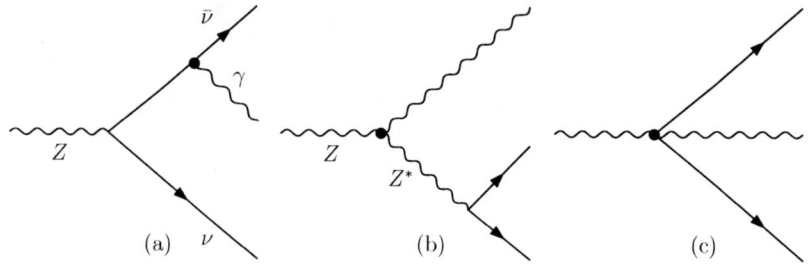

FIGURE 1. Feynman diagrams contributing to the decay $Z \to \nu\bar{\nu}\gamma$ in the effective Lagrangian approach.

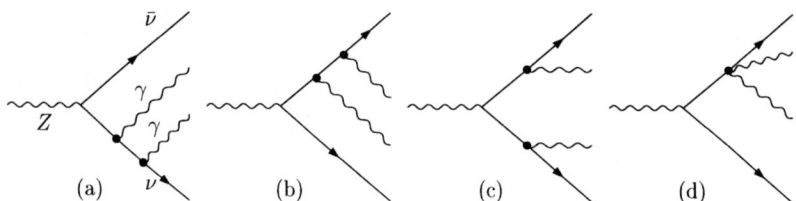

FIGURE 2. Feynman diagrams contributing to the decay $Z \to \nu\bar{\nu}\gamma\gamma$. The crossed diagrams are not shown.

NEUTRINO-ONE-PHOTON INTERACTION

The rare decay $Z \to \nu\bar{\nu}\gamma$ at tree level can be given by the Feynman diagrams shown in Fig. 1.

We are not interested in the contributions from diagrams 1(b) nor 1(c) here. For details of these contributions we refer the reader to Ref. [4, 7]. As for the diagram 1(a), the $\nu\bar{\nu}\gamma$ vertex can be parametrized by the following effective Lagrangian [1]

$$\mathcal{L}_{\nu\bar{\nu}\gamma} = \frac{1}{2}\mu_\nu \bar{\nu}\sigma_{\mu\nu}\nu F^{\mu\nu}, \qquad (1)$$

where μ_ν is the neutrino anomalous magnetic moment. After computing the contribution from diagram 1(a) and that with the photon emerging from the neutrino, the L3 bound can be used to obtain the following bound on the τ neutrino magnetic moment [3, 4]:

$$\mu_{\nu_\tau} < 2.62 \times 10^{-6}\, \mu_B, \qquad (2)$$

where μ_B stands for the Bohr magneton.

The $\nu\bar{\nu}\gamma$ coupling can also give rise to the rare decay $Z \to \nu\bar{\nu}\gamma\gamma$ via the first three Feynman diagrams shown in Fig. 2. We have obtained an upper bound on the τ neutrino magnetic moment by considering this decay channel[2]:

---

[1] We are assuming massive Dirac neutrinos.
[2] More details of our calculations will be given elsewhere [8].

$$BR(Z \to \nu\bar{\nu}\gamma\gamma) \simeq 22.5\mu_{\nu_\tau}^4 \leq 3.1 \times 10^{-6}. \tag{3}$$

It follows that

$$\mu_{\nu_\tau} \leq 6.5 \times 10^{-5} \mu_B, \tag{4}$$

which is one order of magnitude below than the bound of Eq. (2). This fact is not surprising; comparing with the single photon decay, in this process the anomalous coupling enters twice.

NEUTRINO-TWO-PHOTON INTERACTIONS

The interaction of a neutrino pair with a photon pair can be parametrized by the following effective Lagrangian [9]

$$\mathcal{L}_{\nu\bar{\nu}\gamma\gamma} = \frac{1}{4\Lambda^3} \bar{\nu}_j \left(\alpha_L^{ij} P_L + \alpha_R^{ij} P_R \right) \nu_i F^{\mu\nu} F_{\mu\nu}, \tag{5}$$

where $P_{L,R} = (1 \pm \gamma^5)/2$ and $\alpha_{L,R}^{ij}$ are dimensionless coupling constants, and Λ is the new physics scale.

We now use this coupling to obtain the contribution to the decay $Z \to \nu_i \bar{\nu}_j \gamma\gamma$ from diagram 2(d) (plus the crossed one):

$$\Gamma_{Z \to \nu_i \bar{\nu}_j \gamma\gamma} = 1.092 \times 10^3 \left(|F_L^{ij}|^2 + |F_R^{ij}|^2 \right) \leq 3.1 \times 10^{-6} \text{GeV}^{-6}, \tag{6}$$

where we have defined $F_{L,R}^{ij} = \alpha_{L,R}^{ij}/\Lambda^3$. This can also be written as

$$\frac{1}{\Lambda^6} \left(|\alpha_L^{ij}|^2 + |\alpha_R^{ij}|^2 \right) \leq 2.85 \times 10^{-9} \quad \text{GeV}^{-6}. \tag{7}$$

We note that this bound is weaker than the bound obtained in Ref. [10] from the nuclear process $\nu_i Z \to \gamma \nu_j Z$.

Let us now consider the double radiative decay $\nu_i \to \nu_j \gamma\gamma$. Neglecting the mass of the final neutrino ν_j one obtains the following decay width [10]

$$\Gamma(\nu_j \to \nu_i \gamma\gamma) \sim m_j^7 \left(|\alpha_L^{ij}|^2 + |\alpha_R^{ij}|^2 \right) \times 10^{-3} \text{s}^{-1}, \tag{8}$$

where m_j should be expressed in MeV and $F_{L,R}^{ij}$ in GeV^{-6}. From the above expression and the constraint of Eq. (7) we can obtain the following bound for the τ neutrino lifetime

$$\tau_{\nu_\tau \to \nu_\mu \gamma\gamma} > \frac{10^{11}}{m_{\nu_\tau}^7} \text{ s.} \tag{9}$$

where again m_{ν_τ} is to be given in MeV. This bound is two orders of magnitude below the one found in [10]. This is a consequence of the fact that the experimental bound on branching ratio of the rare decay $Z \to \nu\bar{\nu}\gamma\gamma$ is not very strong. In fact, this bound is of the same order of magnitude than the bound on the three body decay $Z \to \nu\bar{\nu}\gamma$. We can hope for some improvement with measurements made in a future linear collider.

ACKNOWLEDGEMENTS

We want to thank CONACYT and SNI (México) for support.

REFERENCES

1. Super-Kamiokande Collab., Y. Fukuda, *et al.*, Phys. Rev. Lett.**81**, (1998) 1562.
2. B. M. Pontecorvo, Zh. Eksp. Teor. Fiz. **36** (1959) 1615; H. Y. Chiu and P. Morrison, Phys. Rev. Lett. **5** (1960), 573.
3. L3 Collab., M. Acciari *et al.*, Phys. Lett. **B**412 (1997) 201.
4. M. Maya, M. A. Pérez, G. Tavares-Velasco and B. Vega, Phys. Lett.**B**434 (1998) 354.
5. J. M. Hérnandez, M. A. Pérez, G. Tavares-Velasco and J. J. Toscano, Phys. Rev. **D** 60 (1999) 013004.
6. L3 Collab., O. Adriani *et al.*, Phys. Lett. **B**295 (1992) 337; OPAL Collab., P. Acton *et al.*, Phys. Lett. **B**311 (1993) 391.
7. F. Larios, M. A. Pérez, G. Tavares-Velasco and J. J. Toscano, Phys. Rev. **D** 63 (2001) 113014.
8. F. Larios, M. A. Pérez and G. Tavares-Velasco, work in progress.
9. J. Liu, Phys. Rev. **D** 44 (1991) 2879; S. Dodelson and G. Feinberg, Phys. Rev. **D** 43 (1991) 913.
10. S. N. Gnienko and N. V. Krasnikov, Phys. Lett. **B**450 (1999) 165.

MSSM Higgs Bosons Production at e^+e^- Colliders

M. A. Hernández-Ruíz*, A. Gutiérrez-Rodríguez* and O. A. Sampayo[†]

*Facultad de Física, Universidad Autónoma de Zacatecas Apartado Postal C-580, 98060 Zacatecas, Zacatecas México.
[†]Departamento de Física, Universidad Nacional del Mar del Plata Funes 3350, (7600) Mar del Plata, Argentina.

Abstract. We study the production of the Higgs bosons predicted in the Minimal Supersymmetric extension of the Standard Model ($h^0, H^0, A^0, H^\pm$), with the reactions $e^+e^- \to b\bar{b}h^0(H^0, A^0)$, and $e^+e^- \to \tau^-\bar{\nu}_\tau H^+, \tau^+\nu_\tau H^-$, using the helicity formalism. We evaluate cross section of h^0, H^0, A^0 and $H^\pm$ in the limit when $\tan\beta$ is large. The numerical computation is done considering two stages of a possible Next Linear e^+e^- Collider: the first with $\sqrt{s} = 500$ GeV and design luminosity 50 fb^{-1}, and the second with $\sqrt{s} = 1$ TeV and luminosity 100-200 fb^{-1}.

INTRODUCTION

Higgs bosons [1] play an important role in the Standard Model (SM) [2]; they are responsible for generating the masses of all the elementary particles (leptons, quarks, and gauge bosons). However, the Higgs-boson sector is the least tested one in the SM. If Higgs bosons are responsible for breaking the symmetry from $SU(2)_L \times U(1)_Y$ to $U(1)_{EM}$, it is natural to expect that other Higgs bosons are also involved in breaking other symmetries.

One of the more attractive extensions of the SM is Supersymmetry (SUSY) [3], mainly because of its capacity to solve the naturalness and hierarchy problems while maintaining the Higgs bosons elementary.

The theoretical frame work of this paper is the Minimal Supersymmetric extension of the Standard Model (MSSM), which doubles the spectrum of particles of the SM and the new free parameters obey simple relations. The scalar sector of the MSSM [4] requires two Higgs doublets, thus the remaining scalar spectrum contains the following physical states: two CP-even Higgs scalar (h^0 and H^0) with $m_{h^0} \leq m_{H^0}$, one CP-odd Higgs scalar (A^0) and a charged Higgs pair ($H^\pm$), whose detection would be a clear signal of new physics. In this paper, we focus on the phenomenology of the neutral CP-even and CP-odd scalar (h^0, H^0, A^0) and charged ($H^\pm$) [5].

PRODUCTION OF THE MSSM HIGGS BOSONS AT POSITRON-ELECTRON COLLIDERS

A Higgs boson h^0, H^0, A^0, and $H^\pm$ can be produced in scattering e^+e^- via the following processes:

$$e^+e^- \to b\bar{b}h^0, \qquad (1)$$
$$e^+e^- \to b\bar{b}H^0, \qquad (2)$$
$$e^+e^- \to b\bar{b}A^0, \qquad (3)$$
$$e^+e^- \to \tau^-\bar{\nu}_\tau H^+, \tau^+\nu_\tau H^-. \qquad (4)$$

The diagrams of Feynman, which contribute at tree-level to the different reaction mechanisms, are depicted in Figs. 1-3, Ref. [5].

We use the Breit-Wigner propagators for the Z^0, h^0, H^0, A^0 and $H^\pm$ bosons. For the SM parameters we adopted the following: $m_b = 4.25$ GeV, $m_t = 175$ GeV, $m_\tau = 1.78$ GeV, $m_\nu = 0$, $m_{Z^0} = 91.2$ GeV, $\Gamma_{Z^0} = 2.4974$ GeV, $\sin^2\theta_W = 0.232$, which are taken as inputs. The widths of h^0, H^0, A^0 and $H^\pm$ are calculated from the formulas given in The Higgs Hunter's [6]. In the next sections we present the numerical computation of the processes $e^+e^- \to b\bar{b}h$, $h = h^0, H^0, A^0$ and $e^+e^- \to \tau^-\bar{\nu}_\tau H^+, \tau^+\nu_\tau H^-$.

In this work, we evaluate total cross section of neutral and charged MSSM Higgs bosons at next generation e^+e^- colliders, including three-body mode diagrams [Figs. 1.1-1.3, 1.5, and 1.6; Figs. 2.1-2.3, 2.5 and 2.6; Figs. 3.2, 3.3, and 3.5, see Ref. [5]] besides the dominant mode diagram [Fig. 1.4; Fig. 2.4; Figs. 3.1, and 3.4, see Ref. [5]] consider two stages of a possible Next Linear e^+e^- Collider: the first with $\sqrt{s} = 500$ GeV and design luminosity 50 fb^{-1} and the second with $\sqrt{s} = 1$ TeV and luminosity 100-200 fb^{-1}. We consider the complete set of Feynman diagrams (Figs. 1-3) at tree-level and utilize the helicity formalism for the evaluation of their amplitudes.

Total Production of the Higgs Bosons

We present in Tables I, II our results for the total production of h^0, H^0, A^0, $H^\pm$ Higgs bosons, taking different values of the center-of-mass energy $\sqrt{s}$, fundamental supersymmetry parameter $\tan\beta$, luminosity $\mathcal{L}$, and the mass of the pseudoscalar m_{A^0}. We take the following values representative of the supersymmetric parameters $m_{A^0} = 100$ GeV and $\tan\beta = 10, 30$. From these results we observed a strong dependence of the supersymmetric parameter $\tan\beta$, in particularly for $\tan\beta$ large, as well as, of the center-of-mass energy $\sqrt{s}$, in the production of the different Higgs bosons $(h^0, H^0, A^0, H^\pm)$. These results make apparent the big importance of investigate the possibility of detecting the Higgs bosons with the reactions at three-body $e^+e^- \to b\bar{b}h^0(H^0, A^0)$ and $e^+e^- \to$

$\tau^-\bar{\nu}_\tau H^+, \tau^+\nu_\tau H^-$, at next generation linear e^+e^- colliders.

Total Production of Higgs Bosons	$\mathcal{L}=50\ fb^{-1}$	
	$\tan\beta = 10$	$\tan\beta = 30$
h^0	1600	1800
H^0	700	470
A^0	900	935
H^+H^-	7000	6500

Table I. Total production of Higgs bosons $h^0, H^0, A^0, H^\pm$ of the MSSM for $\sqrt{s} = 500$ GeV and $m_{A^0} = 100\ GeV$.

Total Production of Higgs Bosons	$\mathcal{L}=100\ fb^{-1}$		$\mathcal{L}=200\ fb^{-1}$	
	$\tan\beta = 10$	$\tan\beta = 30$	$\tan\beta = 10$	$\tan\beta = 30$
h^0	960	1150	1920	2300
H^0	370	220	740	440
A^0	560	615	1120	1230
H^+H^-	4700	4900	9400	9800

Table II. Total production of Higgs bosons $h^0, H^0, A^0, H^\pm$ of the MSSM for $\sqrt{s} = 1$ TeV and $m_{A^0} = 100\ GeV$.

CONCLUSIONS

In this paper, we have calculated the production of the neutral and charged Higgs bosons in association with b-quarks and with $\tau\nu_\tau$ leptons via the processes $e^+e^- \to b\bar{b}h$, $h = h^0, H^0, A^0$ and $e^+e^- \to \tau^-\bar{\nu}_\tau H^+, \tau^+\nu_\tau H^-$ using the helicity formalism. We find that these processes could help to detect the possible neutral and charged Higgs bosons at energies of a possible Next Linear e^+e^- Collider when $\tan\beta$ is large.

In summary, we conclude that the possibilities of detecting or excluding the neutral and charged Higgs bosons ($h^0, H^0, A^0, H^\pm$) of the Minimal Supersymmetric Standard Model in the processes $e^+e^- \to b\bar{b}h$, $h = h^0, H^0, A^0$ and $e^+e^- \to \tau^-\bar{\nu}_\tau H^+, \tau^+\nu_\tau H^-$ are important and in some cases are compared favorably with the dominant mode $e^+e^- \to (A^0, Z^0) + h$, $h = h^0, H^0, A^0$ and $e^+e^- \to H^+H^-$ with $\tan\beta$ large. The detection of the Higgs boson will require the use of a future high energy machine like the Next Linear e^+e^- Collider.

ACKNOWLEDGMENTS

This work was supported in part by *Consejo Nacional de Ciencia y Tecnología* (CONA-CyT), *Sistema Nacional de Investigadores* (SNI) (México) and Programa de Mejoramiento al Profesorado (PROMEP). A.G.R. would like to thank the organizers of the Summer School in Particle Physics and Sixth School on non-Accelerator Astroparticle Physics 2001, Trieste Italy for their hospitality. O. A. S. would like to thank CONICET (Argentina).

REFERENCES

[1] P. W. Higgs, Phys. Lett. **12**, (1964) 132; P. W. Higgs, Phys. Rev. Lett. **13**, (1964) 508; P. W. Higgs, Phys. Rev. Lett. **145**, (1966) 1156; F. Englert, R. Brout, Phys. Rev. Lett. **13**, (1964) 321; G. S. Guralnik, C. S. Hagen, T. W. B. Kibble, Phys. Rev. Lett. **13** (1964), 585.

[2] S. Weinberg, Phys. Rev. Lett. **19**, (1967) 1264; A. Salam, in *Elementary Particle Theory*, ed. N. Southolm (Almquist and Wiksell, Stockholm, 1968), p.367; S.L. Glashow, Nucl. Phys. **22**, (1967) 257.

[3] H. P. Nilles, Phys. Rep. **110**, (1984) 1; H. Haber and G. L. Kane, Phys. Rep. **117**, (1985) 75.

[4] J. F. Gunion and H. E. Haber, Nucl. Phys. **B272**, (1986) 1; Nucl. Phys. **B278**, (1986) 449; Nucl. Phys. B307, (1988).

[5] U. Cotti, A. Gutiérrez-Rodríguez, A. Rosado and O. A. Sampayo, Phys. Rev. **D59**, (1999) 095011; A. Gutiérrez-Rodríguez and O. A. Sampayo, Phys. Rev. **D62**, (2000) 055004; A. Gutiérrez-Rodríguez and O. A. Sampayo, hep-ph/9911361; A. Gutiérrez-Rodríguez, M. A. Hernández-Ruíz and O. A. Sampayo, hep-ph/0005050.

[6] For a recent review see J. Gunion, H. Haber, G. Kane and S. Dowson, *The Higgs Hunter's Guide* (Addison-Wesley, Reading, MA,1990).

High Mountain Cosmic Ray Observatory

Oscar Martínez[1], Humberto Salazar[1], Luis Villaseñor[2], Epifanio Ponce[1], Eucario Pérez[1], Marcelino Anguiano[1], Pedro Bello[1], Javier Hernández[1], Alexei Silaev[3].

[1] Facultad de Ciencias Físico-Matemáticas BUAP Puebla, Pue, Mexico; [2] Universidad Michoacana, Morelia, Michoacán, Mexico; [3] Moscow State University, Moscow, Russia.

Abstract. In this paper, we present a progress report of the Pico de Orizaba's cosmic ray observatory. The scientific rationale, and a brief description of the present status of each component are discussed.

SCIENTIFIC GOALS

There are some high mountain observatories in the world, as Chacaltaya in Bolivia, ARGO and Tien Shan in China. The main advantage of this kind of facilities is to measure the extended air showers nearby to their maximum development, and should allow us better energy determination of the primary particles. Some of the objectives of this array are: To contribute to determine: the chemical composition, the energy spectrum shape and arrival anisotropy of the primary cosmic ray in the energy range from 1×10^{15} to 1×10^{18} eV. Search for possible astrophysical sources of high energy particles for investigation of nuclear interactions. The interest in cosmic ray physics as a division of high-energy physics has essentially increased in the recent years due to the fact that a series of new phenomena, not yet explained in the framework of contemporary theoretical models of nuclear interactions, is observed in the 1–10 PeV energy range beyond the possibilities of the modern accelerators. However, these new phenomena could be related to observations of some new objects of astrophysical origin [1]. Finally, a series of new unusual phenomena are observed in various cosmic ray experiments, mainly detected by high altitude observatories. Is needed, in order to have a statistical reliable sample, to increase the total number of detections of this kind of phenomena.

SIMULATIONS

To design and characterize the array [4,5], we have made simulations of the EAS using the CORSIKA program, developed by the KASKADE collaboration. In our case, we use the latest version (6.2) installed in a computer cluster under LINUX to build a library of EAS produced by protons and Fe nuclei within the energy range and arrival angles between 0 and 60 degrees. This archive allows us to evaluate the lateral and longitudinal distribution function of the secondary particles produced in each case.

The program include the SYBILL and the QGSJET hadronic interaction model, and we use both in the library in order to compare it. As the atmospheric depth of the array is 620 gr/cm^2 (4300 masl) those simulations confirm that we are nearby the maximum development of the EAS.

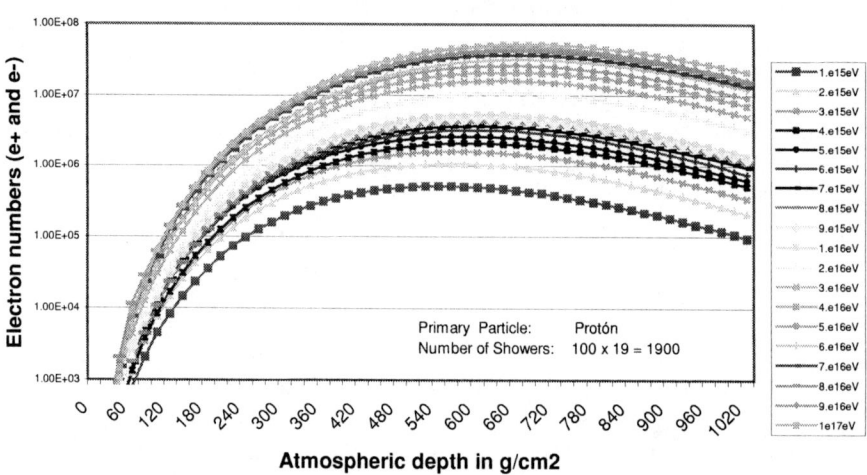

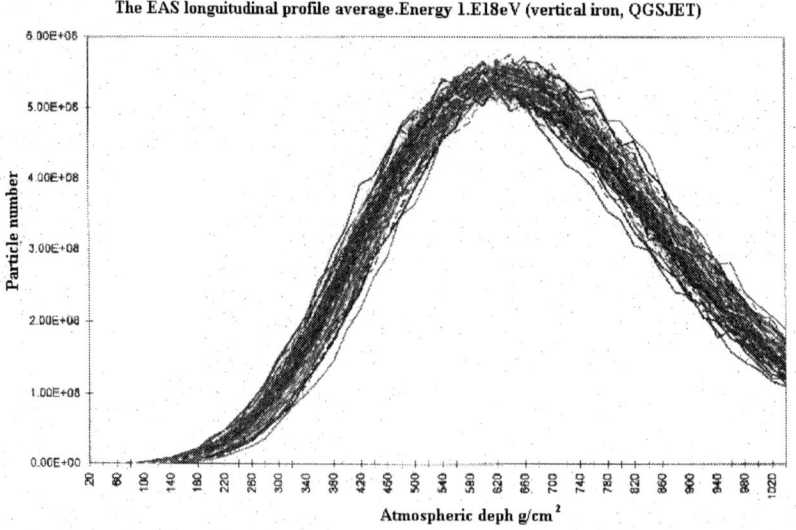

Figure 1 Simulations made with the CORSIKA program to show the atmospheric depth of the maximum development of EAS. Left panel average behavior of the EAS components, right panel, plot of the longitudinal development of 100 vertical showers produced by Fe nuclei with energy 1×10^{18}eV.

ELECTRONICS

The complete observatory will have 21 containers in a hexagonal distribution, the first inner ring schematics is shown in the figure 2.

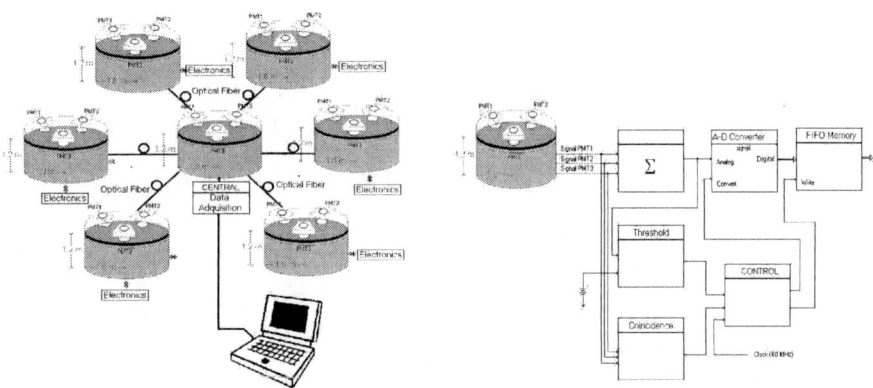

FIGURE 2 Left, diagram of the data acquisition of the first inner ring of the array, right block diagram of the electronics of a single tank.

Signal produced by the PMT's is managed by using a non-invertors operational amplifier. After that, we use a trigger module to get finally the coincidence of events. Schematically a coincidence can be as it shown in the figure 2. Basically, the all PMT's have to detect an event into a small time window, as it shown in the left panel. In the right panel shows the electronic circuit.

The calibration process consists in the transformation of the measured quantities into physical ones. This process allow us to do: statistical studies of the pulses generated by the array; determine the operation voltage of each PMT in order to match their gain; and to have a long term monitoring of the overall performance of the array.

The calibration parameter is related with the vertical flux of muons [2], to determine it accurately, we use two scintillation paddles located on the top and below tank. To discriminate between muons and other particles, the lower paddle have an iron slab working as a filter, in our case this slab is 7 cm thick.

The muon rate measured in the University of Puebla's Campus is 0.011 s^{-1} and 0.085 s^{-1} without and with filter respectively. The rate measured in the Pico de Orizaba is 2.5 times higher than the measured in the Campus.

PRESENT AND FUTURE

The first stage of the observatory consisting of 7 particle detectors [3], and the full array will have 21. In the present, 19 tanks are located in the Pico de Orizaba and one of them was filled with water and have PMT and electronics installed, in order to permit us to obtain the preliminary results presented here. In the next year, the first ring of tanks will be filled with purified water, and install on them PMT's and the acquisition and communication electronics. The design and tests of the discriminator, coincidence and acquisition modules are completed, and the production process is now in progress. The central station will have the computer dedicated to collect and to store the data, linked via optical fiber to the others detectors. The installation and first test of this communication system will be carried out early in the next year.

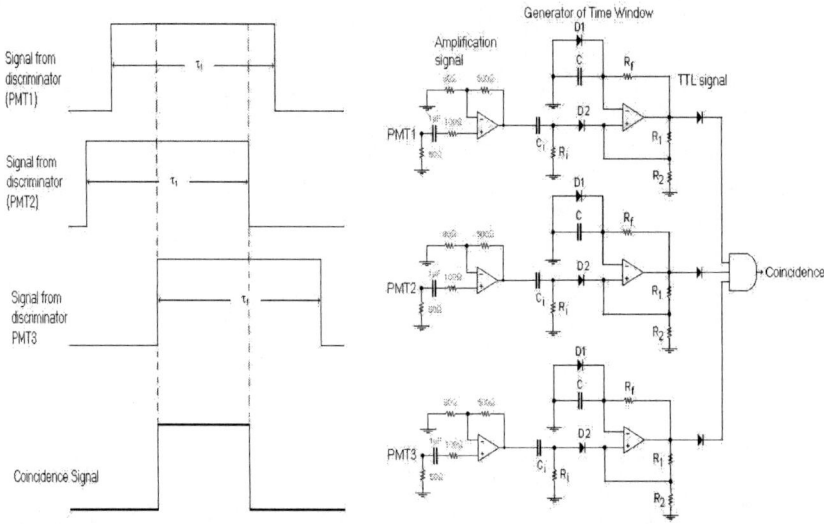

FIGURE 3. The left panel shows schematically the coincidence of the signal of the PMT's in a tank, right, the electronic diagram of the circuit developed.

The development and operation of this kind of facilities represent an important opportunity to involve us with the astroparticle physics and, at the same time to form new researchers and students in one of the most interesting areas of the modern physisc.

ACKNOWLEDGMENTS

This work has partial support of CONACyT, grant G32739-E.

REFERENCES

1. A.A. Watson and J.G. Wilson, J. Phys A, 7, 1199, 1974
2. Anguiano, A., et.al., NIM (submitted) ,2001
3. The Pierre Auger Design Report, 1996
4. V.P. Antonova, et.al., Lebedev Physical Institute, preprint 33, 2000
5. C. Pryke et.al., Astroparticle Physics, 263, 270, 1997

Extensive Air Shower Detector Array at the Universidad Autonoma de Puebla

Cotzomi J.[1], Moreno E.[1], Aguilar S.[1], Palma B.[1], Martínez O.[1], Salazar H.[1], Villaseñor L.[2]

[1] *Facultad de Ciencias Físico Matemáticas, BUAP, Puebla Pue., México*
[2] *Instituto de Física y Matemáticas, UMSNH, Morelia Mich., México*

Abstract. We describe the operation of an Extensive Air Shower Array located at the campus of the FCFM-BUAP. The array consists of 8 liquid scintillation detectors with a surface of 1 m^2 each and a detector spacing of 20 m in a square grid. The array was designed to measure the energy and arrival direction of primary particles that generate extensive air showers (EAS) in the region of 10^{13} eV – 10^{16} eV. The angular distribution measured with this array, $Cos^8(\Theta) \times Sin(\Theta)$, agrees very well with the literature. We also present the measured energies of a number of vertical showers in the range of 5×10^{12} eV to 5×10^{13} eV.

Experimental setup

The array, located at the campus of the FCFM-UAP (19° N, 90° W, 800 g/cm^2), consists of 8 detectors distributed uniformly on a square grid with spacing of 20 m, as shown in figure 1a. Each detector is composed of a light-tight cylindrical container with inner reflective walls filled with liquid scintillator up to a height of 13 cm; each container of 1 m^2 cross section has one 5" photomultiplier (PMT) (EMI model 9030A) facing down 70 cm above the surface of the liquid, as shown in figure 1b. The array has been upgraded from four to six and finally to eight detectors during the last year and it has been operated in a quasi-continuous way. The trigger requires the coincidence of signals in the four central detectors which form a rectangular sub-array with an area of 20×40 m^2. The measured trigger rate is 150 hour^{-1}. The data acquisition (DAQ) system consists of a set of digital oscilloscopes that digitize the signals from the eight PMT's; the system is controlled by a PC running a custom-made acquisition program written in LabView. Commercial NIM and CAMAC modules were used to produce the trigger signal and to measure the individual single-particle rates of each detector; these rates are used to monitor the performance of the array. The DAQ system makes on-line measurements of the integrated charge, arrival time, amplitude and width of the eight PMT signals of each trigger event. Figures 2 and 3a show the front panels of the LabView programs that acquire the PMT traces and measure the single-particle rates of the eight detectors, respectively. As can be seen from figure 2, the typical width of extensive air shower signals detected by each detector is around 200 ns. Figure 3b shows a typical charge distribution of single-muon signals of one detector; this information allows us to calibrate the array by means of the natural flow of single muons and henceforth convert the integrated charge of a given shower into the number of equivalent particles. The two peaks of this charge histogram correspond to the pedestal's peak and the single-muon peak, respectively; for the case of the detector shown, the pedestal-subtracted single-muon charge is 45 pC.

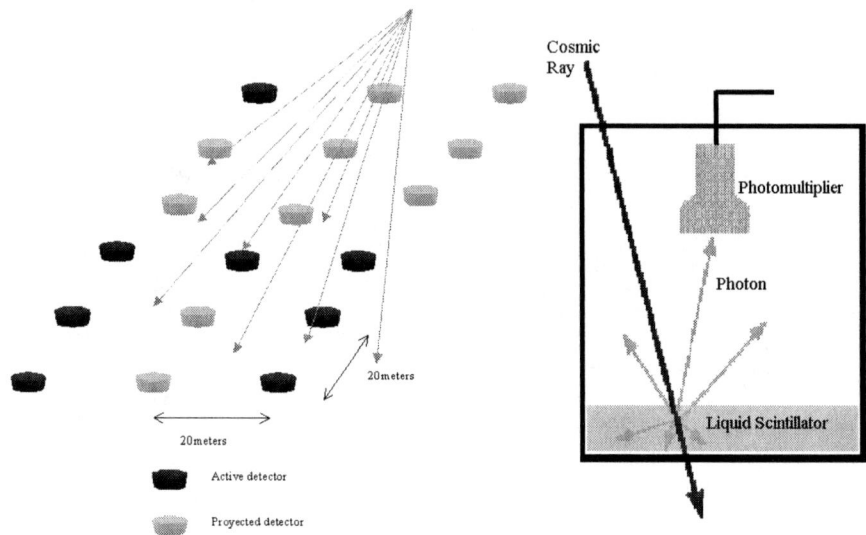

FIGURE 1. a) Extensive Air Shower array layout and b) description of a detector unit.

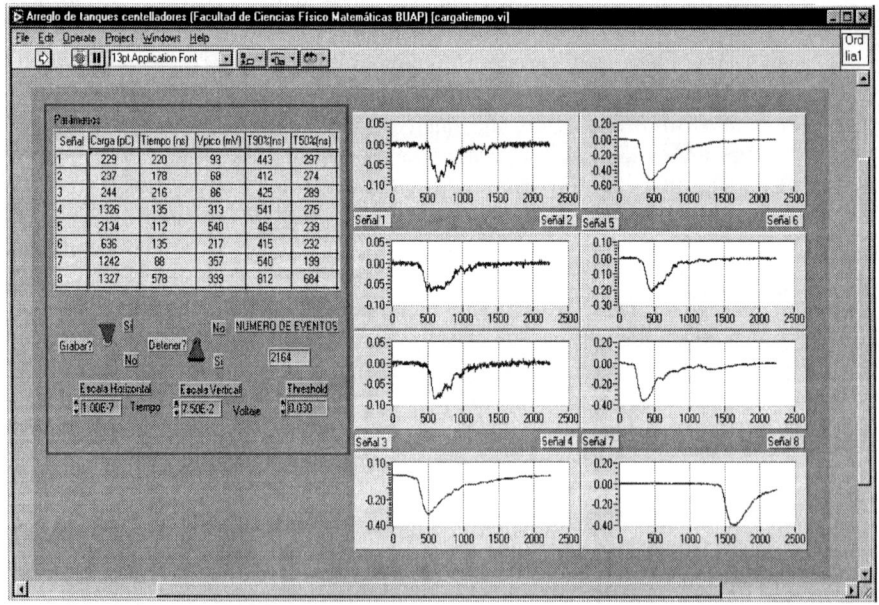

FIGURE 2. Traces of the 8 PMT signals of a shower as displayed by the DAQ program.

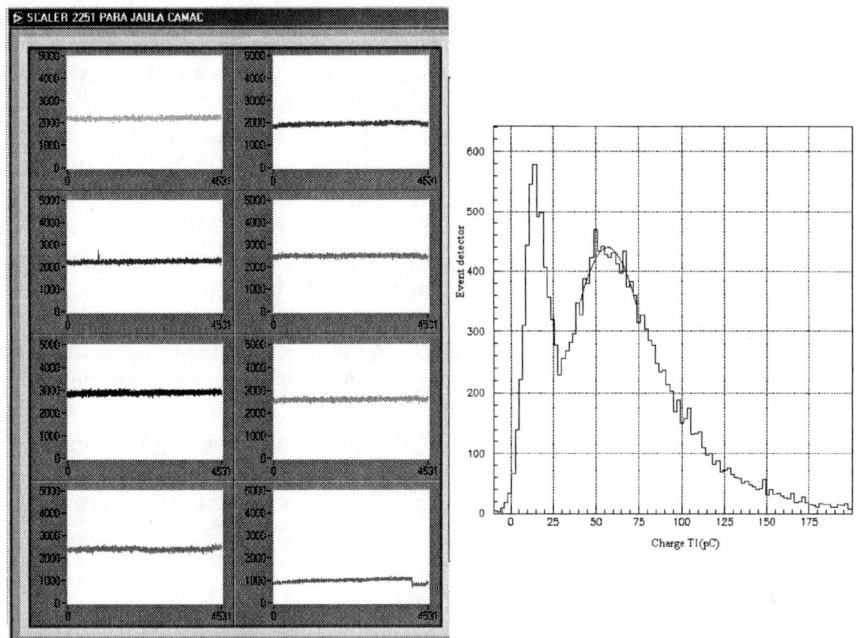

FIGURE 3. a) Single-muon rate monitoring of the 8 detectors and b) Charge distribution of single muons used to calibrate the detector array.

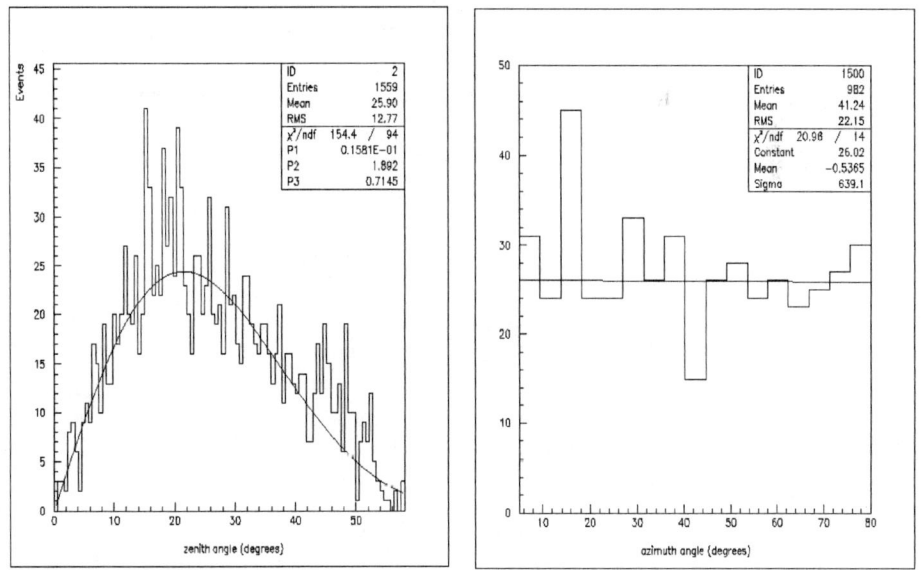

FIGURE 4. Zenithal and azimuthal angular distributions of EAS as measured with the Puebla array.

Discussion and results

During the first phase of operation we measured mainly the arrival direction of showers and the stability of the detectors. The single-muon rate of the detectors varies slightly during day time due to temperature variations. The direction of the primary cosmic ray is inferred directly from the relative arrival times of shower fronts at the different detectors. Figures 4a and 4b show the experimental zenithal and azimuthal distributions, respectively. The solid line in figure 4a corresponds to the function $\cos^8\Theta \times \mathrm{sen}\Theta$ obtained as the best fit to the data, while the azimuthal distribution is best described by the horizontal solid line of figure 4b. In order to obtain the lateral distribution function of the showers, we determine the number of particles in each detector using the single-particle charge spectrum discussed above; then we calculate the position of the shower core by fitting the experimental particle distribution in our array to the NKG function. Finally, the shower energy is obtained by comparing the lateral experimental data with the lateral distribution coming from simulations based on a program called Corsika. Table 1 shows the particle densities at 100 m from the shower core position and the measured shower energy for 15 vertical showers in the range of energy of 5×10^{12} eV to 5×10^{13} eV.

Conclusion

After one year of intermittent operation of the array, we conclude that the detector array and the data acquisition system are stable and work well, while the analysis method we use provides results in good agreement with the literature. The next phase of operation of this detector array will consist of longer and more continuous data-taking periods; the data from this new phase will allow us to make systematic measurements of the lateral distribution function, the primary energy and the composition of cosmic rays around shower energies of 10^{14} eV.

TABLE 1. Shower energies obtained experimentally by the method described based on $\rho(100)$

$\rho(100)$	Energy (eV)	$\rho(100)$	Energy (eV)	$\rho(100)$	Energy (eV)
3.1	1×10^{13}	1.66	5×10^{12}	13.9	5×10^{13}
18.01	5×10^{13}	14.38	5×10^{13}	2.36	1×10^{13}
13.19	5×10^{13}	1.09	5×10^{12}	0.82	5×10^{12}
1.24	5×10^{12}	1.55	5×10^{12}	1.25	5×10^{12}
1.6	5×10^{12}	2.99	1×10^{13}	1.02	5×10^{12}
2.9	1×10^{13}	1.88	5×10^{12}	3.88	1×10^{13}

ACKNOWLEDGEMENTS

The authors are very grateful to Oscar Saavedra for sharing experience and equipment. This work was done with partial support of the CONACY-G32739-E.

REFERENCES

1. Ameev, S, et. Al. *Proceeding of the ICCR*, 1997, pp. 257-260. Vol 7.

Fluorescence detector for Extensive Air Showers in the region of $10^{17} - 10^{21}$ eV

Cuautle M.[1], Moreno E.[1], Pedraza I., Murrieta T.[1], Garipov G[2], Khrenov B.[2], Salazar H.[1], Martinez O.[1] and Villaseñor L.[3]

[1]*Facultad de Ciencias Físico Matemáticas, BUAP, Puebla Pue., México*
2 *Moscow State University, Moscow, Russia.*
[3]*Instituto de Física y Matemáticas, UMSNH, Morelia Mich., México*

Abstract. We present preliminary results on the operation of a fluorescence detector for ultra energetic cosmic rays. This detector was designed to measure the energy of primary particles that generate extensive air showers in the region of $10^{17} - 10^{21}$ eV. We also discuss the main features of the electronic cameras that will be used by the fluorescence detectors of the Kosmotepetl and Pierre Auger projects and present results of simulations appropriate to these cameras.

DESCRIPTION OF THE DETECTOR AND SIMULATIONS

The scientific goal of the Kosmotepetl program [1] is to observe and study ultra high-energy cosmic rays thorough the fluorescent tracks that they produce in the Earth atmosphere with the help of satellite-based optical cameras. The performance of a prototype camera for Kosmotepetl will be tested at 4500 masl (Cerro la Negra) in a hybrid mode, i.e., in conjunction with an array of surface detectors located 5 km away form the telescope at an atmospheric depth of 620gr/cm^2. From simulations we know that this camera can be used to detect longitudinal profiles and positions of shower maximum of showers with primary energy in the range of 5×10^{16} to 10^{18} eV.

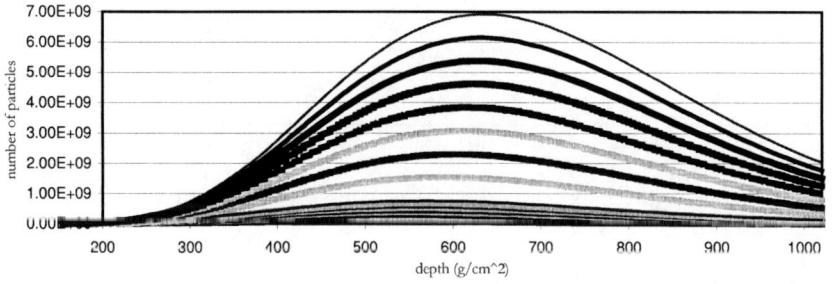

Figure 1. Simulated longitudinal profiles of vertical showers initiated by iron nuclei.

Figure 1 shows the simulated longitudinal profiles of vertical showers initiated by iron nuclei for 10^{17} –9×10^{18} eV. As we can see the shower maximum occurs above 620 gr/cm². Regarding proton-initiated showers, we need to trigger on inclined showers with zenithal angles greater or equal than 30° in order to have their shower maximum above the Cerro la Negra surface array.

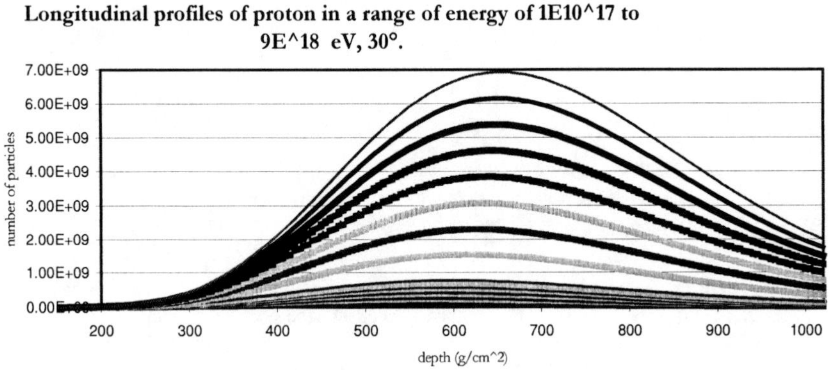

Longitudinal profiles of proton in a range of energy of 1E10^17 to 9E^18 eV, 30°.

Figure 2. Plots of simulated longitudinal profiles of proton-initiated showers with 30° inclination.

Figure 2 shows the simulated longitudinal profiles of proton-initiated showers with 30° inclination. Figure 3 shows the simulation of maximum depth versus the shower maximum for showers with energy lower than 4×10^{18} eV. All the simulation were done with the Corsika program using the QGSJET model [2], which is well fitted with the GIL analytical formula.

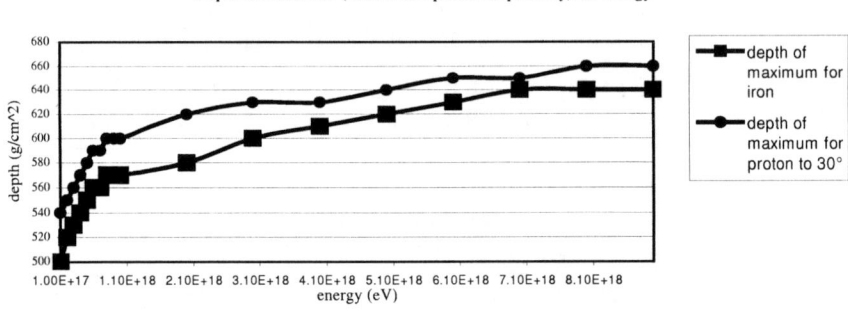

Figure 3. Graphs of the maximum depth versus the shower maximum, for showers with energy lower than 4×10^{18} eV.

The fluorescence detector at Cerro la Negra will have 9 mirrors of diameter 0.31 m with a focal distance of 0.10 m and 16 pixels. We obtain the number of photoelectrons per pixel, Q, at shower maximum by using the Kalmikov approximation [3] given by:

$$Q(S,R,E,\delta) := \frac{\dfrac{E}{1.3} \cdot 5 \cdot S \cdot R \cdot \delta \cdot 0.2}{4 \cdot \pi \cdot R^2} \cdot \exp\left(-\frac{R}{90000}\right)$$

Where S=0.078 m^2 is the mirror area, R=5 km is the distance from the detector to the EAS maximum, the number of fluorescence photons per m is 5, E/1.3 with E in GeV is the number of particles (electrons and positrons) at shower maximum, δ is the pixel FOV (field of view) equal to 0.2 rad, 90 km is the attenuation length of light at an altitude of 4 km, and 0.2 is the quantum efficiency of the PMTs. At a distance of 5 km from the EAS axis the field of view per pixel corresponds to 1 km of vertical track. Therefore the camera with 12x12 pixels allows the detection of up to 12 km of vertical tracks. Some values for the signal, measured in number of photoelectrons per pixel, corresponding to various energies of the primary cosmic ray are:

Energy (GeV)	Signal Q (photoelectrons)	Energy (GeV)	Signal Q (photoelectrons)
5.00E+07	35.81964597	5.00E+08	358.1964597
1.00E+08	71.63929195	1.00E+09	716.3929195

The voltage supplier for the PMT cluster and the block diagram of the electronics of the Cerro la Negra camera are shown in figure 4.

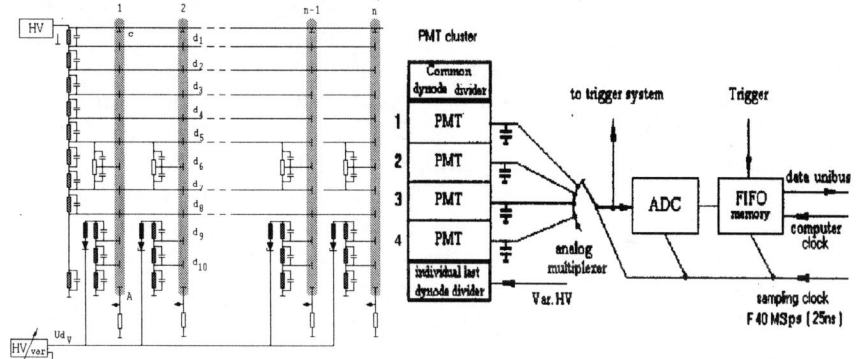

Figure 4. Voltage supplier for the PMT cluster and block diagram of the electronics of the Cerro la Negra prototype camera for Kosmotepetl.

The Pierre Auger Observatory was designed to puzzle out the origin of cosmic rays above the GZK cutoff [4]. The Auger engineering array (EA) consists of a hybrid system where two fluorescence telescopes work in coincidence with an array of 40 surface detectors. Each telescope covers a field of view of 30°x30° and has a camera with 20x22 photomultipliers located at the focal plane. The collected signal is processed with a large dynamical range (15 bit) to detect showers from 7 to 30 km away and with energies in the range of 10^{18}-10^{21} eV. This EA is operational since November 2001, and it is being used to test all the sub-systems of the fluorescence detector system.

Preliminary Results

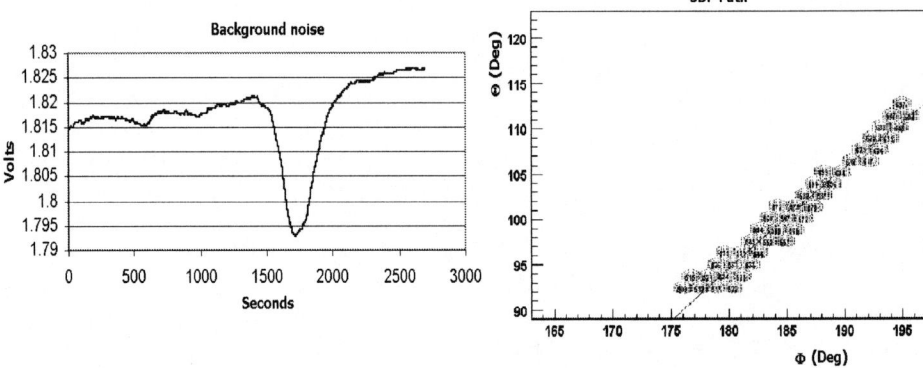

Figure 5. a) Signal produced by a star, crossing the field of view of the prototype camera for Kosmotepetl observed at the campus of the University of Puebla. b) Detection of an extensive air showers with the Auger camera in Argentina.

Figure 5a shows the signal produced by a star, crossing the field of view of the prototype camera for Kosmotepetl observed at the campus of the University of Puebla. We can observe the increase of noise level, as the moon gets higher in the sky. Figure 6b shows the detection of extensive air showers as detected with the Auger camera, in Argentina using the EA. The reconstruction was made using the "Flores" software that provides a tri-dimensional shower axis out of its two dimensional projection on the camera focal plane by using the arrival times of the signals at the different camera pixels. This particular track corresponds to a shower arriving at an angle of 16 degrees.

ACKNOWLEDGEMENTS

This work was done with partial support of the CONACYT grant G32739-E.

REFERENCES

1. B.A. Khrenov et al., Space program Kosmotepetl for the study of extremely high-energy cosmic rays, Proc. of the Intnal. Workshop on Observing Ultrahigh Energy Cosmic Rays from Space and Earth, Eds. H.Salazar, L. Villaseñor, A. Zepeda, AIP **566**, 2000, p 57-75.
2. D. Heck et al., Report FZKA, Forschungszentrum, **6019**, 1998, http://www-ik3.fzk.de
3. .NP. Iljina, N. N. Kalmykov and V.V. Prosin., Russian Nucl. Phys., **55**, 2756, 1992.
   Ameev, S, et. Al. *Proceeding of the ICCR*, 1997, pp. 257-260. Vol 7.
4. The Pierre Auger Project Design Report, November 1996.

Lepton Flavor Violation in the Two Higgs Doublet Model using $g-2$ muon factor

Rodolfo A. Diaz, R. Martinez, J-Alexis Rodríguez and E. Tuiran

Departamento de física, Universidad Nacional de Colombia, Bogota

Abstract. Current experimental data from the $g-2$ muon factor, seems to show the necessity of physics beyond the Standard Model (SM), since the difference between SM and experimental predictions is approximately 2.6σ. In the framework of the General Two Higgs Doublet Model (2HDM), we calculate the muon anomalous magnetic moment to get lower and upper bounds for the Flavour Changing (FC) Yukawa couplings in the leptonic sector.

Current muon anomalous magnetic moment a_μ data have challenged Standard Model (SM) and seem to open a window for new physics. Due to the high precision in a_μ value, it gives very restrictive bounds on physics beyond the SM. Although a_e measurement is about 350 more precise [1], a_μ is much more sensitive to New Physics because contributions to a_l are usually proportional to m_l^2.

The most accurate measurement of a_μ hitherto, has been provided by the Brookhaven Alternating Gradient Syncrotron [2]. Their data have an error one third that of the combined previous data [3], ref [2] reports

$$a_{\mu^+} = 11659202\,(14)\,(6) \times 10^{-10}. \tag{1}$$

On the other hand, SM predictions for a_μ have been estimated taking into account the contributions from QED, Hadronic loops and electroweak corrections. The final current result is [1, 2]

$$a_\mu^{SM} = 11659159.7\,(6.7) \times 10^{-10}. \tag{2}$$

Taking into account (2) is obtained

$$\Delta a_\mu^{NP} = a_\mu^{\exp} - a_\mu^{SM} = 42.6\,(16.5) \times 10^{-10}, \tag{3}$$

where $a_\mu^{\exp}$ is the world average experimental value [1]. Consequently at 90% C.L.

$$21.5 \times 10^{-10} \leq \Delta a_\mu^{NP} \leq 63.7 \times 10^{-10}. \tag{4}$$

Δa_μ^{NP} gives the room available for New Physics, so a_μ^{exp} differs from a_μ^{SM} approximately in 2.6σ. Therefore, physics beyond the SM is needed to achieve an acceptable theoretical experimental agreement. The most studied contributions to a_μ have been carried out in the framework of radiative muon mass models as well as the Minimal Supersymmetric

Standard Model (MSSM), E_6 string-inspired models, and extensions of MSSM with an extra singlet [4].

However, there are several mechanisms to avoid FCNC at the tree level. Glashow and Weinberg [5] proposed a discrete symmetry to supress them in the Two Higgs Doublet Model (2HDM) which is the simplest one that exhibits these rare processes at the tree level. There are two kinds of models which are phenomenologically plausible with the discrete symmetry imposed. In the model type I, one Higgs Doublet provides masses to the up-type and down-type quarks, simultaneously. In the model type II, one Higgs doublet gives masses to the up-type quarks and the other one to the down-type quarks. But the discrete symmetry [5] is not compulsory and both doublets may generate the masses of the quarks of up-type and down-type simultaneously, in such case we are in the model type III [6]. It has been used to search for FCNC at the tree level [7], [8].

Recently, the 2HDM type III has been discussed and classified [9], depending on the way in which the flavor mixing matrices are rotated, showing that there are two types of rotations which generate two lagrangians in the leptonic sector.

In this paper, we calculate the contributions to Δa_μ^{NP} coming from the 2HDM, which includes FCNC at the tree level. We will constrain the FC vertex involving the second and third charged leptonic sector by using the result for Δa_μ^{NP}, equation (4).

The Yukawa's Lagrangian for the 2HDM type III, is as follow

$$\begin{aligned}-\mathcal{L}_Y &= \eta_{ij}^U \overline{Q}_{iL}\widetilde{\Phi}_1 U_{jR} + \eta_{ij}^D \overline{Q}_{iL}\Phi_1 D_{jR} + \eta_{ij}^E \overline{l}_{iL}\Phi_1 E_{jR} \\ &+ \xi_{ij}^U \overline{Q}_{iL}\widetilde{\Phi}_2 U_{jR} + \xi_{ij}^D \overline{Q}_{iL}\Phi_2 D_{jR} + \xi_{ij}^E \overline{l}_{iL}\Phi_2 E_{jR} + h.c.\end{aligned} \quad (5)$$

where $\Phi_{1,2}$ are the Higgs doublets, η_{ij} and ξ_{ij} are non-diagonal 3×3 matrices and i, j are family indices.

In the present report, we calculate Δa_μ^{NP} in the 2HDM with FC interactions. If we assume that $m_\mu^2 << m_\tau^2$ and $m_\mu^2 << m_{h^0,H^0,A^0}^2$ in the calculation of the Feynman integrals, the contribution at the one loop level from all neutral Higgses is given by

$$\Delta a_\mu = \frac{m_\mu m_\tau}{16\pi^2} \sum_i b_i^2 \left[F(m_{H_i}) + \frac{m_\mu}{3m_\tau}G(m_{H_i})\right] + a_i^2 \left[F(m_{H_i}) - \frac{m_\mu}{3m_\tau}G(m_{H_i})\right], \quad (6)$$

where

$$\begin{aligned}G(m_{H_i}) &\equiv \left[\frac{2 + 3\widehat{m}_{H_i}^2 + 6\widehat{m}_{H_i}^2 \ln(\widehat{m}_{H_i}^2) - 6\widehat{m}_{H_i}^4 + \widehat{m}_{H_i}^6}{m_{H_i}^2\left(1 - \widehat{m}_{H_i}^2\right)^4}\right] \\ F(m_{H_i}) &= \frac{[3 + \widehat{m}_{H_i}^2(\widehat{m}_{H_i}^2 - 4) + 2\ln\widehat{m}_{H_i}^2]\widehat{m}_{H_i}}{m_{H_i}\left(1 - \widehat{m}_{H_i}^2\right)^3}\end{aligned} \quad (7)$$

with $\widehat{m}_{H_i} = m_\tau/m_{H_i}$. The sum is over the index $i = m_{h^0}, m_{H^0}, m_{A^0}$. a_i, b_i are the coefficients of the Feynman rules for Scalar and Pseudoscalar Higgses respectively. We have neglected the contribution of the charged Higgs, because of two reasons: on one side, the contribution involve the neutrino mass and on the other hand, the LEP bound on its

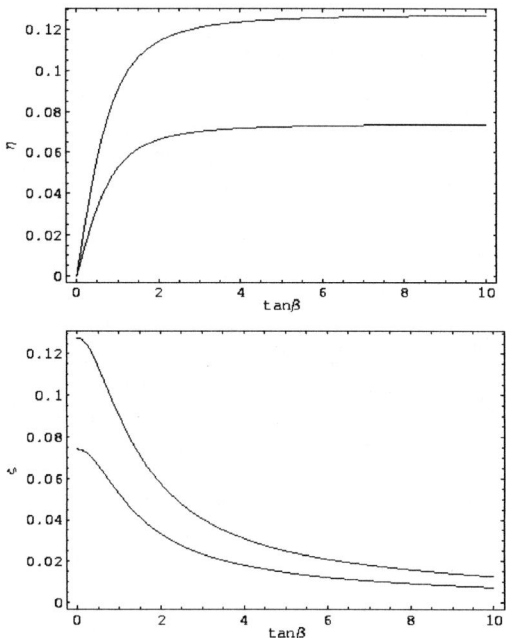

FIGURE 1. Lower and upper bounds for $\eta_{\mu\tau}(\xi_{\mu\tau})$ vs $\tan\beta$, for rotations I and II using $m_{h^0} = m_{H^0} = 150$ GeV and $m_{A^0} \to \infty$.

mass is $m_{H^\pm} \geq 80.5$ GeV. Additionally, we can notice that $m_\mu/3m_\tau G(m_{H_i}) \ll F(m_{H_i})$ and its contribution is negligible. However, we put it for completeness.

If we take into account the experimental data (4), we get some lower and upper bounds on the mixing vertex $\eta(\xi)_{\mu\tau}$ for the rotations of type I (II). In figure 1, we display lower and upper bounds for the FC vertices as a function of $\tan\beta$ for both types of rotations with $m_{h^0} = m_{H^0} = 150$ GeV and $m_{A^0} \to \infty$. In the first case, rotation type I, the allowed region for $\eta_{\mu\tau}$ is $0.07 \lesssim \eta_{\mu\tau} \lesssim 0.13$ for large values of $\tan\beta$. Meanwhile, for rotation type II, the allowed region for small $\tan\beta$ is the same. From Lagrangian (11), which describes rotation type I, we can see that when $\tan\beta \to 0$, $\eta_{\mu\tau}$ should go to zero as well to mantain a finite contribution to Δa_μ. This behaviour can be seen from figure 1. For rotation type II occurs the same but in the limit $\tan\beta \to \infty$.

In conclusion, we have found lower and upper bounds for the FC vertex $\eta(\xi)_{\mu\tau}$ in the context of the general 2HDM by using the allowed range for Δa_μ^{NP} at 90% CL and utilizing several sets of values for the parameters of the model. Additionally, in the limit $m_{A^0} \to \infty$, we get that for small (large) values of $\tan\beta$ the allowed range for the FC vertex $\eta_{\mu\tau}(\xi_{\mu\tau})$ becomes narrower, and both upper and lower bounds go to zero in the rotation of type I (II).

This work was supported by COLCIENCIAS, DIB and DINAIN.

REFERENCES

1. A. Czarnecki and W. Marciano, hep-ph/0102122.
2. H.N. Brown et. al.; hep-ex/0102017
3. C. Caso *et.al.* (Particle Data Group), Eur. Phys. J. C3, 1 (1998).
4. J. A. Grifols and A. Mendez, Phys. Rev. D **26**, 1809 (1989); M. Frank and C. S. Kalman, Phys. Rev. D **38**, 1469 (1988); J. A. Grifols, J. Sola and A. Mendez, Phys. Rev. Let. **57**, 2348 (1986); M. Krawczyk and J. Zochowski, Phys. Rev. D **55**, 6968 (1997); U. Chattopadhyay and P. Nath, hep-ph/0102157; D. Choudhry, B. Mukhopadhyaya and S. Rakshit, hep-ph/0102199.
5. S. Glashow and S. Weinberg, Phys. Rev. D **15**, 1958 (1977).
6. W.S. Hou, Phys. Lett B **296**, 179 (1992); D. Chang, W. S. Hou and W. Y. Keung, Phys. Rev. D **48**, 217 (1993); J.L. Diaz-Cruz, J.J. Godina and G. López Castro, Phys. Lett B **301** (1993) 405.
7. D. Atwood, L. Reina and A. Soni, Phys. Rev. D **53**, 1199 (1996); Phys. Rev. D **54**, 3296 (1996); Phys. Rev. Lett. **75**, 3800 (1993); D. Atwood, L. Reina and A. Soni, Phys. Rev. D **55**, 3156 (1997); G. Cvetic, S. S. Hwang and C. S. Kim., Phys. Rev. D **58**, 116003 (1998).
8. Marc Sher and Yao Yuan, Phys. Rev. D **44**, 1461 (1991).
9. Rodolfo A. Diaz, R. Martinez and J.-Alexis Rodriguez, hep-ph/0010149. To appear in Phys. Rev. D.
10. For a review see J. Gunion, H. Haber, G. Kane and S. Dawson, *The Higgs Hunter's Guide*, (Addison-Wesley, New York, 1990).

List of Participants

Alexis Aguilar Arevalo, ICN-UNAM, *alexis@nuclecu.unam.mx*
Erika Alvarez, IF-UNAM, *erika@ft.ifisicacu.unam.mx*
Jorge Amaro Reyes, IF-UASLP, *jamaro@dec1.ifisica.uaslp.mx*
Paolo Amore, FC-UColima, *paolo@ucol.mx*
Alfredo Aranda, Boston University, *aranda@physics.bu.edu*
Jorge Isidro Aranda, IFM-UMSNH
Juan Carlos Arteaga Velazquez, Cinvestav, *jarteaga@fis.cinvestav.mx*
Alejandro Ayala, ICN-UNAM, *ayala@nuclecu.unam.mx*
Mateo Alejandro Barkovich, ICN-UNAM, *mateo@nuclecu.unam.mx*
Juan Barranco Monarca, Cinvestav, *jbarranc@fis.cinvestav.mx*
Adnan Bashir, IF-UMSNH, *adnan@itzel.ifm.umich.mx*
Jaime Besprosvany, IF-UNAM, *bespro@fisica.unam.mx*
Jochen Bonn, Universität Mainz, *Jochen.Bonn@uni-mainz.de*
Antonio Bouzas, Cinvestav-Merida, *abouzas@mda.cinvestav.mx*
Andrzej J. Buras, TU München, *aburas@ph.tum.de*
Luis Gustavo Cabral-Rosetti, ICN-UNAM, *luis@nuclecu.unam.mx*
Carlos Camargo, Universidad de Colima
Marco Carrasco, Cinvestav, *mcarrasco@fis.cinvestav.mx*
Alexandra Carreño , ICN-UNAM, *alexc@nuclecu.unam.mx*
Mario Castillo, *trichis@hotmail.com*
Victor Castillo-Vallejo, Cinvestav Merida, *azogue@aruna.mda.cinvestav.mx*
Carlos Chavez, *carlosch@fis.cinvestav.mx*
Felix Antonio Cid, UA Santo Domingo
Guillermo Contreras, Cinvestav Merida, *jgcn@moni.mda.cinvestav.mx*
John Conway, Rutgers Univ./CDF, *conway@physics.rutgers.edu*
Peter Cooper, Fermilab, *pcooper@fnal.gov*
Eleazar Cuautle, ICN-UNAM, *ecuautle@nuclecu.unam.mx*
Maritza de Coss, Cinvestav-Merida, *mdecoss@jade.mda.cinvestav.mx*
Lorenzo Diaz Cruz, IF-BUAP, *ldiaz@sirio.ifuap.buap.mx*
Jürgen Engelfried, IF-UASLP, *jurgen@ifisica.uaslp.mx*
Robin Erbacher, Fermilab, *robine@fnal.gov*
Maria C. Espinoza-Hernandez, IF-UNAM, *catalina@ft.ifisicacu.unam.mx*
Olga G. Felix Beltran, IF-BUAP, *olga@sirio.ifuap.buap.mx*
Eden V. Figueroa Barragán, ITESM Monterrey, *al766837@mail.mty.itesm.mx*
Jose Luis Flores, IF-UNAM
Julio Flores, Cinvestav, *julio@fis.cinvestav.mx*
Ruben Flores, IF-UASLP, *ruben@ifisica.uaslp.mx*

José Luis Flores Silva, *jlfs98@yahoo.com*
Juan Bautista Florez Moreno, Universidad Narino, *jbautista@udenar.edu.co*
Nicolao Fornengo, U.Torino / INFN, *fornengo@to.infn.it*
Ricardo Gaitan, UNAM, *rgaitan@servidor.unam.mx*
Augusto Garcia , Cinvestav, *agarcia@fis.cinvestav.mx*
Jose Luis Garcia Luna, Cinvestav, *jlgarcia@fis.cinvestav.mx*
Rosa Magdalena Garcia, IF-BUAP, *rgarciah@sirio.ifuap.buap.mx*
Virendra Gupta, Cinvestav Merida, *virendra@aruna.mda.cinvestav.mx*
Andrea Gutierrez, ICN-UNAM, *andreag@nuclecu.unam.mx*
Maria de los Angeles Hernandez, FF-UAZ, *alexgu@ahobon.reduaz.mx*
Alfredo Herrera, IFM-UMSNH, *herrera@zeus.umich.mx*
Gerardo Herrera Corral, CINVESTAV, *gherrera@fis.cinvestav.mx*
Christoph Hofmann, IF-UASLP, *hofmann@ifisica.uaslp.mx*
Rodrigo Huerta, Cinvestav-Merida, *rhuerta@mda.cinvestav.mx*
Elizabeth Jenkins, UC San Diego, *ejenkins@ucsd.edu*
Piotr Kielanowski, Cinvestav, *kiel@physics.utexas.edu*
Mariana Kirchbach, FF-UAZ, *kirchbach@chiral.reduaz.mx*
Francisco Larios Forte, Cinvestav-Merida , *flarios@belinda.mda.cinvestav.mx*
Ildefonso Leon M. Cinvestav, *leonm@fis.cinvestav.mx*
Gabriel Lopez Castro, Cinvestav, *glopez@fis.cinvestav.mx*
Ricardo Lopez, IF-UASLP, *lopezr@ifisica.uaslp.mx*
Jose Antonio Loza
Jose Luis Lucio, IFUG
Alejandro Martínez, IFUG, *alex@ifug3.ugto.mx*
Oscar Mario Martinez Bravo, FCFM-BUAP, *omartin@fcfm.buap.mx*
Jesús Alberto Martínez, Cinvestav, *macj@fis.cinvestav.mx*
Roberto Martínez, Colombia, *romart@ciencias.unal.edu.co*
Julio Heriberto Mata Salazar, IF-UASLP, *jmata@ifisica.uaslp.mx*
Juan Medellin, IF-UASLP, *juan@ifisica.uaslp.mx*
Jose Angel Mendez Gamboa, Cinvestav Merida, *jmendez@mda.cinvestav.mx*
Jorge Mercado Perez, Cinvestav, *jmercado@fis.cinvestav.mx*
Joaquin Gabriel Miranda Mena, Cinvestav Merida, *mimj74@hotmail.com*
Omar Miranda, Cinvestav, *omr@fis.cinvestav.mx*
Alfonso Mondragon, IF-UNAM, *mondra@ft.ifisicacu.unam.mx*
Myriam Mondragon, IF-UNAM, *myriam@ft.fisica.unam.mx*
Luis Manuel Montano Zetina, Cinvestav, *lmontano@fis.cinvestav.mx*
Benjamin Morales, IF-UNAM, *bamr@ft.fisica.unam.mx*
Antonio Morelos, IF-UASLP, *morelos@ifisica.uaslp.mx*
Gerardo Moreno, IFUG, *gerardo@ifug1.ugto.mx*
Jorge Moreno, Cinvestav, *jmoreno@fis.cinvestav.mx*
Matias Moreno, IF-UNAM
Mauro Napsuciale, IFUG, *mauro@feynman.ugto.mx*
Lukas Nellen, ICN-UNAM
Marek Nowakowski, IFUG, *marek@fisica.ugto.mx*
Pedro Ojeda, *pojeda@astrosmo.unam.mx*
Gabriel Pallares, *pallares@nuclecu.unam.mx*

Sergio Pastor, MPI Physik München, *pastor@mppmu.mpg.de*
Isabel Pedraza Morales, FCFM-BUAP, *mpedraza@fcfm.buap.mx*
Gilberto Perea, IFUG, *perea@ifug3.ugto.mx*
Daniel Perez Astudillo, Cinvestav Merida, *daniel@karin.mda.cinvestav.mx*
Miguel Angel Perez, Cinvestav, *mperez@fis.cinvestav.mx*
Ricardo Perez, Universidad de Saltillo, *rperez@fis.cinvestav.mx*
Gabriella Piccinelli, CT-ENEP UNAM, *gabriela@astroscu.unam.mx*
William A. Ponce, U. de Antioquia, *wponce@fisica.udea.edu.co*
Carlos Ramirez, U. Ind. Santander, *cramirez01@andinet.com*
Alfredo Raya, IFM-UMSNH, *raya@itzel.ifm.umich.mx*
Ricardo Robles, ICN-UNAM, *surfercalaverita@hotmail.com*
Alexis Rodriguez, ICN-UNAM
Simon Rodriguez, IFUG, *srodri@ifug1.ugto.mx*
Ricardo Romero Ochoa, IF-UNAM, *ricardor@fisica.unam.mx*
Eduardo Alonso Rosado Vázquez, *eduardonobel@hotmail.com*
Julio Ruiz Tabasco, *elimazanubelth@hotmail.com*
James Russ, Carnegie Mellon U, *russ@cmphys.phys.cmu.edu*
Sarira Sahu, ICN-UNAM, *sarira@nuclecu.unam.mx*
Angel Sanchez, ICN-UNAM, *ansac@nuclecu.unam.mx*
Alberto Sanchez-Hernandez, Cinvestav, *asanchez@fis.cinvestav.mx*
Carlos Javier Solano, *javier@efei.br*
Ulises Solís, ICN-UNAM
Gilberto Tavares, Cinvestav, *gtv@fis.cinvestav.mx*
Ibrahim Torres, IF-UASLP, *itorres@ifisica.uaslp.mx*
Manuel Torres, IF-UNAM, *torres@fisica.unam.mx*
Eduardo Tututi, FCFM-UMSNH
Richard Van de Water, Los Alamos Nat. Lab., *vdwater@lanl.gov*
Eric Vazquez-Jauregui, IF-UASLP, *ericvj@ifisica.uaslp.mx*
Héctor Vucetich, UNAM-UNLP, *vucetich@fisica.unam.mx*
Albrecht Wagner, DESY, *albrecht.wagner@desy.de*
Jochen Wambach, TU-Darmstadt, *wambach@physik.tu-darmstadt.de*
Hans Weber, University of Virginia, *hw@galileo.phys.virginia.edu*
Andreas Wirzba, FZ Jülich, *a.wirzba@fz-juelich.de*
Jose Wudka, UC Riverside, *jose.wudka@ucr.edu*
C.P. Yuan, Michigan State University, *yuan@pa.msu.edu*

Author Index

A

Aguilar, A. A., 337
Aguilar, S., 399
Amore, P., 293
Anguiano, M., 395
Aranda, A., 333
Ayala, A., 149, 297, 301

B

Barkovich, M., 359
Bashir, A., 313, 355
Bello, P., 395
Besprosvany, J., 149, 351
Birse, M. C., 293
Bonn, J., 189
Bouzas, A. O., 277
Buras, A. J., 3

C

Cabral-Rosetti, L. G., 347
Carreño, A., 321
Casini, H., 359
Chamoun, N., 309
Contreras, J. G., 139
Conway, J. S., 252
Cooper, P. S., 156
Cotzomi, J., 399
Cuautle, E., 297
Cuautle, M., 403

D

de Coss, M., 379
Diaz, R. A., 407
Díaz Cruz, J. L., 115
D'Olivo, J. C., 321, 337, 359
Domínguez, G., 373

E

Engelfried, J., 285, 369
Erbacher, R. D., 245

F

Flores-Mendieta, R., 232
Fornengo, N., 182

G

Garipov, G., 403
Giraldo, Y., 341
Gupta, V., 330
Gutiérrez-Rodríguez, A., 391

H

Hernández, J., 395
Hernández-Ruíz, M. A., 391
Herrera, G., 297, 373
Herrera-Aguilar, A., 281
Hofmann, C. P., 305
Huerta, R., 379
Huet, A., 355

J

Jenkins, E., 36

K

Khrenov, B., 403

L

Larios, F., 387
León-Monzón, I., 373
López Castro, G., 268

M

Martínez, O., 395, 399, 403
Martinez, R., 407
Mata, J., 369
Mc Govern, J. A., 293
Medellin Z., J., 285
Meißner, U.-G., 216
Mercado-Pérez, J., 383
Miranda, O. G., 125
Mondragón, A., 117
Montaño, L. M., 297
Montemayor, R., 359
Morelos, A., 285, 369
Moreno, E., 399, 403
Moreno, J., 289
Moreno, M., 347
Murrieta, T., 403

N

Napsuciale, M., 131
Nellen, L., 321
Nowakowski, M., 325

O

Oller, J. A., 216

P

Pallares, G., 149
Palma, B., 399
Pastor, S., 202
Pedraza, I., 403
Pérez, E., 395
Pérez, M. A., 387
Piccinelli, G., 149
Ponce, E., 395
Ponce, W. A., 341

R

Raya, A., 313, 355
Rodríguez, J.-A., 407
Rosado, A., 347

Russ, J. S., 163

S

Sahu, S., 317
Salazar, H., 395, 399, 403
Sampayo, O. A., 391
Sánchez, A., 301
Sánchez, L. A., 341
Silaev, A., 395
Solano, C. J., 365

T

Tavares-Velasco, G., 387
Torres, I., 369
Tuiran, E., 407

V

Van de Water, R. G., 173
Vázquez-Jáuregui, E., 369
Villaseñor, L., 395, 399, 403
Vucetich, H., 309

W

Wagner, A., 239
Walet, N. R., 293
Wambach, J., 223
Weber, H. J., 209
Wirzba, A., 216
Wolf, G., 61
Wudka, J., 195

Y

Yuan, C.-P., 261